Realistic Asset Creation with Adobe Substance 3D

Create materials, textures, filters, and 3D models using
Substance 3D Painter, Designer, and Stager

Zeeshan Jawed Shah

‹packt›

BIRMINGHAM—MUMBAI

Realistic Asset Creation with Adobe Substance 3D

Group Product Manager: Rohit Rajkumar

Publishing Product Manager: Nitin Nainani

Senior Editor: Mark D'Souza

Content Development Editor: Abhishek Jadhav

Technical Editor: Joseph Aloocaran

Copy Editor: Safis Editing

Project Coordinator: Sonam Pandey

Proofreader: Safis Editing

Indexer: Manju Arasan

Production Designer: Alishon Mendonca

Marketing Coordinator: Nivedita Pandey

First published: November 2022

Production reference: 1031122

Published by Packt Publishing Ltd.

Livery Place

35 Livery Street

Birmingham

B3 2PB, UK.

ISBN 978-1-80323-340-6

www.packt.com

To my mother Shadab Jawed and my father Jawed Ali Shah, for the sacrifices that made me who I am and all their prayers that helped me to overcome the hardships in my life. To my wife Mehwish and my daughters Gia and Reem for always being there for me and supporting me continuously, especially during my tough times. I would also like to thank my employer, Prince Mohammad Bin Fahd University, for appreciating my efforts.

– Zeeshan Jawed Shah

Contributors

About the author

Zeeshan Jawed Shah is a teacher of multimedia, filmmaking, and arts. He is also a creative director, photographer, painter, sculptor, graphic designer, 3D modeler, animator, filmmaker, web designer, and visual effects artist. He is currently working as a lecturer in the graphic design department at Prince Mohammad Bin Fahd University, Saudi Arabia. Zeeshan has worked on many projects.

Zeeshan has taught fine arts, web media, filmmaking, 3D modeling and animation, and visual effects courses at the University College of Bahrain, **New York Institute of Technology (NYIT)** Bahrain, and the University of Bahrain. He holds more than fifteen years of extensive teaching experience, and his problem-based and project-based learning pedagogy has made him a popular faculty member in the Middle East. He is also considered the pioneer of student film projects in Bahrain as the first one to produce commercial feature-length films as student projects there.

Zeeshan was born in Saudi Arabia and went to Bahrain in 2006 to work as a lecturer at NYIT Bahrain. He mainly taught web design, 3D modeling and animation, computer graphics, motion design, and 2D animation courses in the beginning and then he included art history, drawing, and digital photography to his teaching list. Gradually, he started to work on film projects with NYIT students, which were based on his project-based learning teaching style in Bahrain.

Besides filmmaking, Zeeshan is also highly involved in interactive media and has organized various workshops related to Virtual Reality and Augmented Reality. He has been a notable member of Bahrain's gaming society where he participated in many game design events such as Gulf Game Jams.

About the reviewers

Chunck Trafagander has been working in video games since 2018, currently working as an Environment Artist in AAA. In his spare time, he runs the *Get Learnt w/ Chunck* YouTube channel where he creates tutorials centered around game art. He has also created several popular courses for the Flipped Normals marketplace, contributed multiple Substance Designer tutorials on the official Adobe Substance YouTube channel and works frequently with the CG Cookie team to deliver high-quality training resources for their YouTube and website.

Käy Vriend is an award-winning passionate material artist and all-round Substance expert, consultant, coach and mentor who is well-known in the game art community and one of the top Substance Artists in the world.

He loves finding and creating synergies, advise and help building studios and contribute to the games, film and design industry, indie and AAA studios alike.

Martin Schmitter is a talented CG generalist and content creator based in Switzerland. In his spare time, he is also a passionate nature photographer, hiker and loves to play video and boardgames. Blender 3D and other open-source software marked the starting point before he dived deep into the Adobe Substance 3D pipeline. While he began to learn 3D Designer, he created the substance beginner quicktip blog on artstation and social media to share his knowledge to the public. Then he participated on various challenges such as Mayterials, Nodevember and Blender 52 contest.

Now he works self-employed for companies and individuals, creates multiple Substance 3D tutorial series, he is a part of a game project and further releases materials and tools.

First of all, I would like to thank my life partner from the bottom of my heart for all the support and love she always shows me. I would also like to thank my parents, sister, grandparents, uncle, families of mine and my partner and friends for their support. And always remember, believe in your dreams and live your life whether others believe in you or not. Even if it means failing and getting up again. You can create what you want to achieve.

Table of Contents

4

Working with Masks in Adobe Substance 3D Painter   103

5

Working with Advanced Tools in Adobe Substance 3D Painter   117

6

Working with Materials and Smart Materials in Adobe Substance 3D Painter   153

7

Getting Started with Adobe Substance 3D Designer   205

8

Nodes in Adobe Substance 3D Designer   247

9

Blending Modes in Adobe Substance 3D Designer   295

10

Creating a Television Shelf in Adobe Substance 3D Designer   321

11

Adobe 3D Sampler at a Glance   367

12

Getting Started with Adobe Substance 3D Stager — 403

13

Models, Materials, and Lights in Adobe Substance 3D Stager — 417

14

Cameras and Rendering inside Adobe Substance 3D Stager   437

Preface

Adobe Substance 3D is a complete suite with everything that artists need to create stunning 3D digital materials. It can be complex to use, though. Thankfully, you've come to the right place for guidance!

With *Realistic Asset Creation with Adobe Substance 3D*, you will get a practical guide that won't bombard you with reams of textual information. Instead, this is an interactive, project-based book that will give you a sound knowledge of Adobe Substance 3D and set you on the road to a career in 3D design.

This comprehensive guide will start you off with the rudiments of Adobe Substance 3D Painter, giving you the skills to work with layers, masks, shelves, textures, and more. Then, we will move on to Adobe Substance 3D Designer; you'll become thoroughly acquainted with this node-based design tool as we progress through the book. The final section is devoted to Adobe Substance 3D Stager, which will allow you to build complex 3D scenes and visualize your edits in real time.

By the end of this book, you'll have a sound knowledge of Adobe Substance 3D and will have the skills to build a comprehensive portfolio of work, setting you up for a career in 3D design.

Who this book is for

This book is for game designers, 3D generalists, and design students. If you are looking to get started in your 3D career or planning to join the 3D generation, then you'll find this book useful. The book requires very little 3D design knowledge – or none whatsoever. Some understanding of the background history, functionality, and purpose of the Adobe Substance 3D package is presumed but is not mandatory. However, the reader must know how to work with Adobe Photoshop.

What this book covers

Chapter 1, *Getting Started with Adobe Substance 3D Painter*, will explore Adobe Substance Painter, how it looks, its user interface, and user experience. You will also gather the knowledge to prepare yourselves for the following chapters here.

Chapter 2, *Working with Assets in Adobe Substance 3D Painter*, will provide information on one of the essential features of Adobe Substance 3D Painter, which is the shelf. Going over this feature will allow you to work on models without worrying about organizing the files.

Chapter 3, *Working with Layers and Maps in Adobe Substance 3D Painter*, will get into the details of how the layers and maps inside Adobe Substance 3D Painter work. This chapter is divided into five main sections, which provide extensive knowledge of the layers and maps, as these are crucial topics to learn about.

Chapter 4, Working with Masks in Adobe Substance 3D Painter, will help you understand how masking is useful when creating dirt maps or curvature-based damage effects inside Adobe 3D Substance Painter. This chapter explains the whole process using easy yet comprehensive methods.

Chapter 5, Working with Advanced Tools in Adobe Substance 3D Painter, will help you understand the importance, purpose, and usage of a variety of advanced tools in Adobe Substance 3D Painter using conceptual and practical knowledge. These tools are easy to use – however, it could be quite confusing if the user was using them for the first time.

Chapter 6, Working with Materials and Smart Materials in Adobe Substance 3D Painter, will teach you how Smart Materials make the texturing process faster and more procedural. Upon going through this chapter, you will be able to create your own Smart Materials, which you can reuse in many projects or even share with other designers.

Chapter 7, Getting Started with Adobe Substance 3D Designer, will explore Adobe Substance 3D Designer and give a detailed overview of its user interface. You will get to know different types of nodes in Adobe Substance 3D Designer and learn how to blend them using various blending modes.

Chapter 8, Nodes in Adobe Substance 3D Designer, will give you a chance to explore the nodes in Adobe Substance 3D Designer. You will understand why this feature is so crucial and what its purpose is. Every node in this chapter will be explained in detail.

Chapter 9, Blending Modes in Adobe Substance 3D Designer, will explain the Blending Modes in Adobe Substance 3D Designer. The chapter will help you understand why this feature is so crucial and what its purpose is. All the Blending Modes will be explained in detail in this chapter.

Chapter 10, Creating a Television Shelf in Adobe Substance 3D Designer, will teach you how to create a television shelf using Adobe Substance 3D Designer. This chapter will enlighten you with extensive practical knowledge of Adobe Substance 3D Designer.

Chapter 11, Adobe Substance 3D Sampler at a Glance, will inform you on how to get started with using Adobe Substance 3D Sampler. The chapter will begin with an explanation of what this program does and then move on to the user interface and panels. This chapter will delve into the various setup panels, as well as how to access the ready-to-use components within the 3D sampler. Then, you will be able to use what you've learned by creating a basic pavement with blending layers and effects. The chapter will conclude with instructions on how to export our material as a Smart Material and individual maps.

Chapter 12, Getting Started with Adobe Substance 3D Stager, will teach you how to utilize Adobe Substance 3D Stager by equipping yourself with the knowledge needed to make creative judgments in context.

Chapter 13, Models, Materials, and Lights in Adobe Substance 3D Stager, will teach you how the composition is being refined and adjusted in real time. It gives a comprehensive knowledge of basic 3D models plus materials and how to use them in real-time projects.

Chapter 14, Cameras and Rendering inside Adobe Substance 3D Stager, will teach you how visualizing and manipulating intricate lighting and shadows works when rendering a scene. This chapter will also look into how to set up any camera inside Adobe Substance 3D Stager and how to render the scene with it.

To get the most out of this book

You need the version of the Adobe Substance 3D suite from 2021 or later.

Software/hardware covered in the book	Operating system requirements
Adobe Substance 3D Painter 2022	Windows or macOS
Adobe Substance 3D Designer 2022	Windows or macOS
Adobe Substance 3D Sampler 2022	Windows or macOS
Adobe Substance 3D Stager	Windows or macOS
Adobe Photoshop 2022	Windows or macOS
Autodesk Maya 2022	Windows or macOS

You can install any version of Adobe Photoshop and Autodesk Maya since they are used as supporting software in this book.

You need to have sound knowledge of 3D applications such as 3D Studio Max or Maya so that you can model any basic mesh and export it to Adobe Substance 3D for texturing and authoring.

Download the example files

You can download the example code files for this book from `https://packt.link/blbeC`.

We also have other code bundles from our rich catalog of books and videos available at `https://github.com/PacktPublishing/`. Check them out!

Download the color images

We also provide a PDF file that has color images of the screenshots and diagrams used in this book. You can download it here: `https://packt.link/SdEZZ`.

Conventions used

There are a number of text conventions used throughout this book.

`Code in text`: Indicates code words in text, database table names, folder names, filenames, file extensions, pathnames, dummy URLs, user input, and Twitter handles. Here is an example: "Select `Retro_Tv.FBX`, keeping in mind Painter only accepts FBX files."

Bold: Indicates a new term, an important word, or words that you see onscreen. For instance, words in menus or dialog boxes appear in **bold**. Here is an example: "Choose **Normal Map Format** depending on your preference."

> **Tips or important notes**
> Appear like this.

Get in touch

Feedback from our readers is always welcome.

General feedback: If you have questions about any aspect of this book, email us at `customercare@packtpub.com` and mention the book title in the subject of your message.

Errata: Although we have taken every care to ensure the accuracy of our content, mistakes do happen. If you have found a mistake in this book, we would be grateful if you would report this to us. Please visit `www.packtpub.com/support/errata` and fill in the form.

Piracy: If you come across any illegal copies of our works in any form on the internet, we would be grateful if you would provide us with the location address or website name. Please contact us at `copyright@packt.com` with a link to the material.

If you are interested in becoming an author: If there is a topic that you have expertise in and you are interested in either writing or contributing to a book, please visit `authors.packtpub.com`.

Share Your Thoughts

Once you've read *Realistic Asset Creation with Adobe Substance 3D*, we'd love to hear your thoughts! Scan the QR code below to go straight to the Amazon review page for this book and share your feedback.

`https://www.amazon.in/review/create-review/error?asin=1803233400`

Your review is important to us and the tech community and will help us make sure we're delivering excellent quality content.

Download a free PDF copy of this book

Thanks for purchasing this book!

Do you like to read on the go but are unable to carry your print books everywhere?

Is your eBook purchase not compatible with the device of your choice?

Don't worry, now with every Packt book you get a DRM-free PDF version of that book at no cost.

Read anywhere, any place, on any device. Search, copy, and paste code from your favorite technical books directly into your application.

The perks don't stop there, you can get exclusive access to discounts, newsletters, and great free content in your inbox daily

Follow these simple steps to get the benefits:

1. Scan the QR code or visit the link below

https://packt.link/free-ebook/9781803233406

2. Submit your proof of purchase
3. That's it! We'll send your free PDF and other benefits to your email directly

1

Getting Started with Adobe Substance 3D Painter

For 3D professionals and hobbyists, **Adobe Substance 3D Painter** is the go-to texturing software. Substance 3D Painter is widely used in in-game and film production, product design, fashion, and architecture.

From product design to realistic games and visual effects to projects such as stylized animation, Painter can help you create the style you want. It comes with many smart materials that can provide realistic surface detail and simulate realistic wear and tear on any object. It has several mask presets that adapt to any shape and has efficient and dynamic painting tools.

Painter's cutting-edge viewport displays all of your artistic decisions in real time. Experiment with advanced lighting and shadows on complicated materials to make texturing even more creative and painless. You may even use the inbuilt path tracing option to preview your model.

Every action and stroke is stored in Painter and can be recalculated at any moment. This means you may change your project's resolution at any time without sacrificing quality, and you can even adjust existing paint strokes.

However, the Substance 3D Painter software can sometimes be overwhelming to a new user, which is why many users find it difficult to learn. Keeping all this in mind, this book provides you with step-by-step simplified lessons that are designed specially to help you learn this software in a short time. After going through all the lessons in this book, you will become confident with using it with some practice and time.

In the first section of this chapter, you will learn what you should know before starting with Adobe Substance 3D Painter. By the end of this chapter, you will have made a 3D mesh that is ready to texture and paint inside Adobe Substance 3D Painter, which is the first and foremost task you need to perform to start your Painter project.

Moreover, you will learn how to divide your 3D mesh into different parts, which makes it easy to organize your project. In the end, you will also be able to bake textures, which will allow you to create various painting and texturing effects.

In this chapter, we'll be covering the following topics:

- What is texel density?
- Creating texture lists and ID maps for Adobe Substance 3D Painter
- Baking textures in Adobe Substance 3D Painter

Technical requirements

The files used in this book can be downloaded from this link: www.xcxvx.xcxvx.com

The following are the three main actions we'll be performing frequently when navigating inside Adobe Substance 3D Painter:

- Press *Alt* (Windows) or *Option* (Mac) + the left mouse button to orbit around a scene.
- Press *Alt* (Windows) or *Option* (Mac) + the right mouse button to zoom in and zoom out; you can also use the mouse scroll wheel for this purpose.
- Press *Alt* (Windows) or *Option* (Mac) + the middle mouse button to pan across the viewport.

The following screenshot shows us the main user interface of Adobe Substance 3D Painter:

Figure 1.1 – Adobe Substance 3D Painter user interface

Letter labels have been used to highlight the main parts of the interface:

- **A**: Tools and plugins toolbars
- **B**: **Assets** panel
- **C**: Main menu and contextual toolbar
- **D**: Application menu bar
- **E**: Viewport 3D/2D
- **F**: Property panels (**Texture Set List**, **Layer Stack** and **Texture Set Settings**, and **Properties**)
- **G**: Dock toolbar

What is texel density?

When designers and photographers work on 2D designs or photographs, they try to keep their quality to a high scale so that they appear attractive, sharp, and crisp. The standard of quality in images is determined by the number of pixels present.

However, when we apply texture to objects using any 3D software, the number of pixels does not help a lot. This is because in the 3D world, the number of pixels does not matter; it's texels that matter. Texel density is the size of the UV island or UV shell; the bigger the size of the UV island or shell, the sharper and crisper the image will look.

> **UV island, or UV shell**
>
> A **UV island**, also known as a **shell**, is a connected group of UVs on a UV map. UVs resemble vertices, except they determine how a 2D texture is transferred onto a 3D model's surface. Your 2D texture is stored in the UV map, together with the layout of your UV islands or shells, which serve as coordinates for where the texture is placed on the 3D model.

For example, to check the quality of a texture, open the `Retro_Tv.fbx` file in any 3D application (we are opening the file in Autodesk Maya 2022) and perform the following steps:

1. Open Autodesk Maya (or any other 3D application that can be used here).
2. Select **Front** from **Outliner**.

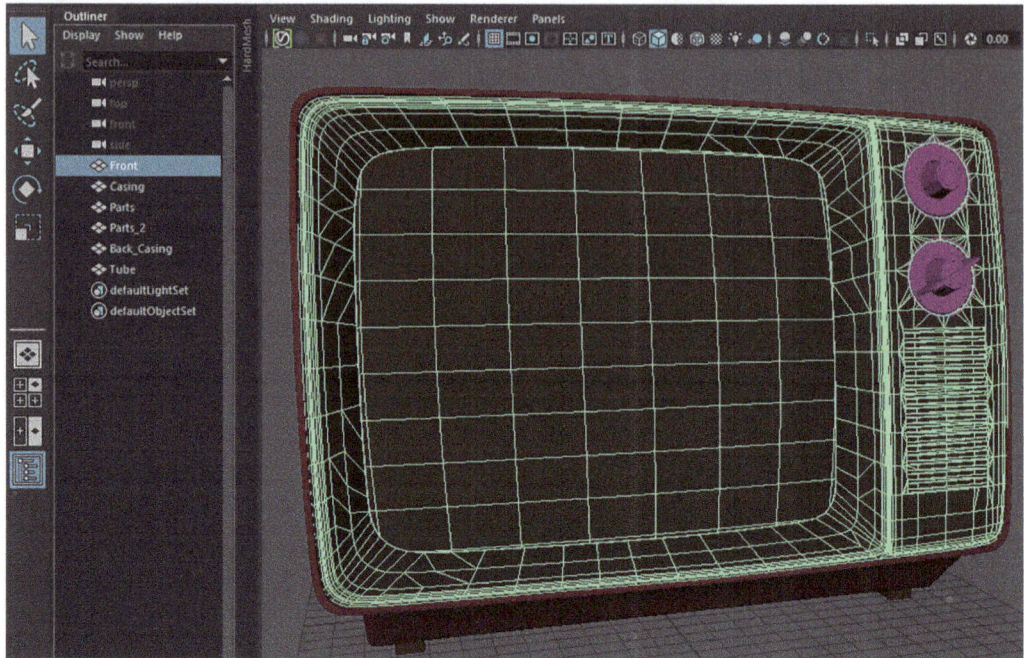

Figure 1.2 – Front of the television

3.   Make sure your menu is set to **Modeling** by choosing it from the drop-down menu indicated by *1* in *Figure 1.3*. Then, choose **UV** from the top menu and click **UV Editor**.

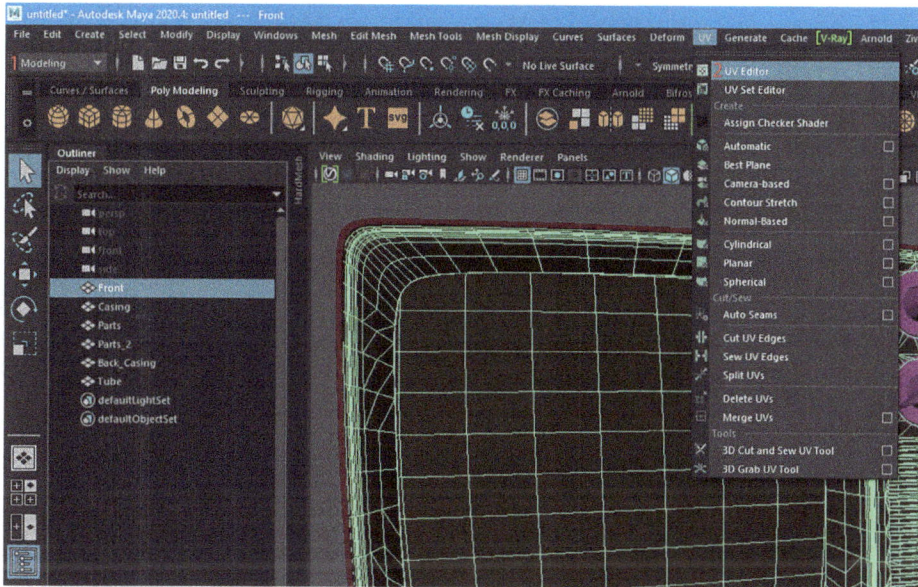

Figure 1.3 – UV Editor menu

4. Observe the UV islands of the TV model.

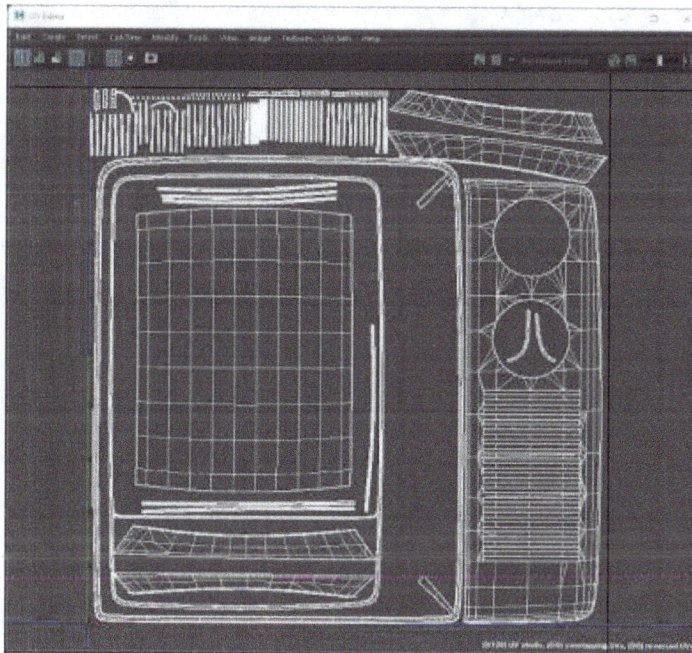

Figure 1.4 – UV island

As seen in *Figure 1.4*, every part of the retro television model is separated into different 3D meshes. This helps in enlarging the UV islands without any spacing issues. Every UV island acts as a container that holds the texel values, which is similar to pixel values.

A **texel**, also known as a **texture element**, is the representation of pixels from a 2D image mapped onto a 3D surface. The larger the UV tile, the more texels it can hold, resulting in sharper and crisper textures on any 3D model.

On the contrary, if the size of the UV islands/shells is ignored and kept small, or instead of keeping the parts of the television separate, the parts are all placed in one single UV tile, as shown in *Figure 1.5*, the UV islands/shells will not be able to hold a much larger texel density. This means that the texture will be low in quality even if the image is originally 8K:

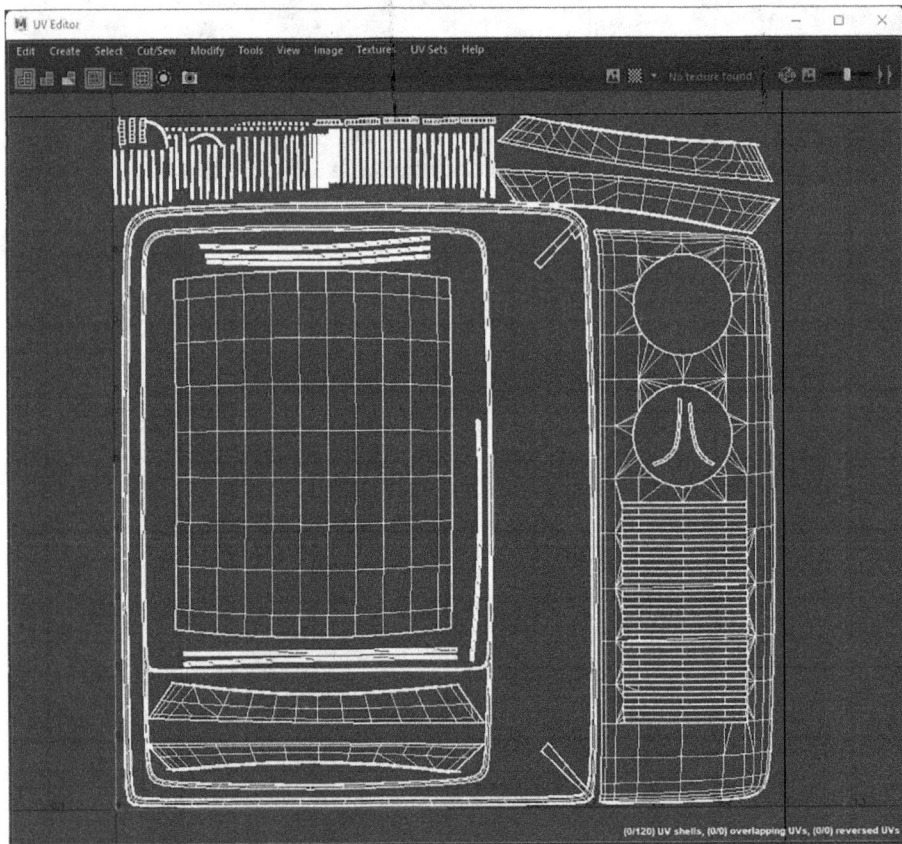

Figure 1.5 – UV islands on one tile

Now that we have a brief understanding of texel density, let's see how we can create texture lists and ID maps.

Creating texture lists and ID maps in Adobe Substance 3D Painter

Once we define the texel density, the 3D mesh is exported to Adobe Substance 3D Painter; however, we must first create the texture sets of the 3D mesh and create the ID maps on it as well. This section demonstrates how texture sets and ID maps are created with the help of Lambert materials and vertex color.

Creating texture sets

The **Texture Set List** pane displays several components of a project's current 3D model. Because the 3D mesh is in the form of layers, Texture Set List allows you to toggle between them and examine the layer stack connected with each material on the model, as well as their specific parameters.

Let's see where to access texture lists from:

1. Open Adobe Substance 3D Painter and click on **Start painting**.

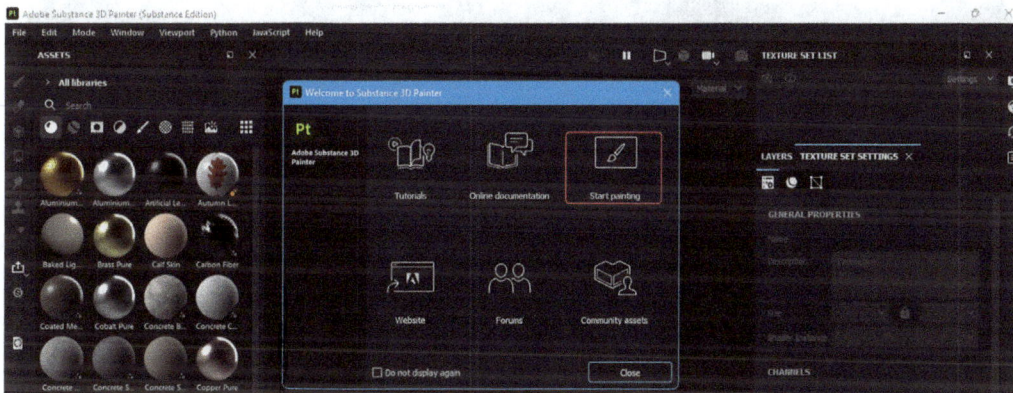

Figure 1.6 – Start painting

2. The **Start painting** option will open the Substance Painter mascot, and you will notice on the right panel that the mascot is divided into different texture sets.

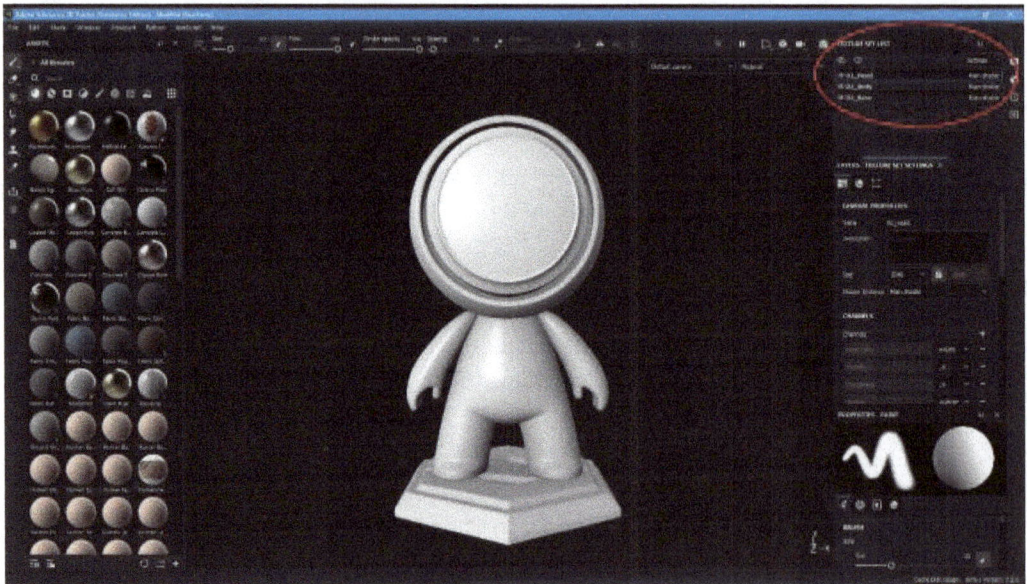

Figure 1.7 – Texture Set List

The mascot is divided into three texture Sets—**Head**, **Body**, and **Base**—and you can easily hide or lock any texture set. For example, if you do not want to work on the body part, you can simply hide it, as shown in *Figure 1.8*, so that it will not distract you while working on your head and base:

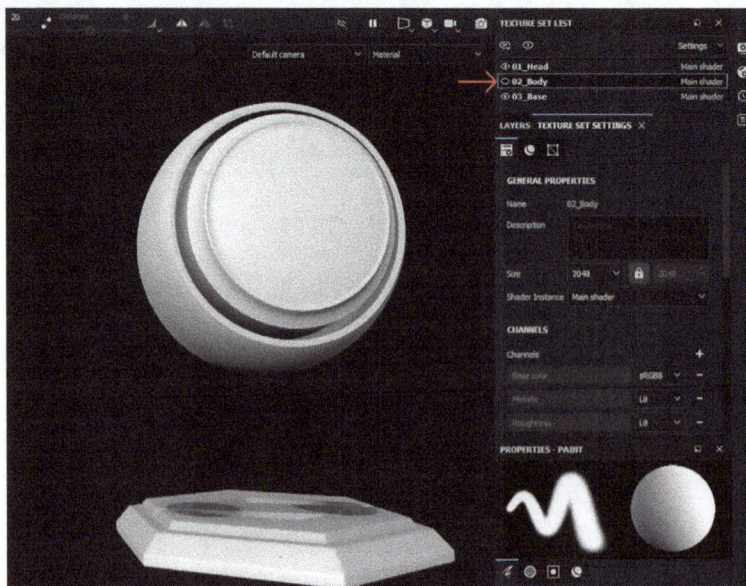

Figure 1.8 – Hiding a texture set

Texture sets can be created in 3D applications such as Autodesk Maya, 3D Studio Max, or Blender.

To create a texture set, we will use Autodesk Maya and perform the following steps:

1. Open Autodesk Maya, go to the **File** option, and click on **Import…**.

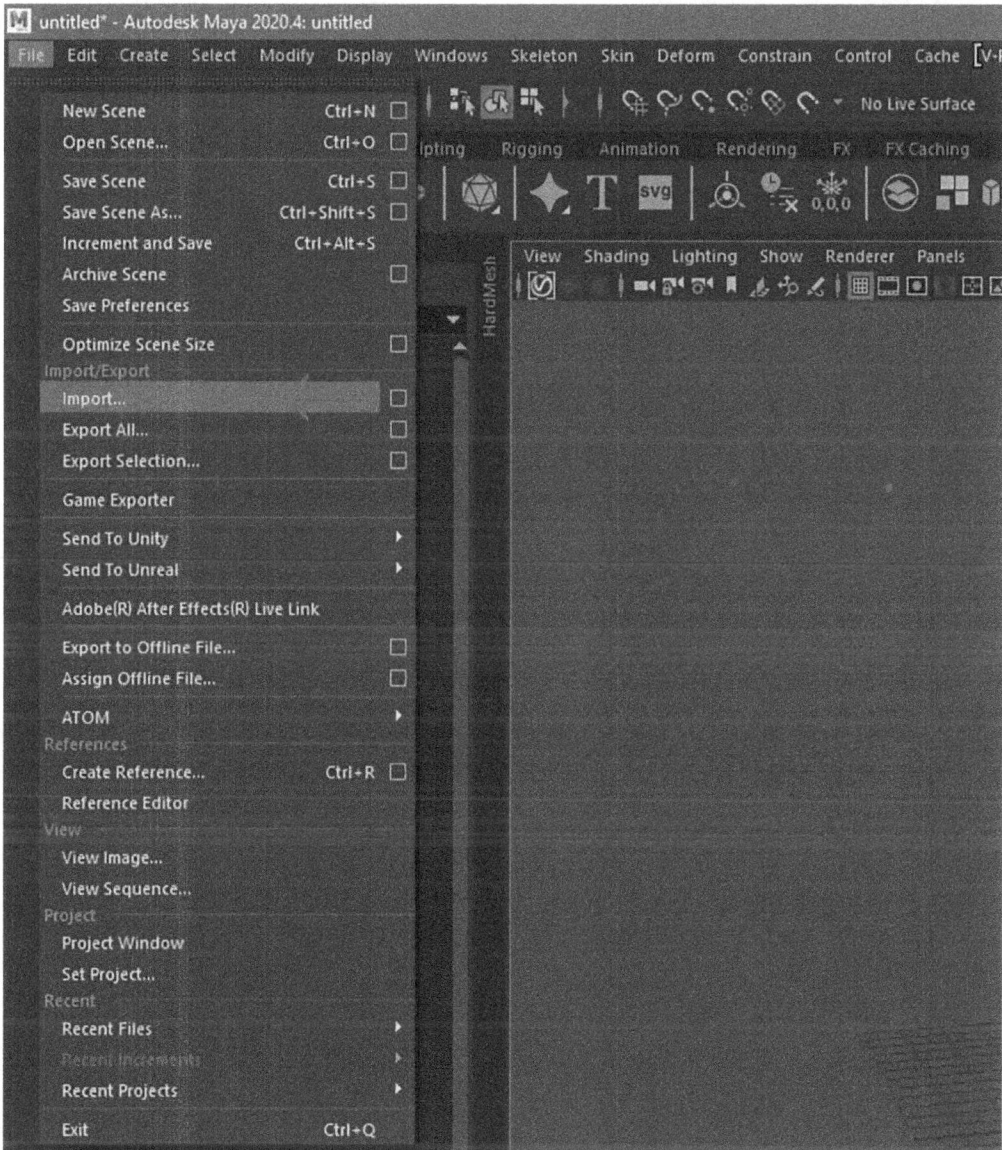

Figure 1.9 – Import… option

2. Choose the Basic_Model.fbx file to import into Maya's new scene.

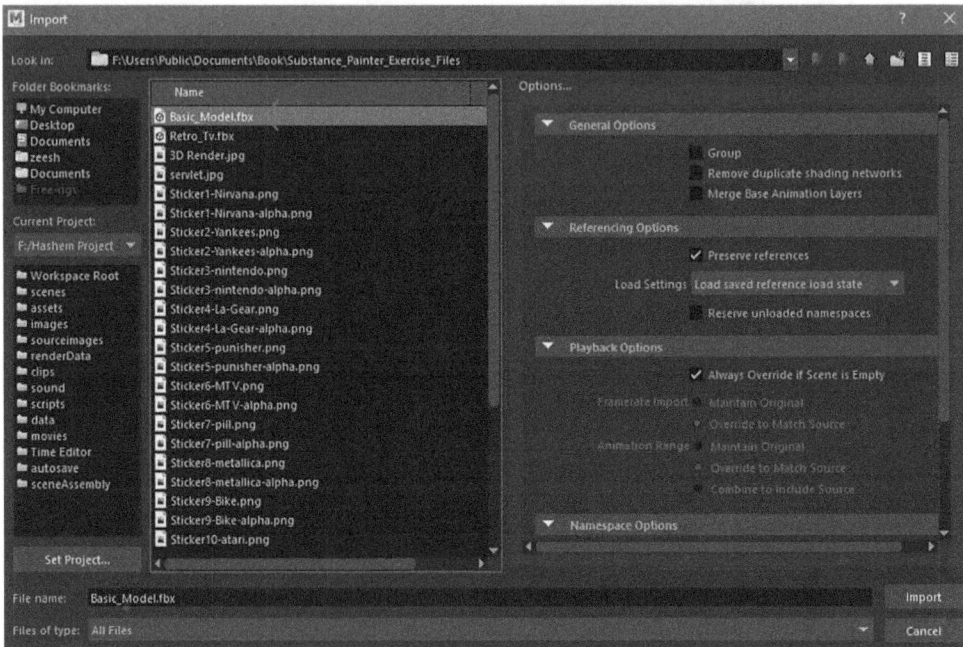

Figure 1.10 – Basic_Model.fbx

3. You will notice, in the Maya **Outliner** pane, that Basic_Model.fbx comprises three parts (**Handle**, **Cap**, and **Base**).

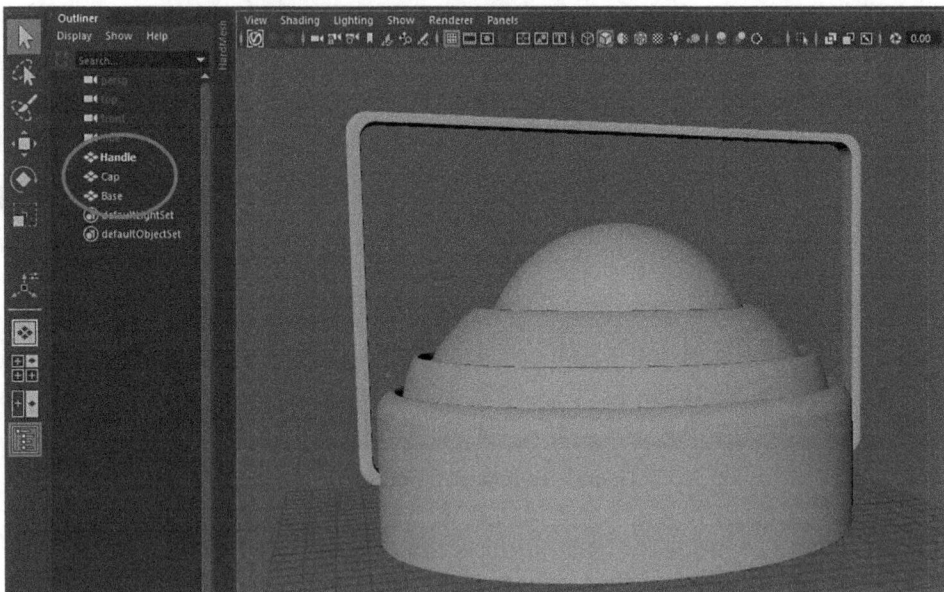

Figure 1.11 – Parts of Basic_Model.fbx

> **Note**
>
> If you export the current model to Adobe Substance Painter, it will be exported as a single object; therefore, you will need to apply different color materials to each part of the 3D model.

4. Select the base of the model (*step 1* in *Figure 1.12*) and click on the **Attribute Editor** tab on the right panel (*step 2*); click the **Material** tab (*step 3*), change the **Lambert** material name to Model_Base (*step 4*), and change the **Color** of the Lambert material (*step 5*):

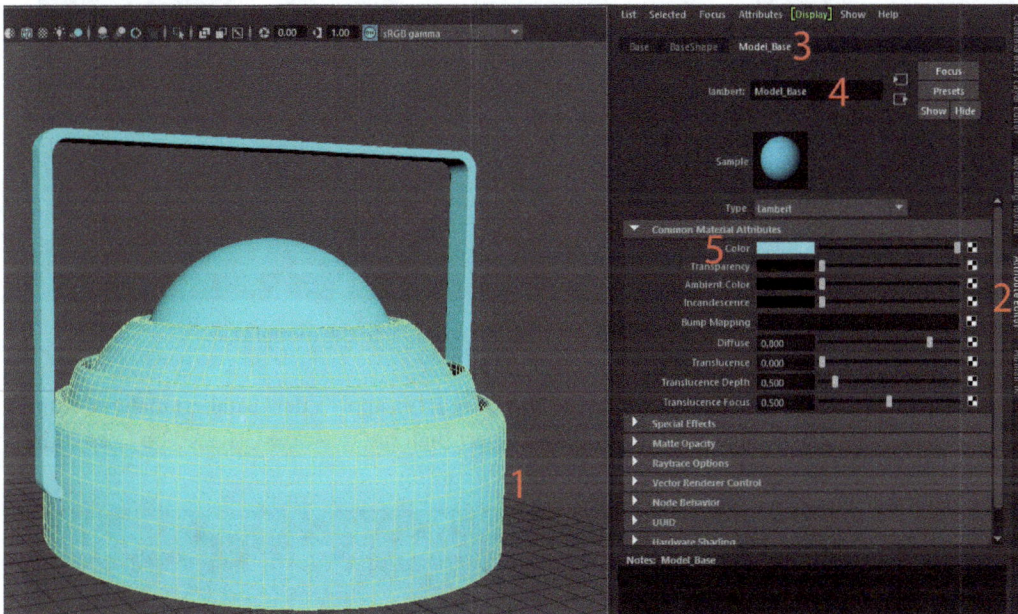

Figure 1.12 – Customizing the Lambert material of the model's base

Congratulations! You have created your first texture set.

5. Next, keep **Attribute Editor** open and select the basic model's cap (*step 1* in *Figure 1.13*). Click on the **Rendering** shelf (*step 2*), choose **Lambert Material** (*step 3*), and in **Attribute Editor**, change the **Lambert** material name to Model_Cap (*step 4*), and change the material color (*step 5*):

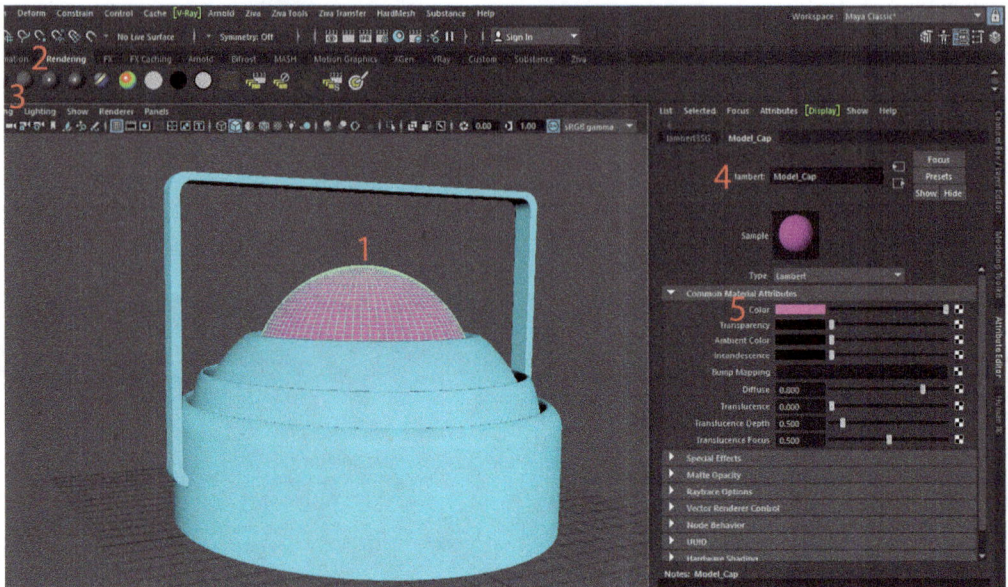

Figure 1.13 – Applying a different Lambert material to the model's cap

6. Next, select the basic model's handle (*step 1* in *Figure 1.14*), apply a new Lambert material from the shelf (*step 2*), change the name of the Lambert material to `Model_Handle` (*step 3*) in **Attribute Editor**, and then change the color (*step 5*):

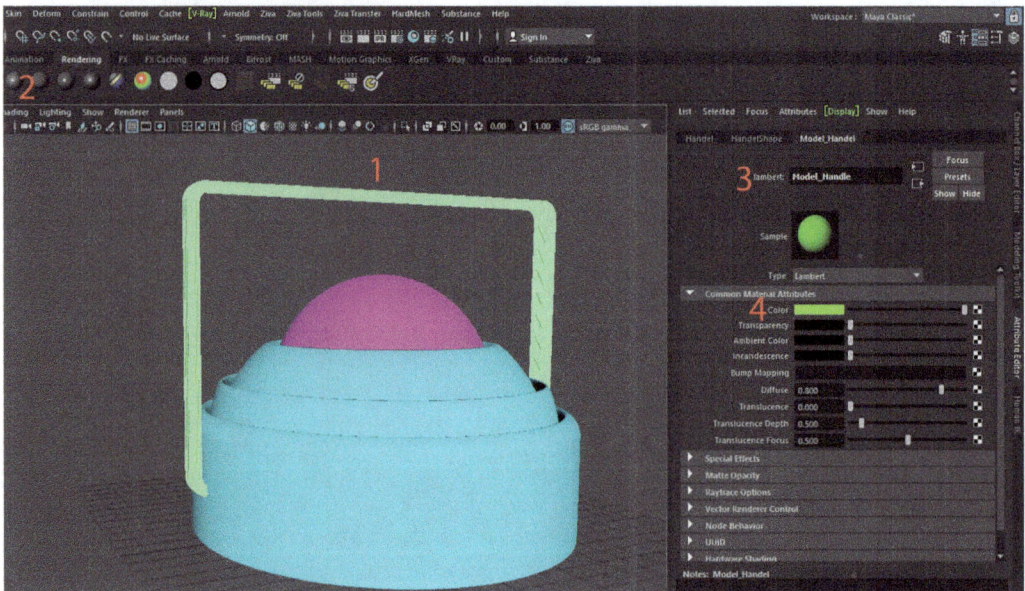

Figure 1.14 – Applying a different Lambert material to the model's handle

7. Now, as the model has a different Lambert material, it's time to export it. Go to the **File** menu and choose **Export All…**.

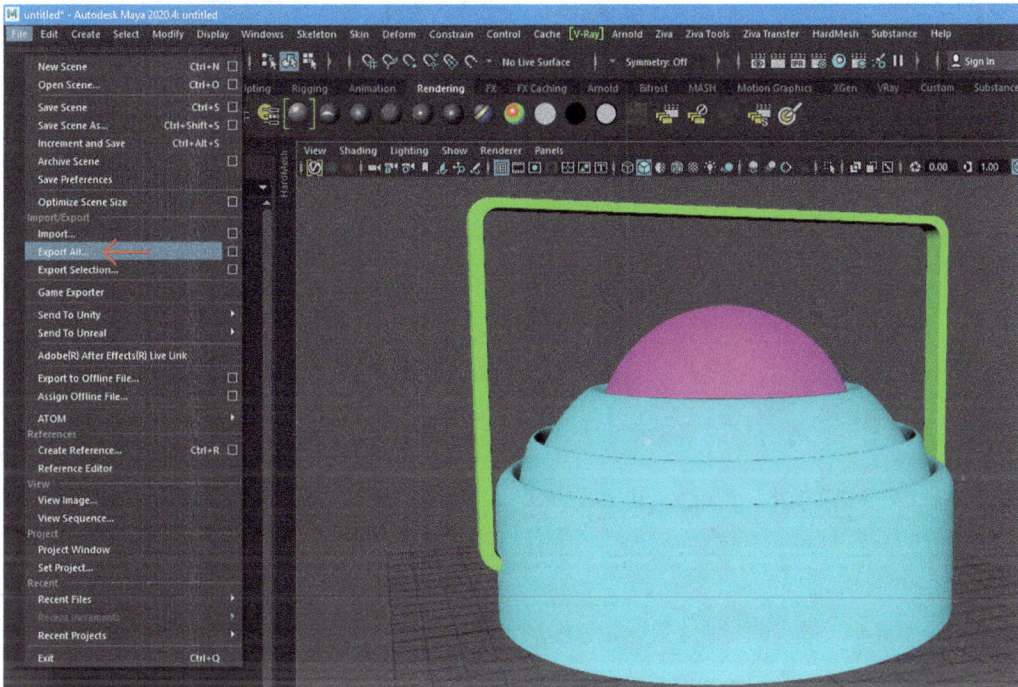

Figure 1.15 – Exporting the basic model

8. Export the basic model as an FBX file, and rename the file to `Basic_Model_Texture_Set.fbx`.

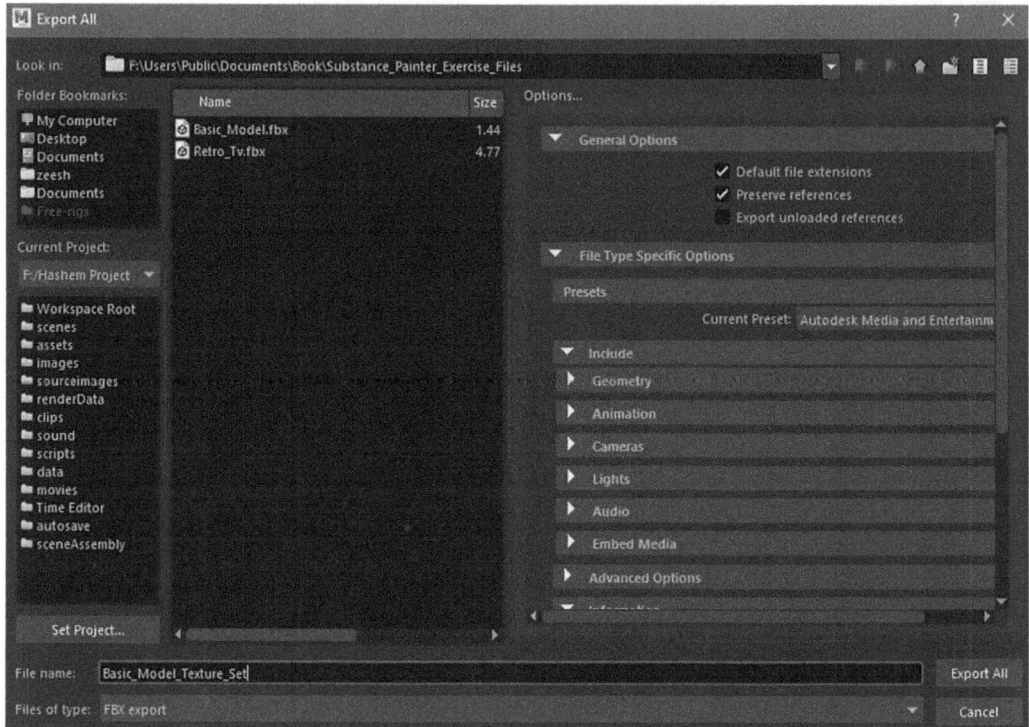

Figure 1.16 – Exporting the basic model as FBX

9.   Now, it's time to import the `Basic_Model_Texture_Set.fbx` file in Adobe Substance 3D Painter. So, open Adobe Substance 3D Painter and from the **File** menu, choose **New…**.

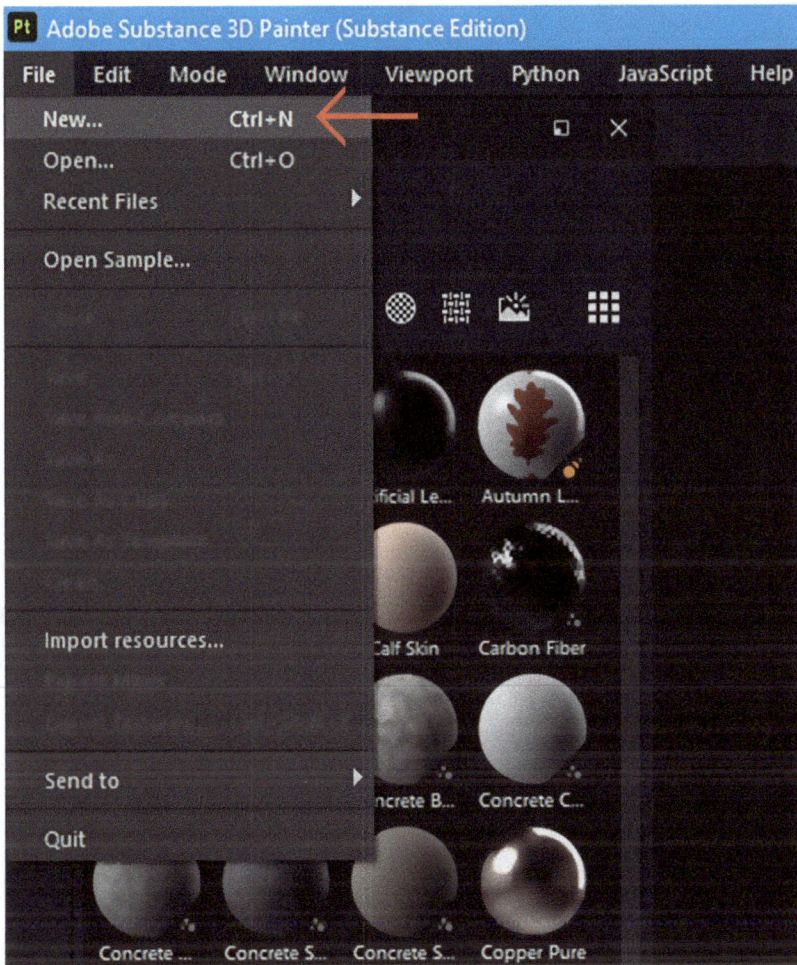

Figure 1.17 – Creating a new file in Adobe Substance 3D Painter

10. Once the **New project** window opens, set **Template** to **PBR - Metallic Roughness Alpha-blend (starter_assets)**. This is the most common template as it holds all the necessary texture elements. For **File**, select `Basic_Model_Texture_Set.fbx`; then, click **OK**.

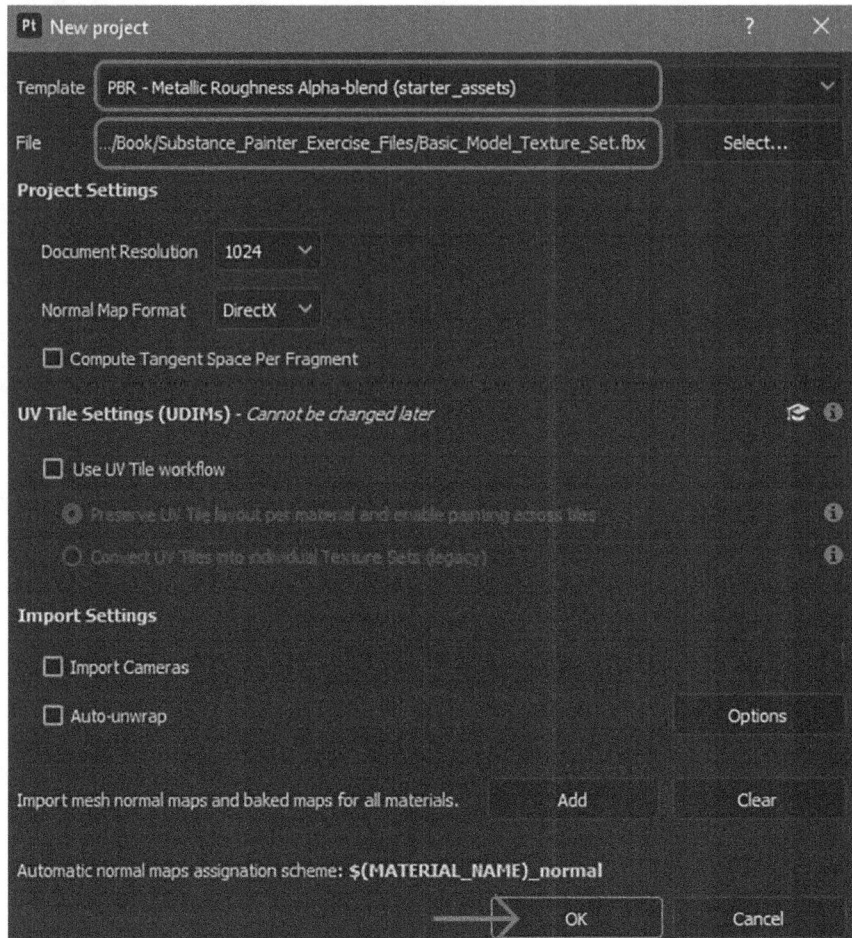

Figure 1.18 – Choosing the template and selecting the 3D model

11. Once the Basic_Model_Texture_Set.fbx file is opened in Adobe Substance 3D Painter, you will notice that the Lambert materials that were applied to each part of the model are now converted into a texture set.

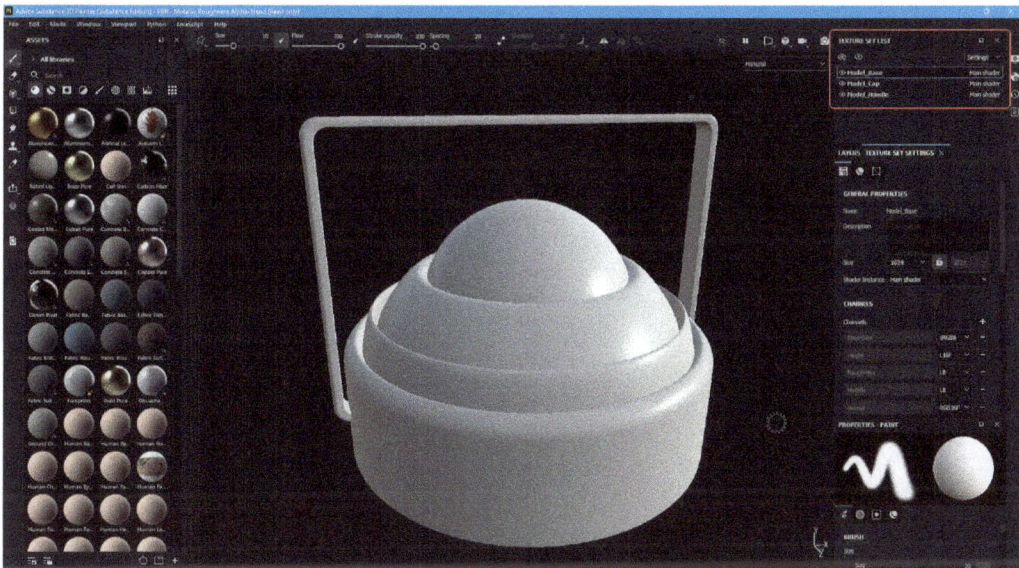

Figure 1.19 – Lambert material as a texture set

With the help of the texture set, you can now have more control over your paintings and material application.

Now that you know how a 3D mesh's different parts can be divided into texture sets, it's time to learn how we can create ID naps on a single mesh using the vertex color option.

Creating ID maps with vertex color

Sometimes, you do not want to separate a 3D model into different texture sets, but you still want to apply different texture maps to different polygon faces of that model without separating them into small models.

This is where the ID maps come in:

1. To see an example of an ID map, open Adobe Substance 3D Painter and choose **Start painting**; once the Painter mascot is opened, hide the body and base of the mascot.

2. Then, from the view modes, choose **ID** under **Mesh maps**, as shown in *Figure 1.20*. In this menu, you can see all the different maps, and if you want to see the material again, you can choose it in the lighting category:

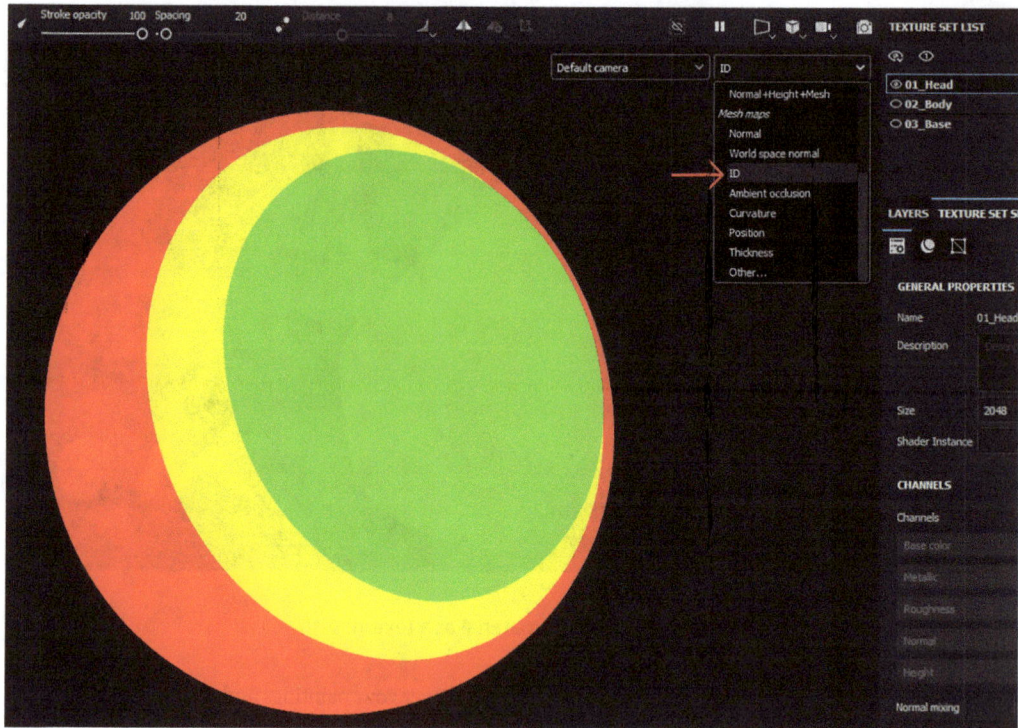

Figure 1.20 – ID map

Notice in *Figure 1.20* that the head of the mascot is divided into three ID maps, which will allow you to apply three different materials to a single one-piece model.

3. To apply a material to the choice of your ID map, you have to press *Ctrl* (Windows) or *Command* (Mac) while dragging and dropping the material onto the head of the mascot.

4. You will notice that once you do that, Adobe Substance 3D Painter highlights the ID map before the material is dropped on it, as shown in *Figure 1.21*:

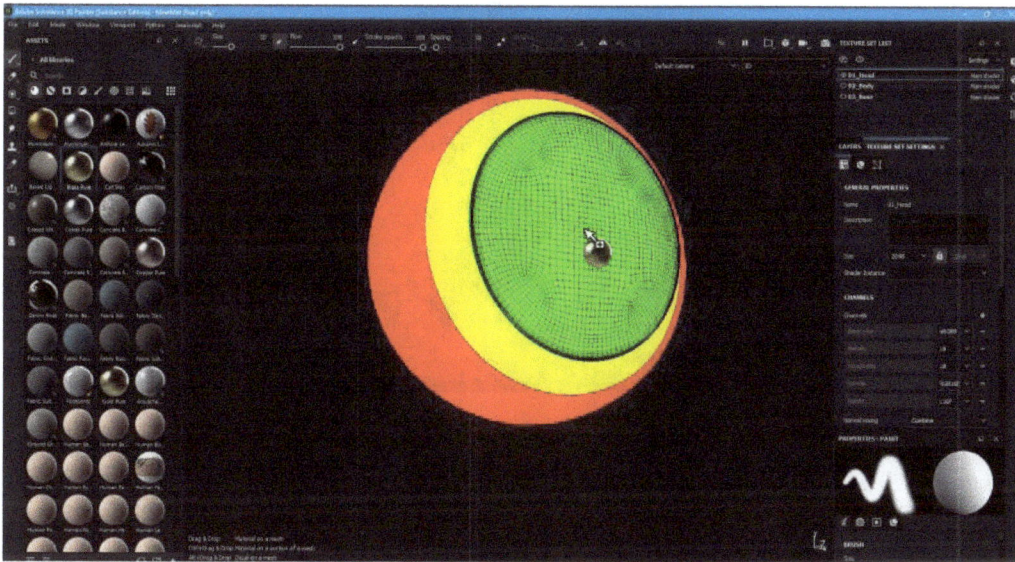

Figure 1.21 – Applying different materials to different ID maps

It then changes back to its previous state, as shown in *Figure 1.22*:

Figure 1.22 – Material applied on a selected ID map

With the help of the ID map, you can apply different materials to each ID map, as shown in *Figure 1.23*:

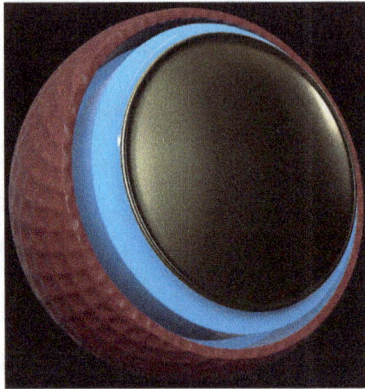

Figure 1.23 – Different materials applied to different ID maps on a one-piece mascot head model

ID maps can be created in any 3D application; for this project, we will create ID maps on `Basic_Model_Texture_Set.fbx` through the following steps using Autodesk Maya:

1. Open Autodesk Maya and import `Basic_Model_Texture_Set.fbx`, then select and hide the handle and the cap of the basic model from **Outliner** by pressing *Ctrl + H* (Windows) or *Command + H* (Mac).

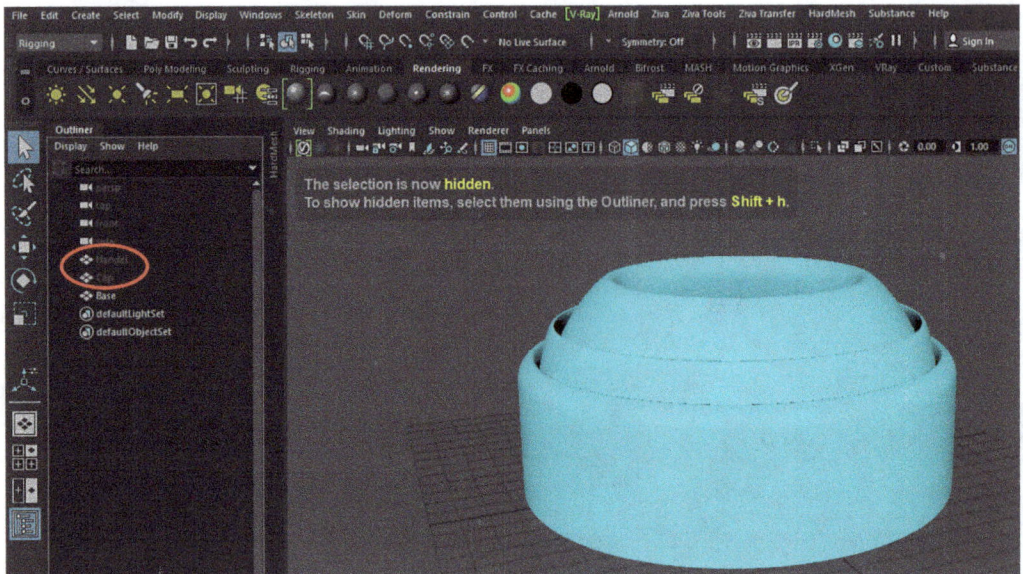

Figure 1.24 – Hiding the handle and cap

2. Right-click the base of the 3D model and choose **Face**.

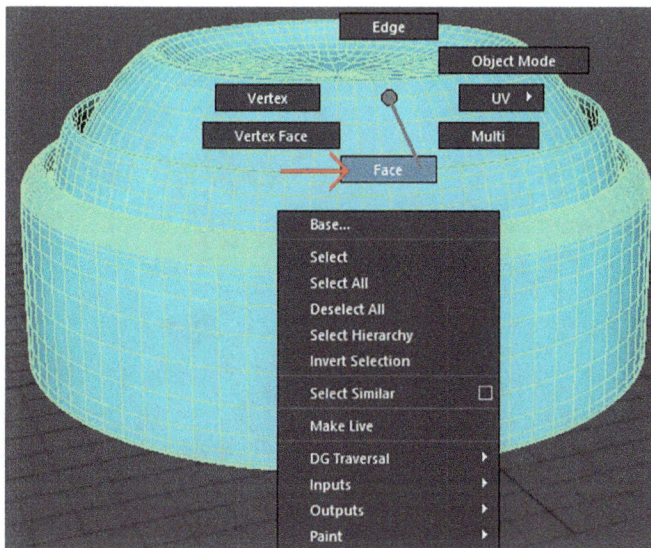

Figure 1.25 – Right-click on the model and choose Face

3. Now, select the bottom polygon faces of the model, as shown in *Figure 1.26*:

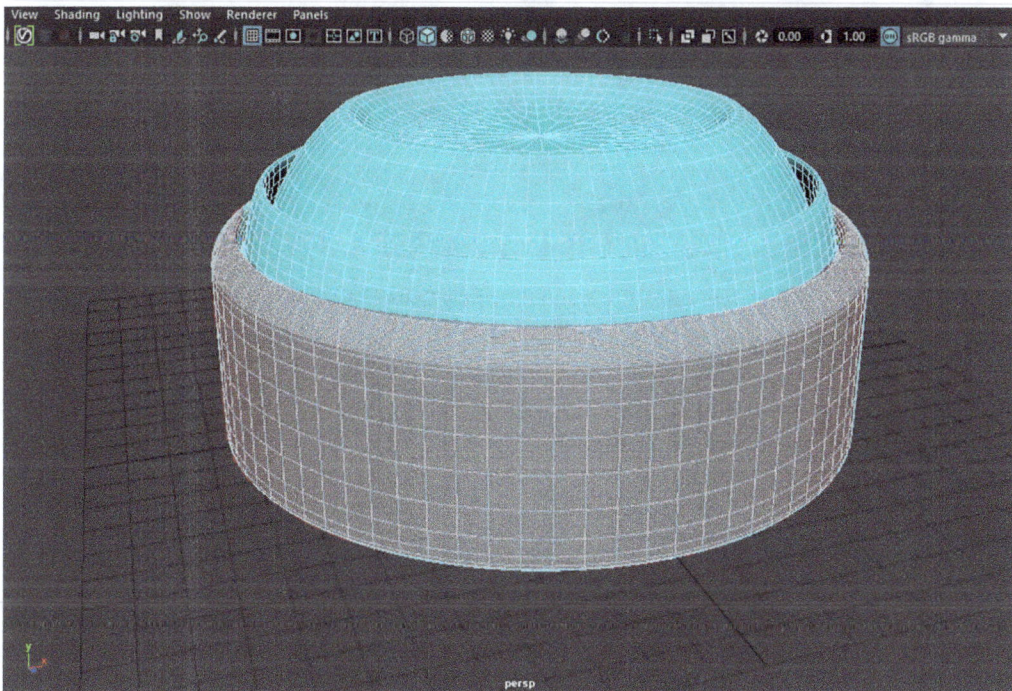

Figure 1.26 – Select the bottom polygon faces

4. Make sure you are in **Modeling** mode, then go to the **Mesh Display** menu and check the square box next to **Apply Color**.

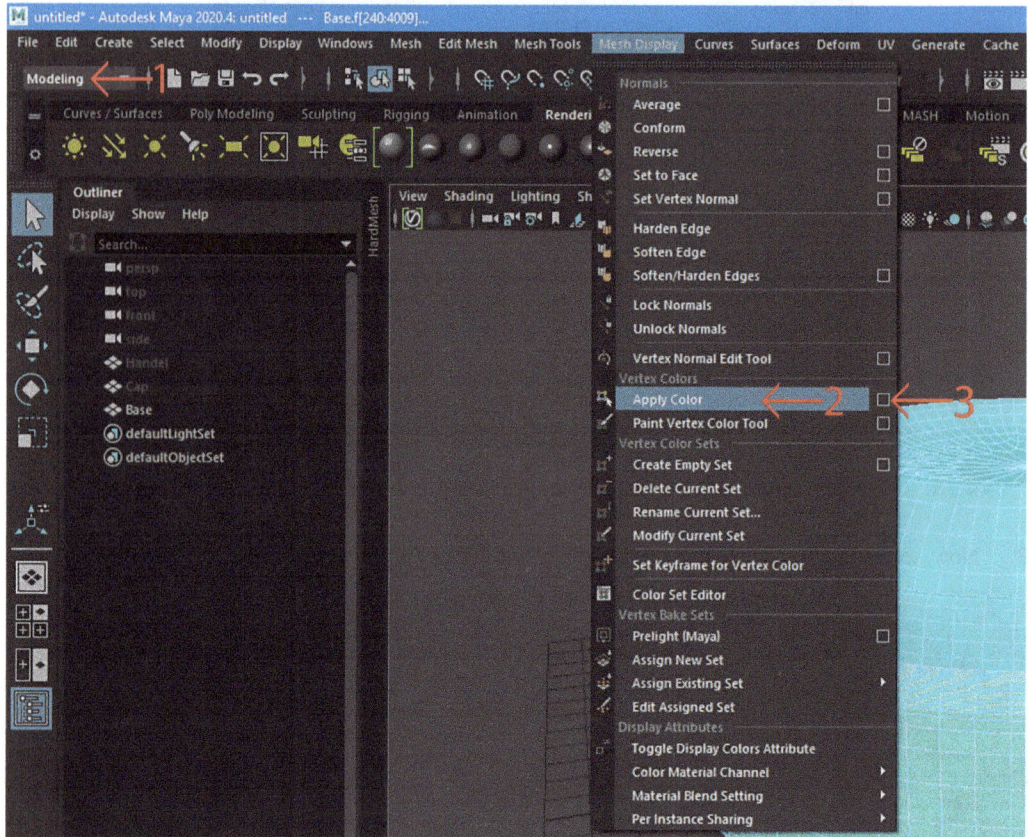

Figure 1.27 – Choosing the Apply Color option

The **Apply Color Options** window will open as soon as you click the box next to **Apply Color**.

5. Before anything else, make sure you click the **Edit** menu of **Apply Color Options** and choose **Reset Settings** to reset any old settings to avoid errors.

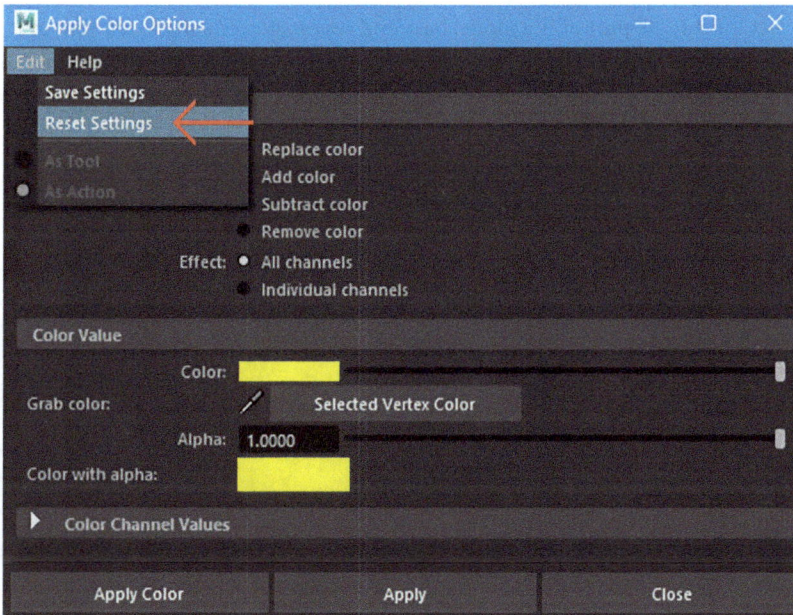

Figure 1.28 – Reset Settings

6. After resetting the settings, choose a unique color under the **Color** value, click **Apply**, and click **Close** to close the **Apply Color Options** window.

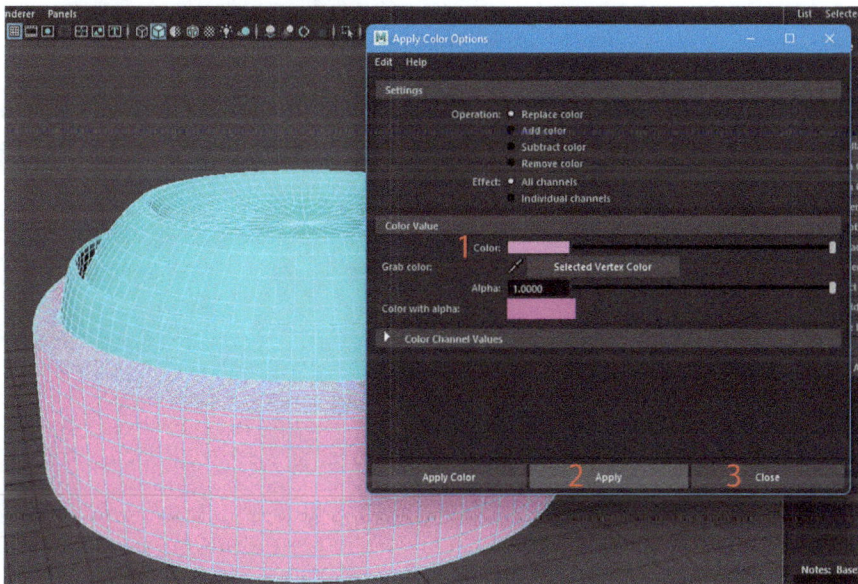

Figure 1.29 – Applying a unique color to selected faces of the base of the model

7.   Then, select the middle part of the model base and repeat step 6 using a different unique color. After that, select the top part of the model base and repeat step 6. In the end, close the **Apply Color Options** window, and your model base should look like that shown in *Figure 1.30*:

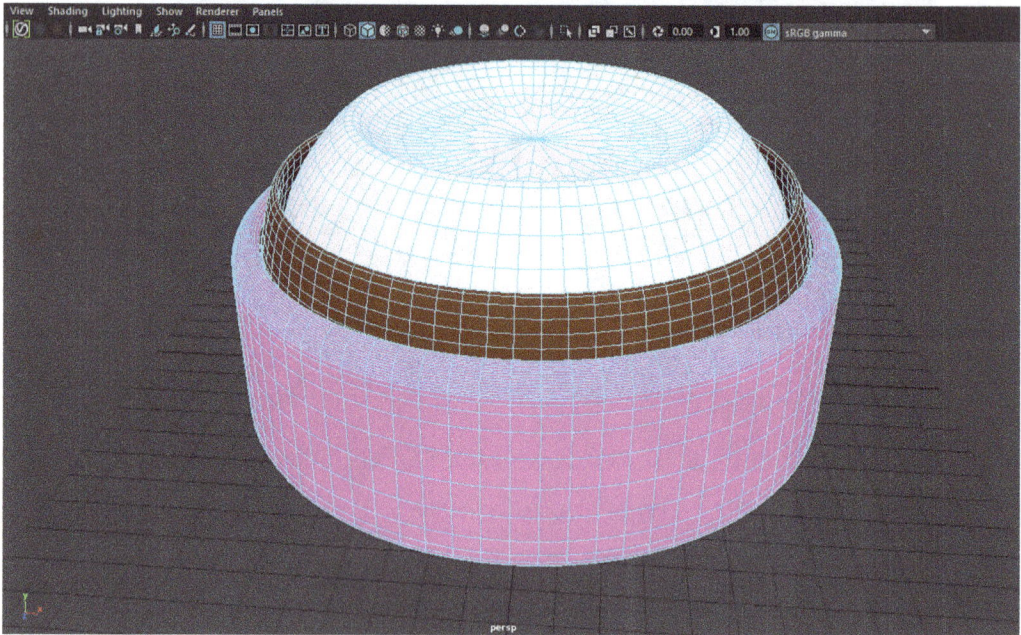

Figure 1.30 – ID map-ready model

8.   Right-click on the model again and choose **Object Mode**, then unhide the handle and the cap of the model by selecting their names in **Outliner** and pressing *Shift + H* (for both Windows and Mac).

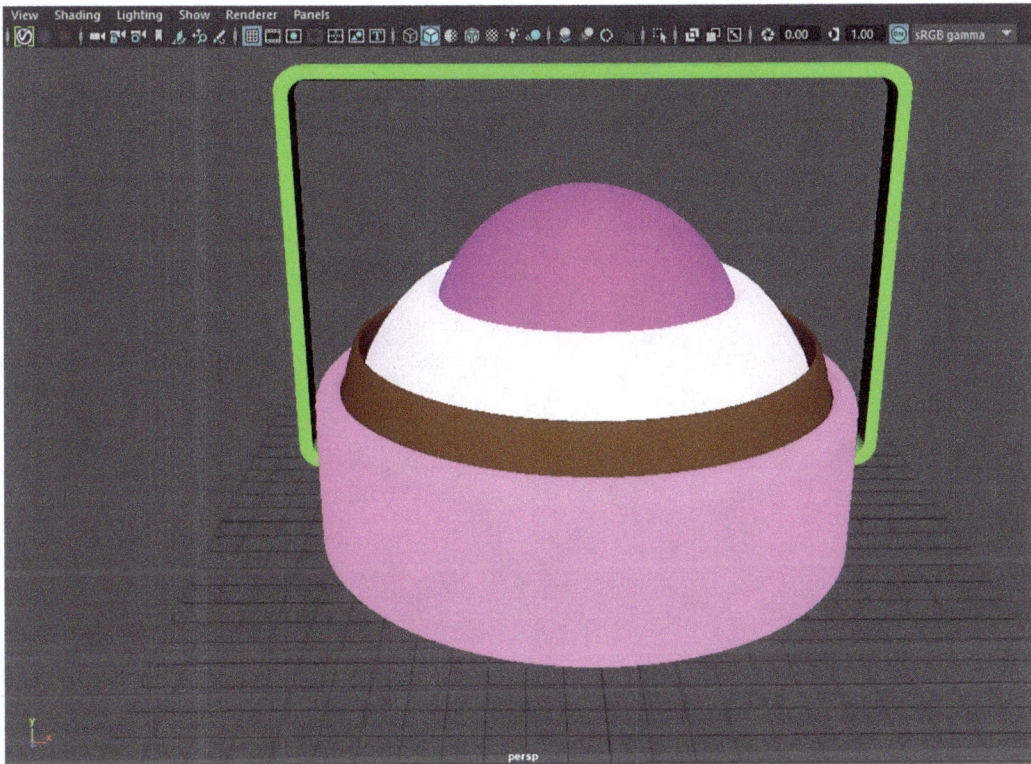

Figure 1.31 – ID map-ready model

9. Click on the **File** menu and choose **Export All…**. Export the file as `Basic_Model_ID_ Map_Ready.fbx`; this file can be now opened in Adobe Substance 3D Painter with the ID maps and previously created texture sets.

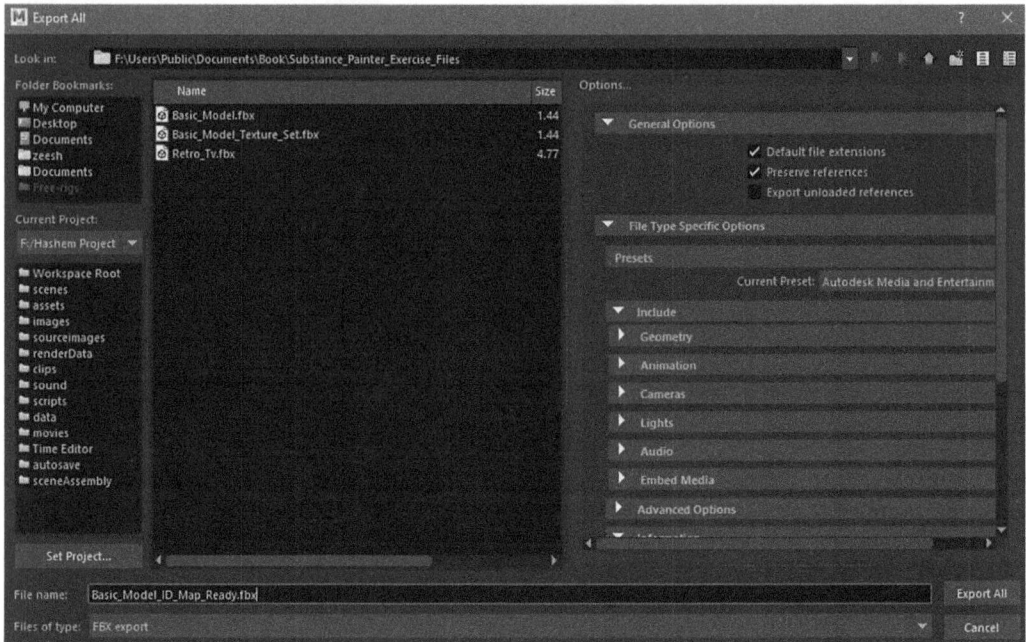

Figure 1.32 – Exporting as Basic_Model_ID_Map_Ready.fbx

However, to work on ID maps, we must bake the textures. We shall see how to do so in the next section.

Baking textures in Adobe Substance 3D Painter

Once the 3D model is exported from any 3D software, it can be easily imported inside Adobe Substance 3D Painter; however, you will not be able to paint over it or apply any material to it until or unless you bake its textures.

> **Baking textures**
>
> The process of storing information from a 3D geometry to a texture file is known as **baking** (bitmap). This procedure usually entails two meshes, a high-poly mesh and a low-poly mesh. Because a high-poly mesh includes a large number of polygons (sometimes millions), it can display high-resolution 3D detail. Because a low-poly mesh has a smaller number of polygons (typically only a few thousand), it is less expensive to store and render. Baking textures gives you the best of both worlds: the high degree of detail provided by a high-poly mesh and the low performance costs provided by a low-poly mesh. The information from the high-poly mesh is transferred to the low-poly mesh and preserved as a texture during the baking process.

First, let's take a look at the different types of mesh maps that can be baked in Adobe Substance 3D Painter:

- **Normal:** This function creates a tangent space. A normal map uses a normal angle at the object's surface. This type of normal map is frequently used to shade and replicate details on a low-poly 3D mesh's surface.

- **World space normal:** This type of normal map can tell which side of an object is up in the environment and which side is down. Mask generators in Adobe Substance 3D Painter, for example, can use this data to deposit dust on any surfaces that are facing up.

- **ID:** This is the mesh map we learned how to create in Autodesk Maya in the previous section. This map may be used to create colored areas, which makes masking and selecting with other tools much easier.

- **Ambient Occlusion:** This function creates a texture with ambient shadows. With this function, a white surface may appear to be overcast from the sky. Moreover, Ambient Occlusion helps to darken enclosed and sheltered areas and provide a more natural look.

- **Curvature:** This function creates a texture with edges and voids. To imitate edge wear, the curvature is frequently utilized to harm the geometry's edges.

- **Position:** Creates a texture with the coordinates of each point on the 3D mesh. To put it another way, this tells Painter where the mesh's top and bottom, front, and rear, and left and right sides are.

- **Thickness:** Creates a texture with the thickness of the 3D mesh in it.

Now that you are familiar with the various mesh maps, it's time to learn how to import any 3D mesh in FBX format and bake its texture so that we can apply materials and paint on it:

1. Open Adobe Substance Painter, go to the **File** menu, and choose **New…**; select **PBR - Metallic Roughness Alpha-blend (starter_assets)** under the **Template** option. We are choosing a **physically based rendering (PBR)** material because it produces photorealistic high-quality results.

2. Select `Retro_Tv.FBX` under the **File** option, keeping in mind Painter only accepts FBX files.

3. Set **Document Resolution**; if you are looking for high-quality results, then choose a higher value. For this project, we will use `2048`.

4. Set **Normal Map Format** depending on your preference; if you are not sure about your choice, then keep it as **DirectX**.

5. Click **OK**.

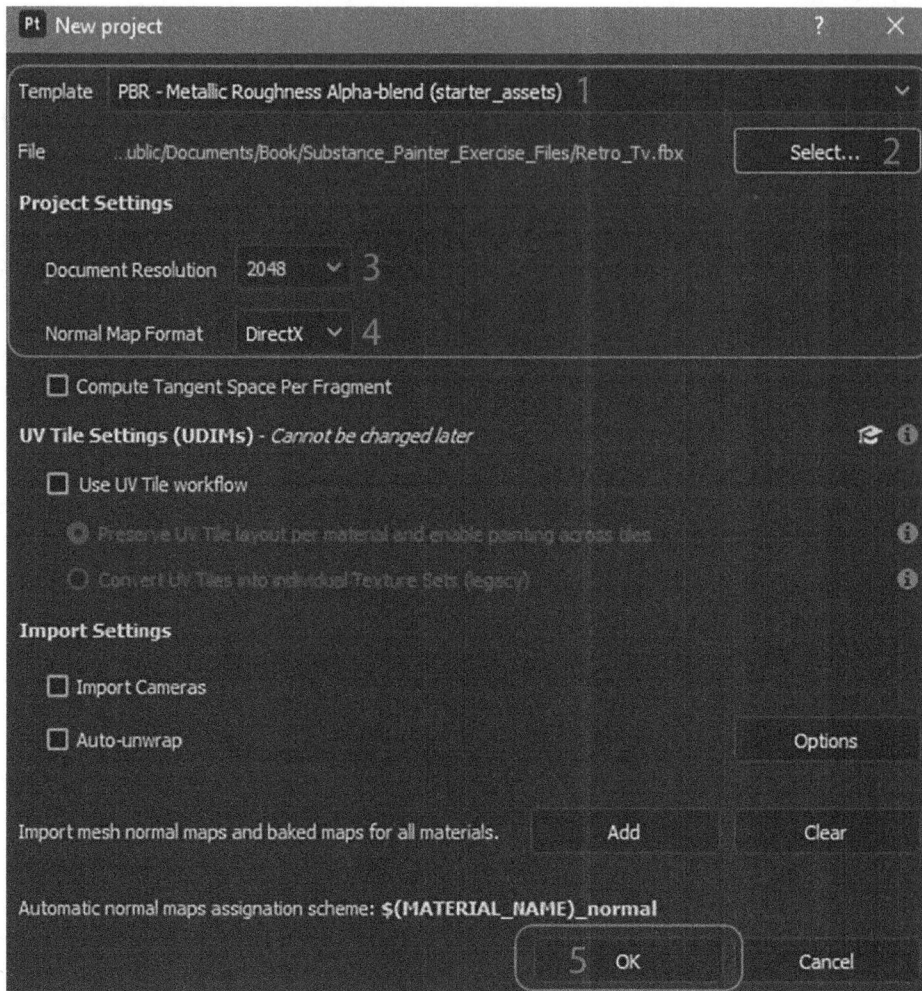

Figure 1.33 – Opening Retro_Tv.FBX in Adobe Substance 3D Painter

6. If your 3D model's UVs are divided into UDIMs, you can select **Use UV Tile workflow**, under **UV Tile Settings (UDIMs)**, as shown in *Figure 1.33*; then, choose **Preserve UV Tile layout per material and enable painting across tiles**, which will allow you to paint on the 3D model by assuming its multiple UDIMs are one single UV shell.

    This way, you will be able to paint continuously; otherwise, choose **Convert UV Tiles into individual Texture Sets (legacy)**, because it works in most 3D applications.

7. Select any texture set from **Texture Set List**, go to the **TEXTURE SET SETTINGS** then click on the **Bake Mesh Maps** option.

Figure 1.34 – Bake Mesh Maps

8. Keep the **Output Size** setting of the texture as **2048** for high-quality results. When you work on a project in a lower resolution, such as 512/1024, you can still bake with higher sizes, for example, 2048/4096 to produce high quality results.

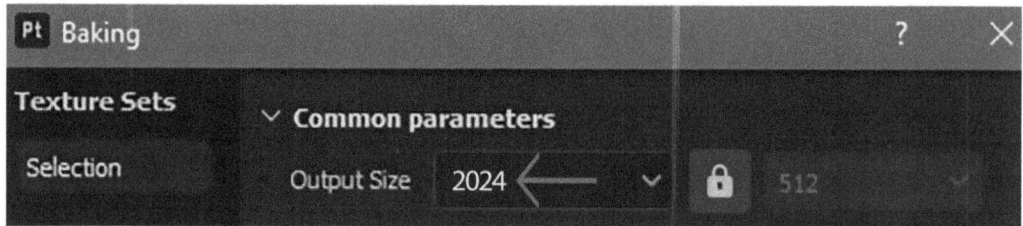

Figure 1.35 – Texture Output Size

9. Click on the **Selection** option and make sure all the texture sets are selected to bake all the textures on all the model parts.

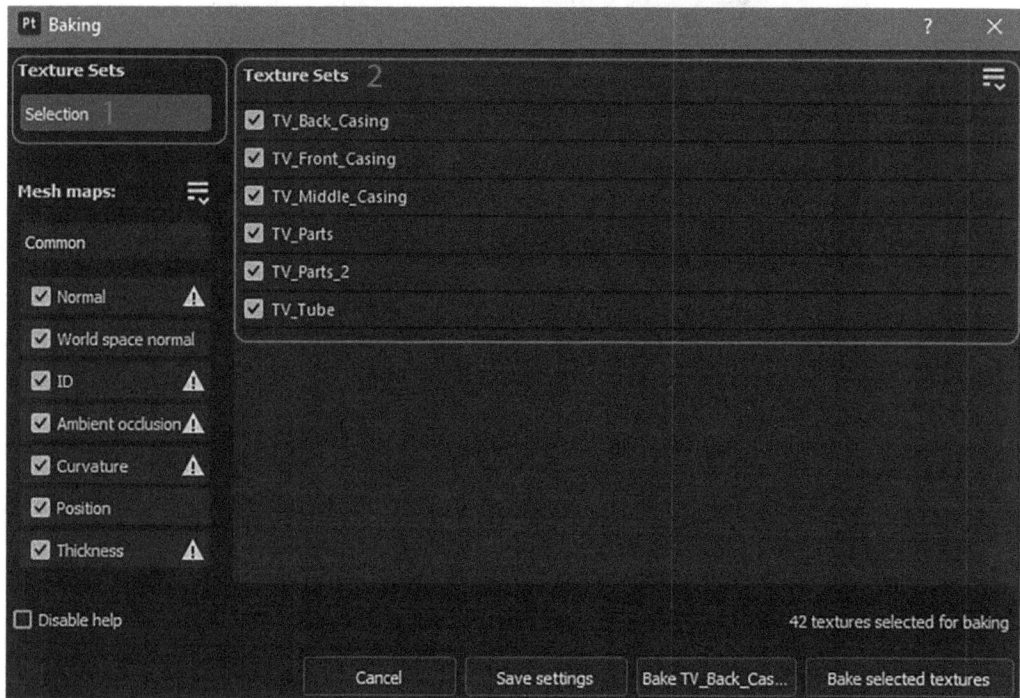

Figure 1.36 – Selected texture sets

10. For this project, we will deselect the **Normal** mesh map, because we do not need a normal map in our model. We will also deselect the **Thickness** mesh map as it is used for sub-surface scattering, which is ideal for skin and wax types of material but we do not have any of these in our material applied to our model.

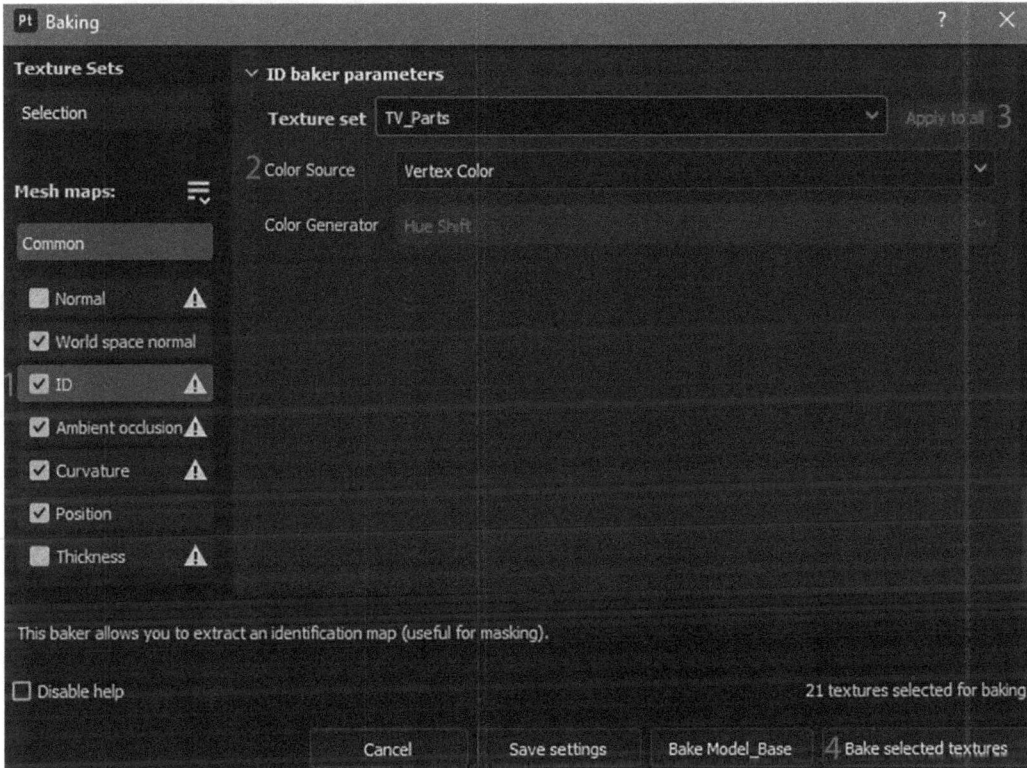

Figure 1.37 – Baking settings

11. Select the **ID** map, and for **Color Source**, choose **Vertex Color** as we have divided the base of the model by vertex color. Then, click on **Apply to all** to apply the settings to all the selected texture sets, and click on **Bake selected textures** to bake all the selected textures.

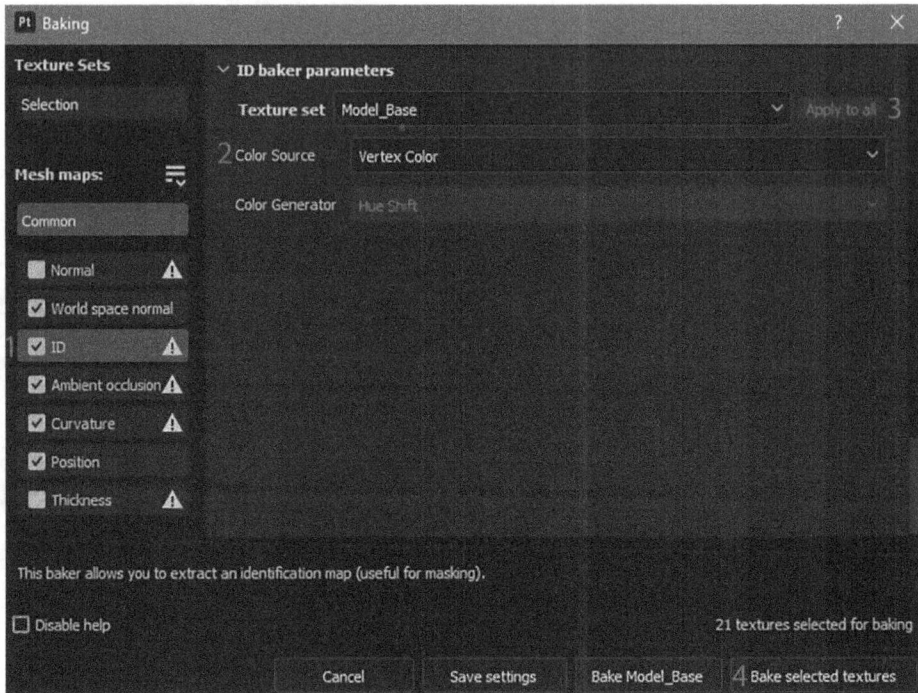

Figure 1.38 – Bake selected textures

12. As soon as you click on the **Bake selected textures** option, the baking process will start. When the process finishes, click the **OK** button.

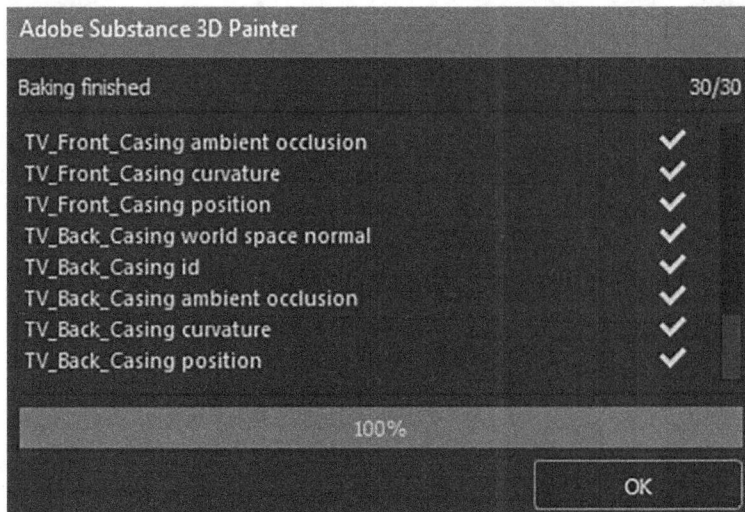

Figure 1.39 – Baking process completed

13. Click on the view modes and choose **ID**. You can see now that the model base has three ID maps represented by three different colors, which will allow you to apply a myriad of materials, textures, or paints to it.

Figure 1.40 – ID maps

Summary

In this chapter, we have gone through many important concepts and steps. First, we learned the concept of texel density, which allowed us to create high-quality texture maps; without proper texel density, even your 8K texture maps will look blurry. Second, we learned how to divide a multi-part 3D mesh into different texture sets.

Without texture sets, your multi-part 3D mesh will appear as a 3D model that has instances instead of independent parts, which makes make it impossible to paint over any 3D model. Lastly, we learned how to divide a single 3D mesh into different ID maps, which will make it easier for us to apply different assets and paint to that 3D model. We have also learned that ID maps on a 3D mesh can be easily created using vertex colors in any 3D application and then imported into Adobe Substance 3D Painter.

In the next chapter, we will go through the **Assets** panel, which contains all the important resources that are essential to creating textures. We will also learn how we can create custom tabs in the **Assets** panel, which will be helpful in organizing your project. We will also get the hang of importing your own external assets and study different types of assets in Adobe Substance 3D Painter and how to apply them.

2
Working with Assets in Adobe Substance 3D Painter

Adobe Substance 3D Painter is owned by Adobe, and therefore, there is a strong connection between Painter and Adobe products, especially Adobe Photoshop. A lot of artists create resources inside Adobe Photoshop and import them into Painter.

In this chapter, we will first go through conceptual knowledge of the **Assets** panel, which is essential for creating any Adobe Substance 3D Painter project as it contains important resources. Then, we will go through methods to create our custom tabs in the **Assets** panel, which will help us organize our project and make our texturing and painting process agile.

When you will move to the real world, you will see for yourself that organizing a project is one of the most crucial steps because if you start working on your project without organizing it, you will be lost halfway through and miss out on important things. Once we have a comprehensive understanding of the **Assets** panel, we will learn how to apply these assets.

In this chapter, we will be covering the following topics:

- Familiarizing ourselves with the **Assets** panel
- Importing resources and creating custom tabs in the **Assets** panel
- Understanding the different types of assets in the **Assets** panel
- Applying assets to 3D meshes

Familiarizing ourselves with the Assets panel

You will find the **Assets** panel on the left side of the interface. The panel has a dock and undock option on the panel's top-right side, as shown in *Figure 2.1*:

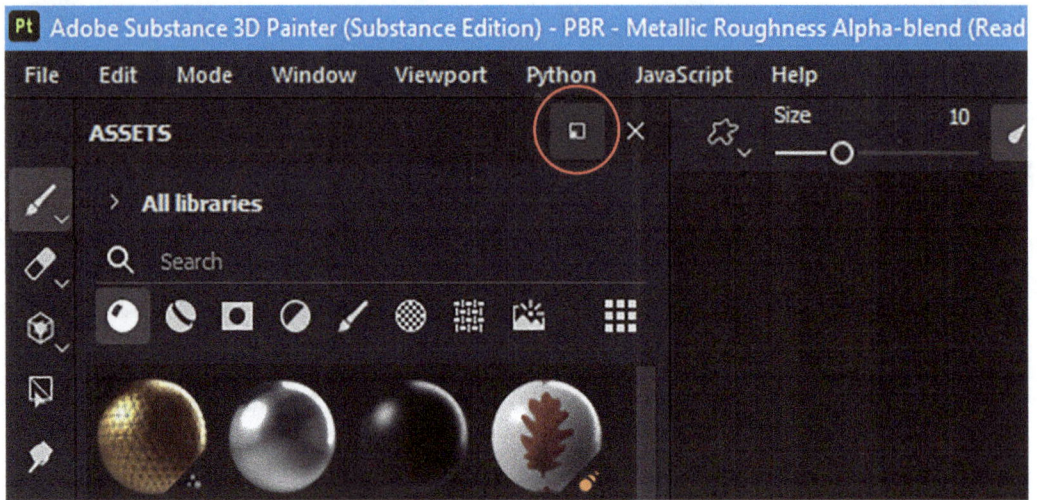

Figure 2.1 – Docking and undocking option

With the help of this option, you can undock the panel, resize it, and place it anywhere you want, as shown in *Figure 2.2*; you can dock the panel back to its original place by simply moving it to its original position and waiting till it docks back by itself.

In case you change your UI by mistake, you can select **Window** and then click on **Reset UI**, which will restore your interface to its default state. Whenever you try to dock back any panel, Adobe Substance 3D Painter makes space for the panel by moving other panels, as shown in *Figure 2.2*:

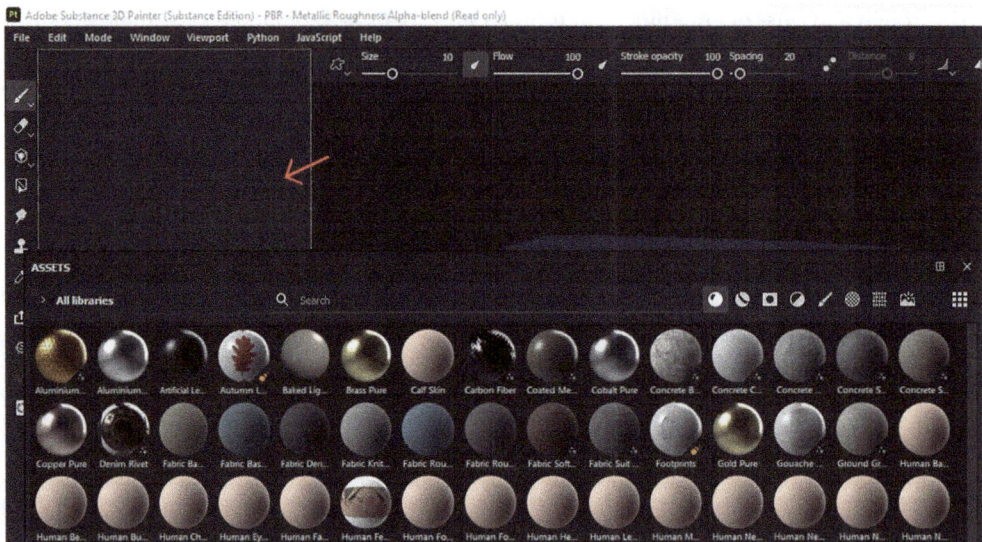

Figure 2.2 – Docking back the panel

Let us now familiarize ourselves with all the elements in the **Assets** panel shown in *Figure 2.3*:

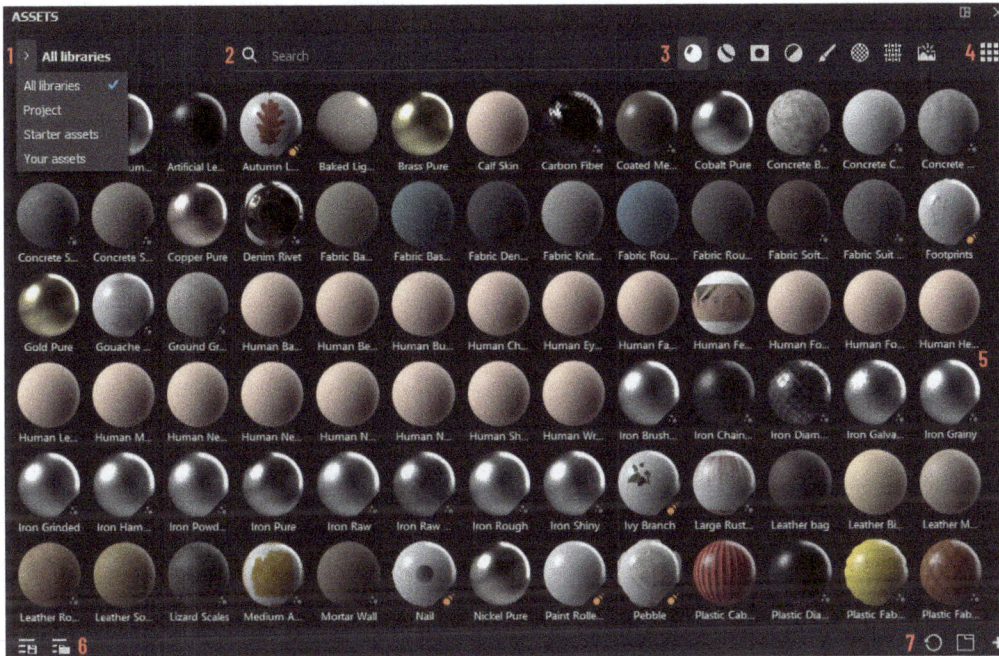

Figure 2.3 – Assets panel

Important elements in the preceding figure have been labeled with numbers, and their descriptions are as follows.

1. **Breadcrumbs**: There are four different ways you can view libraries inside the **Assets** panel, which you can access through a dropdown:

    -   **All libraries**: This will show you all asset libraries.

    -   **Projects**: This will show you all assets that belong to your project mesh and the assets that are created after baking maps. However, you need to clear all the search results to view this option.

    -   **Starter assets**: This is the assets library that contains assets that are used most commonly in any project. Once you choose **Starter assets**, you will notice that there is an arrow next to its name, which will further filter the choice of your starter assets.

    -   **Your assets**: This is the library that shows the assets you created or imported from other files.

2. **Search field**: The search field allows you to quickly search any material, smart, mask, imported assets, and so on.

3. **Assets**: The icons here represent various assets (from left to right): Materials, Smart Materials, Smart Masks, Filters, Brushes, Alphas, Textures, and Environment maps. These assets are the most frequently used features in Adobe Substance 3D Painter to generate textures.

4. **Asset thumbnail size**: This helps users to customize the way they want to see their assets.

5. **Assets view**: This part of the interface displays all the assets based on the type you have chosen to display.

6. **(From left to right) Saved searches** and **Filter by path**: The **Saved searches** option allows you to save your recent searches, while the filter path gives you various filters to customize your search. With these options, you can save time if you are frequently searching for your assets.

7. **(From left to right) Reset all queries, Open a new sub-library tab,** and **Import resources**: **Reset all queries** will help you to remove all the saved searches, **Open a new sub-library** will allow you to open a tab in which you can show assets of your choice, and the **Import resources** option is a shortcut to import resources outside the Adobe Substance 3D Painter. We will learn about all of them in detail in later sections.

Now that we are familiar with the components of the **Assets** panel and what each can do, let us learn about importing resources into the panel to work on our project.

Importing resources and creating custom tabs in the Assets panel

Resources are the items that you import from external files; these can be created in any third-party software, such as Adobe Photoshop or Illustrator, and can comprise Smart Materials, Alpha Maps, Textures, and so on. On the other hand, tabs are used to access nested panels for quick access.

Importing resources

To import any resources in the **Assets** panel, use the following steps:

1. Click on the **File** menu and select the **Import resources...** option:

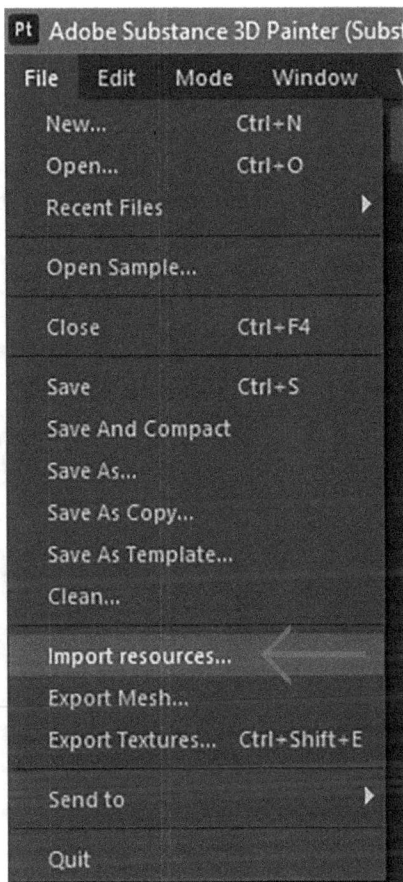

Figure 2.4 – Import resources

2. Once the **Import resources** panel is opened, click on the **Add resources** option:

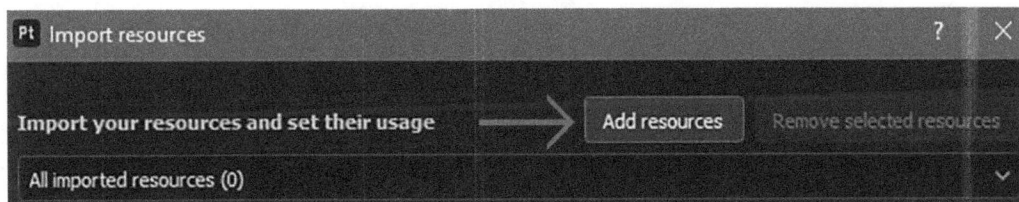

Figure 2.5 – Add resources

3. Select the resources (these are the same resources whose download link was given to you in *Chapter 1, Getting Started with Adobe Substance 3D Painter*). Select all the files from the resources folder except for any `.fbx` file and click **Open**.

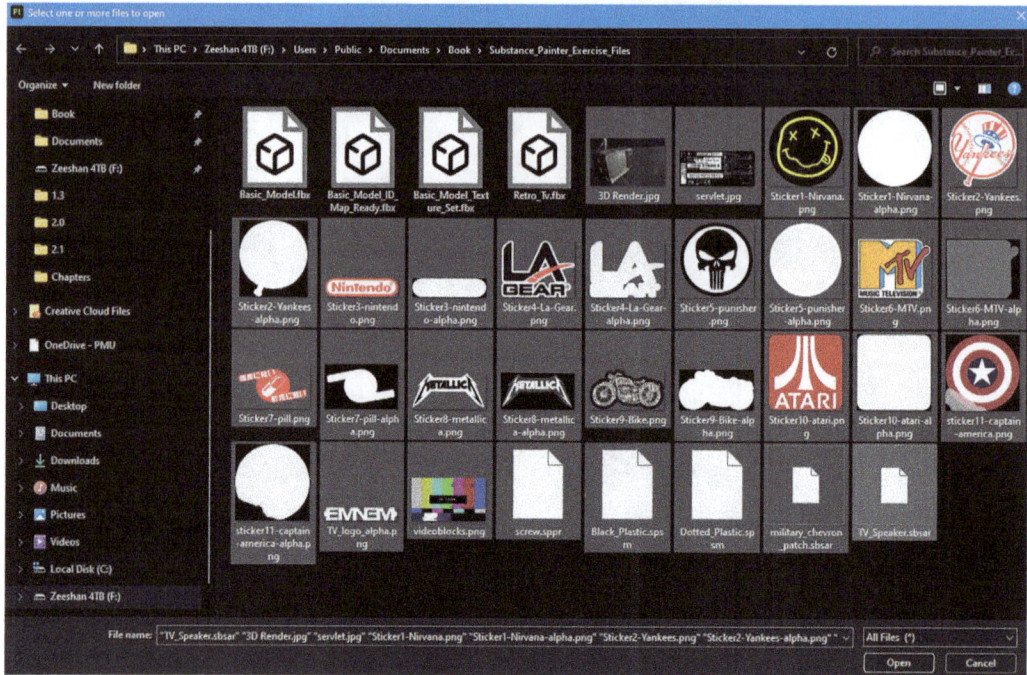

Figure 2.6 – Select resources

4. Once you click **Open**, all your resources will be collected in the resources list; however, you will not be able to import them yet because the type of these resources must be defined first:

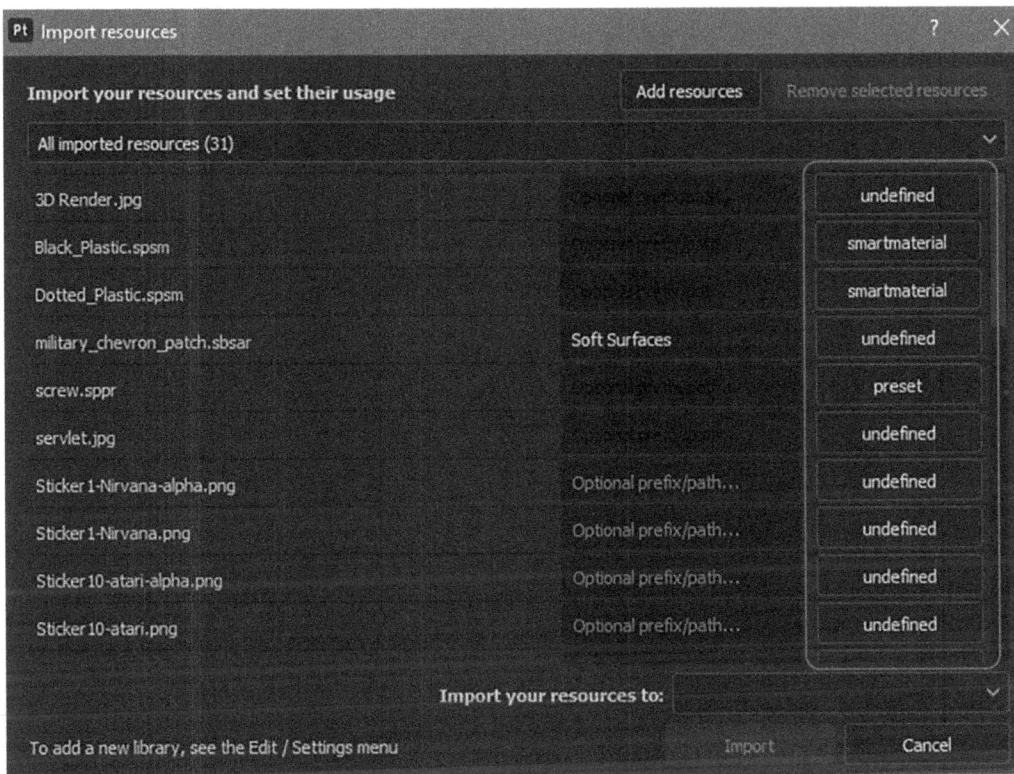

Figure 2.7 – Imported resources list

5. Select the **basematerial** type for `military_chevron_patch.sbsar` and `TV_Speaker.sbsar`. Define all the JPEG and PNG files without the word `alpha` as **texture**, and define all the PNG files with the word `alpha` as **alpha**. You will notice that Adobe Substance 3D Painter has already defined all the **Substance Painter Smart Material (SPSM)** files as **smartmaterial** because it is automatically programmed to detect files directly related to Painter.

Note

Texture

A texture is a two-dimensional picture that may be used in three dimensions. Textures might be grayscale, which means just one channel is utilized, or colorful, which means many channels are used. Materials are often made up of a variety of textures, each of which serves a distinct purpose, such as color, roughness, and metalness.

Alpha

Alpha is a mask that may be used to paint complicated shapes or details, such as a barcode, decals, stickers, or a logo.

---

6.   Next to the type of resource, you will find **Prefix**. This option helps you to create common prefixes in case you want to search for any asset. So, select all the resources with the *Shift* key and, for example, type `Retro`. The result will be as follows:

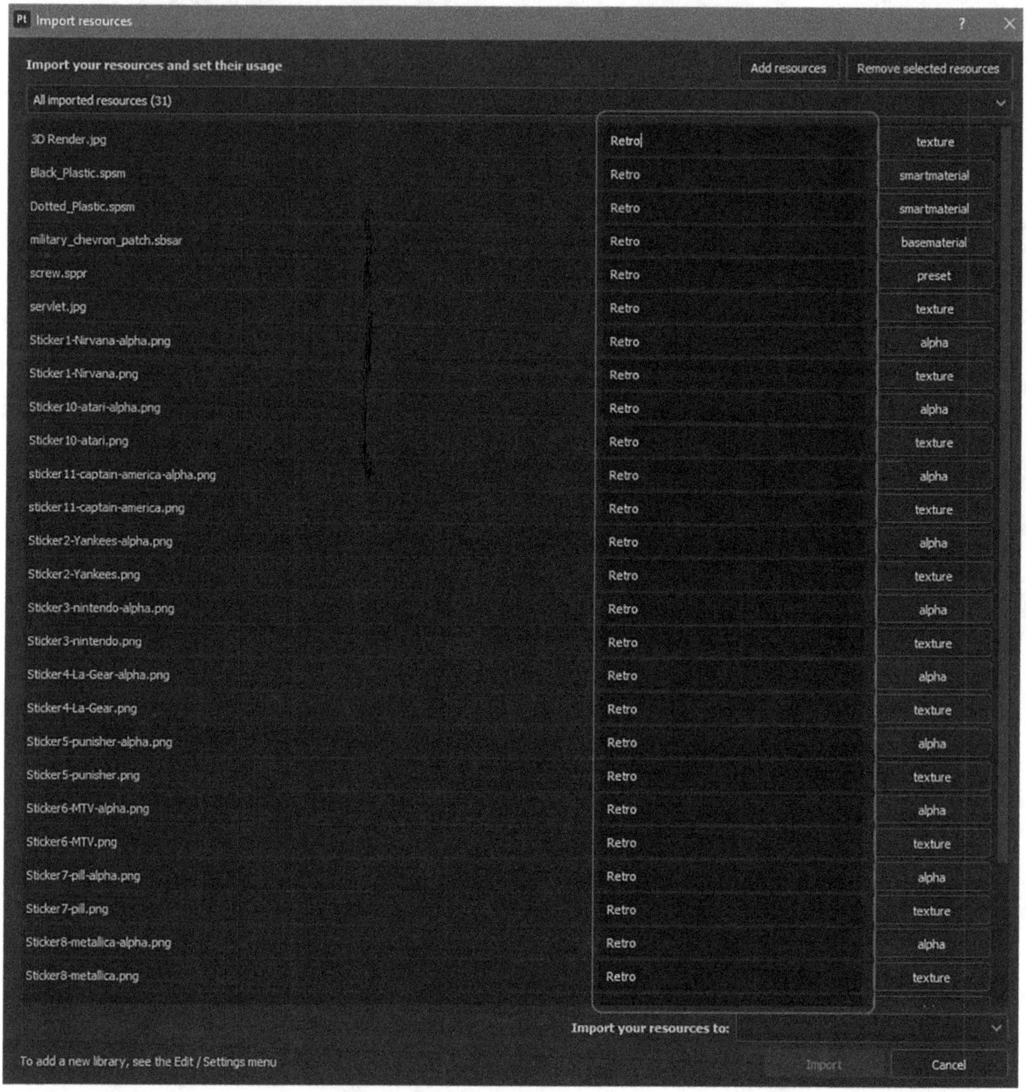

Figure 2.8 – Typing prefix

Now, whenever you search in the **Assets** panel with the word `Retro`, all these resources will show up. This will make your search more flexible, and you will be able to search for assets more quickly, especially if you have a lot of resources.

7. Now you need to choose the project type from the **Import your resources to:** option, which is located at the bottom right of the panel:

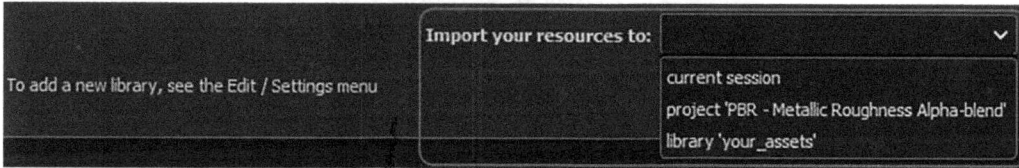

Figure 2.9 – Project types

You will find three different options in the project type, as shown in *Figure 2.9*. These options can be explained as follows:

- **current session**: This type will import the resources to the current session of the project, and the resources will be available only for that session. Once the project is closed and then reopened, all the previously imported resources will be gone.

- **project 'PBR - Metallic Roughness Alpha-blend'**: This type will import the resources to the project you are working on, and the resources will be available throughout the project. No matter how many times you close and reopen the project, the resources that you imported previously will be always available. However, you have to save the project before closing it. The PBR - Metallic Roughness portion is actually the name of your project until you save it, so if you are importing resources after you have already named and saved a file, it can look different than what we have here.

- **library 'your_assets'**: This type will import the resources to the assets library; therefore, these resources will be available for all the new and old projects, unless you remove them manually from the your_assets folder.

8. Select **project 'PBR - Metallic Roughness Alpha-blend'** and click **Import**:

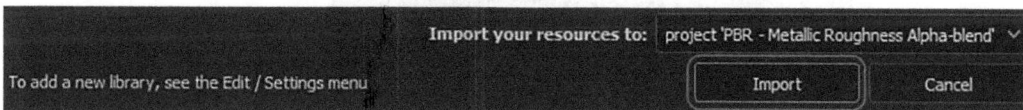

Figure 2.10 – Import your resources to project 'PBR-Metallic Roughness Alpha-blend'

9. Once you will click on **Import**, all the resources will be imported into the **Assets** panel. These resources will be available under the **Project** library. Moreover, a search filter called **Imported resources** will also be generated automatically, as you can see in *Figure 2.11*. However, you don't need that filter as you have already created the prefix. So, you can delete that filter by clicking on the cross next to it:

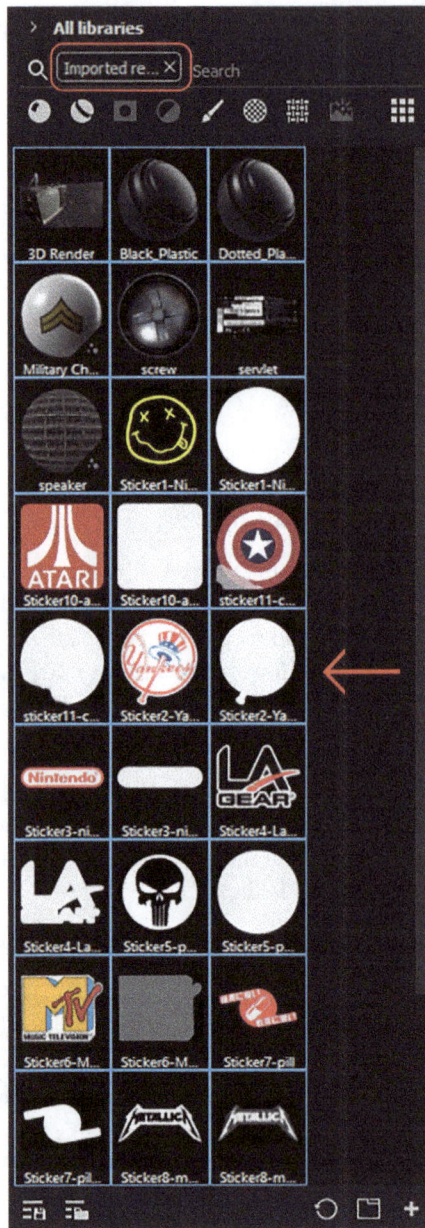

Figure 2.11 – Imported resources

Now, our resources are ready to use. Let us move on to creating custom tabs inside our **Assets** panel to help ourselves work faster.

Creating custom tabs inside the Assets panel

You can create custom tabs inside the **Assets** panel to work on your project faster and more efficiently. However, before you can create custom tabs, you must create a search keyword.

Creating a search keyword

You can create search keywords by using a filter or the **Saved searches** option.

Creating a search keyword using a filter

To create a search keyword using a filter, go through the following steps:

1. Click on **All libraries** and choose **Project**. You will notice that your imported resources are there along with the textures that you previously baked:

Figure 2.12 – Project library

2. It's quite overwhelming sometimes when you have a lot of textures showing, especially the baked textures. So, to only see the imported resources, you can click on the small arrow next to the **Project** option and choose a filter. For our example, you can choose the **Retro** filter, which will show you all the resources assigned with the `Retro` prefix:

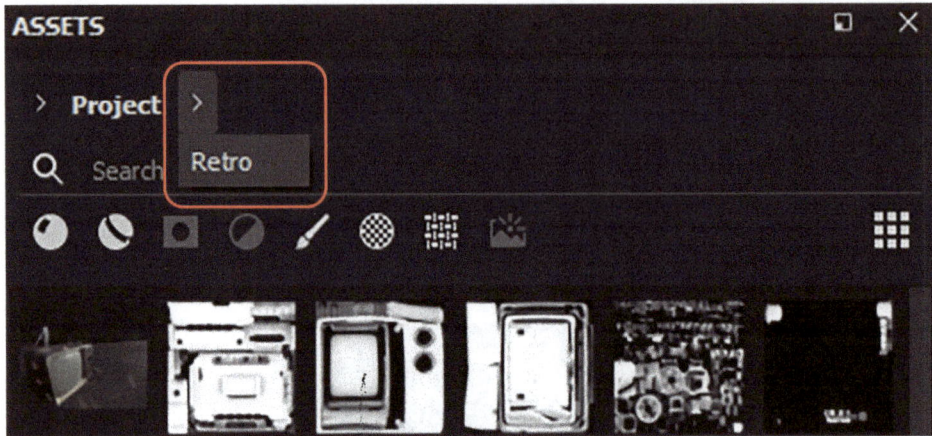

Figure 2.13 – The Retro prefix filter

You can also perform the same operation by clicking the **Filter by path** option (**1**) at the bottom of the **Assets** panel, as shown in *Figure 2.14*:

Figure 2.14 – The Filter by path option

3. Then, click on **Project** (**2**) and choose the prefix you are looking for. In our case, we will select **Retro** (**3**). Now, you have successfully created a search keyword to represent a filter.

Creating a search keyword using Saved searches

You can also create a search keyword using **Saved searches**:

1. You can also save searches in the **Assets** panel. Click on the **Saved searches** option at the bottom of the control panel (circled in *Figure 2.15*):

Figure 2.15 – Saved searches

2. Then, type any search keyword, for example, `silver`:

Figure 2.16 – Searching using typed keywords

3. To save the keyword, click on the *Save* icon:

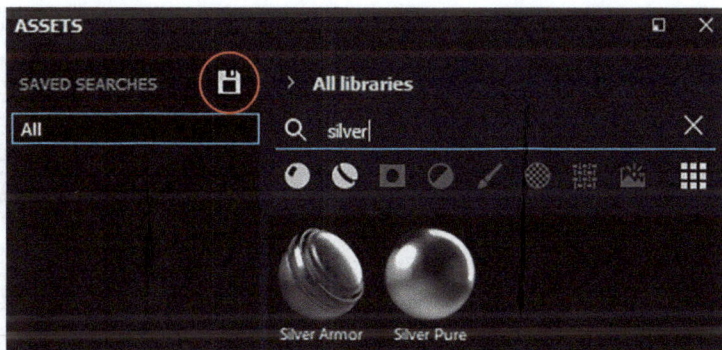

Figure 2.17 – Saving search keywords

4.   After saving the keyword, you will notice that the search keyword is saved in the list. Now you can come back to this option whenever you want to search for the same keyword:

Figure 2.18 – Saved searches list

However, it's quite time-consuming if you are going to search for filters using the **Search path** option, or the **Saved searches** option. To save time, it's best to create a tab that will help you to switch between keywords much faster.

Creating a custom tab

Let's see how to create custom tabs that will help you switch between keywords faster, and as a result, help you save time:

1.   Select the keyword **silver** that you saved and then click on the **Open new query** option at the bottom of the **Assets** panel, as highlighted in the following figure:

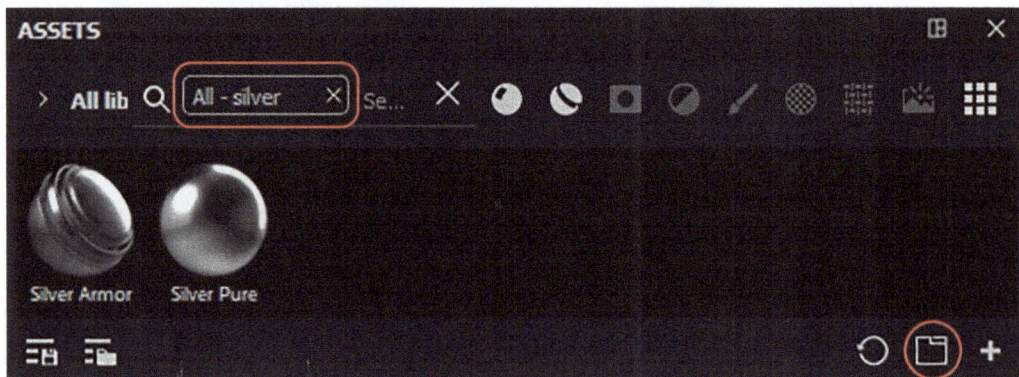

Figure 2.19 – Open new query option

2. As soon you click the **Open new query** option, you will notice that a new tab with a **ALL - SILVER** keyword is created, as shown in *Figure 2.20*:

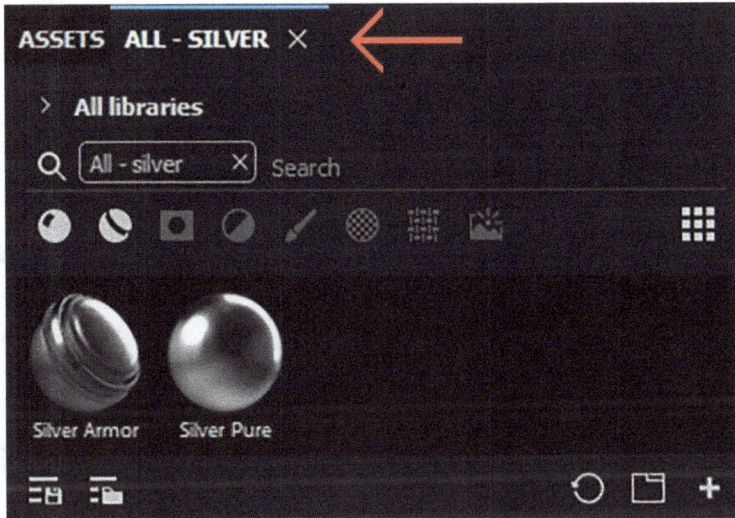

Figure 2.20 – Created SILVER tab

3. Now you can go to the **Assets** tab and delete the filter from there, as shown in *Figure 2.21*, but do not delete it from the **SILVER** tab; otherwise, you will not be able to see the results.

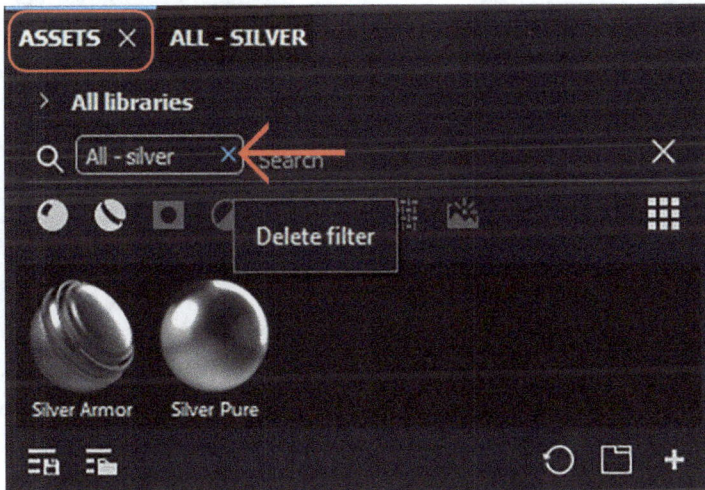

Figure 2.21 – Delete the filter from the Assets tab

4. Now you can easily switch between your **Assets** panel and the **SILVER** panel. In this way, you can create as many tabs as you like, with the keywords of your choice, and easily access them:

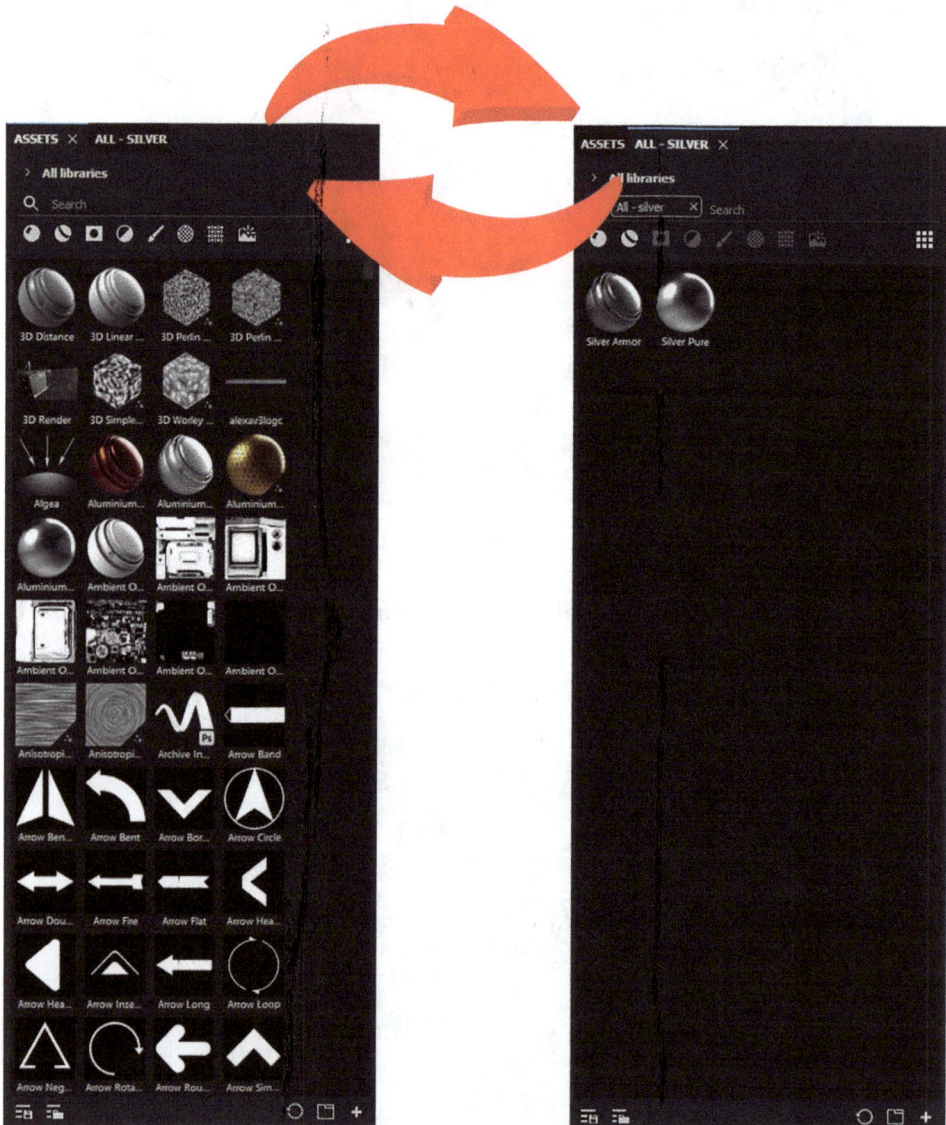

Figure 2.22 – Quick switch between tabs

5. Now, go to the **File** menu and choose **Save As** and save the file in the `Substance_Painter_Exercise_Files` folder as `Retro_Television.spp`. Bear in mind that your `Painter` file can be huge depending on mesh details:

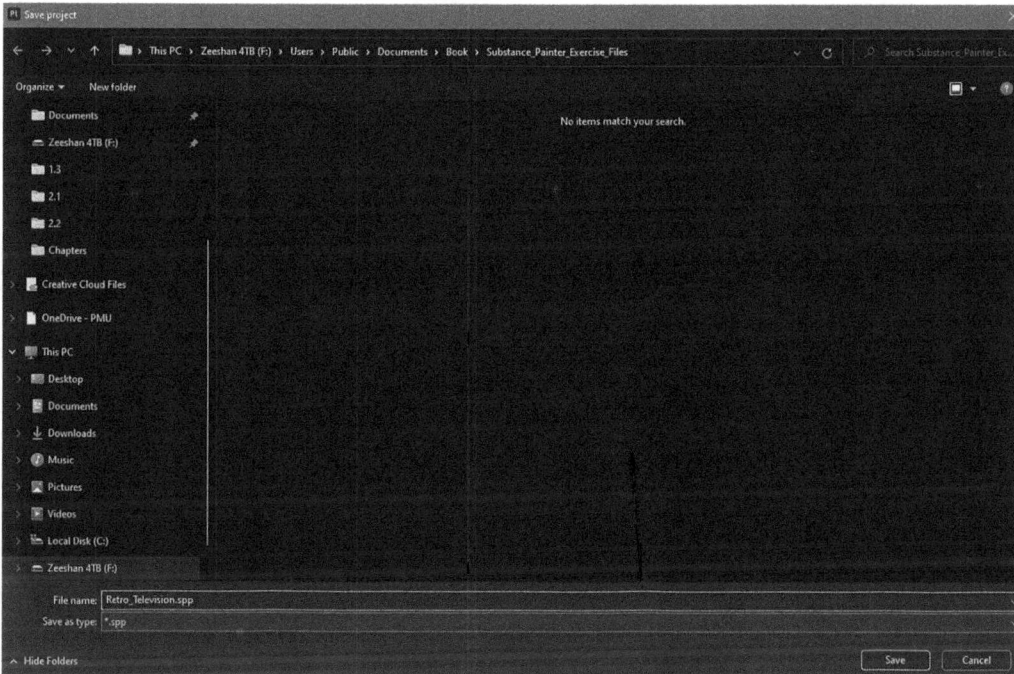

Figure 2.23 – Saving the project

As you have learned how to create your own custom tabs, we will now study different types of assets so you can comprehend their correct use. You may create more custom tabs based on your choice of assets.

Understanding the different types of assets in the Assets panel

Adobe Substance Painter has a variety of asset types that you can simply drag and drop on the model and customize according to your preferences. In this section, we will go through each one of them and learn about their usage.

In the following figure, the asset -type icons have been labeled with numbers from **1** to **8**:

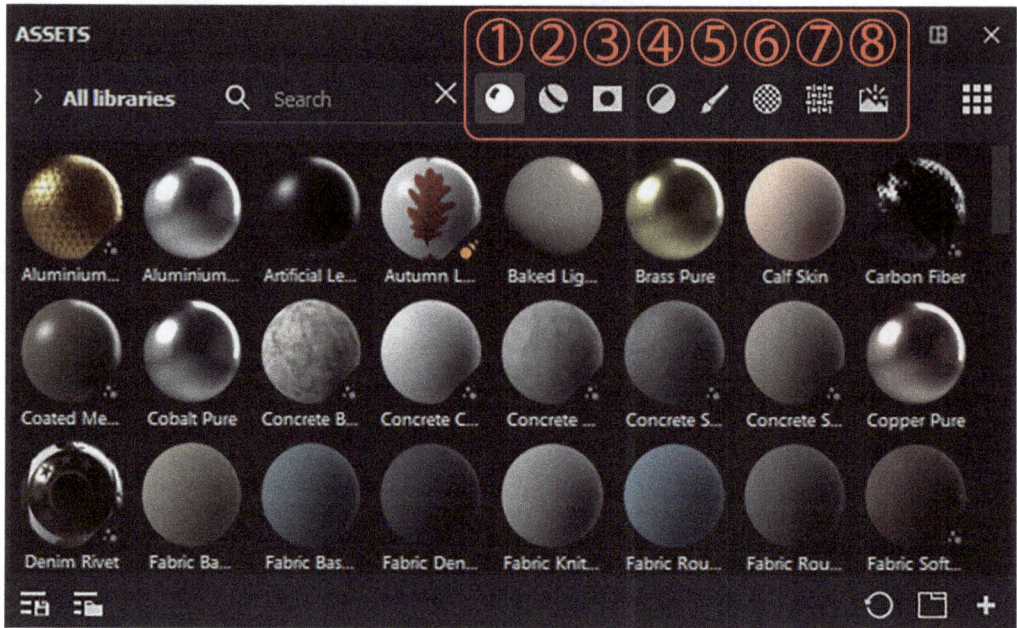

Figure 2.24 – Various types of assets

Let us now explore what each icon labeled with the number entails:

1. **Materials**: Materials comprise tiling materials that are standard and consistent. They don't require baking and don't have any mesh-specific details. Materials can be created using a fill layer in Painter, but for additional control, use Substance 3D Designer or Substance 3D Sampler.

   If you type Human Skin in the search field, you will find human skin under the materials. Human skin is widely used for humans, cartoon characters, and creatures.

2. **Smart materials**: Painter's Smart materials are a one-of-a-kind feature. They feature mesh-specific characteristics, in addition to tiling and uniform detail, that are automatically customized to your mesh. You must first bake your maps for this to work. Only Substance 3D Painter can create and use Smart materials.

3. **Smart masks**: Smart materials can be applied to any layer in the stack. However, Smart masks can only be applied to the effect stack because these are effect presets (for masks specifically). We will learn about effect stacks in *Chapter 4, Working with Masks in Adobe Substance 3D Painter*, and *Chapter 6, Working with Materials and Smart Materials in Adobe Substance 3D Painter*.

4. **Filters**: These are effects like the filters you find in Adobe Photoshop, such as blur, color correction, and emboss. However, these are specially designed for Adobe Substance 3D Painter.

5. **Brushes**: These provide you with brush types; you can also customize available brushes:

    -   **Particles**: This falls under the brushes category and can provide you with physical effects that can act as physical particles on your meshes, for example, dripping oil, spray paint, rust, and so on.

6. **Alphas**: Alphas are used for the shape of the brushes, text, decals, and stamps, and as a height map.

7. **Textures**: These are generally JPEG or PNG files or other formats of colored images such as LUTs, which are lookup tables and are most commonly used in color grading. All the imported texture files can also be accessed under this category. The details of texture subtypes are given here:

    -   **Grunges**: These can be used as masks or as an effect to create some dirt, damage, or wear and tear effects.

    -   **Procedurals**: These are procedural effects that you can apply on meshes. You can customize them to get random results that will prevent a tiled or a computer-generated look and feel, and output more realistic and natural effects.

    -   **Hard Surfaces**: These contain a variety of normal maps that will allow you to create an effect with an engraved or height map (displacement) effect. These are specially used for effects such as screws, bolts, grills, and so on.

8. **Environment maps**: The environment is a representation of the infinite space that surrounds the three-dimensional scene. These are HDRI files that give you realistic lighting results that are derived from the HDRI files themselves. There are a variety of HDRI files under this category.

Hopefully, you have understood the different types of assets and their usage, so let's learn how we can apply them in our next section.

Applying assets to 3D meshes

There are many ways to apply assets to 3D meshes; this section will cover some of those methods and discuss them in detail.

Using normal maps to apply assets

Normal mapping is a texture mapping technique used in 3D computer graphics to simulate the lighting of bumps and dents. To apply a normal map in Adobe Substance 3D Painter, you need to do the following:

1. Select the texture set from **TEXTURE SET LIST**; for this project, let's choose TV_Front_Casing:

Figure 2.25 – Applying a normal map

2. After selecting the texture set, click on the **Textures** asset in the **Assets** panel.

3. Then type the keyword `hard surface`.

4. Once the hard surfaces are shown, then select a basic brush from the **brushes** menu in the **Assets** panel, and go to the **Paint** property on the right side.

5. Click the **Material** tab under **PROPERTIES - PAINT**. You should see the **Normal** map further down.

6. Drag the **Screw Cross Round** normal map from the left panel to the **Normal** map area on the right panel.

7. **Screw Cross Round** is loaded in your **Normal** map.

    Now, you can take the brush and click on the front of the TV screen using the left mouse button. You need to only tap on the TV screen; do not drag over the screen; it will create unwanted results.

Changing the Brush attributes

To increase the size of the brush, you can press], and to decrease the size, you can press [on your keyboard; however, these shortcuts might not work if you are using a different keyboard layout such as the Swiss keyboard; therefore, you can change the shortcuts.

Also, you can press *Ctrl* (Windows) or *Command* (Mac), keep the right mouse button pressed, and drag the mouse left or right to change the size of the brush, and up or down to change the softness of the brush. You can also rotate the brush by keeping *Ctrl* (Windows) or *Command* (Mac) pressed and dragging the mouse up or down while the mouse's left button is pressed.

Moreover, you can change the flow of the brush by keeping *Ctrl* (Windows) or *Command* (Mac) pressed and dragging the mouse left or right while the left mouse button is pressed.

Figure 2.26 – Screw stamped on the front screen of the television

However, you will notice that the screw is drawn in a white color. If you want to change the brush color, then choose the **STENCIL** tab in the **PROPERTIES - PAINT** panel and choose a different color under **Base color**:

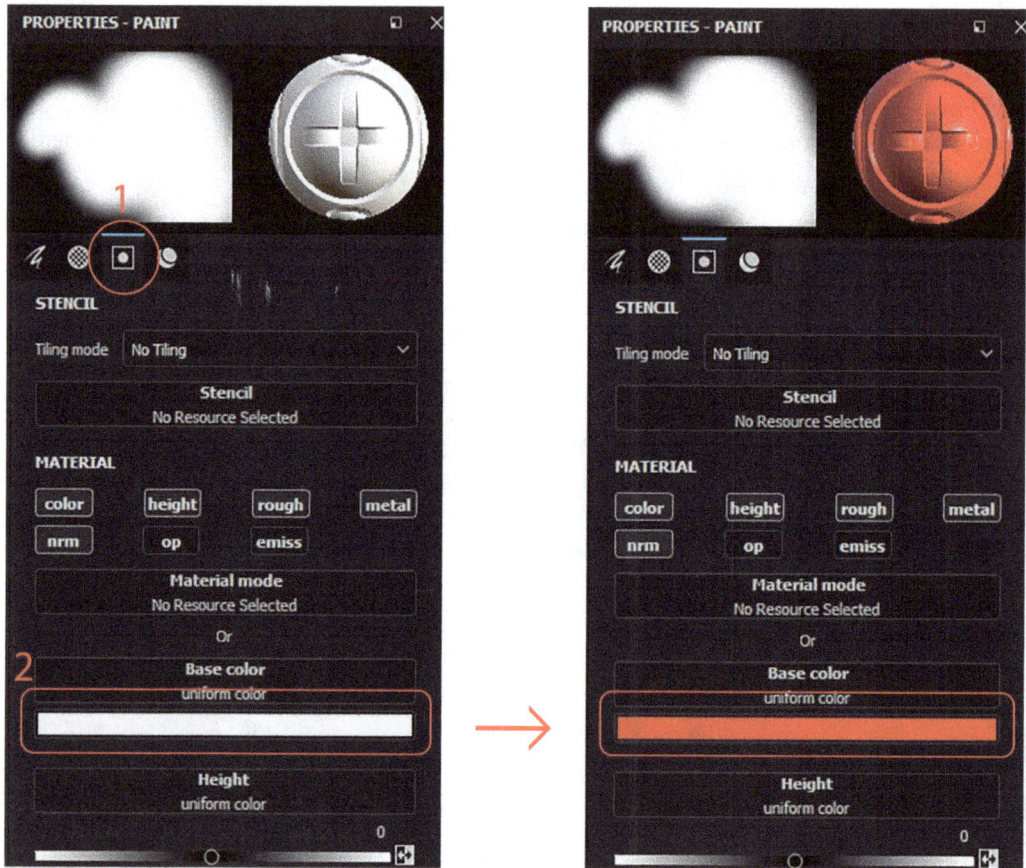

Figure 2.27 – Changing brush base color

Now, if you click with the brush using the left mouse button on the TV screen, you will get a red screw:

Figure 2.28 – Red screw

Suppose you do not want color at all. All you have to do is follow these steps:

1. Deselect the color option under the **STENCIL** tab.

2. Now, click on the TV screen using the left mouse button; you will not notice any color on the screw:

Figure 2.29 – Painting without color

Now let's create the same Screw Shape brushstroke but with the material. Go to the previously created **SILVER** tab and drag **Silver Pure** to **Material mode** in the Paint properties, as shown in *Figure 2.30*:

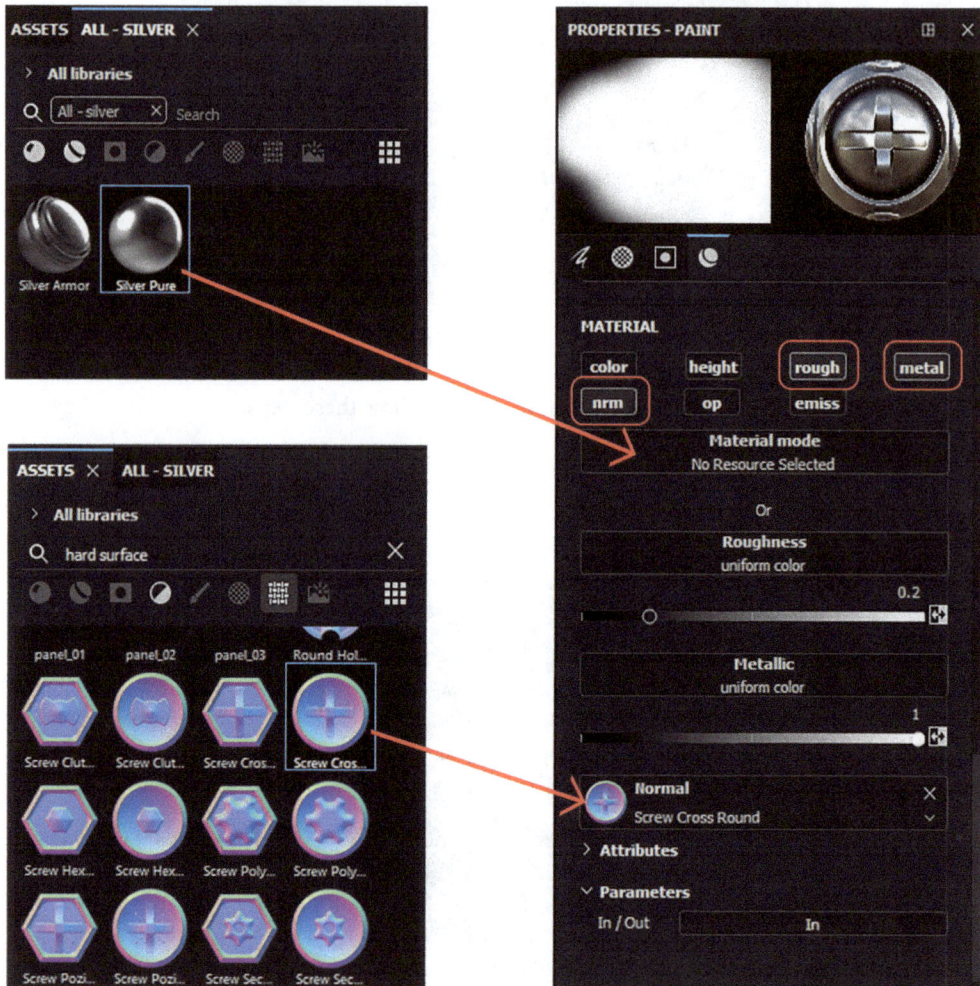

Figure 2.30 – Creating Silver Screw

Make sure your normal (**nrm**) option is selected under **MATERIAL** and drag **Screw Cross Round** across to the **Normal** map, as shown in *Figure 2.30*.

Using alpha maps to apply assets

Previously, you might have noticed that the edges of the painted screw were blurred or feathered; this was because the alpha that the brush was using had soft edges. To create a solid screw, go to the **ALPHA** tab under **PROPERTIES - PAINT**.

In case you don't have any solid round alpha inside Adobe Substance 3D Painter, you can use the alpha that we imported. So, go to the **Assets** panel and choose **Project library** and then choose **Retro**. Then, drag any solid round alpha from the **Assets** panel to the **ALPHA** section on the right **Properties** panel and replace it, as shown in *Figure 2.31*:

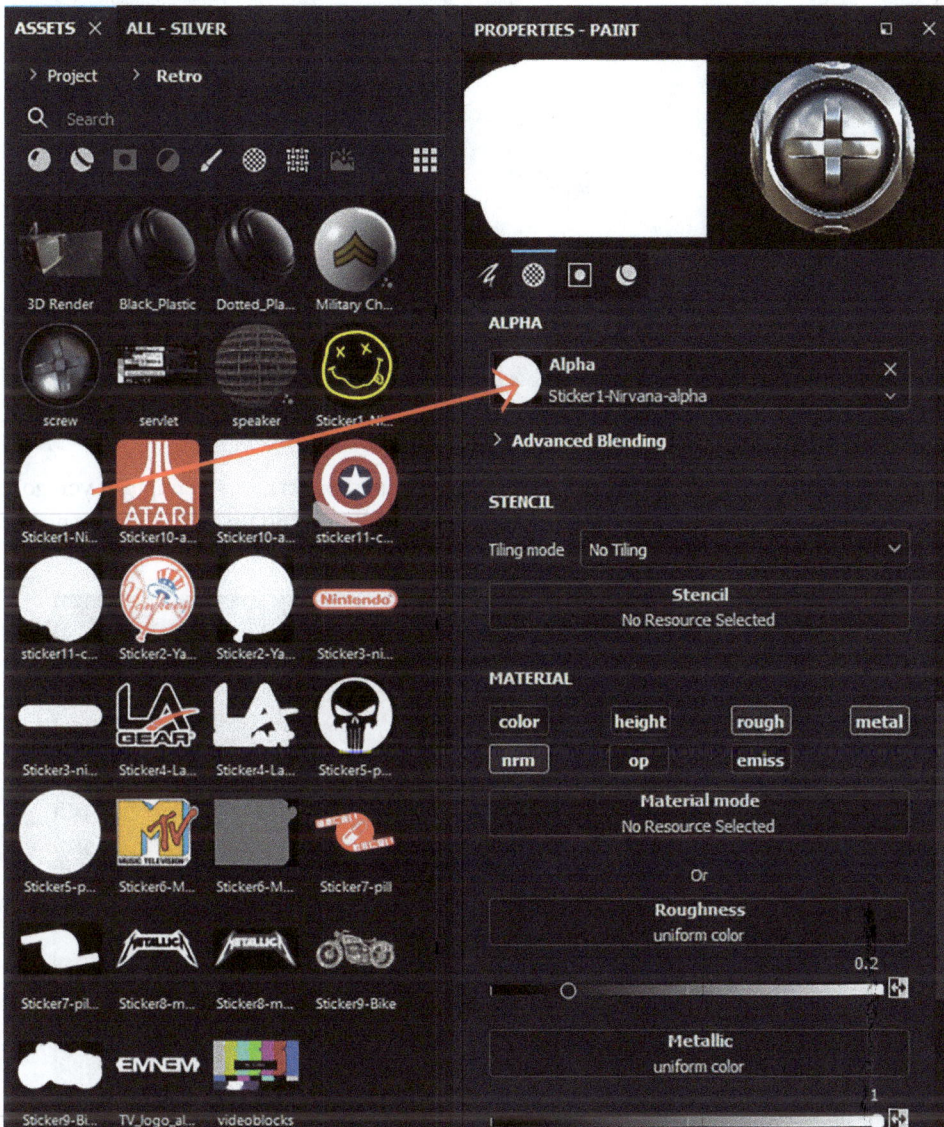

Figure 2.31 – Alpha

Now, click on the front screen of the TV with the left mouse button, and you will get a solid silver screw:

Figure 2.32 – Solid silver screw

Every material, normal map, alpha, and so on, has different parameters. For example, if you go to the **Normal** map in **Attributes**, as shown in *Figure 2.33*, you will find **In/Out**, which creates normal maps by flipping the tangents:

Figure 2.33 – Parameters

Undo the previous screw brush strokes, so we can perform some other experiments.

You can also increase the height of the normal map, but first, make sure the **height** map option is selected:

Figure 2.34 – Height map

To increase the height of the normal map, you can move the slider to the white area, and if you want to reduce it, then move the slider to the black area. White means 100% positive height, and black means 100% negative height. You can also invert the height by clicking the invert height map button:

Figure 2.35 – Positive and negative height

Now, click on the TV screen with positive and negative heights and see the results:

Figure 2.36 – Positive and negative heights

Now, you can undo and remove the screw or use the **Eraser** option from the left toolbar to erase it. And practice using the brush with different **Alpha**, **Material**, **Height**, **Roughness**, **Metal**, **Emission**, **Opacity**, and **Color** combinations.

Summary

In this chapter, we covered some vital areas of Adobe Substance 3D Painter. First, we learned about the use of the **Assets** panel. We studied the **Assets** panel's interface and its usage. Second, we learned how to import resources from the outside of Adobe Substance 3D Painter into the **Assets** panel and create custom tabs inside it. Third, we learned in detail how to apply assets to 3D meshes using different methods. Last, we studied different types of assets in Adobe Substance 3D Painter and learned how to apply them.

In the next chapter, you will gain basic to advanced practical knowledge of layers in Adobe Substance 3D Painter. Layers will help you to keep your textures and materials organized in your future projects. You will also learn how to apply materials in more detail, which is a key skill in Adobe Substance 3D Painter, and in this part, you will master this technique.

Once you become a master in applying materials, you will be able to create and apply highly complex and in-demand materials and textures. You will be learning techniques showing how to apply a material through ID maps, which will quicken the workflow in Adobe Substance 3D Painter.

You will also learn how to create transparent materials in Adobe Substance 3D Painter, such as a television screen and LED lights. By learning about the aforementioned techniques and skills, you will be able to create materials for professional gaming assets, architecture models, and visual effects for movies.

3
Working with Layers and Maps in Adobe Substance 3D Painter

Layers are one of the vital features of any graphic design software. The layers inside Adobe Photoshop are renowned for their organized and simple usage, and they are quite similar to layers inside Adobe Substance 3D Painter.

Layers help you stack different textures you create inside Painter; they allow you to stack masks, effects, and paint layers on top of each other to create an interesting realistic texture.

In this chapter, we will learn about layers inside Adobe Substance 3D Painter. While Painter is more about textures and materials, creating materials can be quite a tedious task if you do it without the proper knowledge. Therefore, in this chapter, we will also go through a simple and easy step-by-step guide to creating complex materials and applying them directly to, or by using ID maps on, any type of 3D model.

This chapter is crucial because you will learn how to work on layers and apply complex materials to any type of 3D meshes. You will also learn how to create different types of materials, such as television screens and LED lights, which are highly used in film production, 3D models in game design, and so on.

In this chapter, we will cover the following topics:

- Understanding the layer stack in Adobe Substance 3D Painter
- Applying materials in Adobe Substance 3D Painter
- Applying materials using ID maps in Adobe Substance 3D Painter
- Creating LED lights in Adobe Substance 3D Painter
- Creating a television screen in Adobe Substance 3D Painter

Understanding the layer stack in Adobe Substance 3D Painter

The layer stack in Painter is used when you wish to alter the layers of a Texture Set. A layer holds the artwork and effects that will be used to build a texture on a scene's 3D object. Layers can be hidden and unhidden, placed in folders, and changed in terms of their opacity and blending mode.

First, let's take a look at the type of layers available in Painter.

Type of layers

There are three different types of layers inside Adobe, as shown in *Figure 3.1*: **Paint** layer, **Fill** layer, and **Folder**:

Figure 3.1 – Types of layers

Layers with a specific hierarchy are displayed in the layer stack; the layer at the bottom is drawn first on the mesh, followed by the layer at the top. As a result, the item at the very top of the stack is the last one, while the item at the very bottom is the first, as shown in *Figure 3.2*:

Figure 3.2 – Layer stack order

Folders follow the same idea, but the folder's content takes precedence. This means that the content of a folder will be analyzed first, followed by layers on the same level (see *Figure 3.9*).

Paint layer

With this layer, you can use brushes and particles to paint anything on a 3D model or add texture to the models.

Fill layer

Sometimes, you want to fill a 3D model with a pattern or texture. In such a case, you do not want to paint on the model and instead use the **Fill** layer. This option allows you to load material into a layer to fill the channels. You won't be able to paint on the layer. For example, you can use transformation to make the material repeat itself, such as a cheetah skin pattern.

Folder

This layer exists just to contain additional layers. The **Folder** layer is where you can store sub-layers; it is mostly used to organize the layer stack.

Now that you are familiar with the layer stack and the purpose of the different types of layers, let's explore the layer stack view mode, which will allow you to switch between layer channels to produce myriad painting and texturing effects on 3D models.

Viewmode

The layer stack's viewmode is controlled via the top-left drop-down menu. It is not feasible to display all of these attributes at once since a layer might encompass many channels. As a result, the current display context may be specified using the viewmode.

This dropdown allows you to choose which channels should be displayed in the layer thumbnails and control the blending mode and opacity for this channel only. The channels available in the **TEXTURE SET SETTINGS** parameters are used to populate the list in this dropdown:

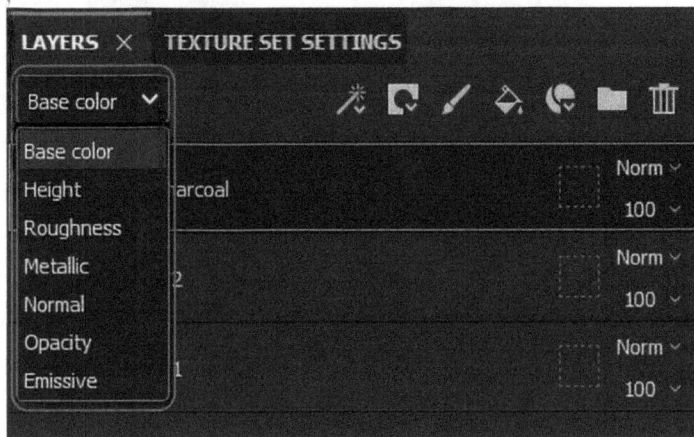

Figure 3.3 – Viewmode

Actions

There are various actions inside Adobe Substance 3D Painter that can be used in the layer stack. In *Figure 3.3*, from left to right, these are as follows: **Add effect**, **Create mask**, **Add new smart materials**, and **Delete layer**:

Figure 3.4 – Actions

Add effect

You can use this action to add a new effect and apply it to the currently chosen layer. Effects are a collection of different actions that can be performed on the content or mask of a layer in Substance 3D Painter's layer stack.

They enable an infinite number of variations, ranging from simple color changes to complex mask creations. Substance 3D Painter comes with multiple effects by default, but you can also create your own in Substance 3D Designer.

Effects can be added to the stack by right-clicking on any layer or mask, or by using the dedicated button on the layer stack window:

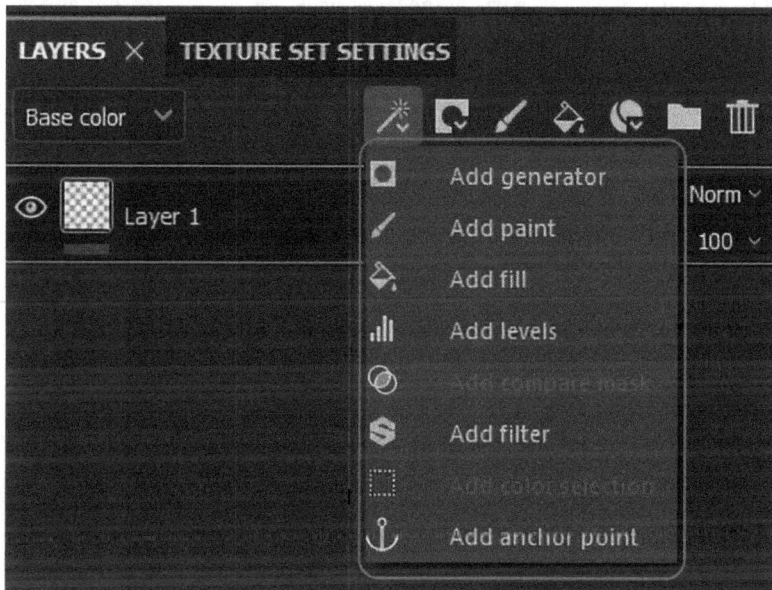

Figure 3.5 – Add effect

Most effects, like ordinary layers, have a blending mode and opacity and may be reordered, allowing you to build a whole stack of effects to make a complicated mask; this is just one of the multiple use cases.

There are a total of eight effects:

- **Add generator**: Generators are substances that use the extra maps available in **TEXTURE SET SETTINGS** to produce a mask or a material, depending on the mesh topology.

- **Add paint**: This effect allows you to paint over other effects. It operates as a layer, so you can use the various blending modes and adjust the opacity.

- **Add fill**: Fill effects work similarly to **Fill** layers, but they can be applied to a layer or a mask, enabling you to create more complicated materials or masks in a single layer.

- **Add levels**: The level effect is used to modify an image's color ranges. It allows you to tone and balance colors as well as grayscale values. You can do so by going to the **Add effect** menu and selecting it.

- **Add compare mask**: This effect allows you to compare two channels fast and efficiently and generate a mask as a consequence. This effect is only achievable when using the layers mask.

- **Add filter**: Filter effects are substances that change the appearance of a layer or mask's content.

- **Add color selection**: This effect allows you to choose or work with a certain color and helps you to mask out areas based on the color you select.

- **Add anchor point**: An anchor point is used to expose any resource or element in the layer stack and reference it in various parts of the layer stack, for various reasons and with various changes. They open up a whole new world of possibilities by letting you link layers or masks together and having a single anchor point to influence several areas of your project, thereby turning Substance 3D Painter into a non-linear experience. We will cover this in more detail in *Chapter 6*, *Working with Materials and Smart Materials in Adobe Substance 3D Painter*.

Create mask

This action activates the mask action menu, which includes the following options. We will work on these in *Chapter 5*, *Working with Advanced Tools in Adobe Substance 3D Painter*, and *Chapter 6*, *Working with Materials and Smart Materials in Adobe Substance 3D Painter*:

- **Add white mask**

- **Add black mask**

- **Add bitmap mask**

- **Add mask with color selection**

- **Add mask with height combination**

Add new smart materials

Inserting a new smart material above the presently selected layer is possible with this action. This button opens a mini-shelf where you can browse the list of smart materials that are currently available in Assets.

Delete layer

You can delete the currently selected item using this operation (layer, folder, or effect).

Now, as you are familiar with layer stack's viewmode and have learned how to add effects, create masks, add new smart materials, and delete layers, it's time to see how to apply different materials to 3D models, which is one of the most significant features of Adobe Substance 3D Painter.

Applying materials to 3D models and meshes in Adobe Substance 3D Painter

There are several ways you can apply materials in Adobe Substance 3D Painter. In this section, we will go through these methods and learn how we can apply these materials easily because when applying materials, you will face different 3D models, and each 3D model requires a different technique. For example, applying material to a ceramic vase will be different from applying material to a broken wooden box.

To learn how to apply materials in Adobe Substance 3D Painter, let's go through the following steps:

1. From **TEXTURE SET LIST**, choose **TV_Middle_Casing**:

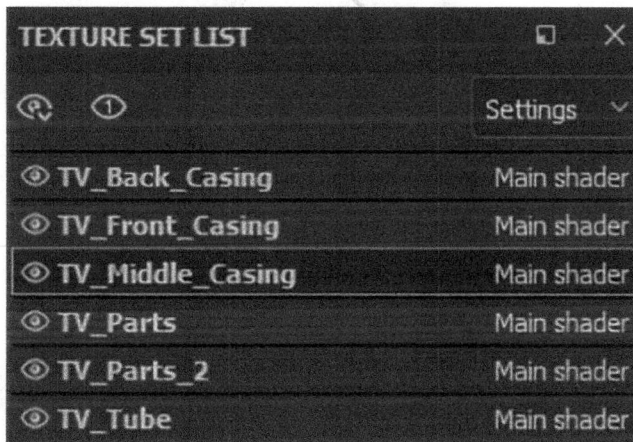

Figure 3.6 – Selecting TV_Middle_Casing

Now, keep in mind that you can only apply textures or materials to the selected Texture Set; therefore, as we have selected **TV_Middle_Casing**, we can only apply textures and materials to this Texture Set.

2. From the left-hand **Assets** panel, choose **Smart materials**, and in the search field, type `plastic`. Once the results show up, select **Black_Plastic**; if you don't have **Black_Plastic**, you can choose any dark plastic material:

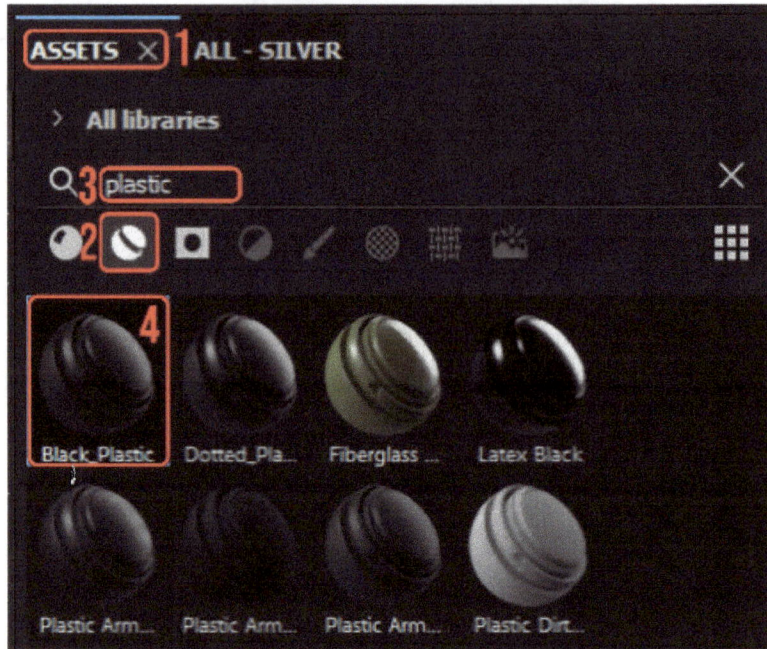

Figure 3.7 – Searching for a plastic smart material

3. Drag the selected **Black_Plastic** material onto the **TV_Middle_Casing** Texture Set:

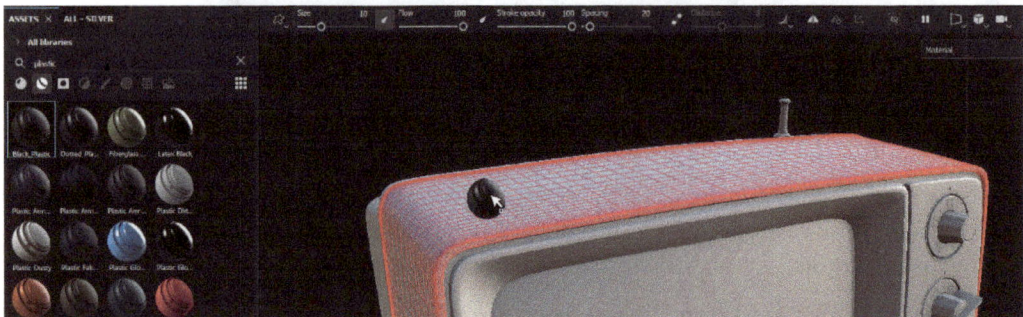

Figure 3.8 – Applying a smart material to the TV_Middle_Casing Texture Set

4. Once you have applied **Black_Plastic** to **TV_Middle_Casing**, go through the following steps:

    I.   Note that in the **LAYERS** panel on the right side, a new layer folder has been created called `Black_Plastic`.

    II.   Inside the `Black_Plastic` folder, you will find the editable base layer.

    III. Under the base layer, you will find **Levels – Roughness**, which is also editable.

IV.   Under the **LAYERS** panel, you will now see **PROPERTIES - FILL**, which now holds attributes related to the **Black_Plastic** base layer:

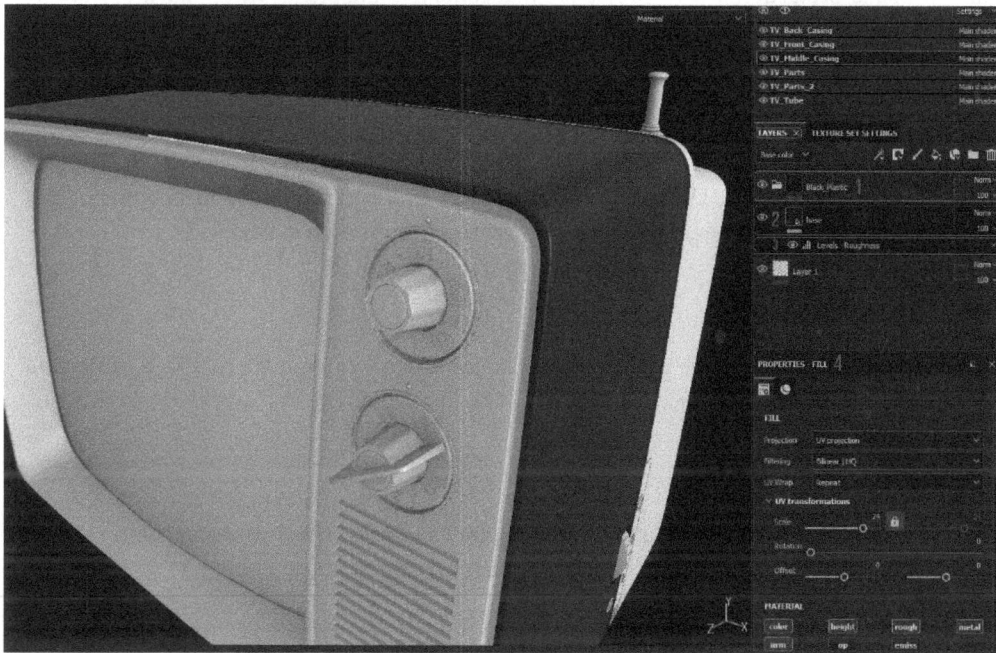

Figure 3.9 – The Black_Plastic layer folder and PROPERTIES - FILL

5.   Now, it's time to adjust the smart material we have applied. First, we will change the **Projection** type from the **PROPERTIES - FILL** panel. The **Fill** property determines how a material is applied to the mesh and projected. The **Projection** option in the **Properties** window can be used to switch projection modes.

Let's take a bit of a detour before we continue creating our material. Let's take a quick glance at the seven different types of projection modes.

UV projection

To all UV tiles, a single picture or the first image from a sequence is applied. This also allows for deformation control – for example, if you want to scale, rotate, or move any texture map.

Fill (match per UV Tile)

A dedicated UV Tile is assigned to each image in a sequence without deformation controls. There are no restrictions for deformation.

Tri-planar projection

The fill's Tri-planar projection is a 3D projection that combines and blends three planar projections to cover an entire 3D mesh. It's a great way to project noises and patterns without having to worry about apparent seams.

Planar projection

The fill's planar projection is a 3D projection that displays an image in a plane-like manner. Signs, decals, and other patterns can be projected on the surface of or through a 3D object.

Spherical projection

Filling images and patterns can be projected completely around an item via spherical projection. It's possible to project on round objects or distort texture into circular patterns with this technique.

Cylindrical projection

The fill's cylinder projection allows you to project graphics and patterns all around an object. Fitting pillars or columns, as well as organic structures such as arms, is possible.

Warp projection

The fill's warp projection is a 3D projection that allows you to modify a texture by changing grid points. On a non-planar surface, it can be utilized to accommodate patterns and logos.

Now that we've had a look at the different types of projection modes, let's get back to the creation of our material:

1. Choose **Tri-planar projection** as the **Projection** type; this will give a seamless texture output with a clean look.

2. Under **UV transformations**, you can control the UV's position, rotation, and size per se. You can control the Tri-planar projection by changing the offset, rotation, or scale values under **3D projection settings** (**A**), or by using the *Manipulator* directly on the screen by moving, rotating, or scaling the texture map from the *Contextual Toolbar* (**B**) (see *Figure 3.10*):

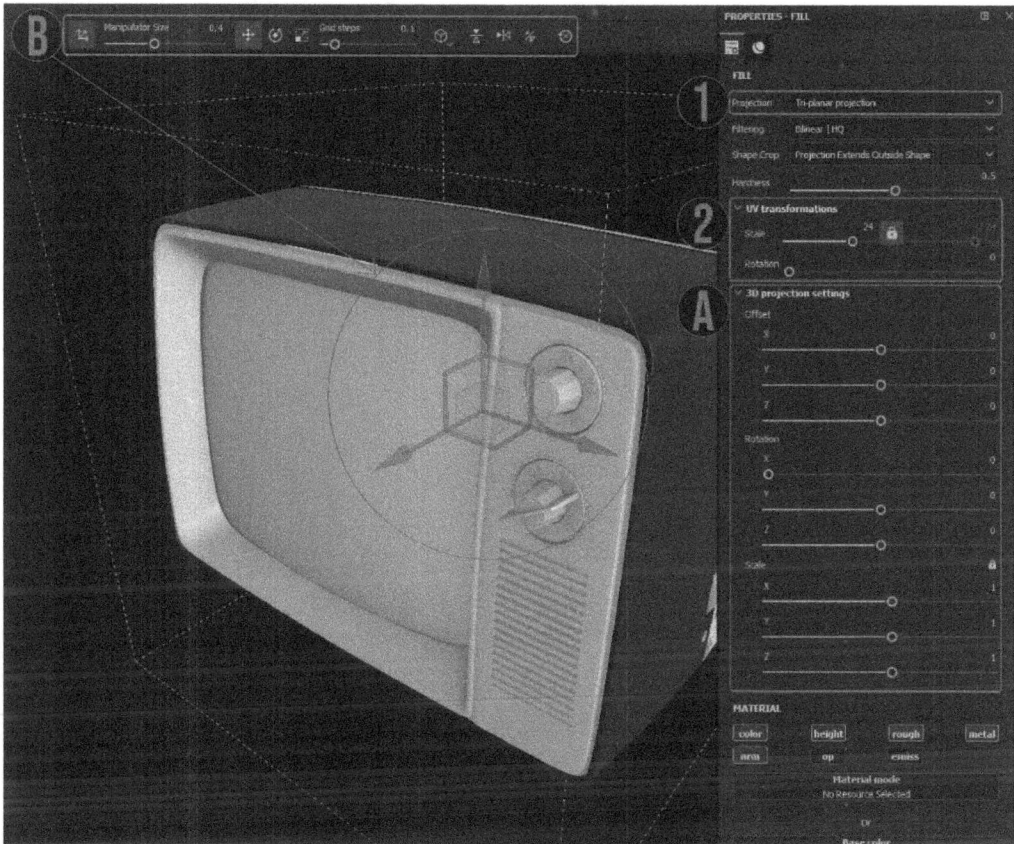

Figure 3.10 – Fill and Tri-planar options

3. To give a more plastic look to **TV_Middle_Casing**, make sure that **Scale** is set to 24 under **UV transformation**. Under **3D projection settings**, lock the **Scale** aspect ratio as shown in *Figure 3.10* and change the **X** value to 0.6. As the aspect ratio is locked, the **Y** and **Z** values will also change to 0.6 automatically.

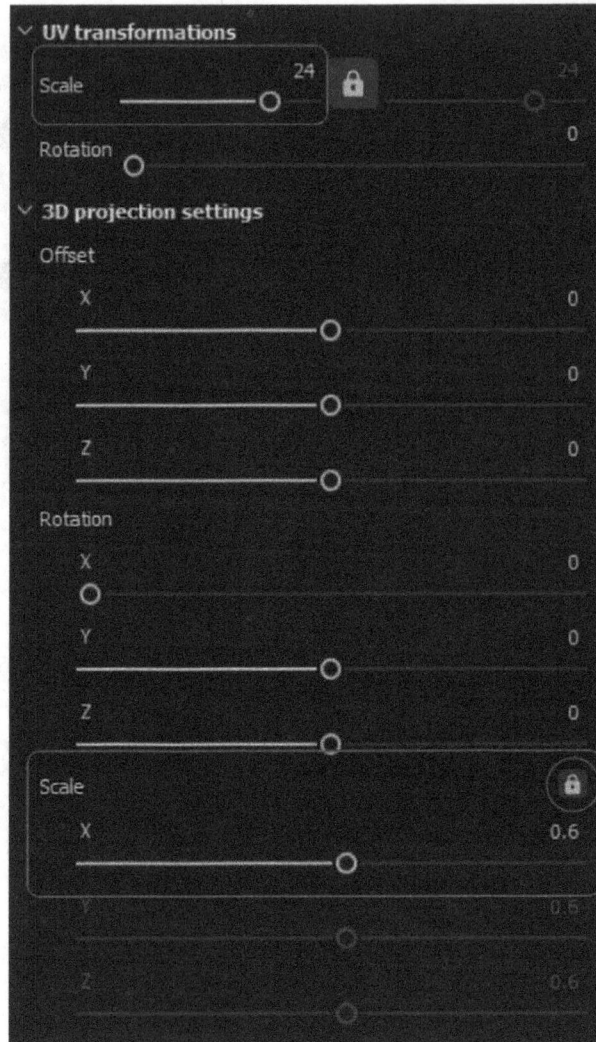

Figure 3.11 – UV transformations and 3D projection settings

Now, you will notice that the plastic material has become more realistic, and you can change any material by adjusting its UV transformations, 3D projection settings, or both. However, you have to decide which project type best suits your project. If you are more comfortable working with gizmos, then use the Manipulator directly from the Contextual Toolbar; otherwise use **3D projection settings**.

Now that we have learned how to apply materials on 3D models directly, it's time to learn how we can apply materials using ID maps. This method is used for models divided into different ID maps, which we learned about in *Chapter 1, Getting Started with Adobe Substance 3D Painter*.

Applying materials on 3D models using ID maps in Adobe Substance 3D Painter

Now that you have learned how to apply a material on any Texture Set and modify its settings, it's time to apply materials using ID maps. Applying materials through ID maps allows you to drop different materials on selected polygon faces, so let's do this!

1. From the **TEXTURE SET LIST** menu, select **TV_Front_Casing**.

    Note that there are a total of 18 speaker vents on the television's front casing; now, if I want to apply a certain material by dragging it over **TV_Front_Casing**, it will get applied over the whole Texture Set:

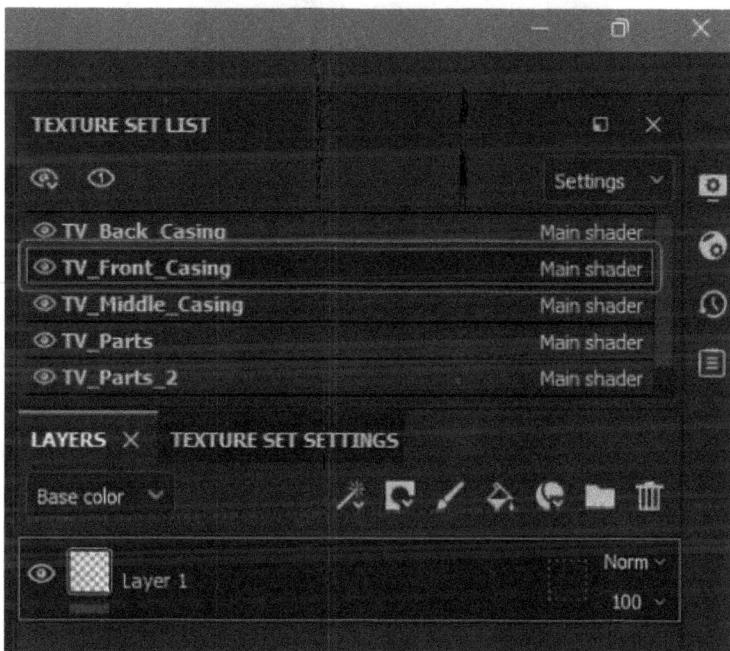

Figure 3.12 – Selecting TV_Front_Casing from TEXTURE SET LIST

The 3D model that we are using in this project already has ID maps created on it using Autodesk Maya, which we learned about in *Chapter 1, Getting Started with Adobe Substance 3D Painter*.

2. To view the created ID maps, click on the viewmodes and choose **ID** under the **Mesh maps** category:

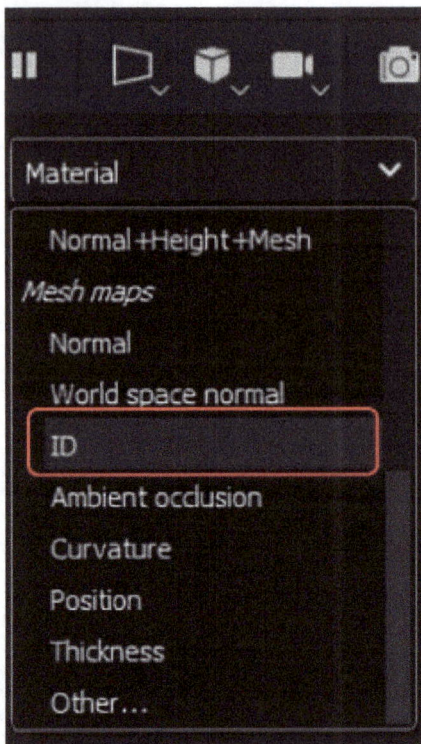

Figure 3.13 – Select ID in the viewmodes under the Mesh maps category

Now, note that each speaker vent is yellow. This is because the yellow vertex color is applied on each polygon face of the speaker vent in Autodesk Maya:

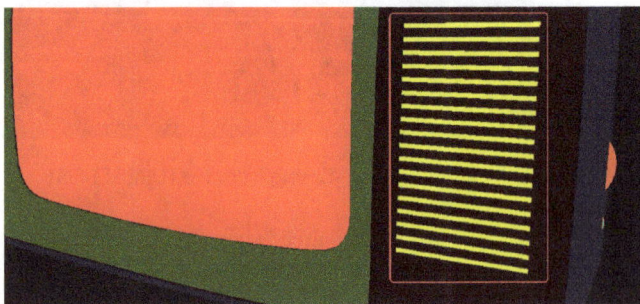

Figure 3.14 – Yellow ID maps on the speaker vents

3.   Once you have seen what the ID maps for the speaker vents look like, go back to the viewmode and choose **Material** under the **Lighting** category:

Figure 3.15 – Selecting the Material viewmode under the Lighting category

4. Now, go to the **Assets** panel on the left side, choose **Project** from the drop-down menu, and then choose the **Materials** asset type:

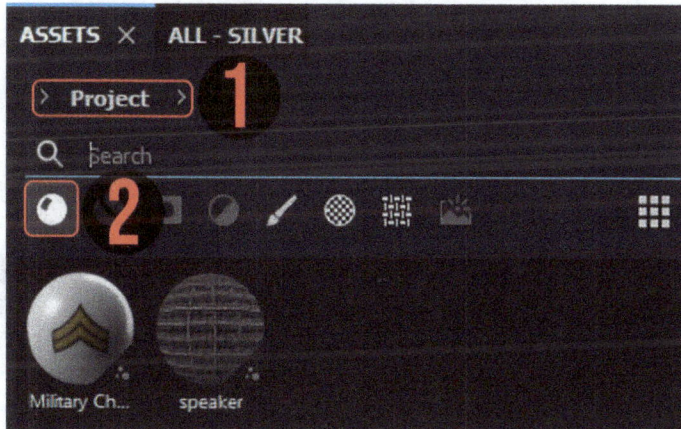

Figure 3.16 – Selecting Project and Materials

5. One of the materials you will see is the **Speaker** material. To apply this **Speaker** material on the speaker vents through the ID map, you need to first press *Ctrl* (Windows) or *Command* (Mac) on your keyboard.

Once you do that, you will notice that the yellow ID maps for the speaker vent become highlighted. While the vents are highlighted with yellow color, just drop the **Speaker** material on the yellow speaker vent:

Figure 3.17 – Applying a material using an ID map

6. Once you will apply the material, the viewmode will automatically go back to **Material** mode:

Figure 3.18 – The applied Speaker material on the speaker vent

Note that the **Speaker** material looks awkward on the speaker vent – it will need some adjustment. You will find all the settings to do so under **PROPERTIES - FILL** on the right panel.

7. First, change the **Projection** type to **Tri-planar projection** so that you can change the texture's size easily. Then, under **UV transformations**, change the scale to 60:

Figure 3.19 – Adjusting PROPERTIES - FILL

8. It is always recommended to apply the same vertex color to the same group of polygon faces using Autodesk Maya, Autodesk 3D Studio Max, or any other 3D software. For example, press *Alt* (Windows) or *Option* (Mac) on your keyboard and click on the eye of the **TV_Parts** Texture Set.

9. This will hide all the Texture Sets except **TV_Parts**. Then, select **ID** under the **Mesh maps** viewmode.

Note that the TV dials and jogs are in orange ID maps. This is because the same vertex color was applied to these using 3D software such as Autodesk Maya or 3D Studio Max. So now, whenever you will apply any material on any of these dials or jogs using an ID map, it will apply that material to all the dials and jogs using the same ID map/vertex color.

Apart from the orange-colored **TV_Parts** Texture Set, you will see green-colored TV parts. These are a different set of groups, which belong to metallic objects such as antennae or pins:

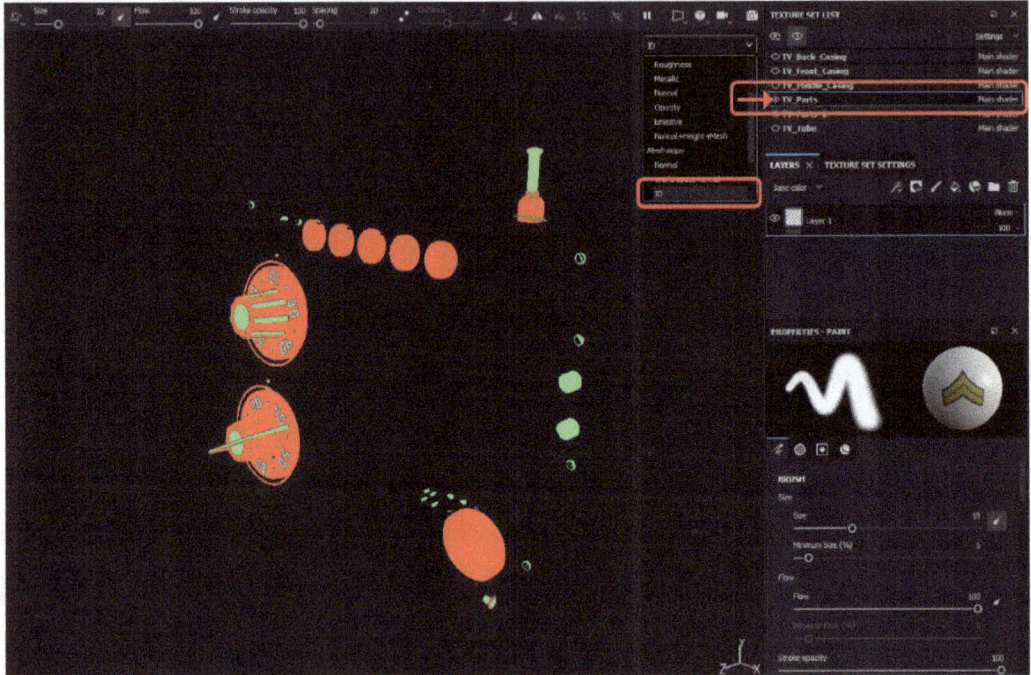

Figure 3.20 – The TV_Parts ID map

10. Now, unhide all the other Texture Sets and switch back to **Material** viewmode:

Figure 3.21 – Unhide all the Texture Sets and switch mack to Material viewmode

11. In the **Assets** panel, choose **All libraries**. Select the **Smart materials** asset and type `Dotted Plastic` in the search field.

You should now see the **Dotted Plastic** smart material; if you don't have this material, you can choose any similar dark plastic material:

Figure 3.22 – Searching for the Dotted Plastic smart material

12. Now, press *Ctrl* (Windows) or *Command* (Mac) on your keyboard, and drag the **Dotted_Plastic** smart material on one of the orange dials or jogs:

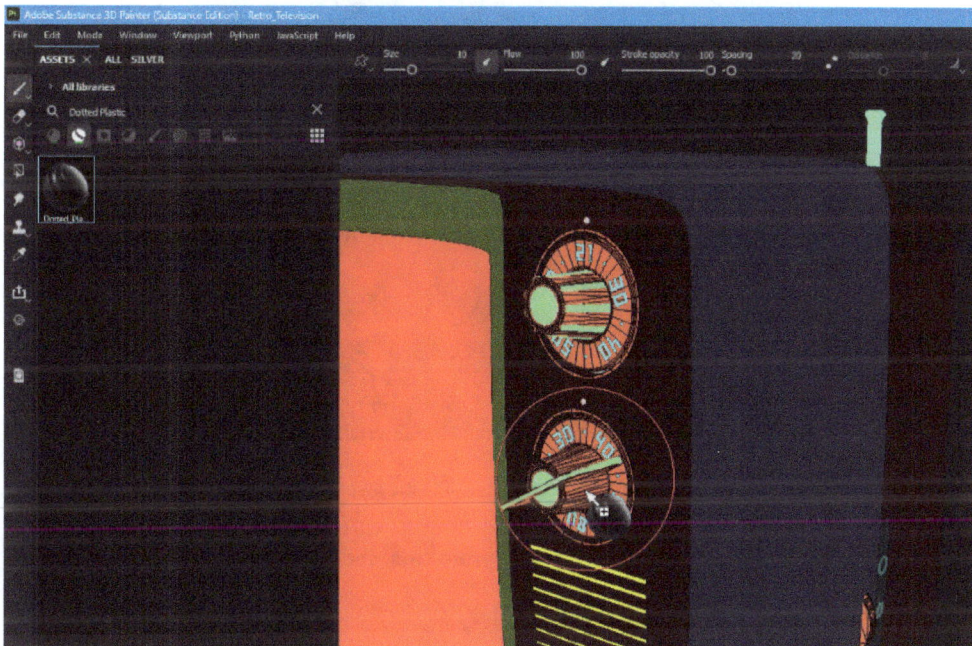

Figure 3.23 – Applying Dotted_Plastic on the dials and jogs ID map

13. On the **LAYERS** panel on the right side, you will see a **Dotted_Plastic** layer with a folder. Click on the folder icon, which will open the **Plastic Matte Pure** sub-layer. Once you select the **Plastic Matte Pure** sub-layer, new attribute options will open under the **PROPERTIES - FILL** panel:

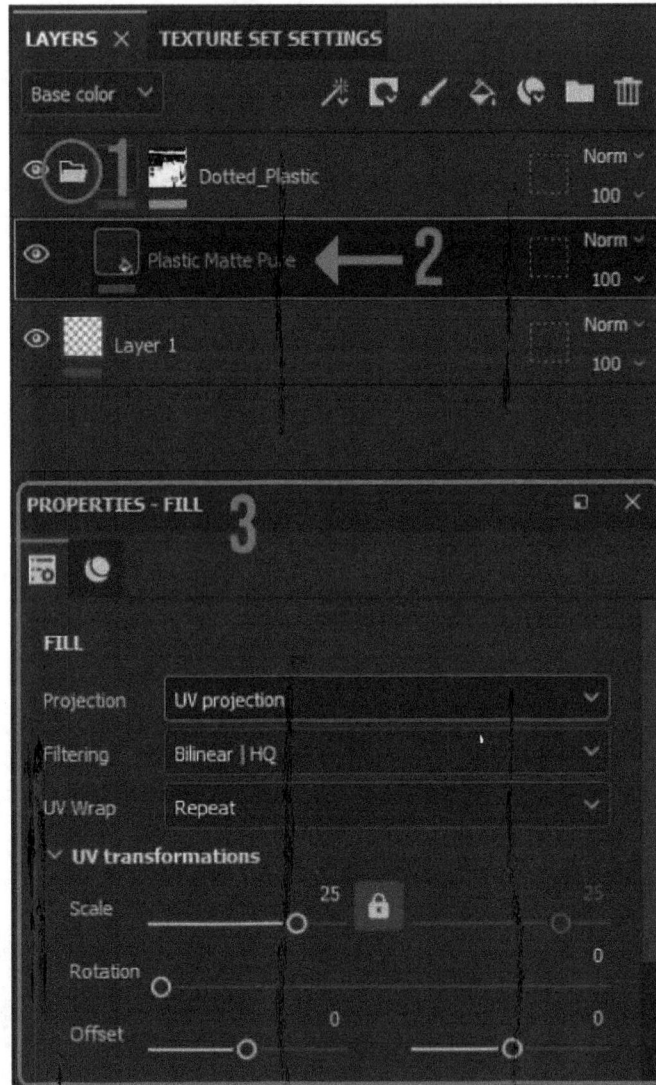

Figure 3.24 – The Dotted_Plastic layer

14. Now, change the **Projection** type to **Tri-planar projection**, and **Scale** under **UV transformations** to 35:

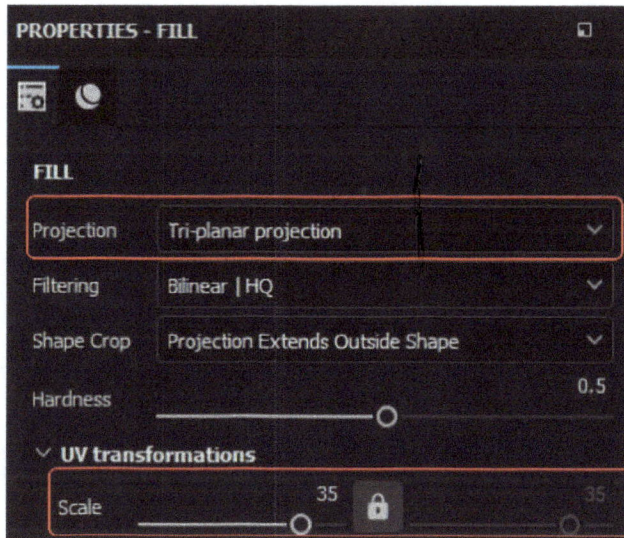

Figure 3.25 – Projection and UV transformations settings

15. Now, scroll down in the **PROPERTIES - FILL** panel till you see **Base color**. Change **Base color** to dark gray, as shown in *Figure 3.26*:

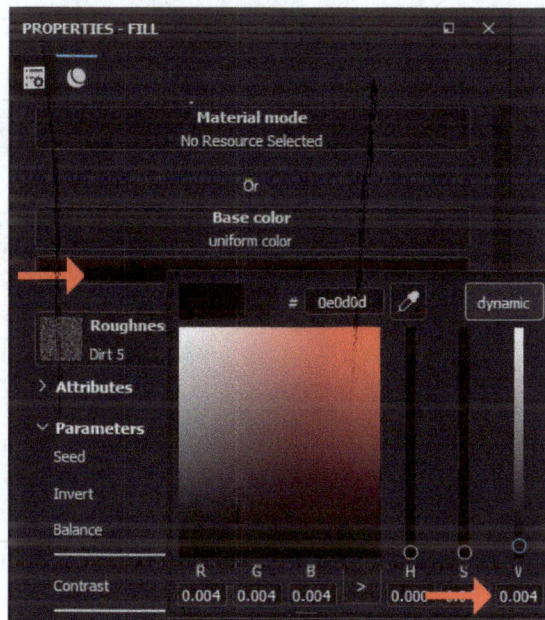

Figure 3.26 – Changing Base color

16. Now, repeat *step 11*; however, this time you will search for Steel instead of Dotted Plastic:

Figure 3.27 – Select the Steel smart material

17. Now, using the same method as we did in *step 12*, apply the **Steel** smart material to the green-colored ID maps:

Figure 3.28 – Applying the Steel smart material to the green-colored ID map

18. Now, you can see that the **Steel** smart material is applied to all the green-colored, ID-mapped **TV_Parts** Texture Sets, as shown in *Figure 3.28*:

Figure 3.29 – The Steel smart material on the green-colored ID-mapped TV_Parts Texture Sets

Note

In *Figure 3.30*, note that the **TV_Front_Casing** and **TV_Back_Casing** Texture Sets are the same-colored ID maps. Now, if I apply any material on any one of them using the ID map method as we did in the previous steps, the material will not get applied to both Texture Sets at the same time because they are different Texture Sets. This means that different Texture Sets will not get the same materials assigned to them at the same time, even if they are the same-colored ID maps. So, you need to select each Texture Set separately and then apply the material.

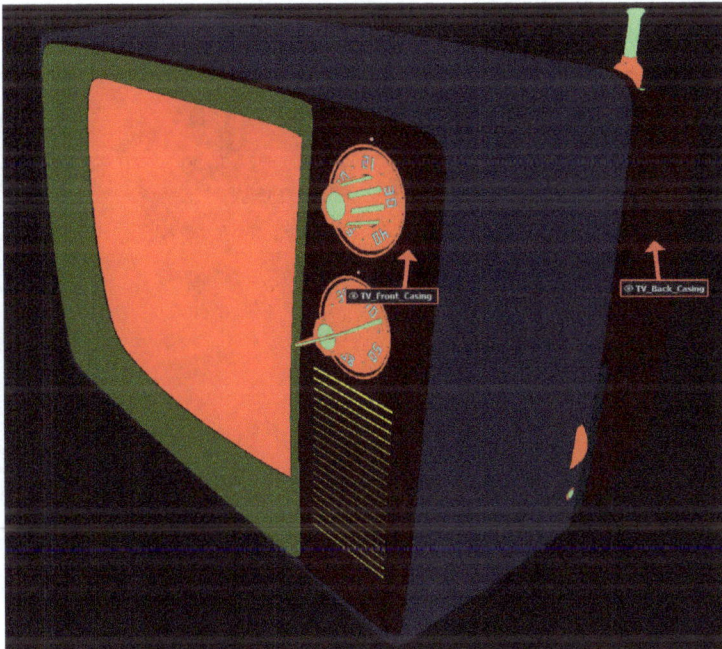

Figure 3.30 – The TV_Front_Casing and TV_Back_Casing Texture Sets

19. Next, we will apply a plastic material to the side labels of the television set; the **Labels** panel is on the **TV_Parts_2** Texture Set, and the labels are on the **TV_Parts** Texture Set, as shown in *Figure 3.31*:

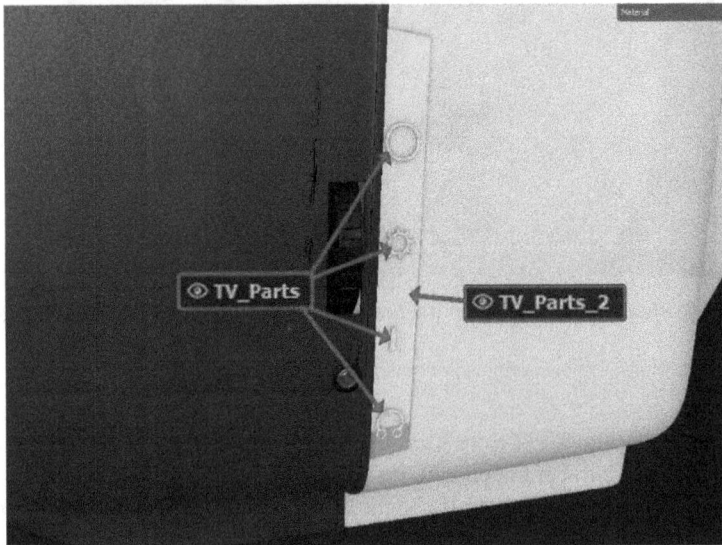

Figure 3.31 – The labels and the Labels panel

20. Go to the **Assets** panel and make sure it is set to **All libraries**. Set the asset to **Materials**, and search for Plastic Grainy in the search field:

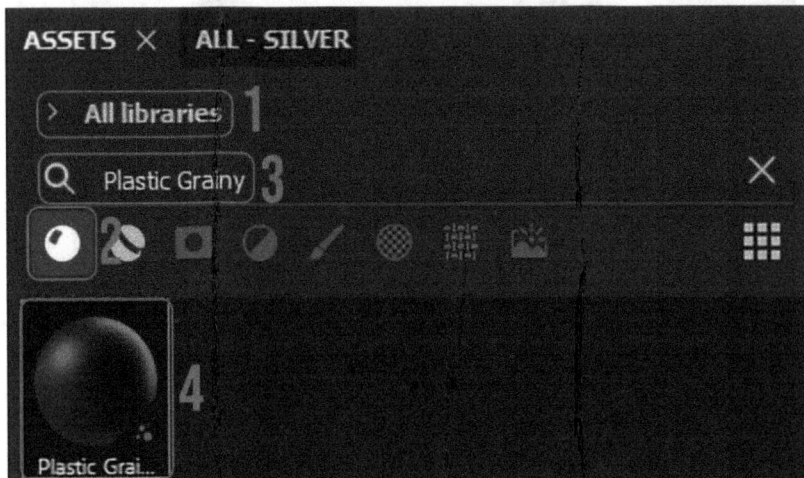

Figure 3.32 – The Plastic Grainy material

21. Now, apply the **Plastic Grainy** material using the same method we used in *step 11* and *step 12* to the labels and the **Labels** panel:

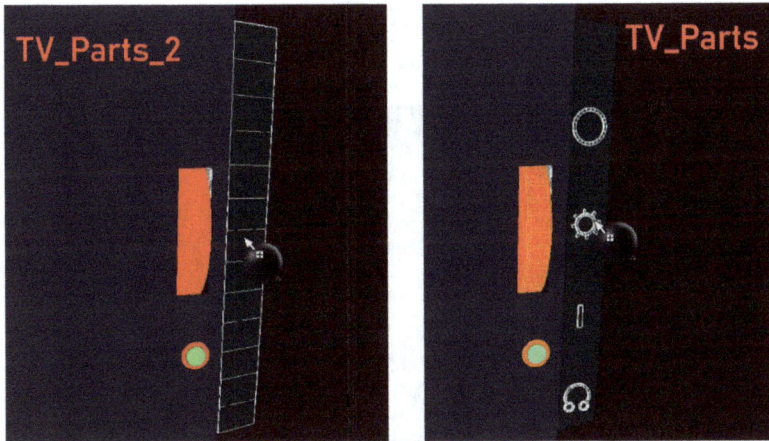

Figure 3.33 – Applying Plastic Grainy to the labels and the Labels panel

22. After applying grainy plastic, note that the material has large grains, which are not suitable for the labels and the **Labels** panel:

Figure 3.34 – Large grains on the labels and the Labels panel

23. Now, we will adjust the **Grainy Plastic** material for both **TV_Parts** and **TV_Parts_2**. First, select **TV_Parts** and select **Plastic Grainy** in the **LAYERS** panel on the right-hand side:

Figure 3.35 – The Plastic Grainy layer

24. Now, change the **Projection** type to **Tri-planar projection** under the **PROPERTIES - FILL** panel on the right-hand side, and then under **UV transformations**, change **Scale** to 20:

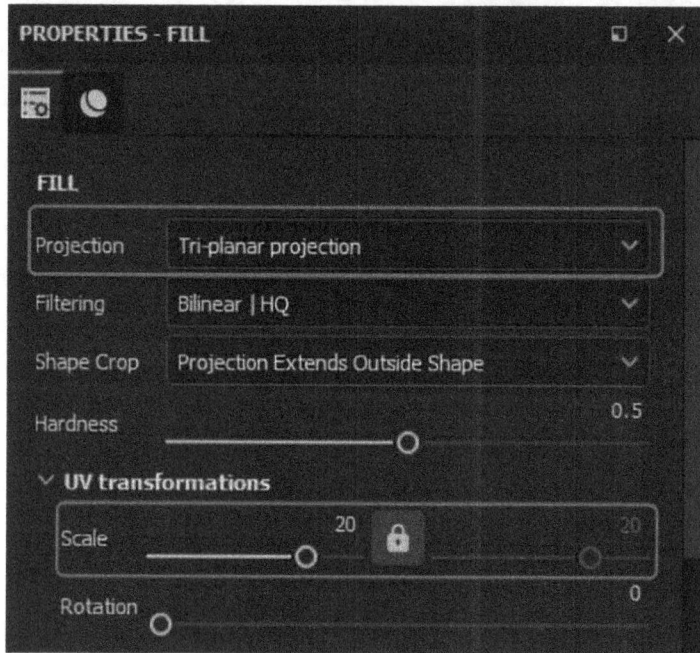

Figure 3.36 – The Plastic Grainy material settings

25. Once the **TV_Parts** material is ready, select **TV_Parts_2** in the **LAYERS** panel, repeat *step 23* and *step 24*, and your labels and **Labels** panel will be finished:

Figure 3.37 – The finished Plastic Grainy material

Now that we know how to apply materials on 3D models directly and by using ID maps, let's try it out practically in our next section.

Creating LED lights in Substance Painter

It's quite tricky to create some materials inside Adobe Substance 3D Painter, such as glass materials, transparent materials, and LED lights, and there are some tricks and shortcuts to create such materials.

So, let's learn how to create LED lights inside Painter:

1. The LED lights on the TV are located in the **TV_Parts** Texture Set; therefore, select **TV_Parts** from the **TEXTURE SET LIST**:

Figure 3.38 – LED lights

2. Make sure that the first layer is selected in the **LAYERS** panel on the right-hand side so that any time you create a new layer, it will be on top of every other layer. Then, select the **Add fill** layer from the **LAYERS** panel and scroll down to **Base color** in the **PROPERTIES - FILL** panel; once you see **Base color**, click on it and add the following values:

    *   **R:** 0.106

    *   **G:** 0

    *   **B:** 0

    *   **H:** 0

    *   **S:** 0.997

    *   **V:** 0.106

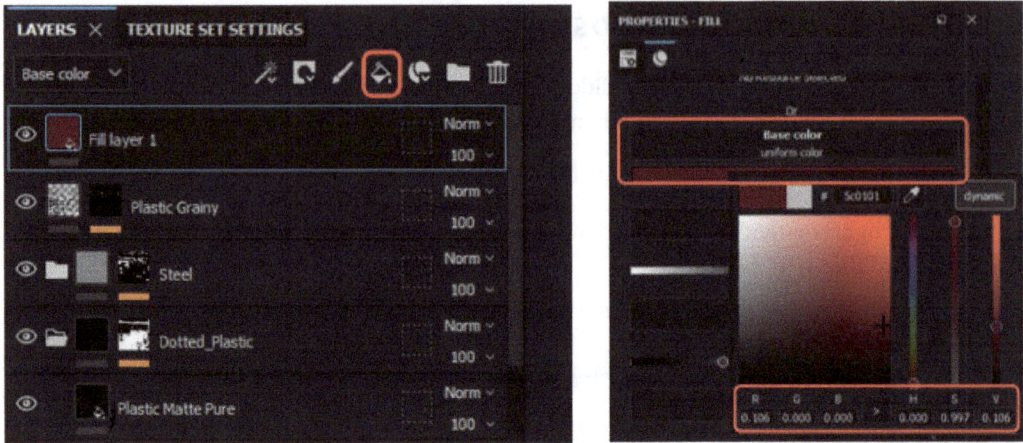

Figure 3.39 – The Add fill layer and Base color

> **Note**
>
> Once you add the fill layer, note that every TV part has turned a maroon color, which we do not want. This is because the **Add fill** layer option applies the **Fill** effect on the whole Texture Set; therefore, we need to create a mask that only affects the LED lights.

3.   Before creating masks for LED lights, switch to **ID** under **Mesh maps** in the viewmode, and note that the ID map color for both LED lights is pink; we will use this pink ID map to create masks easily:

Figure 3.40 – Pink ID map on LED lights

4. Now, switch back to the **Material** viewmode under the **Lighting** category and rename the new **Fill layer 1** layer LED Lights:

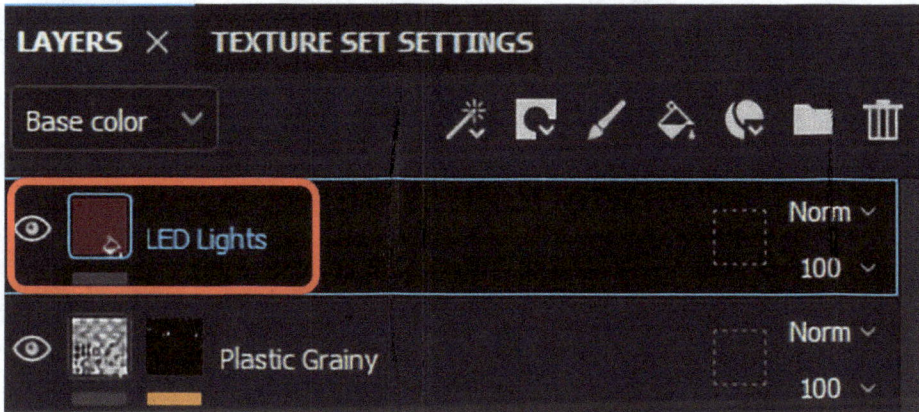

Figure 3.41 – Rename Fill layer 1 LED Lights

5. Now, right-click on the **LED Lights** layer and select **Add mask with color selection**:

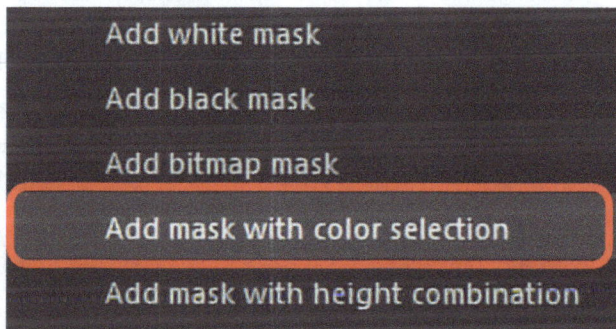

Figure 3.42 – Right-click on the LED Lights layer and select Add mask with color selection

6. As soon you select **Add mask with color selection**, a new **COLOR SELECTION** property will appear under **PROPERTIES - COLOR SELECTION**. Under the **COLOR SELECTION** settings, you will see **Pick color**. Once you click on that, an eyedropper tool will appear, and the viewmode will automatically switch to **ID**. With the help of the eyedropper, click on the LED light's *pink-colored* ID map:

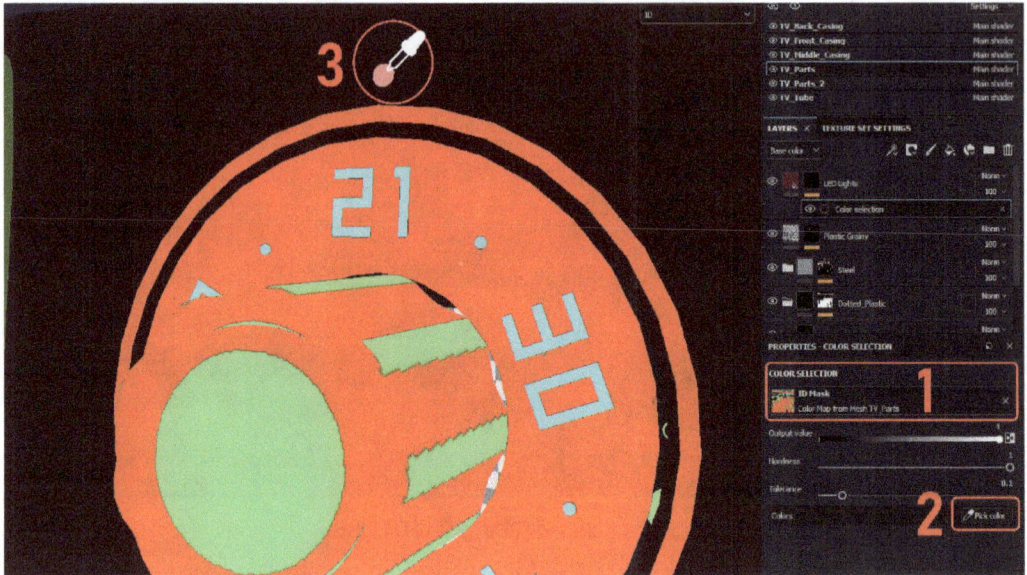

Figure 3.43 – Creating a mask with the Add mask with color selection option

7. Once the LED light's ID map is picked up, the viewmode will switch back to **Material** and the LED light's pink-colored ID map will be set as a mask:

Figure 3.44 – The LED light's ID maps as masks

Now, the maroon color is applied to the LED lights only; however, they still don't resemble LED lights. To give them a realistic look, let's do some adjustments.

8. You have to click on the fill color icon again to switch to **PROPERTIES - FILL**. Scroll down under **PROPERTIES - FILL** until you see the **Metallic** and **Roughness** settings. Change the **Metallic** setting to around 0.3 and **Roughness** to around 0.24. These settings will give you a nice specular highlight because of the low roughness value and a good reflection, owing to the high metallic value:

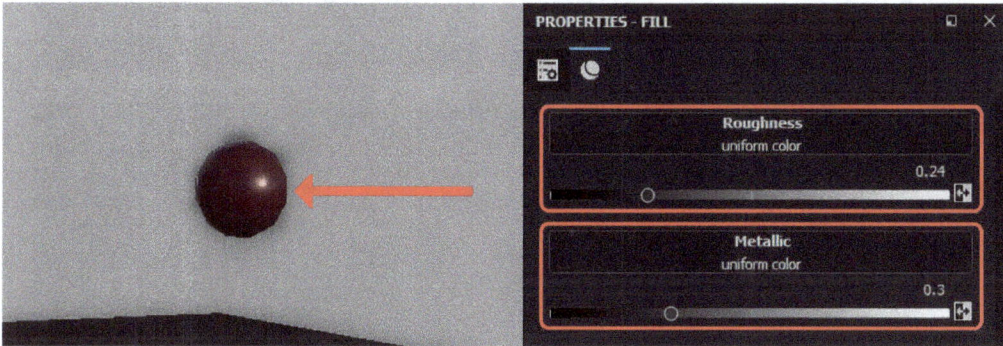

Figure 3.45 – Realistic specular highlight and reflection on the LED lights

In this section, we have learned how to create an LED texture map. Now, let's push ourselves further and learn to create a television screen, which is a bit more complex than an LED light.

Creating a television screen in Adobe Substance Painter

In the previous section, we used some tips and tricks to create LED lights. In this section, we will go a little further and create a TV screen:

1. Select **TV_Front_Casing** from **TEXTURE SET LIST**. After selecting **TV_Front_Casing**, go to the **LAYERS** panel and select **Add fill layer**. Once the fill layer is added, rename it TV Screen. Under the **PROPERTIES - FILL** panel, scroll down to **Base color** and change the color to the following values:

    *   **R:** 0.166

    *   **G:** 0.170

    *   **B:** 0.133

    *   **H:** 0.186

    *   **S:** 0.216

    *   **V:** 0.170

2. Then, scroll even more till you see **Roughness** and **Metallic**. Change the **Roughness** value to 0.05335 and the **Metallic** value to 0.0617:

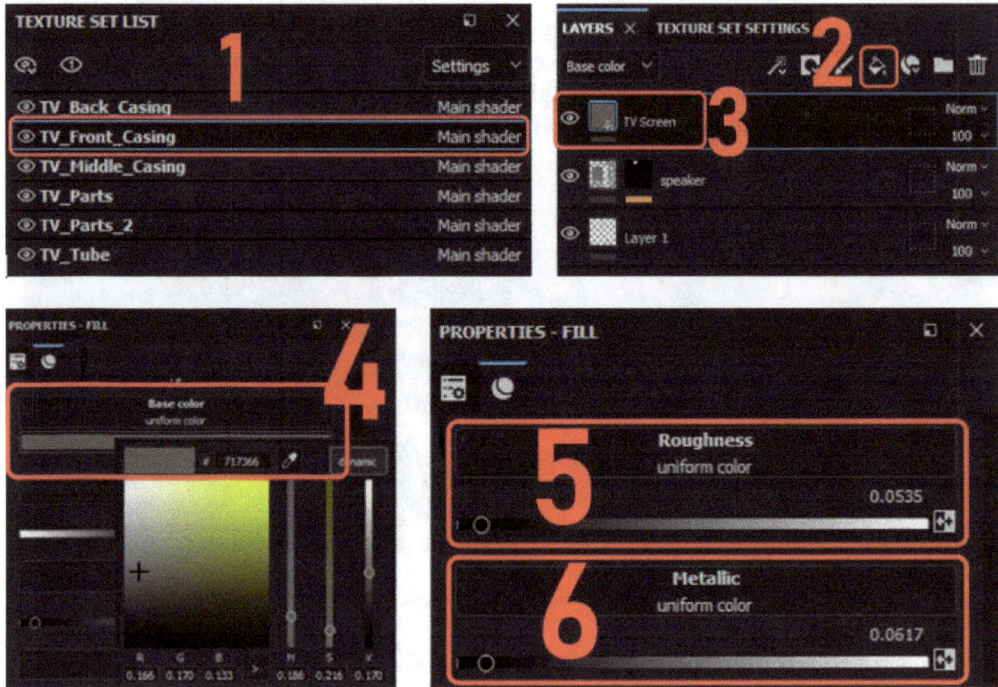

Figure 3.46 – The TV screen settings

After creating the **TV Screen** fill layer, note that the new fill layer is applied to the whole **TV_Front_Casing** Texture Set:

Figure 3.47 – The new fill layer

3. To only affect the TV screen, right-click on the new **TV Screen** layer and choose **Add mask with color selection**. When **PROPERTIES - COLOR SELECTION** appears, select **Pick color** and use the eyedropper to click on the TV screen from the viewport:

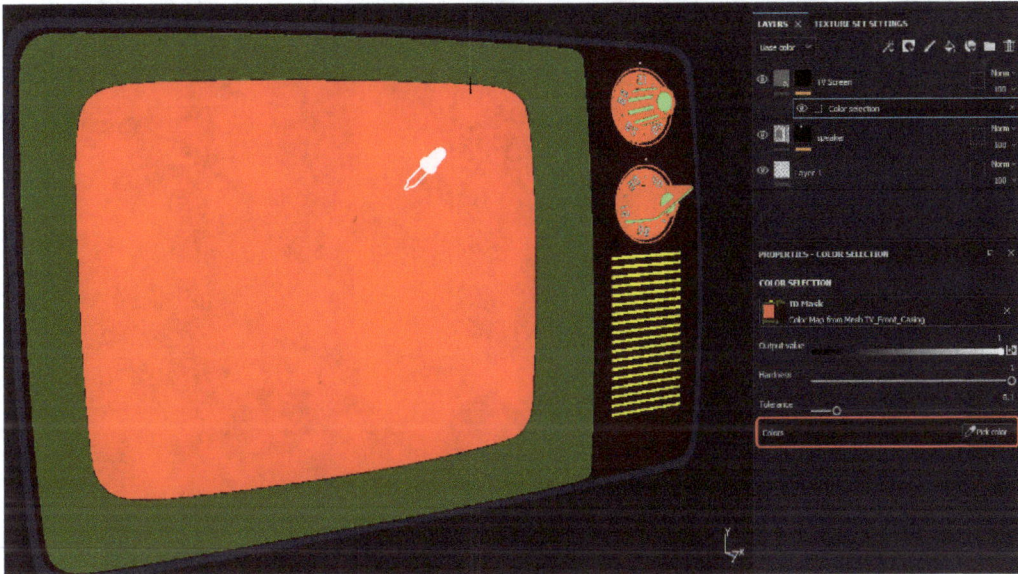

Figure 3.48 – Selecting the screen ID map as a mask

Now, the **Screen** material is created and only affects the TV screen. However, the forest reflection on the screen doesn't make any sense. Nobody leaves their TV set in the middle of a forest.

This reflection is there because it's using the default **Environment Map** setting, which is always there for every project. So, let's change the environment by clicking **Display Settings** from the **Dock** menu on the right-hand side, and then choose **Studio Tomoco**, as shown in *Figure 3.49*, from the list under the **Environment Map** options.

Studio Tomoco

Studio Tomoco is a panoramic high-definition environment map, which is quite famous for its realistic lighting effect.

Figure 3.49 – Selecting Studio Tomoco as an environment map

4. The **Studio Tomoco** environment suits the TV Screen, but it needs some light adjustment. In **Display Settings**, change **Environment Rotation** to 344. You can also rotate the environment by pressing *Shift* (Windows and Mac) on your keyboard and dragging your mouse left or right while holding the right mouse button:

Figure 3.50 – Adjusting Environment Rotation

To make the screen look dirty and overused, let's create some grunge effects on it. In Adobe Substance 3D Painter, there is a generator that can generate various mask effects.

5. So, let's add a new fill layer with a black base color and rename it `Dirt`. Once it is created, right-click on it and select **Add mask with color selection** and repeat *step 3*. Select the color selection sub-layer inside the new **Dirt** layer so that when we add a generator, it appears on top of it. Now, select **Add effect** and click on **Add generator**:

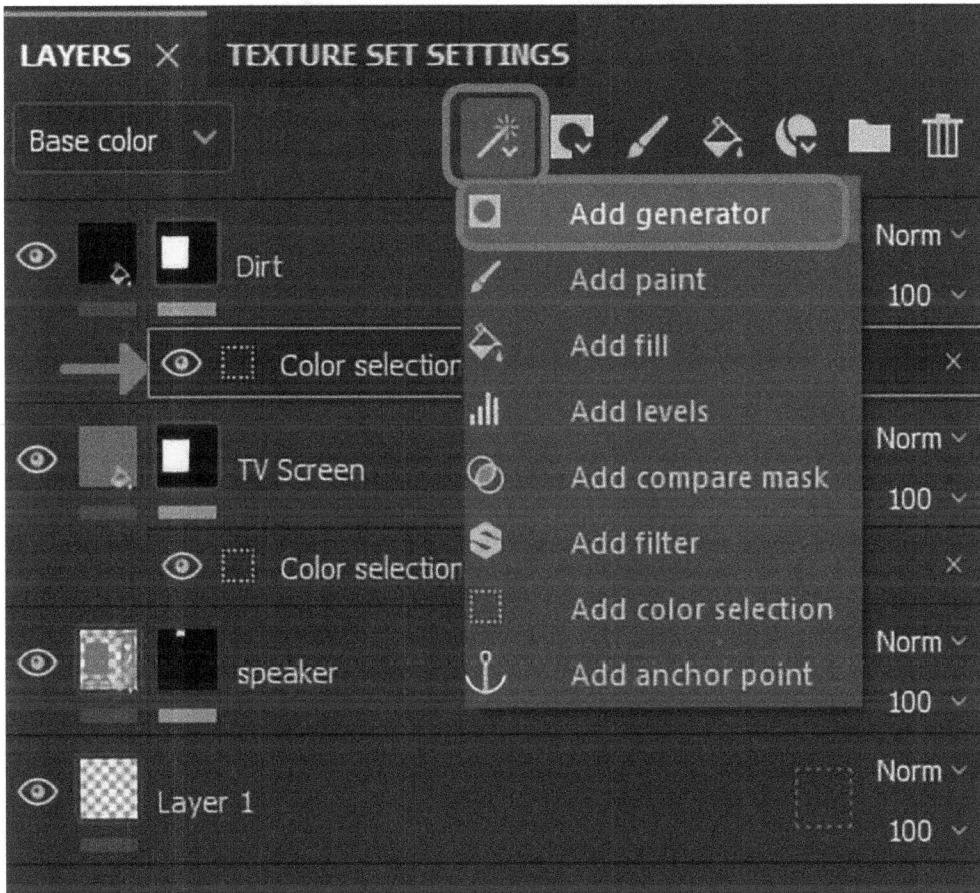

Figure 3.51 – Add generator

6. As soon you click on **Add generator**, the **PROPERTIES - GENERATOR** panel will appear. Select the **Generator** option under this panel and choose **Dirt** from the list:

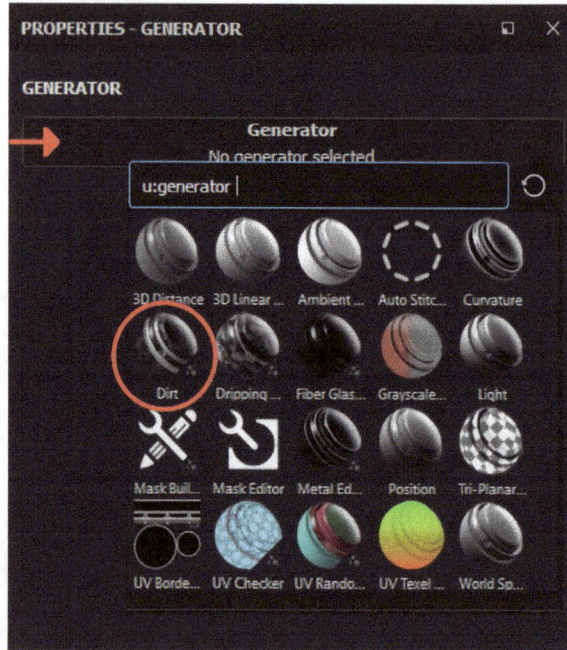

Figure 3.52 – Adding the Dirt generator

Note that the **Dirt** generator is doing its job, but the dirt is all over the front part of the television:

Figure 3.53 – The dirt effect all over the front part of the TV

7. Now, unlike the main layer, you cannot add a mask with color selection to any sub-layer or effects, such as the **Dirt** generator that we just created. Therefore, to apply the dirt effect to the television screen and to the entire front part, we must create a folder first and rename it Screen. Then, move the **Dirt** layer inside it.

    We can also move the **TV Screen** layer inside the Screen folder to keep everything organized; however, you must make sure that the **Dirt** layer is on top and **TV Screen** below it:

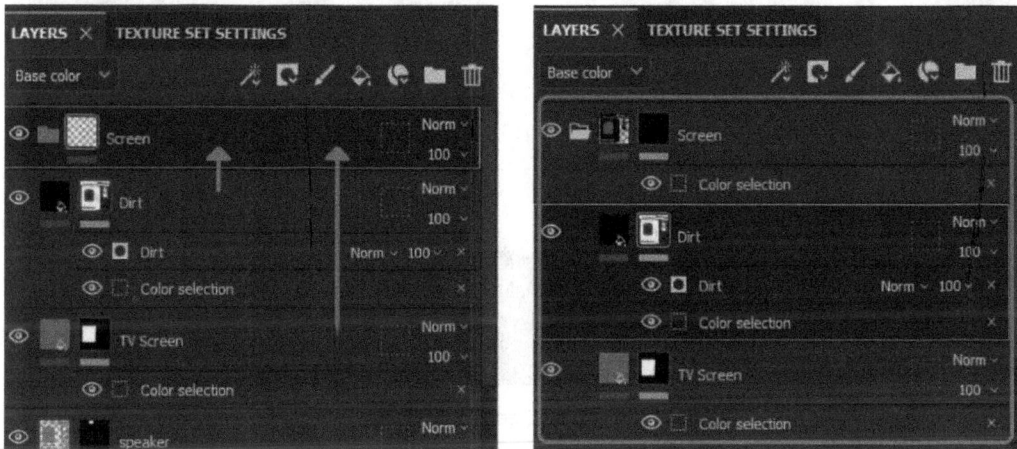

Figure 3.54 – Creating the Screen folder and moving the Dirt and TV Screen layers into it

8. Now, right-click on the Screen folder, select **Add mask with color selection**, and repeat *step 3*. Now, you will see that the dirt effect is only on the screen.

    If your effects or sub-layers are affecting the whole area, to avoid this, you can create a folder and move every affected layer inside it and apply the mask, as we did in this step:

Figure 3.55 – The Screen folder mask

9. Now, select the dirt effect and change the values of the following under **PROPERTIES – GENERATOR**:

    *   **Dirt Level**: 0 . 6

    *   **Dirt Contrast**: 0 . 55

    *   **Grunge Amount**: 0 . 26

    *   **Grunge Scale**: 5

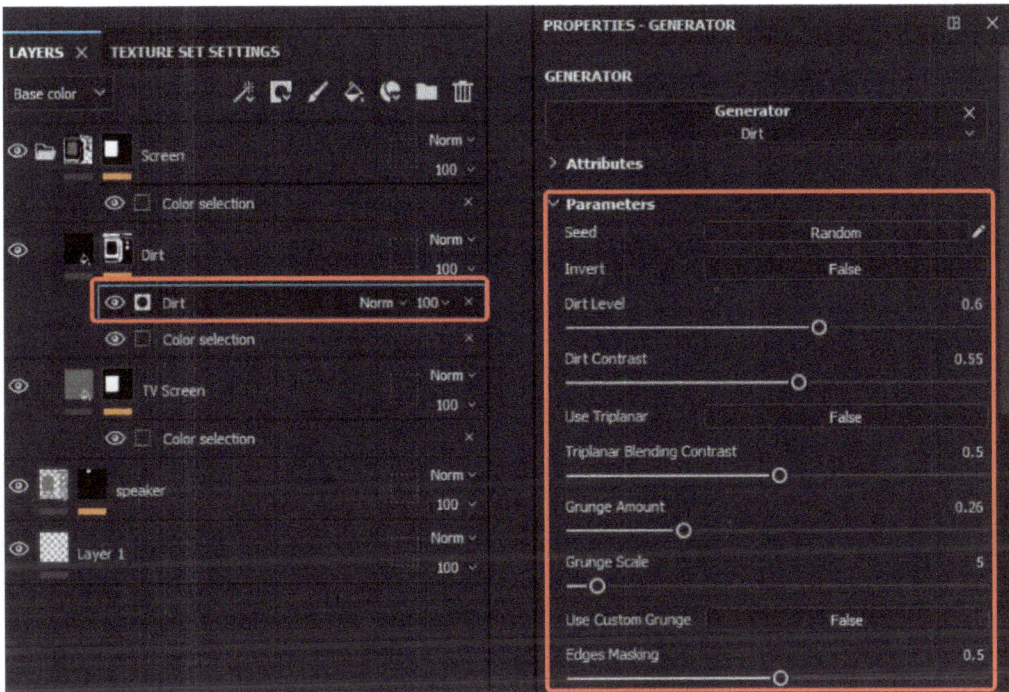

Figure 3.56 – The dirt effect settings

Now, you can see that the screen looks more realistic:

Figure 3.57 – A realistic TV screen

Summary

In this chapter, we have learned about layers, how to apply materials on 3D models and 3D meshes, how to use an ID map to apply materials, and how to create transparent and dirt materials and dirt effects.

In the next chapter, we will go through masks in detail and learn the usage, purpose, and importance of complex masks, as well as the importance of Planar and Tri-planar masks, while working on practical parts in Adobe Substance 3D Painter.

Working with Masks in Adobe Substance 3D Painter

This is chapter is quite important because we will learn how masks are created inside Adobe Substance 3D Painter. You might have noticed that in movies and video games where 3D objects are used, they look quite realistic because of their imperfections, withering, dirt, and the damage effects used on them, and without these effects, the objects will not look realistic and attractive.

Masking is very useful when creating withering, dirt maps, or curvature-based damage effects inside Adobe 3D Substance Painter – this chapter will take you on that journey and explain the whole process with easy yet comprehensive methods.

In this chapter, we will cover the following topics:

- Creating complex masks in Adobe Substance 3D Painter
- Creating planar masks in Substance Painter

Creating complex masks in Adobe Substance 3D Painter

Regular masks are only used to show textures in the white-masked area, while the black area hides the texture, as we studied in the previous chapters. However, complex masks can do more than regular masks – for example, you can create nested masks and each mask can control different **MATERIAL** settings. So let us do that:

1. Select **TV_Middle_Casing** from **TEXTURE SET LIST** – you will notice that the middle casing of the retro television looks new and untouched, whereas it should look dirtier, as it's an old television set.

2. In the **LAYERS** panel, make sure the `Black_Plastic` folder is collapsed so that it will not annoy you and the **LAYERS** panel looks clean and organized. To collapse any folder, you just need to click on the folder icon, as shown in *Figure 4.1*:

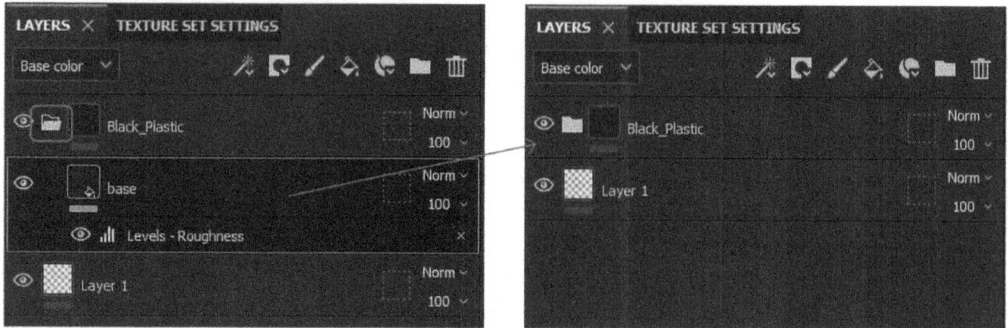

Figure 4.1 – Collapsing a folder

3. Click on the folder icon, as shown in *Figure 4.2*:

Figure 4.2 – Adding a group

4. Rename the newly created folder Dirt Scratches:

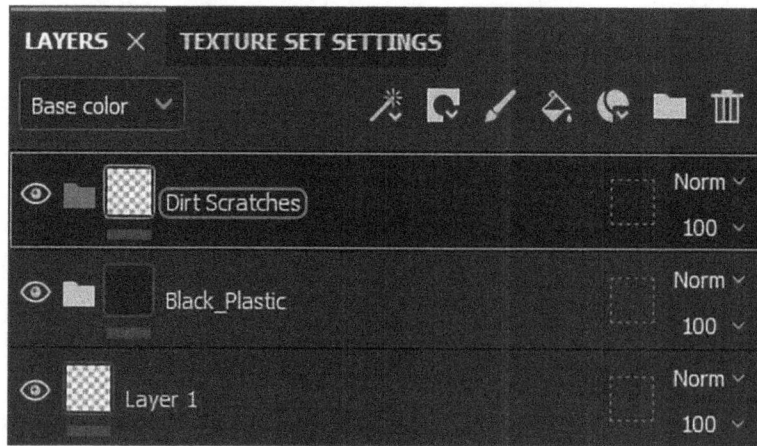

Figure 4.3 – Renaming the folder

5.   We will add a new fill layer – this will cover our previously created **Black_Plastic** layer. However, don't worry about it because **Black_Plastic** was only added to teach you the methods of adding a fill layer. Now, select **Add fill layer** inside the `Dirt Scratches` folder, rename it `Base`, and change its **Base color** property to black:

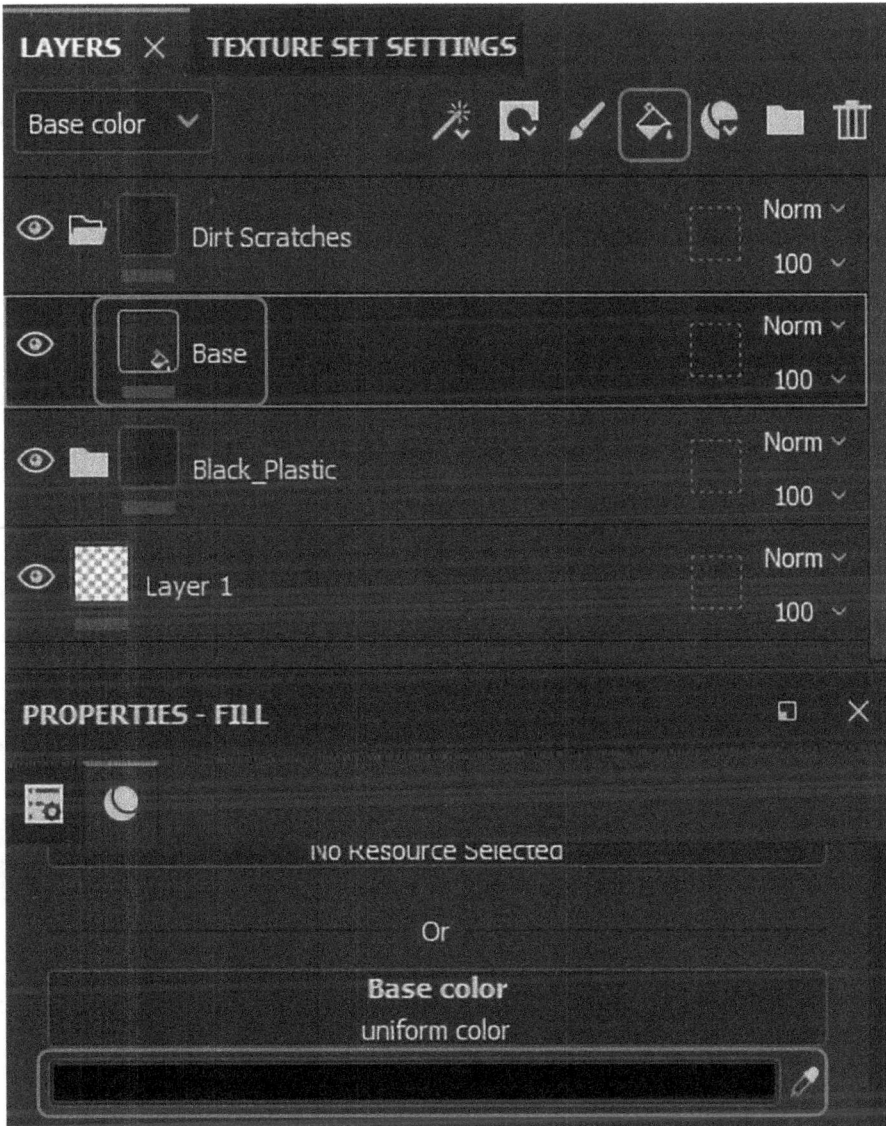

Figure 4.4 – A new fill layer

This new **Base** fill layer will act as a base for the dirt map we will create – creating this base layer will prevent the **Black_Plastic** layer from being affected.

6. Now, change the **Roughness** property under **PROPERTIES HYPHEN FILL** of the **Base** layer to 0.4444, as shown in *Figure 4.5*:

Figure 4.5 – The Roughness value for the Base layer of Dirt Scratches

This will make the base of **TV_Middle_Casing** a bit rougher, as shown in *Figure 4.6*:

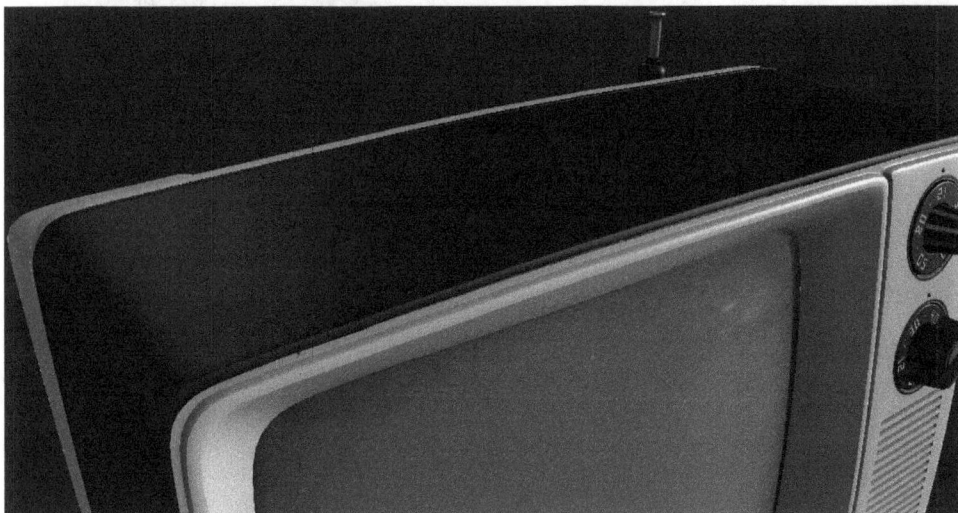

Figure 4.6 – The Roughness effect on the Base layer of Dirt Scratches

7. Now, select **Add fill layer** above the **Base** layer and rename it Dirt Roughness, as shown in *Figure 4.7*:

Figure 4.7 – A new Dirt Roughness layer

8. Now, under **PROPERTIES HYPHEN FILL** of the **Dirt Roughness** layer, deselect all the MATERIAL settings except for **rough**, as shown in *Figure 4.8*:

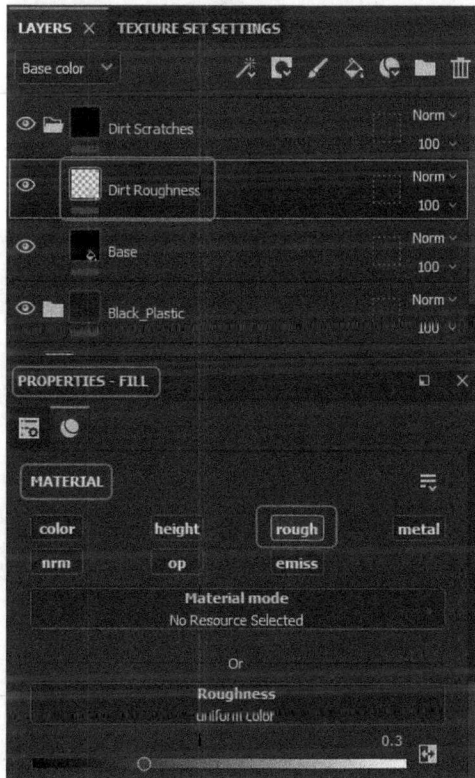

Figure 4.8 – The MATERIAL properties of the Dirt Roughness fill layer

9. Instead of directly applying any materials to the **Dirt Roughness** layer, we will use **Add black mask** on it so that the main **Dirt Roughness** layer stays independent in case we want to change its **Base color** property or any other **MATERIAL** settings:

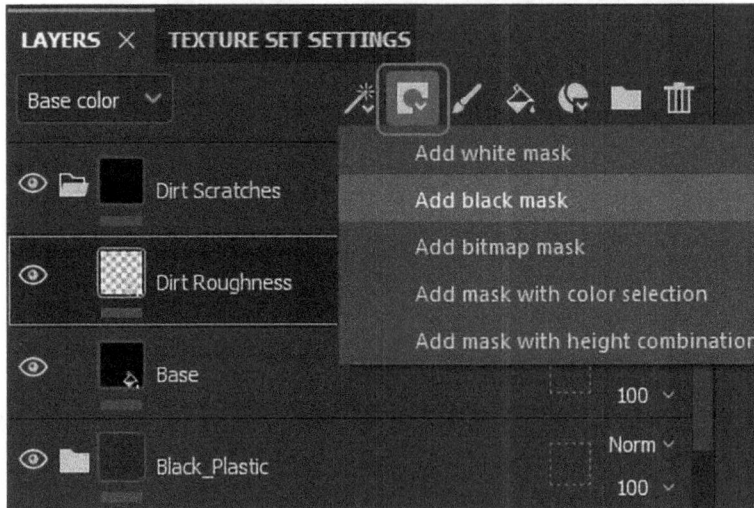

Figure 4.9 – Adding a black mask to the Dirt Roughness layer

10. Now, select the black mask that we have created on the **Dirt Roughness** layer and choose **Add fill** under the **Add effect** option, as shown in *Figure 4.10*:

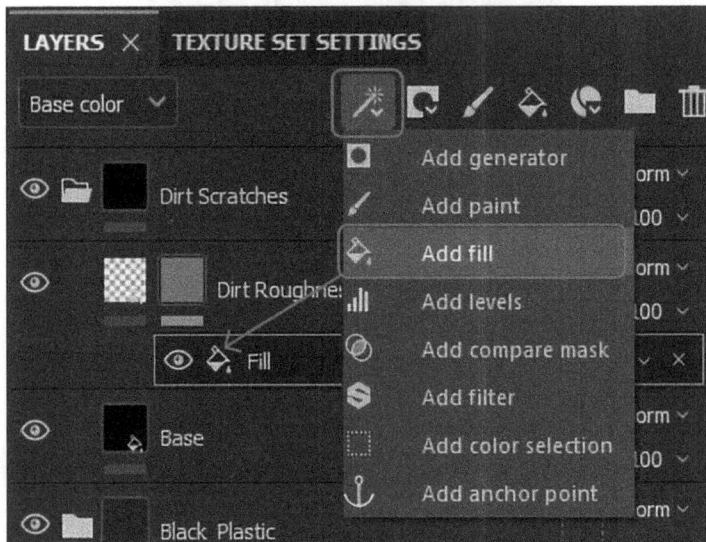

Figure 4.10 – Adding a fill layer to the Dirt Roughness mask

11. After adding the fill layer to the **Dirt Roughness** mask, select the same mask, and under **PROPERTIES HYPHEN FILL**, go to the **GRAYSCALE** settings and click on **grayscale uniform color**. Then, in the **RESOURCES** area, search for the grunge rough dirty mask and select it.

This mask will create an old dirty effect on **TV_Middle_Casing**. The fill layer of the **Dirt Roughness** mask layer will automatically be renamed grunge rough dirty:

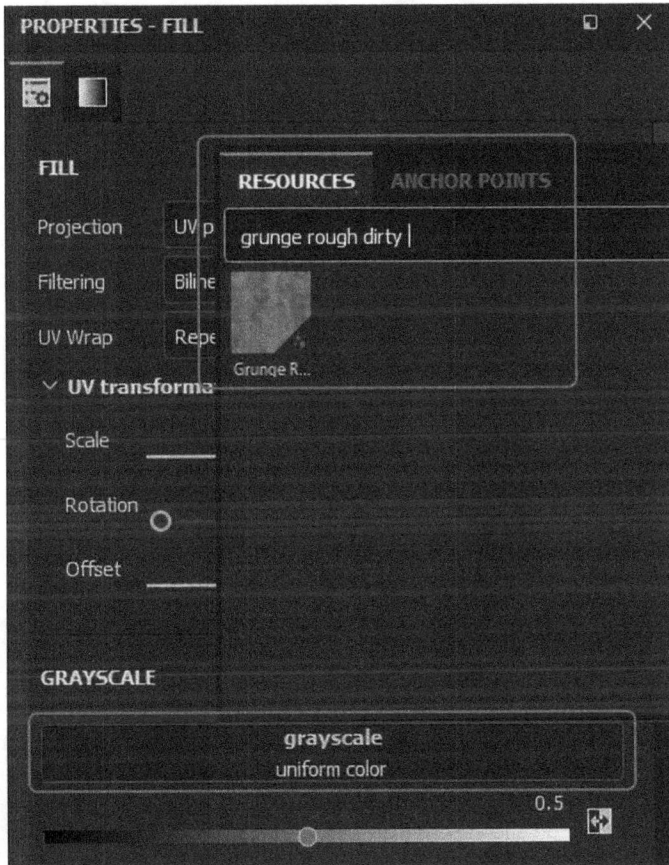

Figure 4.11 – Using a grayscale uniform color on Dirt Roughness

12. Select the **Grunge Rough Dirty** mask layer and apply the following changes:

- **Projection**: **Tri-planar projection**
- **UV transformations Scale**: 2
- **3D projection settings Scale Y**: 0.72
- **3D projection settings Scale Z**: 0.77

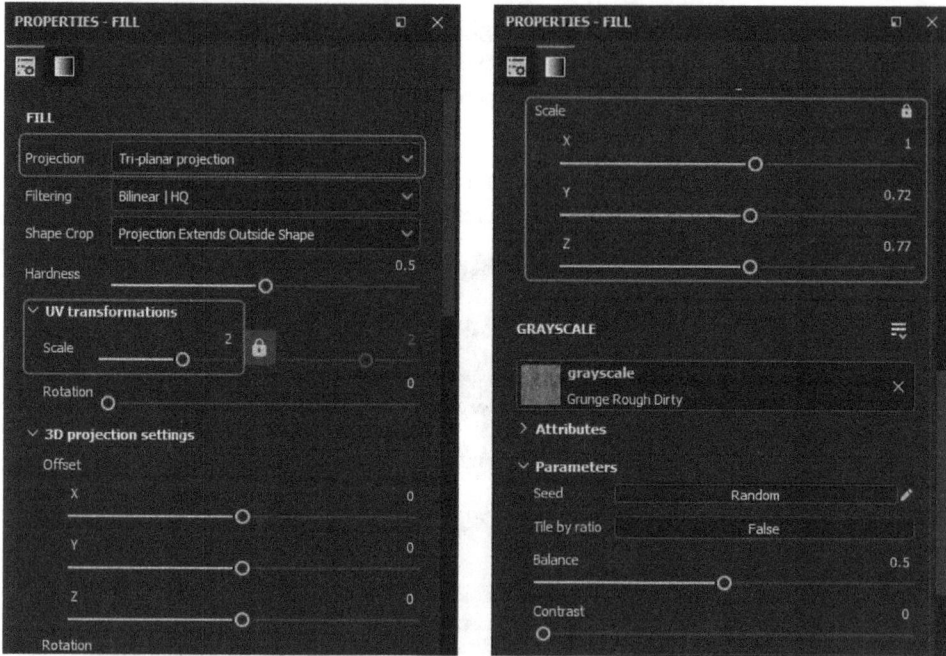

Figure 4.12 – Settings for the Grunge Rough Dirty mask

13. Now, select the main **Dirt Roughness** layer, and under the **PROPERTIES HYPHEN FILL** settings, set **Roughness uniform color** to 0 . 1, and you will notice that **TV_Middle_Casing** looks old and rough, as shown in *Figure 4.13*:

Figure 4.13 – The final Dirt Roughness material

Now that you have hopefully comprehended the usage, purpose, and importance of complex masks, in the next section, we will learn how to create planar masks inside Painter.

Creating planar masks in Substance Painter

Planar masks use **Planar projection** to create uniformly scaled planar textures on 3D models – to avoid stretched and squeezed textures, **Planar projection** uses the **Depth Culling** feature:

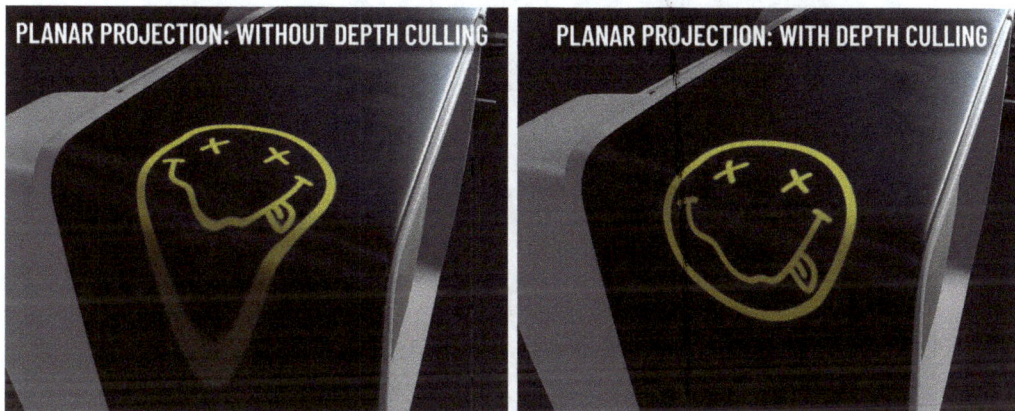

Figure 4.14 – The difference between planar projection with and without the Depth Culling feature

So, let's put this all together and apply a sticker on the television using **Planar projection**:

1. Keep **TV_Middle_Casing TEXTURE SET LIST** selected and go to the **ASSETS** panel. Click on the small arrow next to the **All libraries** breadcrumbs and change it to **Project**. Then, click on the small arrow next to **Project**, select **Retro**, and select the **Textures** asset type:

Figure 4.15 – Viewing the imported textures in the ASSETS panel

2. You will see the **MTV Sticker** inside the **ASSETS** panel. Drag that logo onto the upper-left-hand corner of the television, and after dropping the logo, a menu will appear for selecting the **MATERIAL** type. Choose the **Base color** property from the menu, as shown in *Figure 4.16*:

Figure 4.16 – Applying the MTV logo sticker to the television

3. After applying **MTV Sticker** on the television, a new layer will be created under the **LAYERS** panel called **Sticker6-MTV** – it's the same name as the sticker:

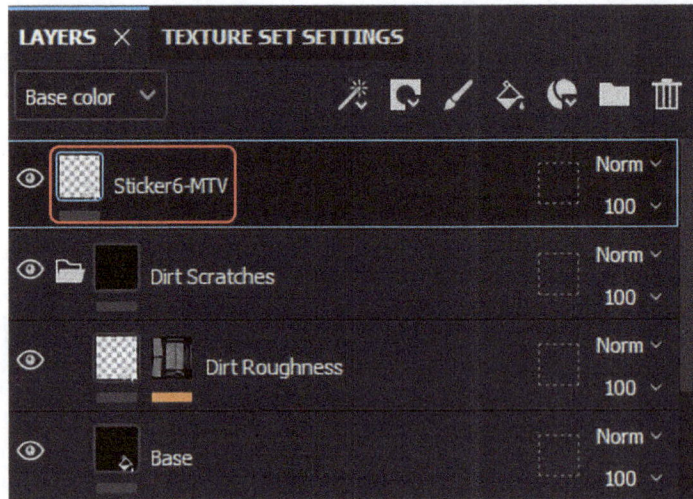

Figure 4.17 – The automatic creation of the Sticker6-MTV layer

4. Keep **Sticker6-MTV** selected. Under **PROPERTIES HYPHEN FILL**, go to the **Projection** type and choose **Planar projection**. Set **Scale** to 1.5 and **Rotation** to 45 under **UV transformations**. Make sure the **Depth Culling** and **Blackface Culling** checkboxes are checked:

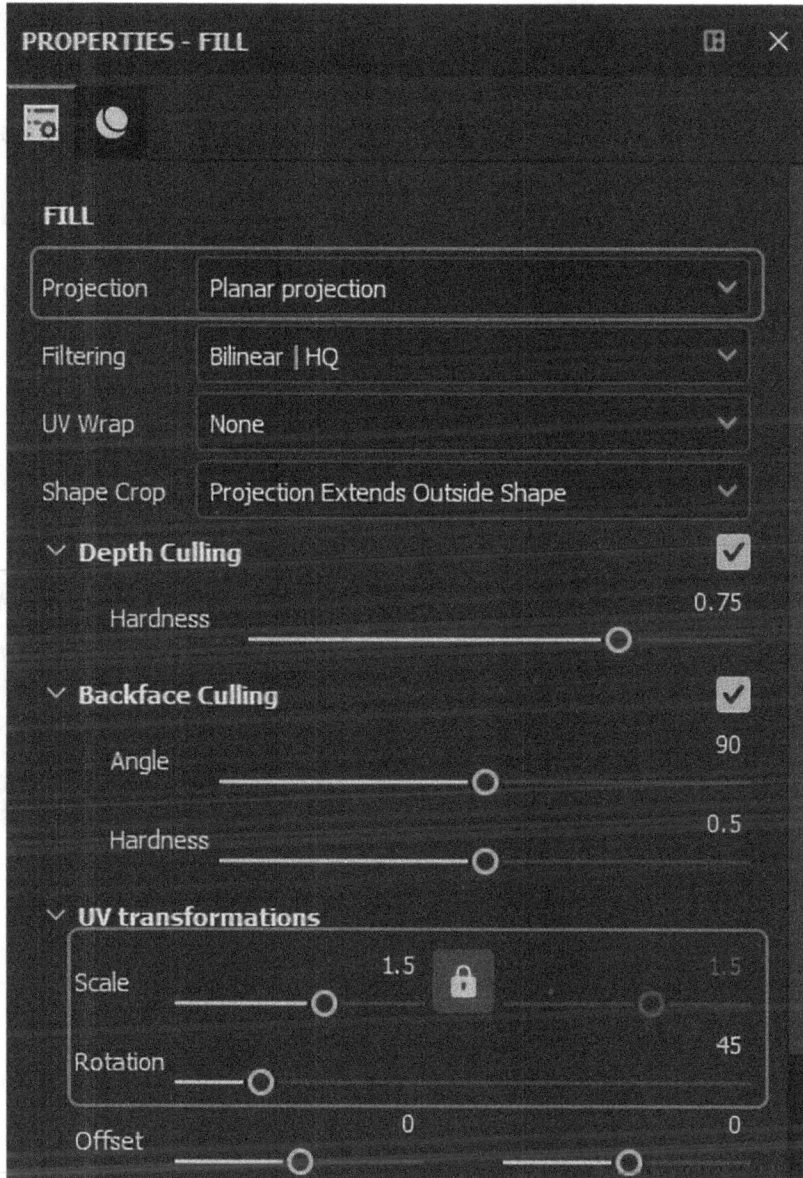

Figure 4.18 – The Sticker6-MTV settings

5. If you notice a black fading feather effect on **MTV Sticker**, as shown in *Figure 4.20*, that means the logo is stretching and that black feather effect is the **Depth Culling** indication, which warns you about stretching or any other sort of deformation:

Figure 4.19 – Depth Culling

6. To avoid stretching, we need to reposition the sticker until we stop seeing any **Depth Culling** indication. To reposition the sticker, you have to change the **3D projection settings** properties for **Sticker6-MTV** under **PROPERTIES HYPHEN FILL** as follows:

- **Offset** X: -0.946 ; **Rotation X**: 92

- **Offset** Y: 0.658 ; **Rotation Y**: -137

- **Offset** Z: 0.38 ; **Rotation Z**: 92

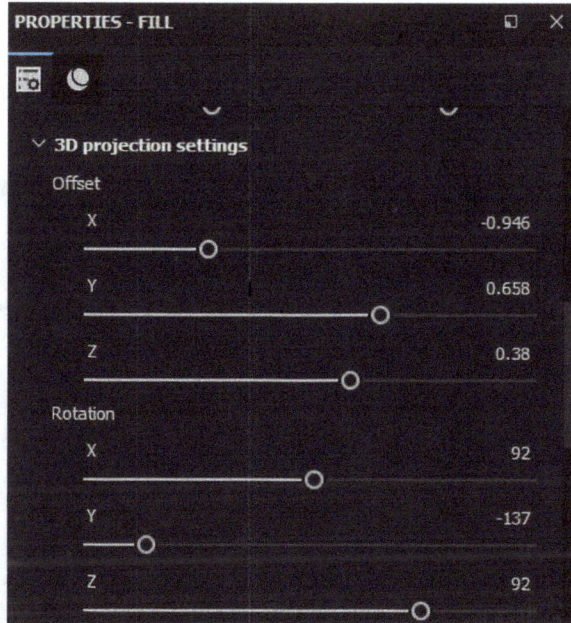

Figure 4.20 – Sticker6-MTV's 3D projection settings

7. After changing **3D projection settings** for **Sticker6-MTV**, the sticker will look cleaner and more uniform, as shown in *Figure 4.22*:

Figure 4.21 – The uniform MTV sticker

8. If you do not want to move the planar textures with **3D projection settings** under **PROPERTIES - FILL**, you can use the **3D projection settings** feature located in the contextual toolbar, as shown in *Figure 4.23*, and with your mouse, you can move, rotate, scale, flip, and warp your planar texture:

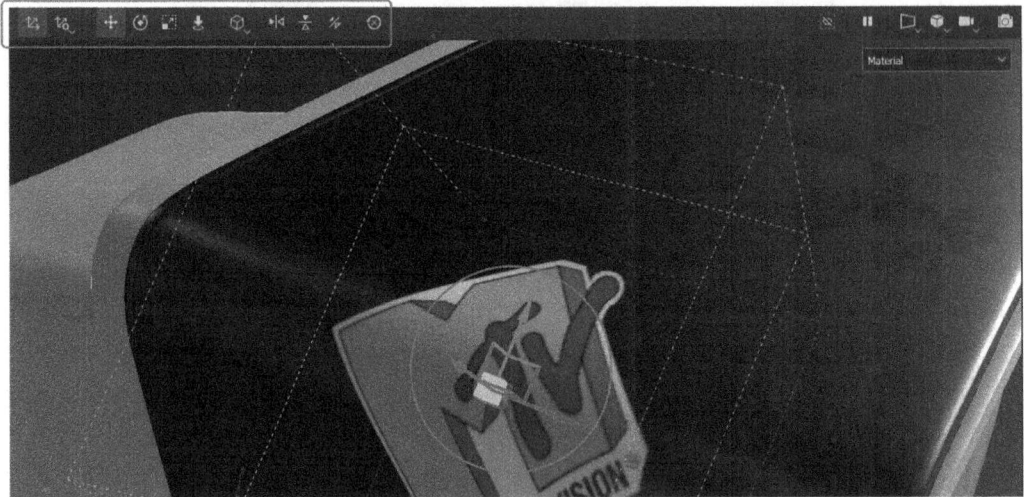

Figure 4.22 – 3D projection settings in the contextual toolbar

Summary

Now that you are familiar with complex masks, you can create nested masks with different types of creative effects. Moreover, we have learned how to apply a uniform planar texture, which prevents any type of stretching or deformation.

In the next chapter, we will learn how to apply bitmap textures, create a 3D logo from scratch, make a custom brush, work with stencils and projection, and use text, fonts, and the clone tool and smudge tool in Adobe Substance 3D Painter.

Working with Advanced Tools in Adobe Substance 3D Painter

There are a variety of advanced tools in Adobe Substance 3D Painter and they are easy to use. However, they can be quite confusing if the user is using them for the first time.

In this chapter, you will learn the importance, purpose, and usage of these tools with conceptual and practical knowledge. The reason these tools are important to learn is that they have the ability to create texture and paint effects that are hard to create with normal tools and procedures. However, the procedure for using these tools is quite tricky, and you have to follow certain steps, but these can make the tools easier to comprehend.

In this chapter, we will cover the following topics:

- Applying bitmap textures in Adobe Substance 3D Painter
- Creating a 3D logo from scratch in Adobe Substance 3D Painter
- Making a custom brush in Adobe Substance 3D Painter
- Working with stencils and projection in Adobe Substance 3D Painter
- Using text, fonts, and the Clone and Smudge tools in Adobe Substance 3D Painter

Applying bitmap textures in Adobe Substance 3D Painter

Previously, you learned the methods to apply planar masks and in this section, we will learn to apply bitmap textures using **Base color**. The bitmap textures are colorful images that can be used as texture maps on any mesh, for example, red floral pattern on sofa and so on. So, let us start the bitmap texture application:

1. Rotate the television so that you can see its back and select **TV_Back_Casing** from **TEXTURE SET LIST**. Go to the **LAYERS** panel and click on **Add fill layer** to create a new layer and rename the layer Servlet Map.

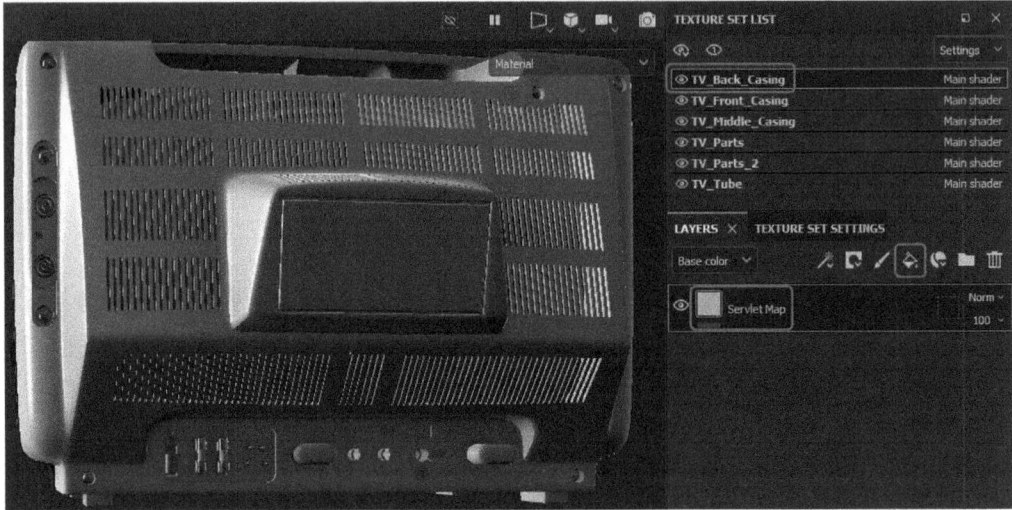

Figure 5.1 – New Servlet Map fill layer

2.   Keep the **Servlet Map** layer selected, go to the **PROPERTIES - FILL** panel, click on **Base color uniform color**, type `servlet` in the **RESOURCES** pop-up menu search field, and the servlet bitmap texture that we imported in the previous chapters will appear.

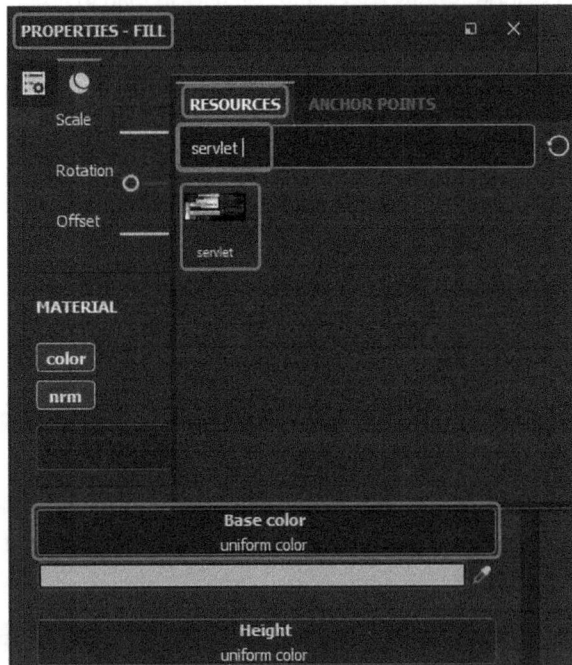

Figure 5.2 – Dropping the servlet bitmap texture into Base color

3. After dropping the **servlet** bitmap texture into the **Base color MATERIAL** area, you will notice the bitmap texture is all over **TV_Back_Casing**, as shown in *Figure 5.3*.

Figure 5.3 – Servlet bitmap texture all over the television

4. To avoid this situation, we must create a mask on the **Servlet Map** layer. Select **Servlet Map**, click on **Add mask**, and select **Add mask with color selection**.

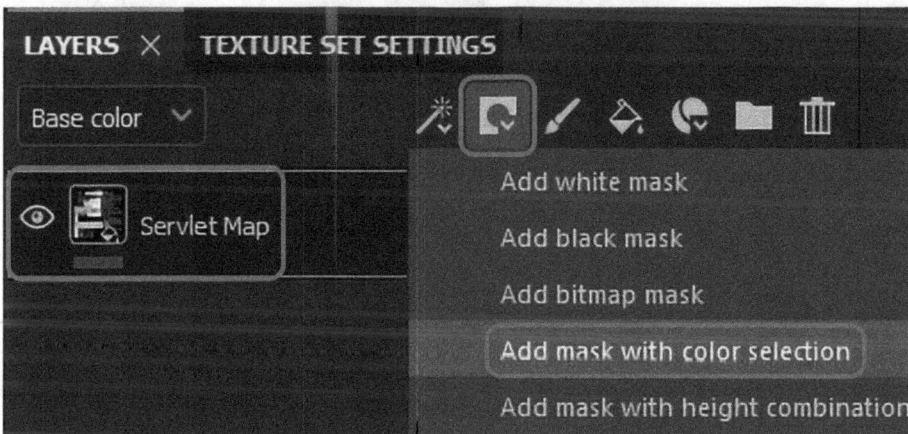

Figure 5.4 – Add mask with color selection

5.   Select the newly created **Color selection** mask in the **Servlet Map** layer.

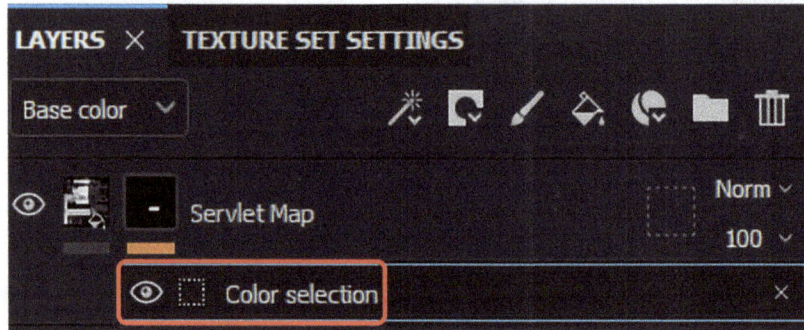

Figure 5.5 – Selecting the Color selection mask

6.   Now, go to **PROPERTIES - COLOR SELECTION**, click on the eye dropper in the **Colors** area, and with the eye dropper click on the green area of **TV_Back_Casing**.

Figure 5.6 – Selecting the servlet ID map with the help of the color selection mask

7.   You will notice that Servelet Texture Map is now appearing only in the designated servlet area, as shown in *Figure 5.6*. However, the Servelet Texture Map appears to be cropped, as shown in *Figure 5.7*; to solve this, we need to go to the next step.

Figure 5.7 – Servlet appearing only in the color selection mask area

8. To fix the servlet bitmap texture, select the main **Servlet Map** layer, then go to the **PROPERTIES – FILL** area, and change the following settings under **FILL** and **UV transformations** respectively:

- **Scale**: `-4.9, -10.5` (make sure the lock icon is not pressed so they stay non-uniformed)

- **Offset**: `0.093, -0.227`

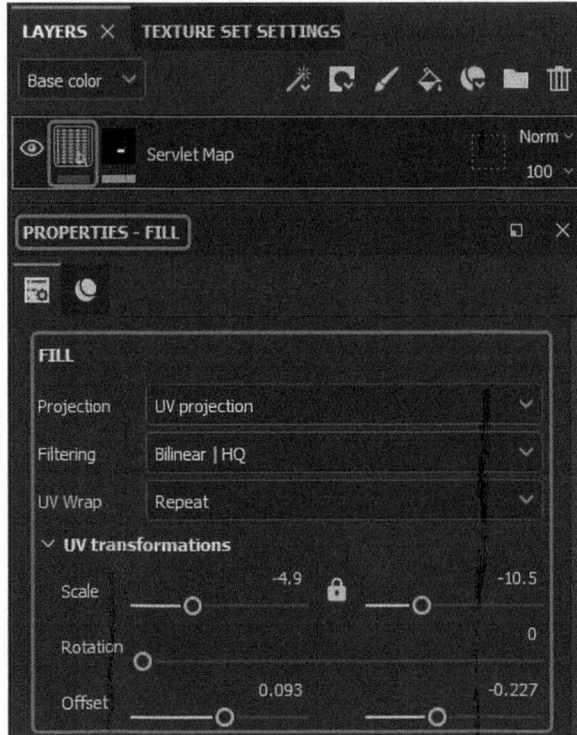

Figure 5.8 – Servlet bitmap texture map settings

9. After applying these settings, you will notice that the servlet bitmap texture sits in its designated area perfectly.

Figure 5.9 – Servlet bitmap texture perfectly sitting in its designated area

Now that you understand the usage, purpose, and importance of bitmap textures while working on practical elements in Adobe Substance 3D Painter, we can move on to the next section in which you will learn how to create a 3D embossed logo.

Creating a 3D logo from scratch in Adobe Substance 3D Painter

So far, we have learned how to apply planar texture maps that have no depth or height, which are good if you want flat textures like stickers and posters on walls, and so on. However, if you want to create a texture that has a certain depth or height effect, you need to go through the following steps in which you will learn how to create a 3D embossed logo:

1. Rotate the 3D viewport so that you can see the front of the television, select **TV_Front_Casing** from the **TEXTURE SET LIST**, and make sure all the folders in the **LAYERS** panel are collapsed to keep the panel neat and organized. We will create the logo in the red-circled area, as shown in *Figure 5.10*.

Figure 5.10 – 3D logo application area

2. In the **LAYERS** panel, click on **Add fill layer** and create a new fill layer, then rename the layer TV Logo. Change the **Base color** value of the **TV Logo** layer to white.

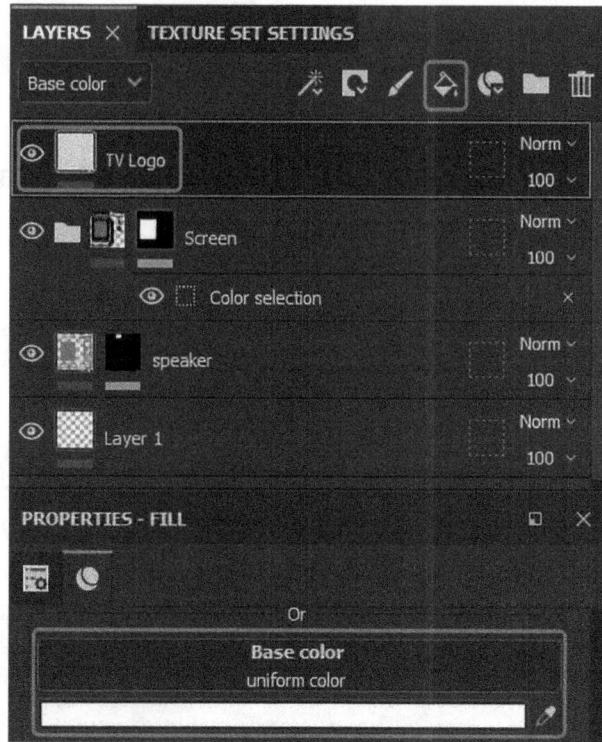

Figure 5.11 – Creating a new TV Logo fill layer

3.   Keep the **TV Logo** layer selected and then click on **Add mask** and select **Add black mask**. We need the mask to put the 3D logo alpha inside so that the texture effects are only on the 3D logo alpha area.

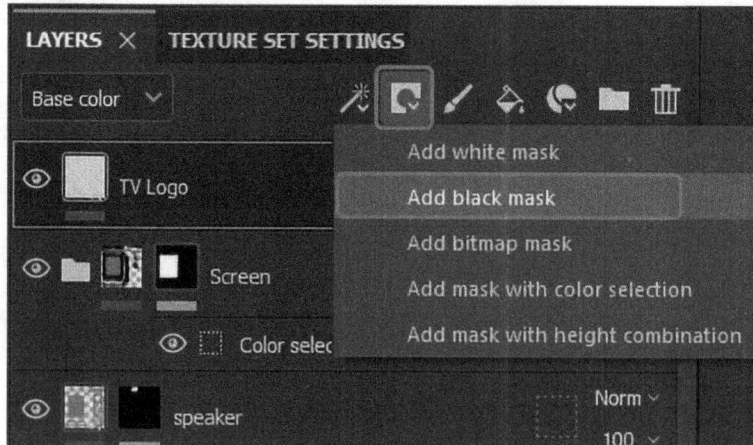

Figure 5.12 – Adding a black mask to the TV Logo layer

4. Select the **TV Logo** mask, and then click on **Add effect** and select **Add fill**.

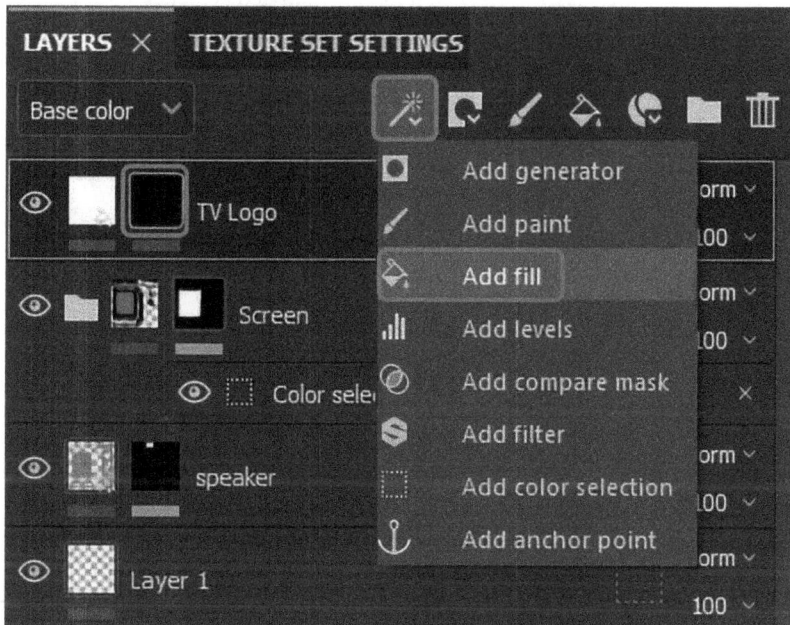

Figure 5.13 – Adding a fill layer to the TV Logo mask

5. Keep the **TV Logo** mask's fill layer selected, go to **PROPERTIES - FILL**, and under the **GRAYSCALE** option, click on **grayscale uniform color**. When the **RESOURCES** menu pops up, type TV_logo_alpha in the search field and the EMNEM TV logo will appear. Select that logo; this is the logo we imported in the previous chapter.

Figure 5.14 – Applying the TV logo alpha to the TV logo mask's Fill layer

6. First, we need to switch the **Projection** mode to **Tri-planar**. Now, you will notice that the TV logo is all over **TV_Front_Casing**. To keep it in its designated area, keep the **TV Logo** mask's **Fill** layer selected, go to **PROPERTIES – FILL**, go to **3D projection** settings, and change the following settings under **Offset** and **Scale** respectively:

- **Offset X**: 0.74, **Scale X**: 0.125

- **Offset Y**: -0.44, **Scale Y**: 0.03

- **Offset Z**: 0.75, **Scale Z**: 0.03

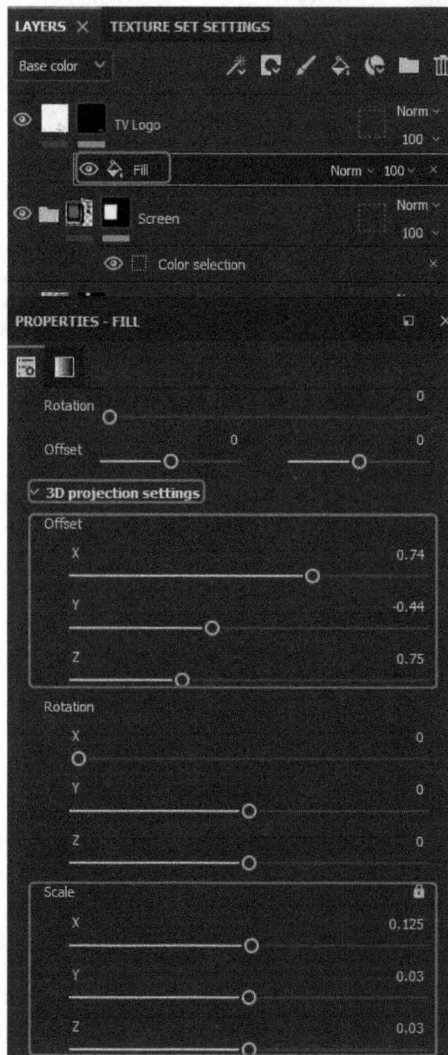

Figure 5.15 – Changing the settings of TV Logo's mask Fill layer

7. Select the main **TV Logo** layer and go to **PROPERTIES – FILL**. Under **MATERIAL**, select **color**, **height**, **rough**, and **metal**, and deselect **normal**, **opacity**, and **emissive**. To give the **TV Logo** layer a 3D look, change the following settings for **Height**, **Roughness**, and **Metallic**.

-   **Height**: 1, **Roughness**: 0.22, **Metallic**: 1

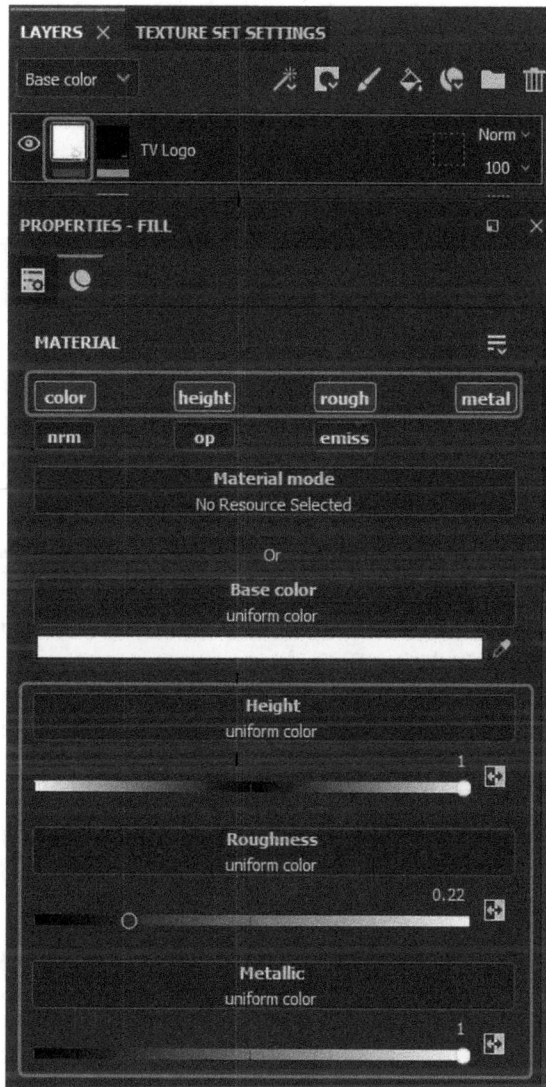

Figure 5.16 – Changing the settings of the main TV Logo layer

As the **TV_logo_alpha** map was applied to the **TV Logo** mask's **Fill** layer, the new settings, as shown in *Figure 5.16*, will only appear in the **TV_logo_alpha** shape as you can see in *Figure 5.17*.

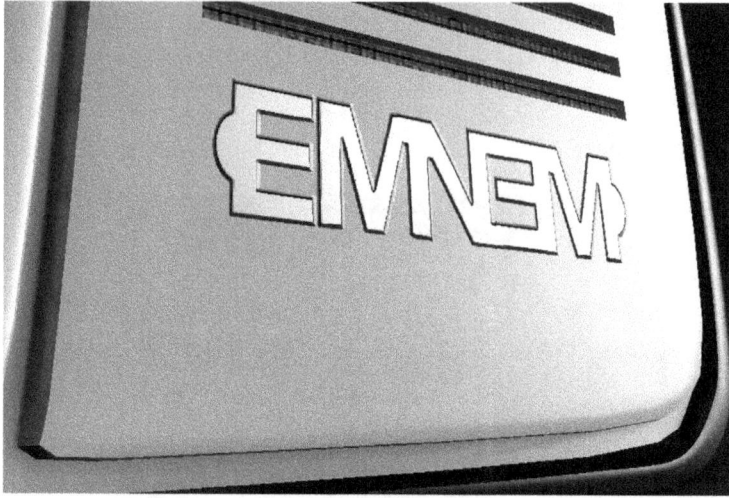

Figure 5.17 – The final look of the 3D TV logo

As you are now familiar with how to give depth and height to a texture map, and how to make it look metallic and control its roughness, in the next section we will learn how to make our own brushes and use them creatively.

Making a custom brush in Adobe Substance 3D Painter

Applying textures and masks, as we learned in the first two sections of this chapter, can be tedious if you have a lot of them to apply. To make it easier, you can create your own custom brush and apply the textures or masks very easily and quickly. However, this method only works for simple textures or masks such as screws and image stamps, because they work like stamps and not like regular texture maps.

So, let's start with the following steps and make a brush with a screw pattern:

1. Select **TV_Front_Casing** so that we can use our custom screw brush on the previously created **TV Logo**, and make sure all the folders in the **LAYERS** panel are collapsed so that the panel looks neat and organized.

2. Click on **Add Layer** to create a paintable layer on top of all layers and rename it Screw Stamp Brush.

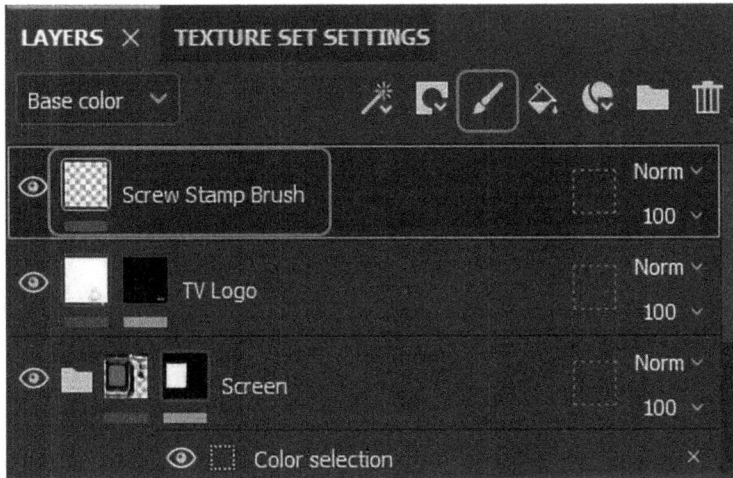

Figure 5.18 – Creating a new paintable layer called Screw Stamp Brush

3. Now, go to the **ASSETS** panel and select the **Brushes** asset type, then in the search field type `Basic Hard`, and select the **Basic Hard** brush when it appears in the panel.

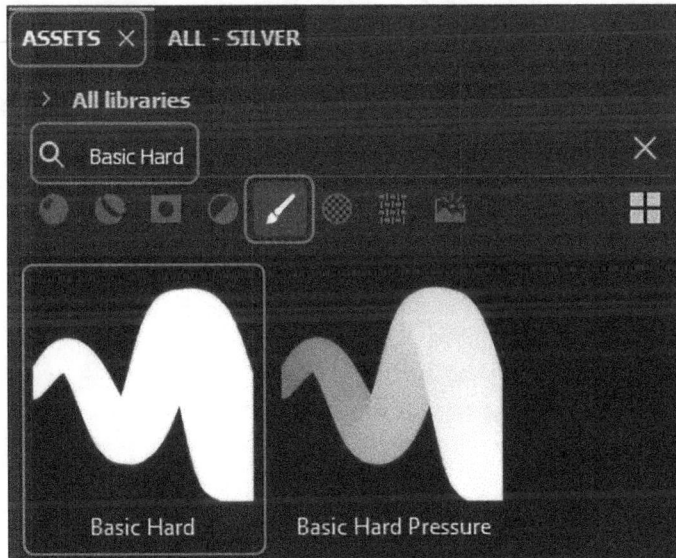

Figure 5.19 – Selecting the Basic Hard brush

4. Now it's time to customize the **Basic Hard** brush to create your own **Screw Stamp Brush**. To do that, go to **PROPERTIES – PAINT**, and under the **ALPHA** settings, usually, **Hardness** is by default set to 1. However, if it is not, then change **Hardness** all the way to 1, so that whatever you create has a solid outline instead of a feathered or faded outline.

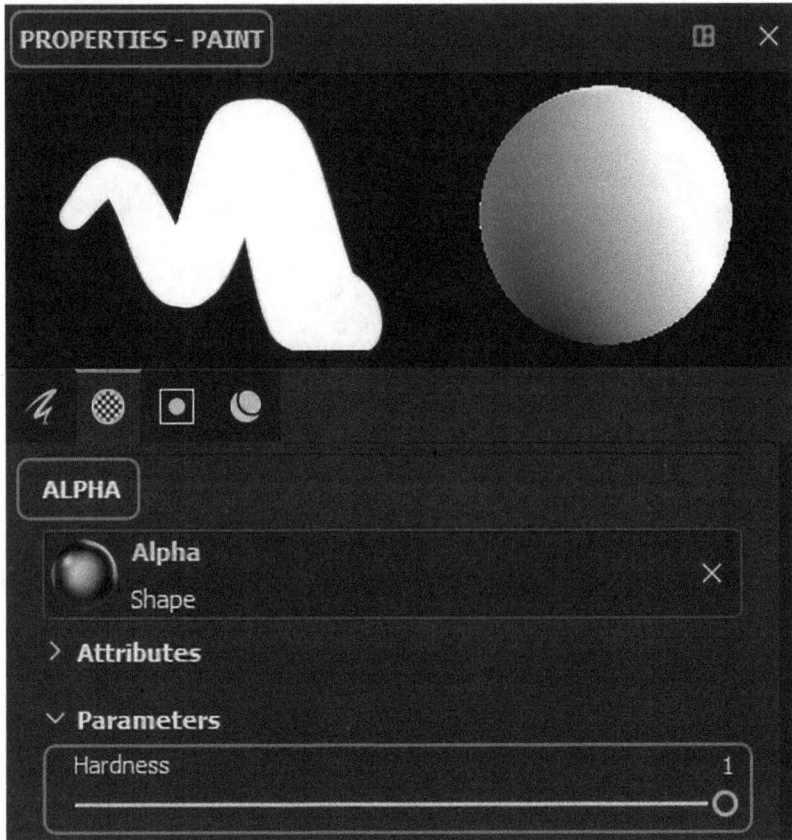

Figure 5.20 – Increasing the ALPHA Hardness value of Screw Stamp Brush

5. Now, to give the brush a metallic and shiny look, change the following settings under the **MATERIAL** settings:

    - **Base color**: Use light gray

    - **Roughness**: 0.25

    - **Metallic**: 1

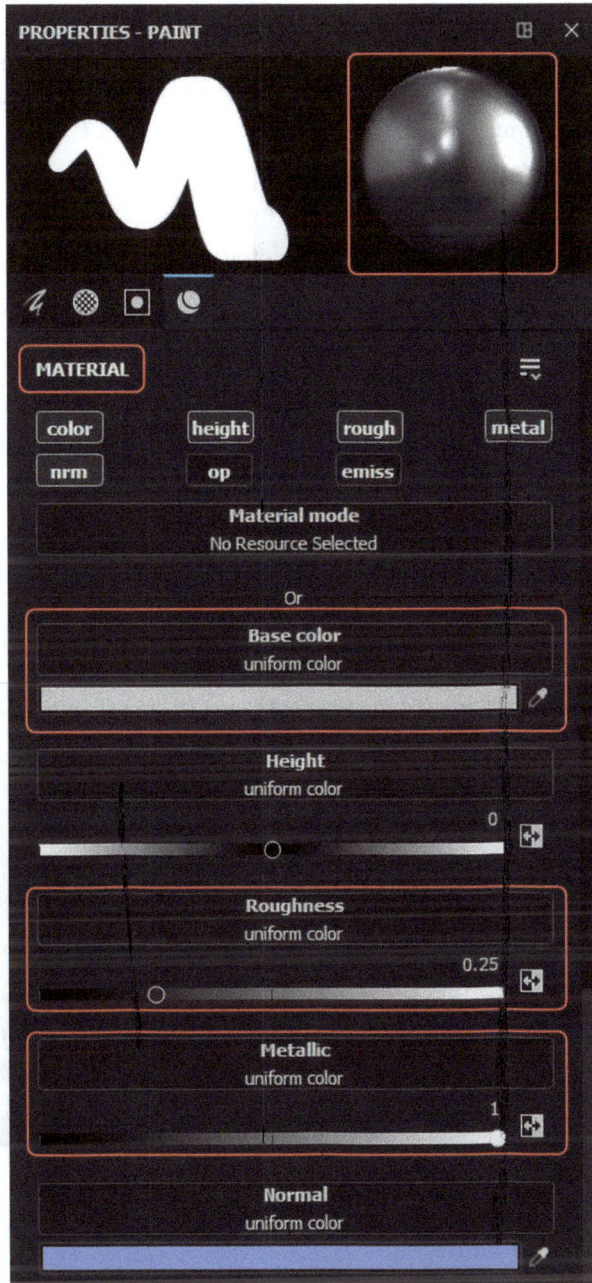

Figure 5.21 – Changing the Base color, Height, Roughness, and Metallic settings

Now, you will notice that the material of the brush looks more metallic like a screw. In the next step, we will give the material a screw-shaped depth.

6. To give the brush a screw-shaped depth, change **Height** to 1 under the **MATERIAL** settings, then click on **Normal**. As soon you click on **Normal**, a menu will pop up with the search field. Type Screw Cross Round in the search field. Select **Screw Cross Round**, which is a built-in normal height map.

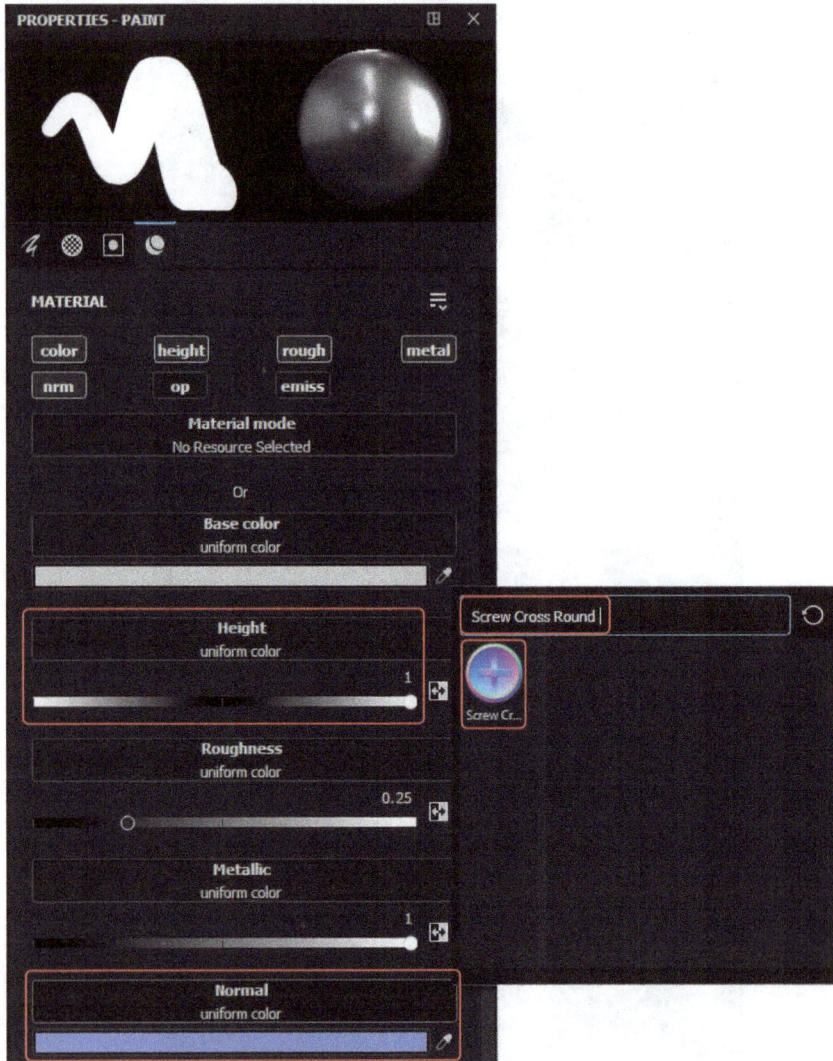

Figure 5.22 – Changing the Height map and adding a Normal map

Normal

Normal mapping is a texture mapping method used in 3D computer graphics to simulate the lighting of bumps and dents.

Now you will notice that the brush material looks more like a screw.

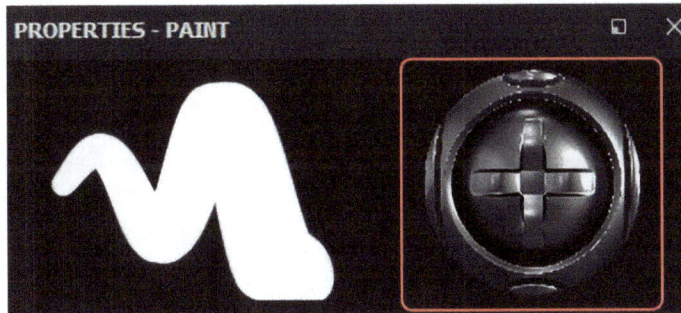

Figure 5.23 – Brush material with a screw shape

7. To decrease the brush size, press [on your keyboard, and to increase the brush size, press] (however, if you are using a different keyboard layout such as a Swiss keyboard, you might need to change the shortcuts using **Edit | Settings | Shortcuts**). We can also use the **Size** option under the **BRUSH** setting in **PROPERTIES - PAINT**, and change **Size** to 3 . 85.or to any size of your choice. The bigger you will make the brush size, the screw map size will also get bigger.

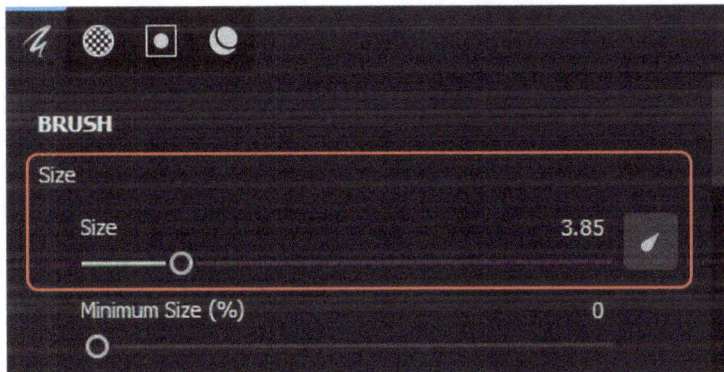

Figure 5.24 – Changing the brush size

Now, if you use **Screw Stamp Brush** with the existing settings, you will get a single impression of the screw by clicking the left mouse button once on the surface, as shown in *Figure 5.25*.

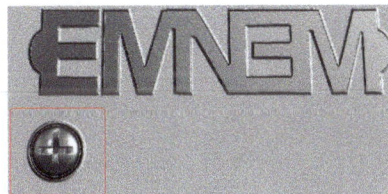

Figure 5.25 – Single-click brush impression

However, if you click the left mouse button and drag the mouse over the surface, you will get a row of screws without any gaps, as shown in *Figure 5.26*.

Figure 5.26 – Brush stroke drag effect

8. To avoid creating a brush stroke without any gaps, we need to increase the gaps between the brush strokes. To do that, go to **PROPERTIES - PAINT** in the **Screw Stamp Brush** layer. Then, under the **BRUSH** settings, change **Spacing** to 200.

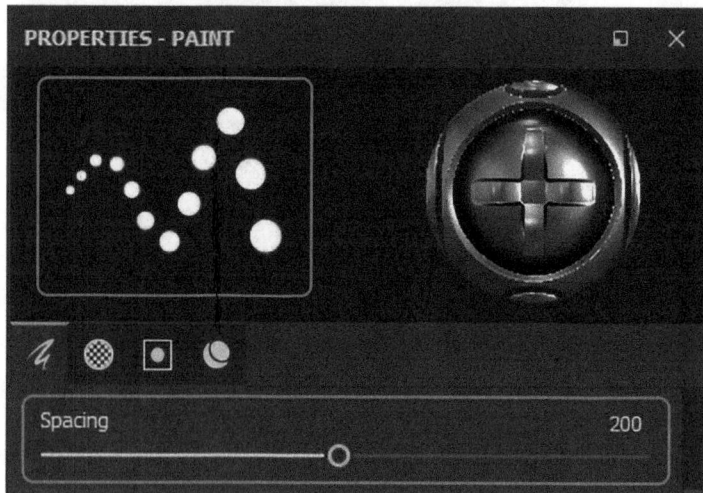

Figure 5.27 – Brush stroke spacing

9. Now if you drag your mouse across the surface with the left mouse button pressed, you will get the same impression, as shown in *Figure 5.28*.

Figure 5.28 – Non-straight brush stroke

10. However, if you want to create a straight row of brush strokes, you need to press *Shift* on your keyboard and drag your mouse across the surface with the left mouse button pressed, as shown in *Figure 5.29*.

Figure 5.29 – Straight brush stroke with the Shift key

Undo and remove these brush strokes if you have created them because *steps 8*, *9*, and *10* were only for demonstration purposes.

11. As we have created our own custom brush, it's time to save this brush. To save the brush, right-click anywhere inside **PROPERTIES - PAINT**, and select **Create tool preset**. The reason you choose **Create tool preset** and not **Create brush preset** is that the brush preset only saves the **BRUSH** settings while the tool preset saves all the settings.

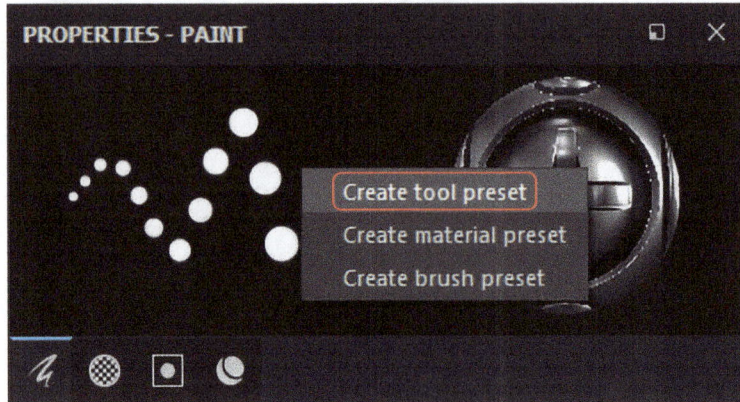

Figure 5.30 – Saving the brush tool preset

12. Once you click on **Create tool preset**, a filter named **tool** will be automatically created, as shown in *Figure 5.31*. You will see a new tool preset created named **new_tool_preset**. Right-click **new_tool_preset** and rename it Screw Stamp Brush. Now, you can click on the cross next to the **tool** filter to close it.

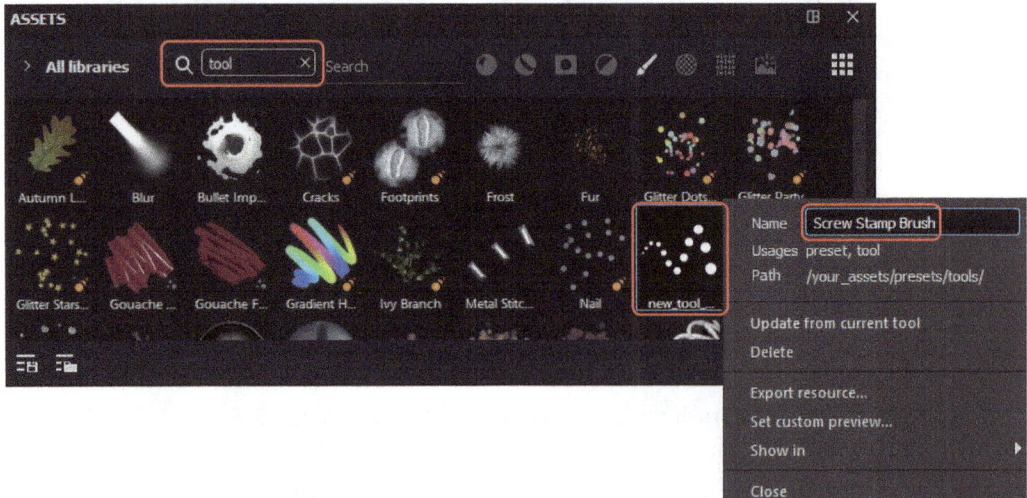

Figure 5.31 – Renaming new_tool_preset as Screw Stamp Brush

13. As we have created a new tool preset, you don't need the **Screw Stamp Brush** layer, so delete **Screw Stamp Brush** under the **LAYERS** panel.

Figure 5.32 – Deleting the Screw Stamp Brush tool

14. Now, click on **Add layer** to create a new paintable layer and rename it Screws.

Figure 5.33 – Creating screw layers

15. Now, resize **Screw Stamp Brush** to 0.75 and apply screws on both ends of the logo, as shown in *Figure 5.34*.

Figure 5.34 – Applying brush strokes to the TV logo

Hopefully, you have understood the process of creating your own custom brush; in the next section, you will learn how to work with projection and stencils. These are widely used to digitally print any texture or bitmap on the 3D model.

Working with stencils and projection in Adobe Substance 3D Painter

Stencils and projection are very widely used tools inside Painter. Through these tools, you can easily project images, textures, or bitmaps onto the surface of your 3D model with the amalgamation of brush and texture effects, whereas you cannot mix the brush and texture effects with the complex mask and planar mask methods that we studied in *Chapter 4, Working with Masks in Adobe Substance 3D Painter*.

So, let's see how it is done in practice:

1. Go to the **ASSETS** panel and select the **Brushes** asset type, then select the brush you want to use. In this project, we will use a **Basic Hard** brush.

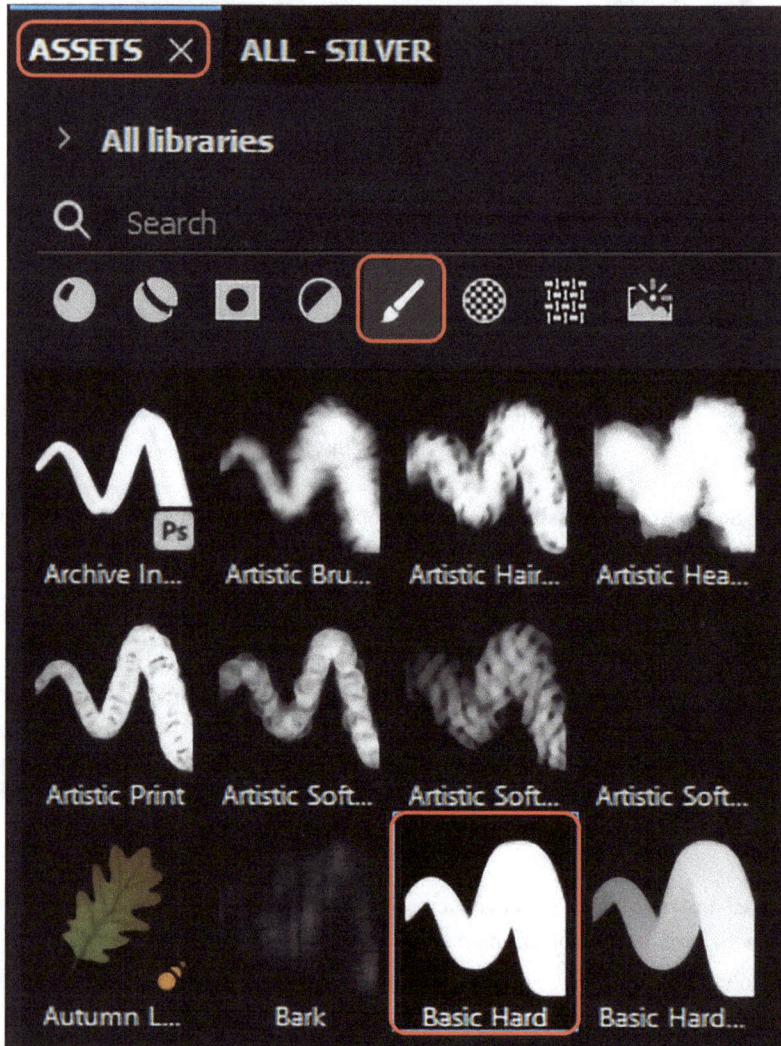

Figure 5.35 – Selecting the Basic Hard brush

2. Now go to **TEXTURE SET LIST** and select **TV_Middle_Casing**, then go to the **LAYERS** panel and click on **Add layer** to create a new layer. Once the new layer is created, rename it Stencil Artwork.

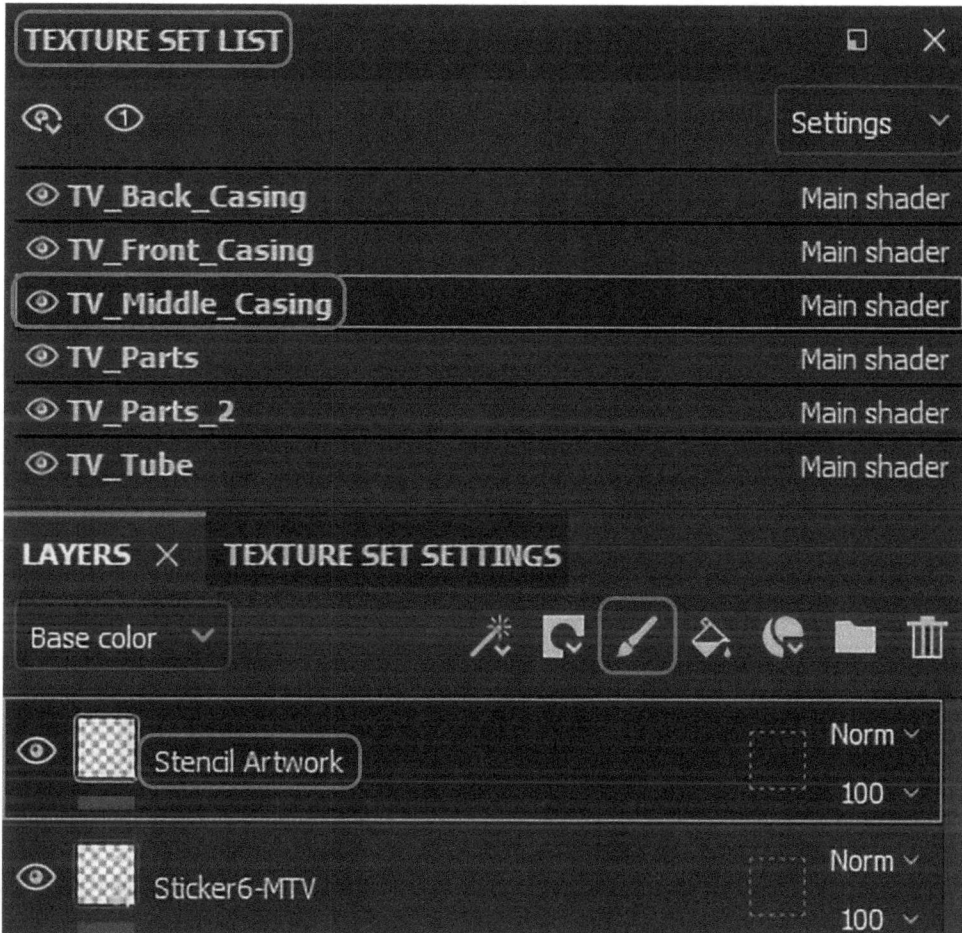

Figure 5.36 – Creating a new paintable layer

3. Make sure the **Basic Hard** brush is already selected, then go to the **PROPERTIES - PAINT**, and click on the **Stencil** option under the **STENCIL** settings. You can choose any alpha stencil you want, so I will choose **Logo Painter**, as shown in *Figure 5.37*.

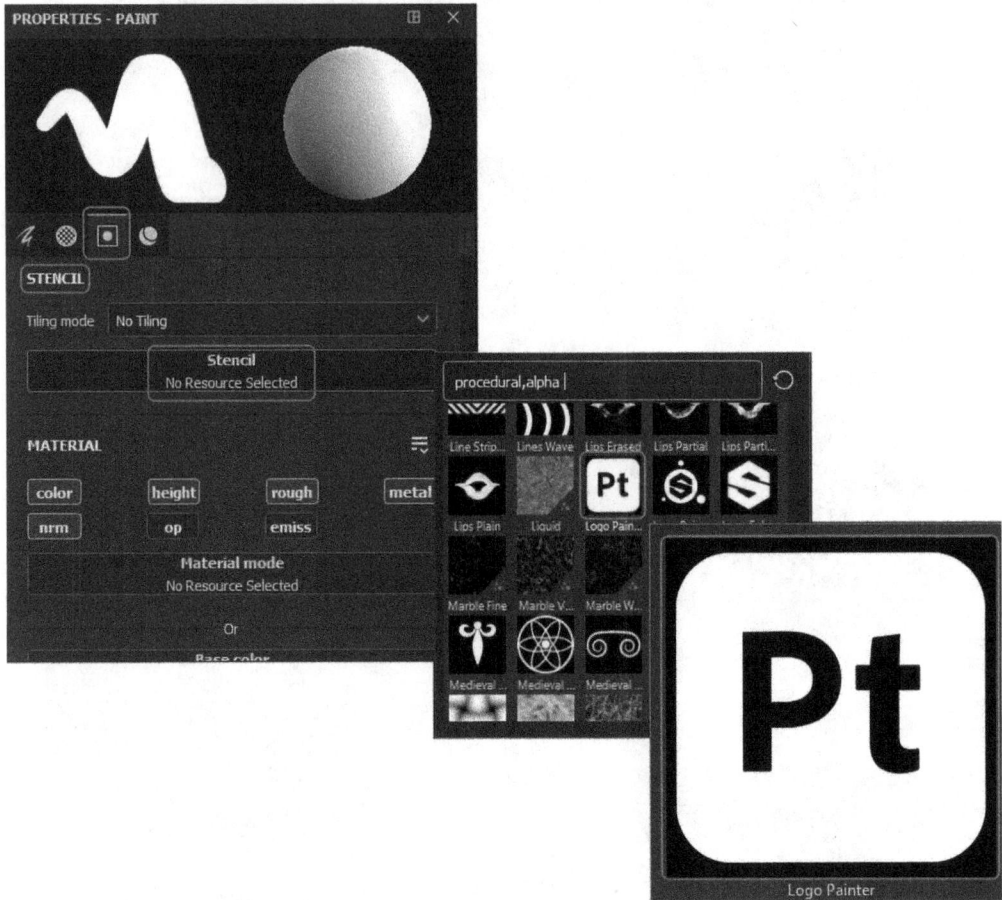

Figure 5.37 – Selecting a stencil alpha

4. Now the stencil is loaded in the brush, your 3D viewport will be as shown in *Figure 5.37*. If you try to rotate, move, or rescale the viewport, only the 3D model and its viewport will be affected. It won't rotate, move, or rescale the stencil.

    To do this, you need to hold the letter *S* on your keyboard and press the left mouse button to rotate the stencil, hold the *S* key and press the right mouse button to rescale the stencil, and finally, hold the *S* key and press the middle mouse button to move the stencil.

Figure 5.38 – Stencil loaded in the viewport

5. You can rescale **Logo Painter** with keys, as explained in *step 4*, and decide on the location you want to paint with the selected **Basic Hard** brush, using the stencil. Make sure the area you want to paint with the stencil is flat; otherwise it will look like *Figure 5.39* after painting.

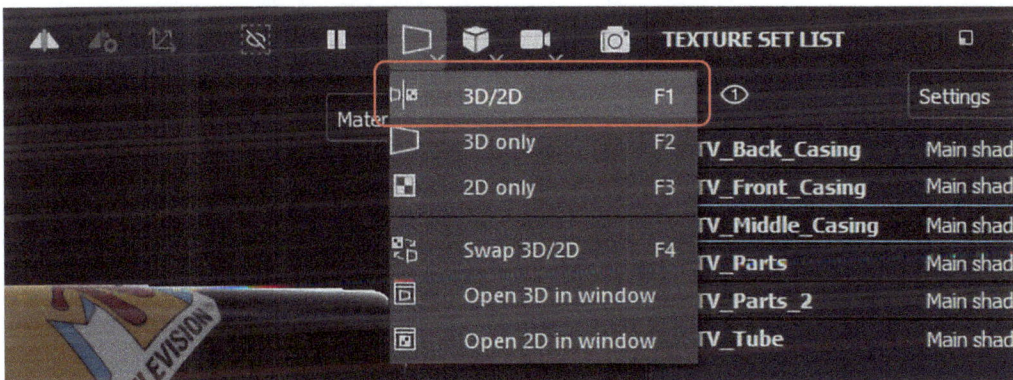

Figure 5.39 – Find a location to paint with the stencil

6. It's always better to use a 2D viewport while painting with the stencil so that the result is always flat and accurate. Therefore, you can choose the 2D viewport from the viewport menu, but for this project, we will use the 3D/2D viewport so that we can compare both viewports while working.

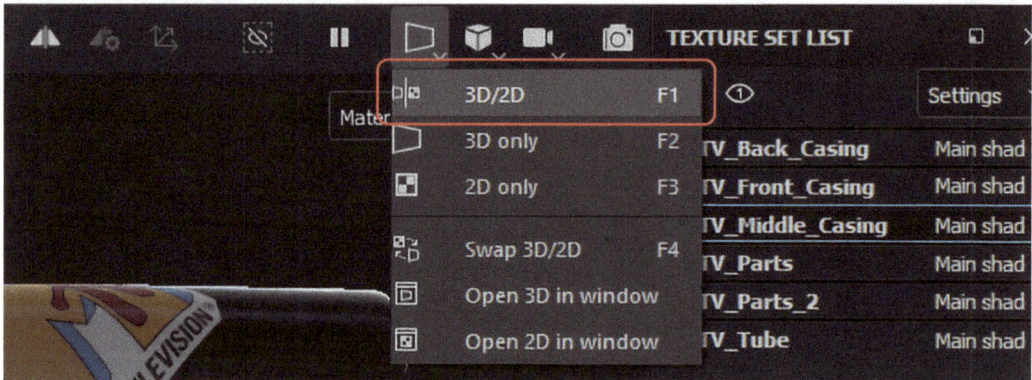

Figure 5.40 – Choosing 3D/2D viewport

7. The 2D viewport is the UV tiles representation of your 3D model; as UV tiles are flat, whatever you paint on them will look accurate. Therefore, adjust the 2D viewport with the same navigation keys you use to adjust the 3D viewport, then adjust the **Logo Painter** stencil with the same keys, as shown in *step 4,* and change the **Stencil Logo Painter Base** color to whatever you want. I will change it to green, as shown in *Figure 5.41.*

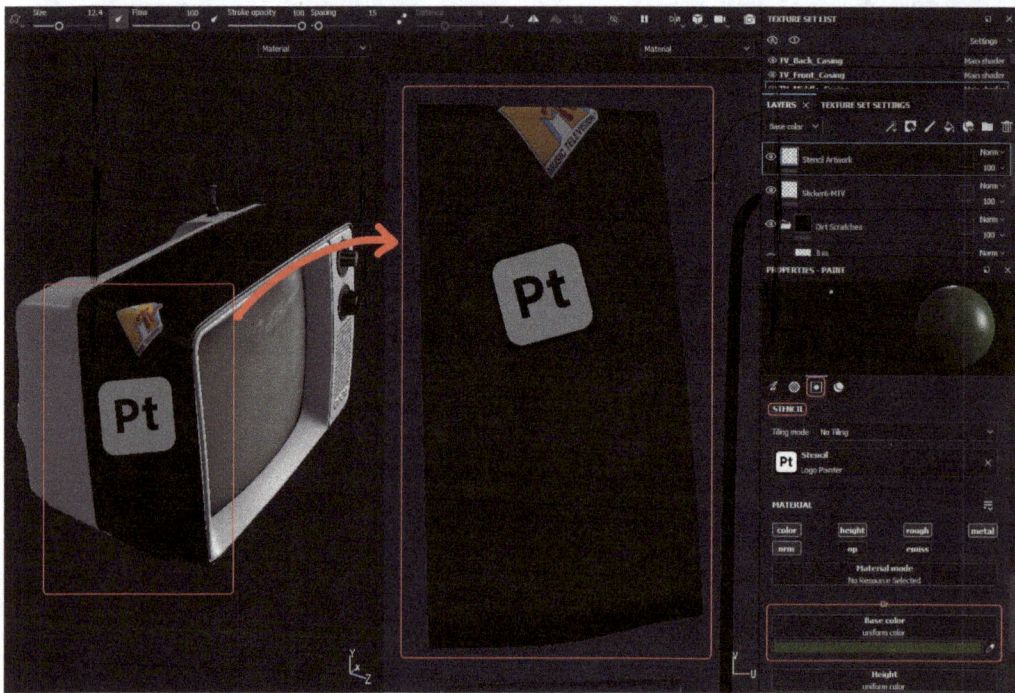

Figure 5.41 – Adjusting the 2D viewport and changing the stencil's base color

8. Now paint over the stencil by dragging using the left mouse button, and once the stencil is fully painted you can click on the cross on the stencil alpha, as shown in *Figure 5.42*. You can also switch back to **3D only** view when you are done.

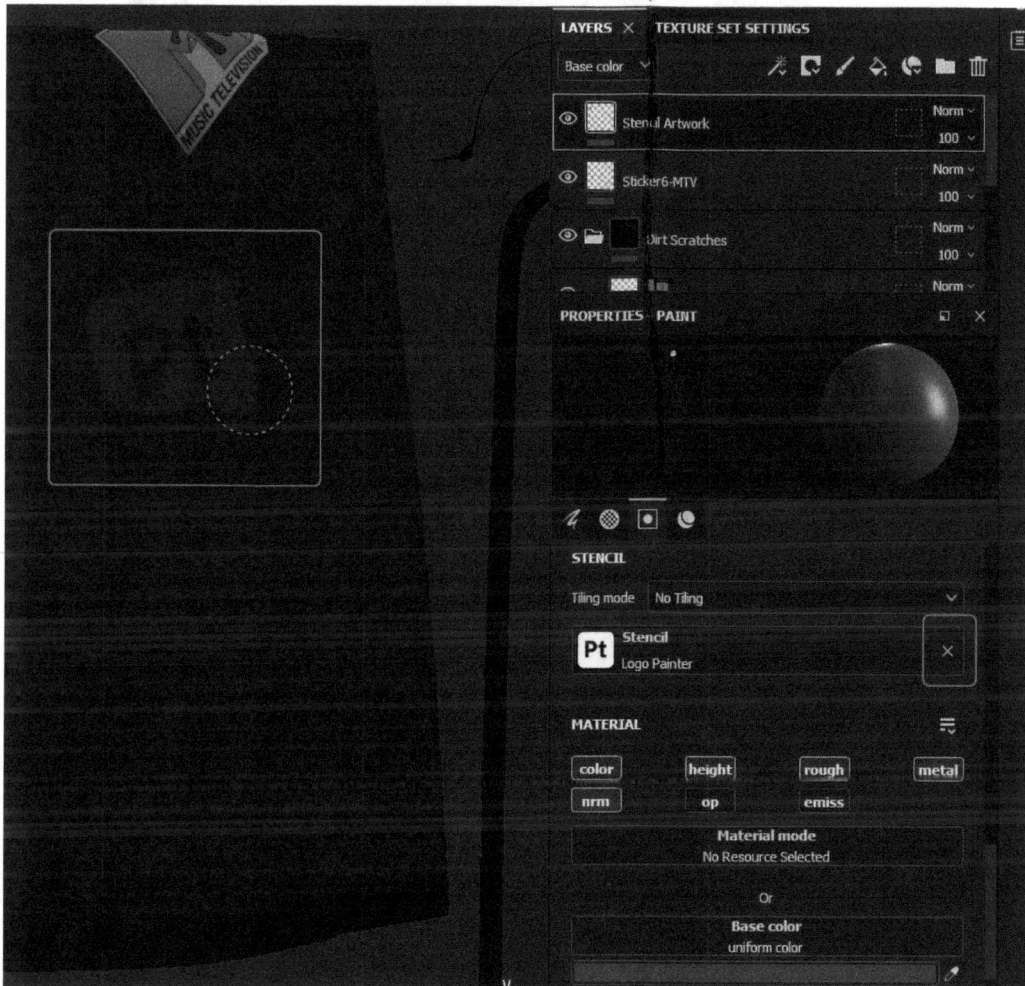

Figure 5.42 – Painting with the stencil

Your final output of the painting will look like *Figure 5.43*.

Figure 5.43 – Final output of the stencil painting

Hopefully, you are now aware of the stencil projection and the way to handle it. In the next section, we will learn how to clone inside Adobe Substance 3D Painter, type with different fonts, and smudge textures.

Using text, fonts, and the Clone and Smudge tools in Adobe Substance 3D Painter

At times you might want to type on your 3D model using different fonts, do some cloning, or smudge some already painted textures to create a different kind of effect. Let us type a *Made by* label, clone an existing texture, and smudge it to give it an aged look:

1.   Click on **Add layer**, create a new paintable layer, and rename it Made in Substance.

Figure 5.44 – Creating a new paintable layer called "Made in Substance"

2. Go to the **ASSETS** panel and select the **Brushes** asset type, then select the brush you want to use; in this project, we will use a **Basic Hard** brush.

3. Go to **PROPERTIES - PAINT**, then jump to the **STENCIL** settings and click on **STENCIL**. When the resources window pops up, type font in the search field, then select **Font Libre Baskerville**.

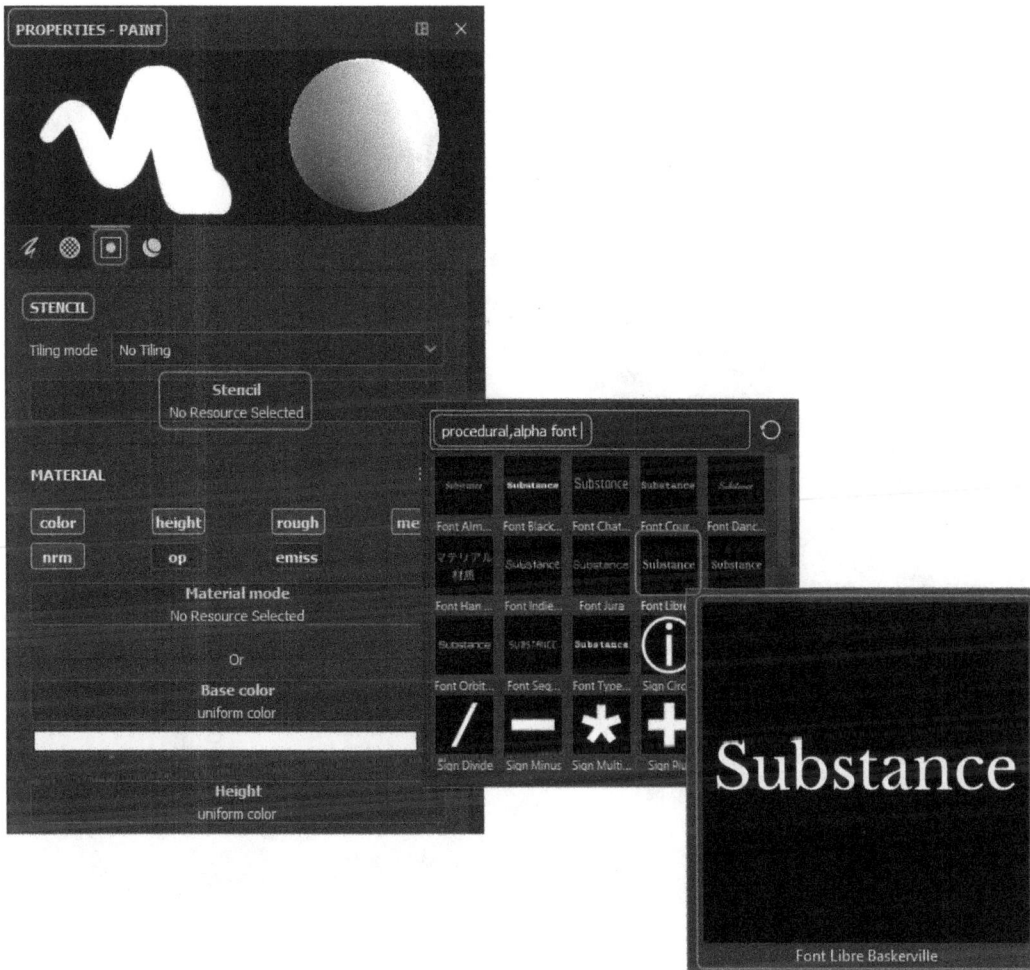

Figure 5.45 – Choosing fonts as alpha

4.   Once you select **Font Libre Baskerville**, you will see some new parameters. You can change the parameter settings as follows and paint over the bottom-right area of the television after adjusting the stencil, as shown in *Figure 5.46*. Moreover, you can also adjust the stencil scale and position by pressing *S* + the middle mouse button and *S* + the right mouse button:

- **Text: Made in Substance, Type: Regular, Size:** 0.12

- **Alignment: Center, Position X:** -0.021**, Position Y:** 0.05

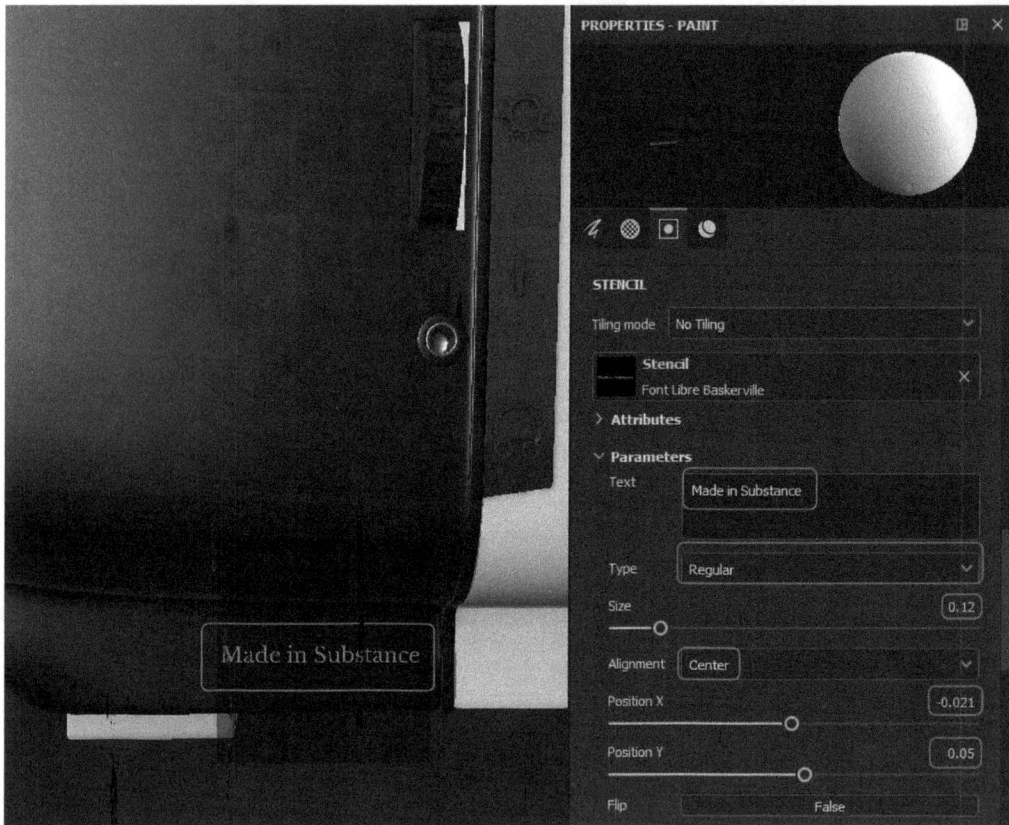

Figure 5.46 – Setting font and area to type

Now, you can click on the cross of the stencil, as shown in *Figure 5.47*.

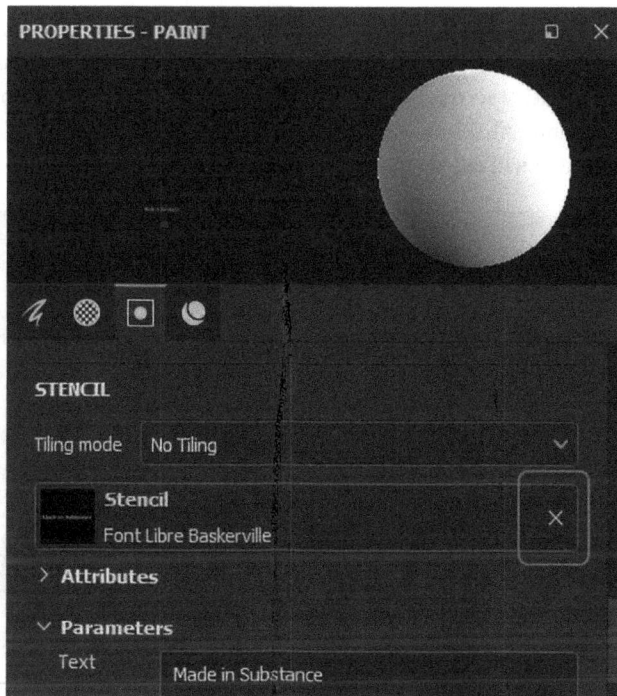

Figure 5.47 – Closing the stencil

Your final output will look like *Figure 5.48*.

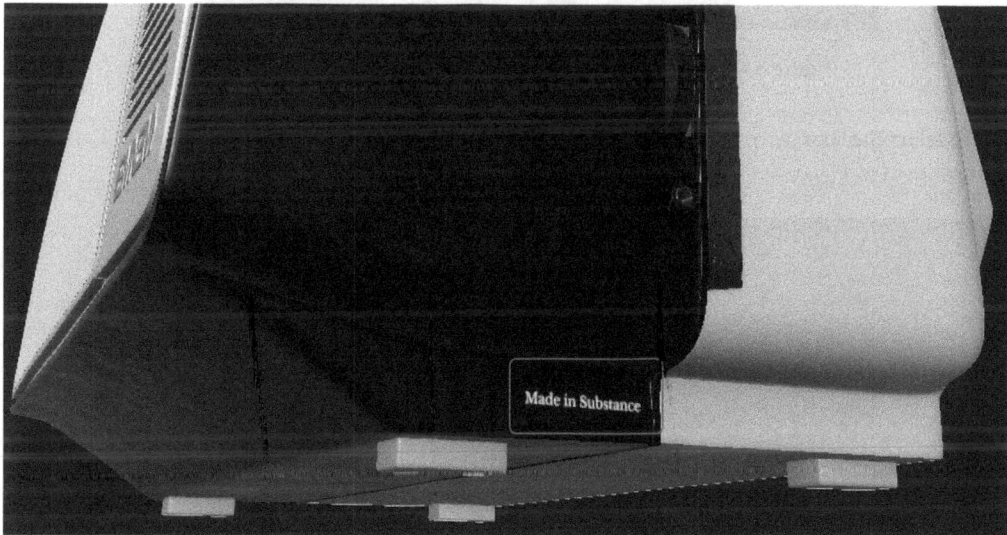

Figure 5.48 – Final output of the font selection

5. Now, let us clone the **Logo Painter** tool that we used in the previous section. To do this, go to the **Stencil Artwork** layer. Now, to clone **Logo Painter**, select the **Charcoal Medium** brush and click on **Clone (relative source)**, as shown in *Figure 5.49*. The difference between **Absolute** and **Relative** is that, while **Absolute** will use any provided resolution and replace the input and parent settings, **Relative** will use the base parent settings as its resolution.

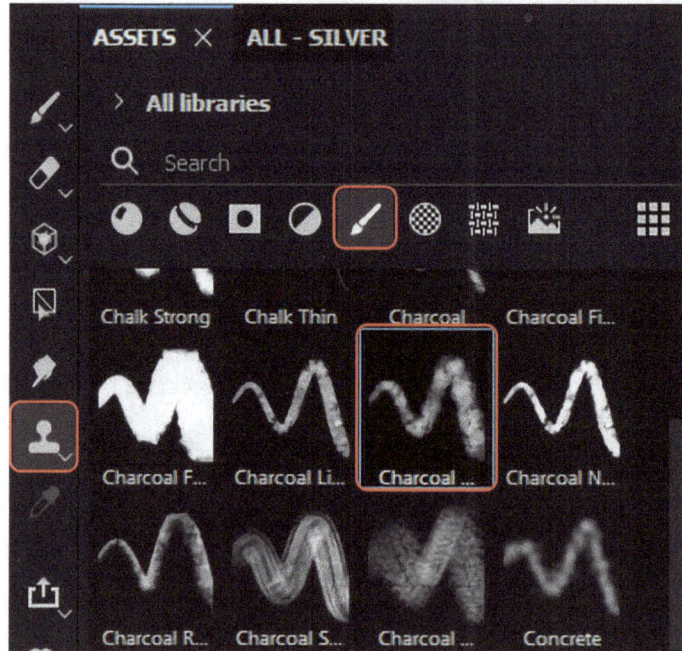

Figure 5.49 – Creating a new layer for the cloned Logo Painter

6. To select the source, you need to press the letter *V* on the keyboard and click on **Logo Painter**, as shown in *Figure 5.50*.

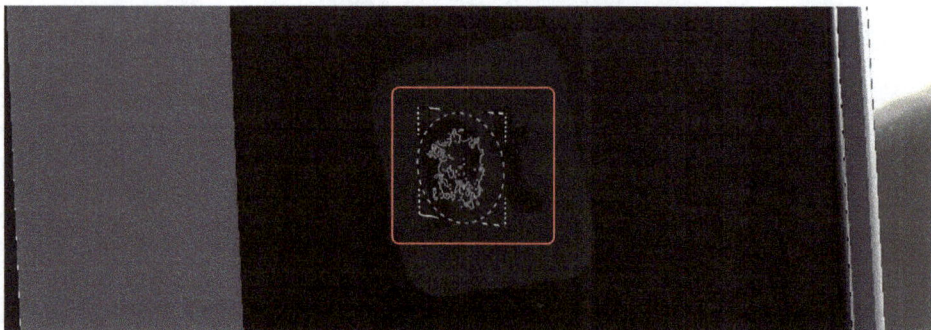

Figure 5.50 – Selecting a clone source

7. Now, you paint with the Clone tool just under the source logo and make it look eroded, as shown in *Figure 5.51*.

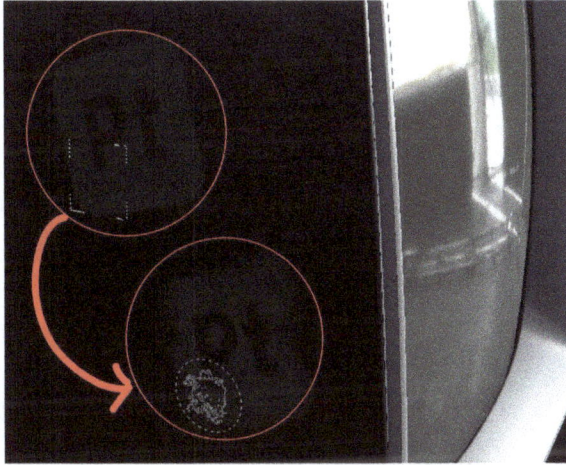

Figure 5.51 – Cloning Logo Painter

8. To make the logo look older, use the Smudge tool, which you will find above the Clone tool; go to **PROPERTIES - PAINT**, and under the **BRUSH** settings change the following settings:

Size: 3.11, **Minimum Size (%)**: 5, **Flow**: 50

Then, use the Smudge tool and try to smudge some edges in different directions, but do not overuse it; otherwise, **Logo Painter** will start looking unrealistic.

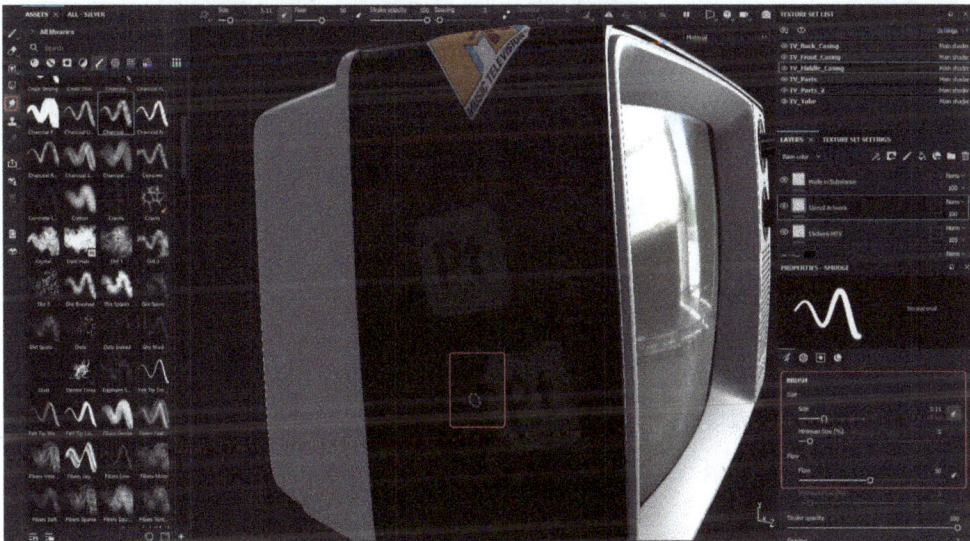

Figure 5.52 – Smudging Logo Painter

9. Now as you have a new old-looking **Logo Painter**, select the Eraser tool, which you will find under the Paint tool, and erase the first **Logo Painter** you created, as shown in *Figure 5.53*:

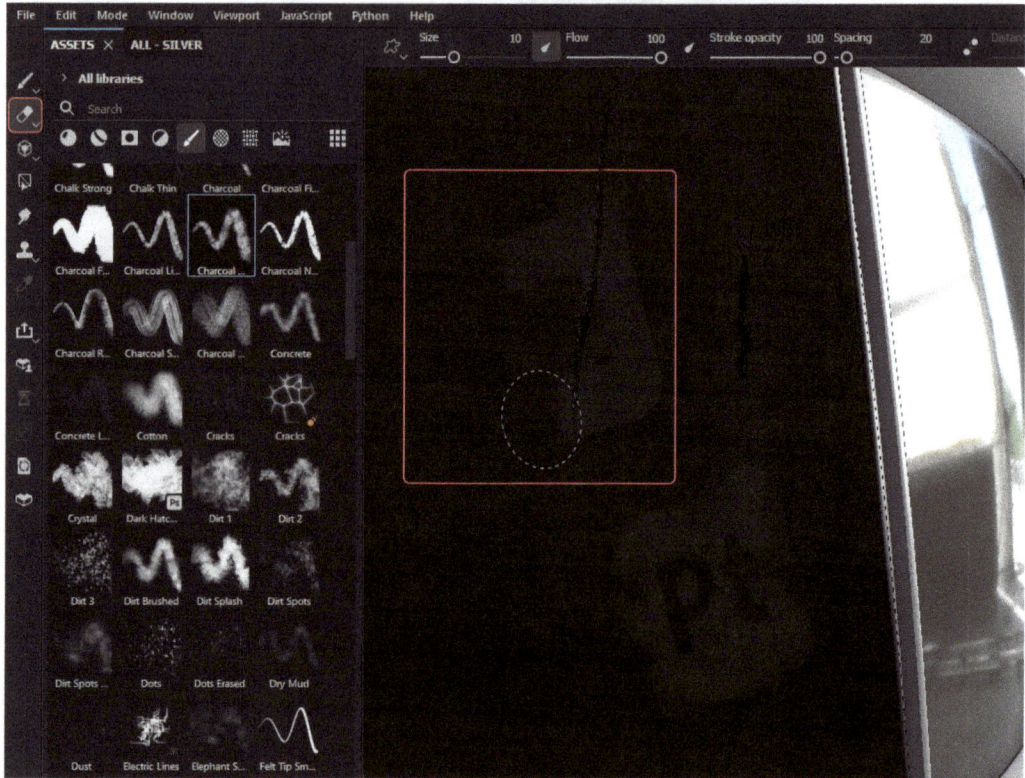

Figure 5.53 – Erasing the original Logo Painter

The final output will look like *Figure 5.54*.

Figure 5.54 – Final output after cloning, smudging, and erasing Painter Logo

Summary

Hopefully, you have comprehended the usage, purpose, and importance of bitmap and 3D-based textures; projection and stencils; text and fonts; and the Clone and Smudge tools, while working on practical parts of Adobe Substance 3D Painter.

In the next chapter, we will learn about Smart Materials. Smart Materials make the texturing process faster and more procedural. Upon going through the next chapter, you will be able to create your own Smart Materials that you can re-use in many projects or even share with other designers.

6

Working with Materials and Smart Materials in Adobe Substance 3D Painter

This is the last chapter covering Adobe Substance 3D Painter; previously you have learned the usage, purpose, and importance of Bitmap and 3D-based textures, projection and stencils, text, fonts, the Clone tool, and the Smudge tool, while working on practical parts in Adobe Substance 3D Painter.

In this chapter, we will learn about Smart Materials, which are vital because they make the texturing process faster and more procedural. After going through this chapter, you will be able to create your own Smart Materials, reuse them in many projects, or even share them with other designers.

Moreover, sometimes you need to apply an overall effect on a whole 3D model. To do this, you need a special map called a **position map**; therefore, we will also cover this in this chapter as well.

In the end, you will learn how to render inside Painter, which is quite crucial, as it produces a rendered version of photo-realistic output that can be shown to clients and companies.

In this chapter, we will cover the following topics:

- Creating a material or Smart Material from scratch in Adobe Substance 3D Painter
- Creating a material or Smart Material from existing material in Adobe Substance 3D Painter
- Applying stickers and decals in Adobe Substance 3D Painter
- Adding an overall layer effect with the position map in Adobe Substance 3D Painter
- Exporting textures from Adobe Substance 3D Painter
- Rendering in Adobe Substance 3D Painter using iRay

Creating a material or Smart Material from scratch in Adobe Substance 3D Painter

The difference between Materials and Smart Materials is that the Materials are tiling texture maps that are standard and consistent. They don't require baking and don't contain any mesh-specific information. A fill layer in Painter can be used to create Materials. Painter, on the other hand, is the only program that has Smart Materials.

They feature mesh-specific details, which are automatically fitted to your mesh, in addition to tiling and uniform detail.

Figure 6.1 – The difference between Materials and Smart Materials

Let's create some Smart Material inside Adobe Substance 3D Painter and apply it to the screen frame of the **TV_Front_Casing**:

1. Select **TV_Front_Casing** from **TEXTURE SET LIST**, click on **Add group** to create a new folder, and rename it TV Frame.

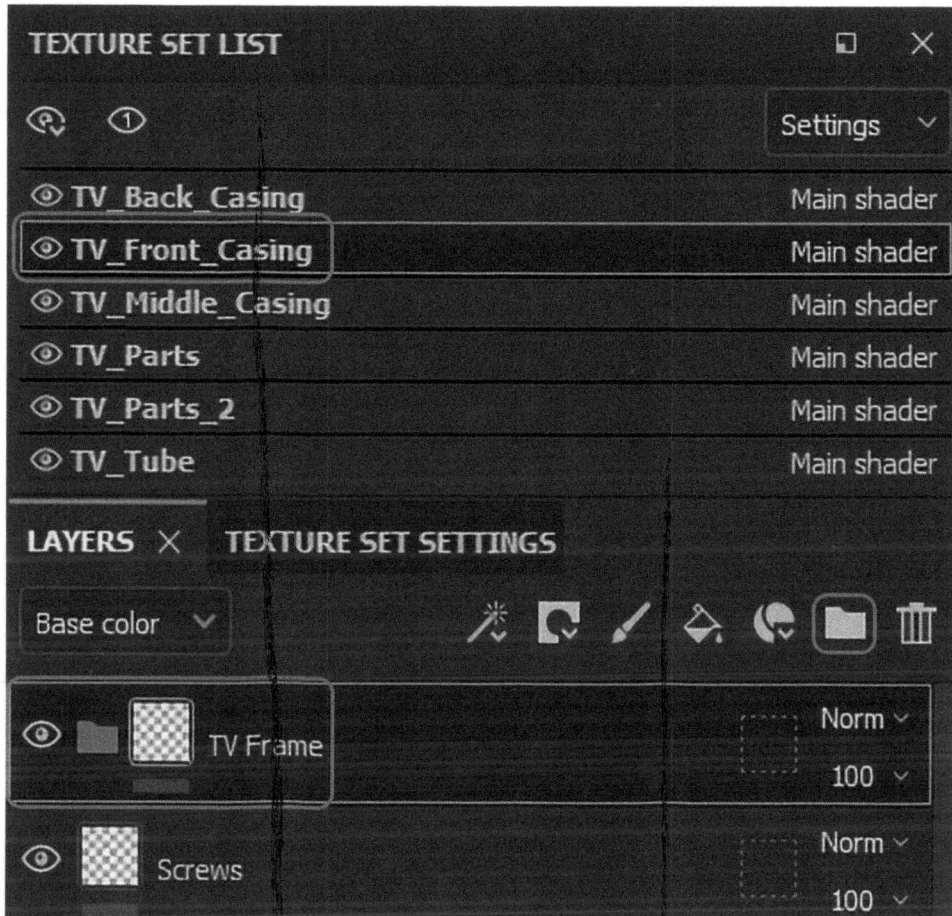

Figure 6.2 – Creating a new folder in the TV_Front_Casing Texture Set

2. Now, we have to create a **TV Frame** base material. Firstly, click on **Add fill layer** to create a new fill layer inside the TV Frame folder. Secondly, rename the fill layer Frame Base Material, and lastly, change **Base color** to 979268 and **Roughness** to 0.27.

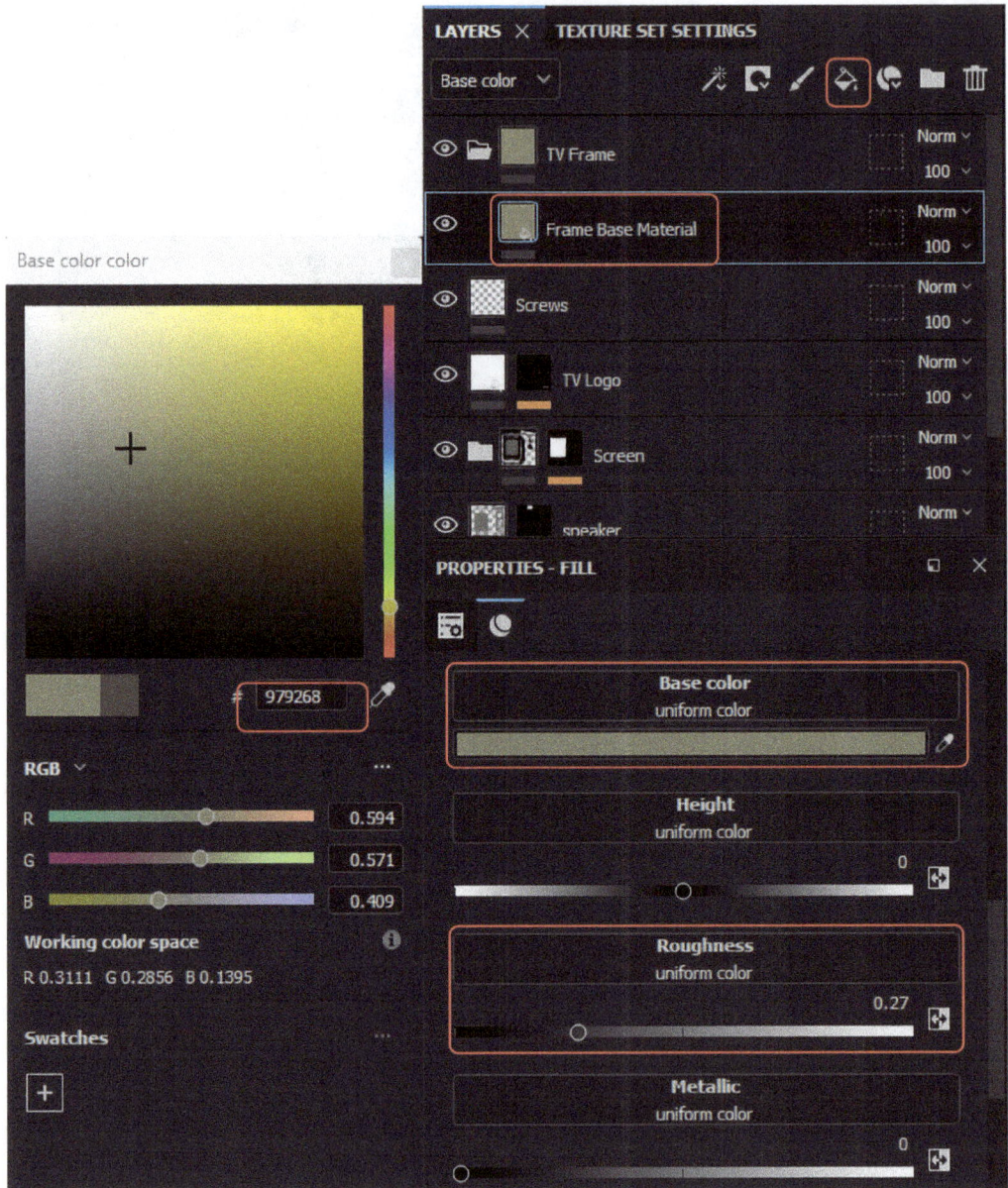

Figure 6.3 – Creating a frame base material

3. Now, create another fill layer, rename it Frame Dirt Material, and change **Base color** to 382F0E and **Roughness** to 0.18.

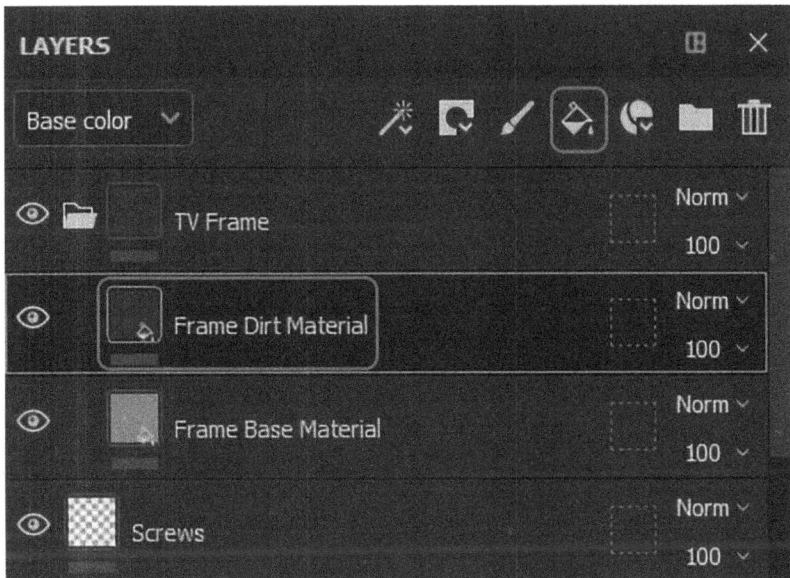

Figure 6.4 – Creating the Frame Dirt Material fill layer

4.   It's time to add a black mask to **Frame Dirt Material** so that when we add the masking effect, the effect appears in the white area only. Therefore, go to **Add mask**, click on it, and choose to **Add black mask**.

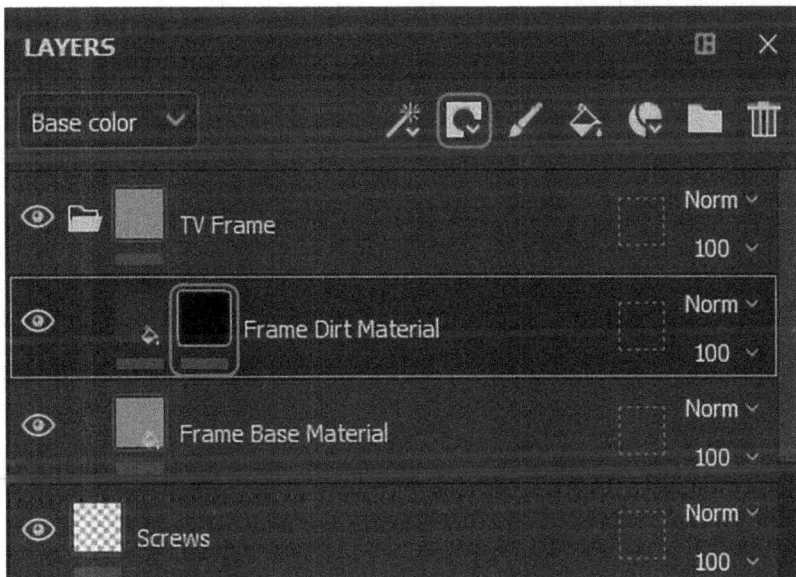

Figure 6.5 – Adding a black mask to Frame Dirt Material

5. Let's add some edge wear effect to the frame to make it look like old material with wear and tear. So, go to **Add effect**, click on it, and then select **Add generator**. Once the generator is added, go to **PROPERTIES - GENERATOR**, click on the **Generator** option, and then select the **Metal Edge Wear** generator.

Figure 6.6 – Adding the Metal Edge Wear generator

6. Once the **Metal Edge Wear** generator is added, you need to change its parameters to make it look more realistic. Change the following settings:

- **Wear Level**: 0.9

- **Wear Contrast**: 0.33

- **Grunge Amount**: 0.42

- **Curvature Weight**: 0.35

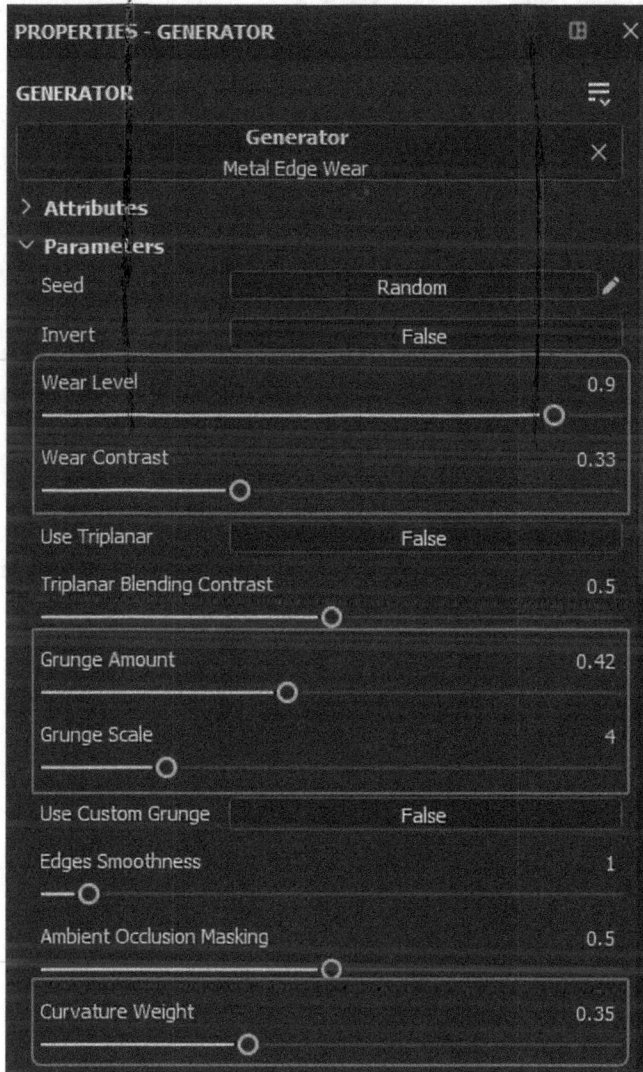

Figure 6.7 – Changing the generator's parameters

7. Now, collapse the **TV Frame** folder so that everything inside the **LAYERS** panel looks neat and organized.

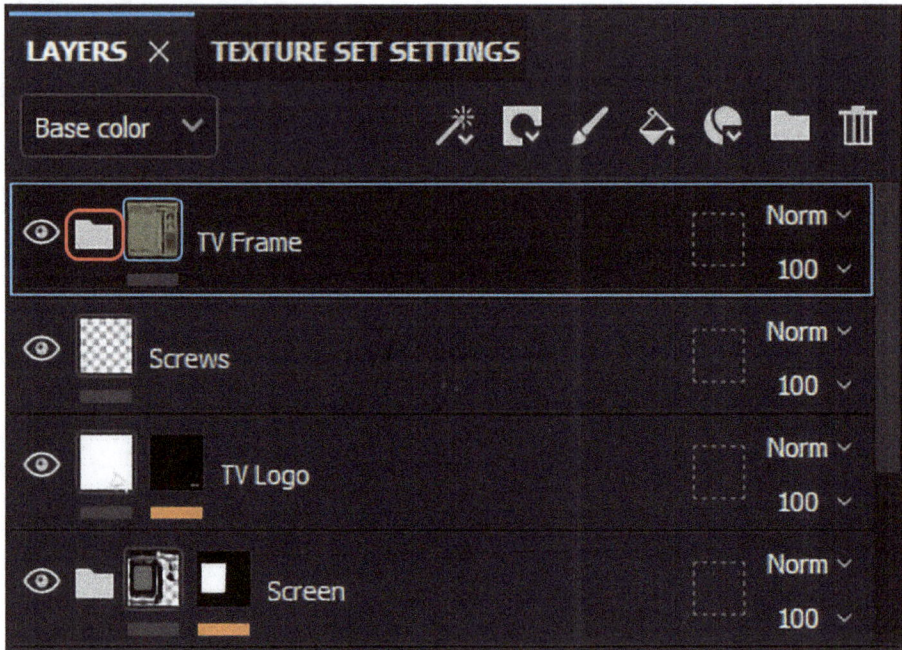

Figure 6.8 – Collapse all folders

8. Now, we will convert this **TV Frame** material to a Smart Material. Right-click on the **TV Frame** layer and select **Create Smart Material**.

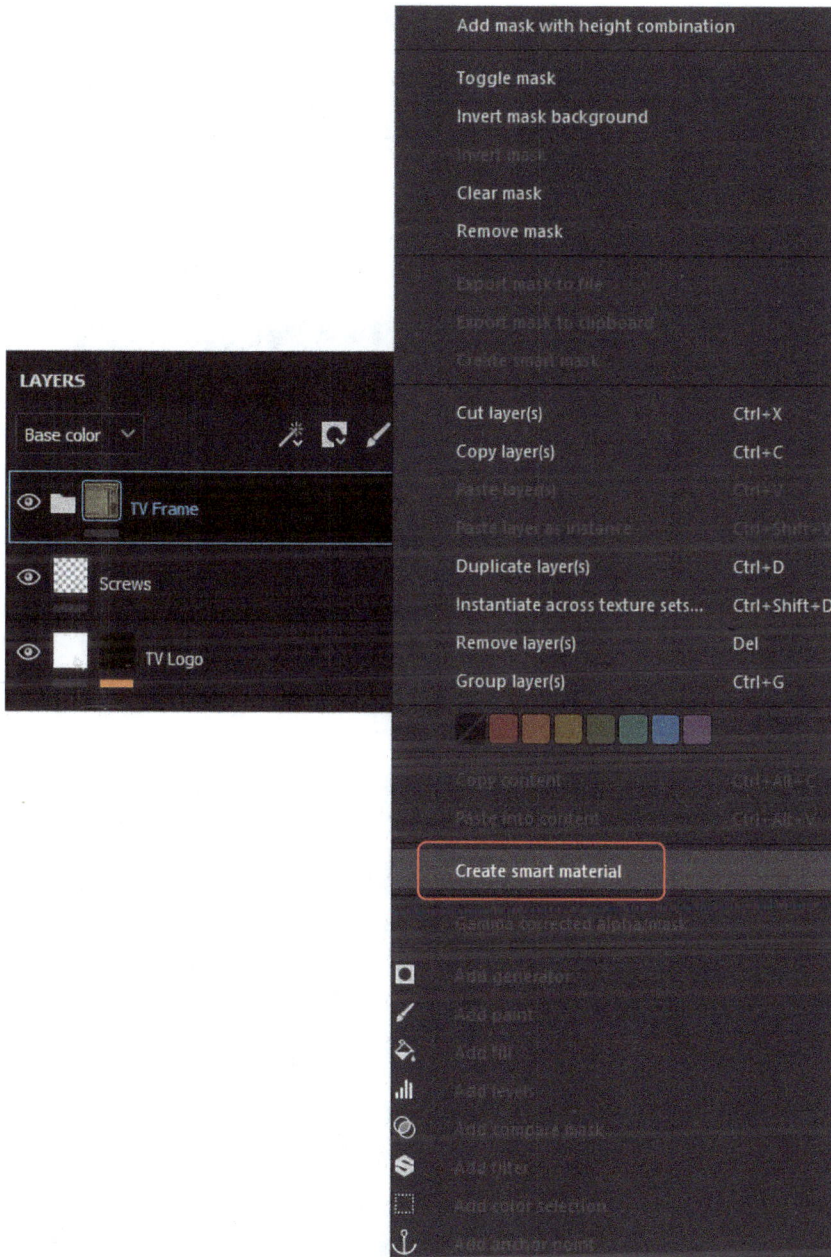

Figure 6.9 – Creating a Smart Material

9. Once you click on **Create Smart Material**, the **TV Frame** material will become a Smart Material and will be shown in the **Smart Material** asset type. This material can be applied at any time in any project to any 3D model by adjusting its settings and parameters.

Figure 6.10 – TV Frame Becomes a Smart Material

10. However, the **TV Frame** material is all over **TV_Front_Casing**, and we need it only on **TV Frame**. Therefore, keep the **TV Frame** folder selected, choose **Add mask**, and select **Add mask with color selection**. Then, select **Pick color** under **COLOR – SELECTION** and click on **TV Frame**, as shown in *Figure 6.11*.

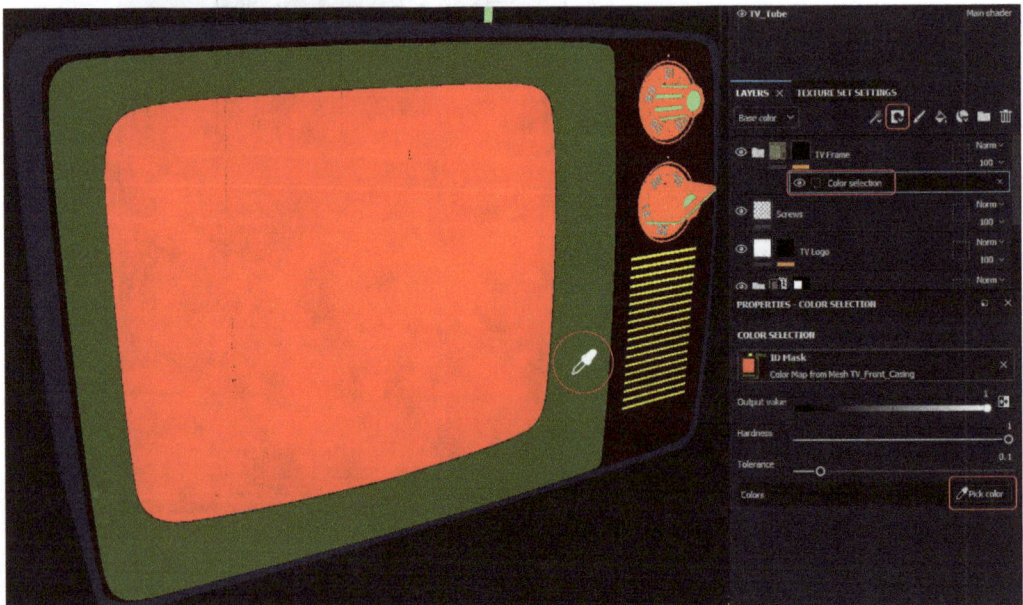

Figure 6.11 – Apply the TV Frame Smart Material to the frame only

11. Now, as the **TV Frame** Smart Material is masked to frame itself, the result will look more realistic, as shown in *Figure 6.12*.

Figure 6.12 – Final output of the TV Frame Smart Material

Now that you know how to create your own Smart Material from scratch, in the next section, we will learn how to create our own Material or Smart Material from existing materials.

Creating a material or Smart Material from an existing material in Adobe Substance 3D Painter

Sometimes, it is easier and faster to create a Material or Smart Material from an existing Smart Material or material. In this section, we will create a dark plastic material and apply it to everything with a maroon ID map.

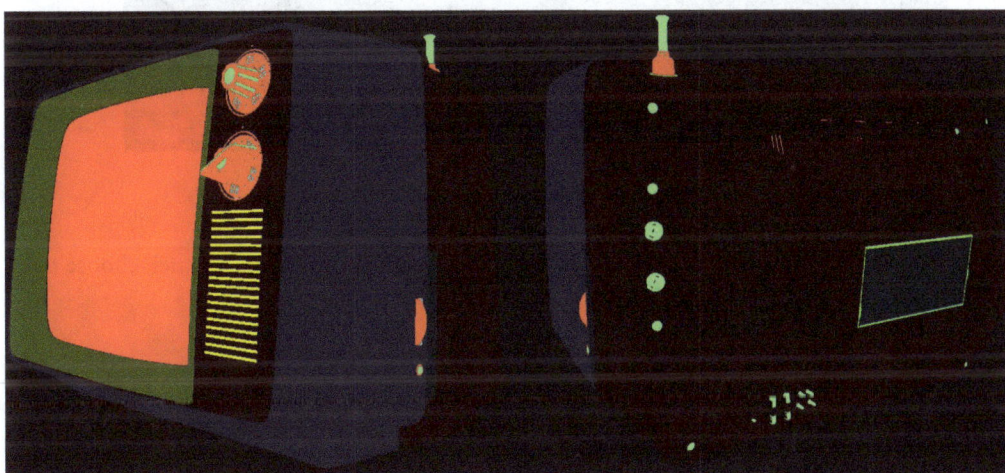

Figure 6.13 – A maroon ID map

12. Select **TV_Front_Casing** from **TEXTURE SET LIST**, click on **Add group**, and rename it `Dark Plastic Parts`. Then, inside the folder, create a new fill layer by clicking **Add fill layer**, and rename it `Plastic Material`. Make sure the new folder is under the **TV Logo** layer so that it doesn't hide the logo layer and the layers above it.

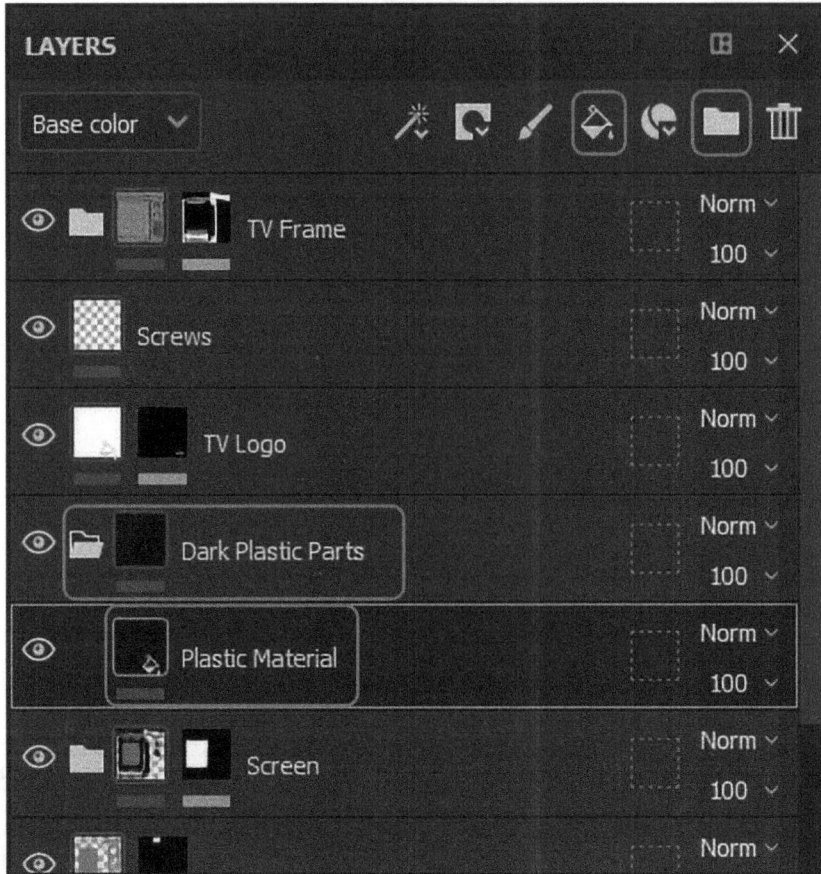

Figure 6.14 – Creating a new fill layer

13. Make sure that the **Plastic Material** layer is selected, go to the **ASSETS** panel, choose the **Material** asset type, and select **Plastic Matte Pure**.

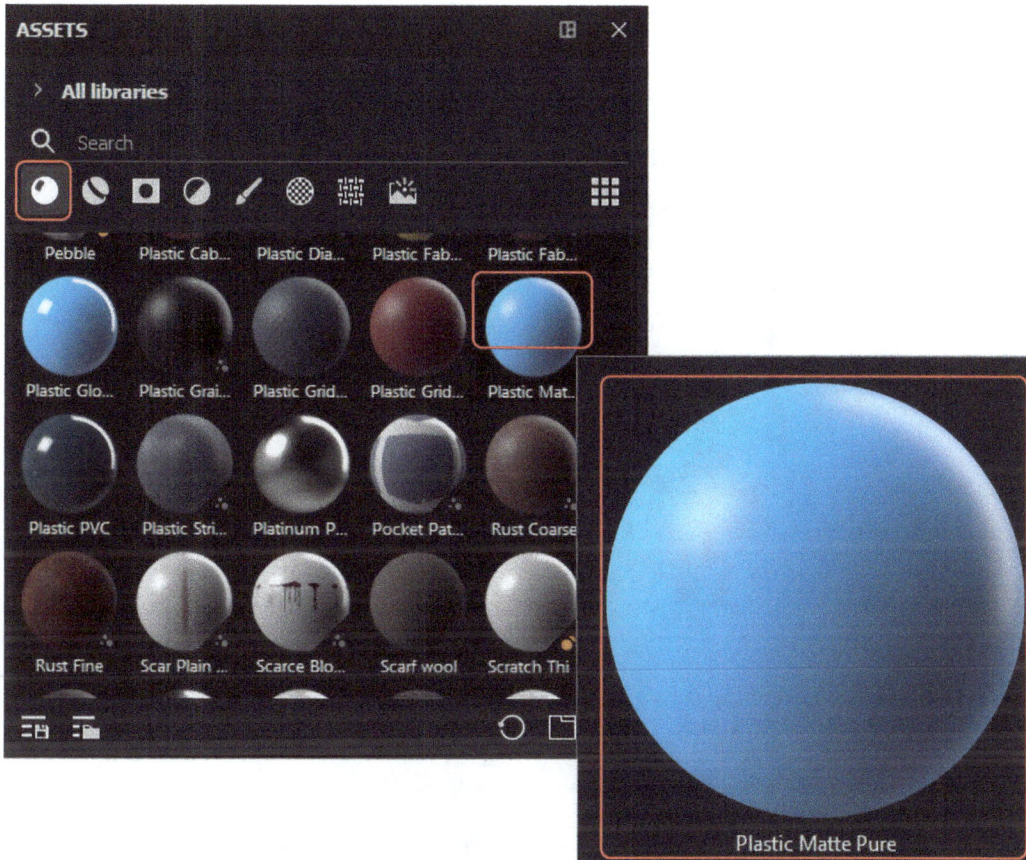

Figure 6.15 – Selecting the Plastic Matte Pure material

14. Now, go to the **PROPERTIES - FILL** window of the **Plastic Material** fill layer and change the following settings under the **MATERIAL** settings.

- **Base color**: 111111
- **Roughness**: 0.17
- **Metallic**: 0

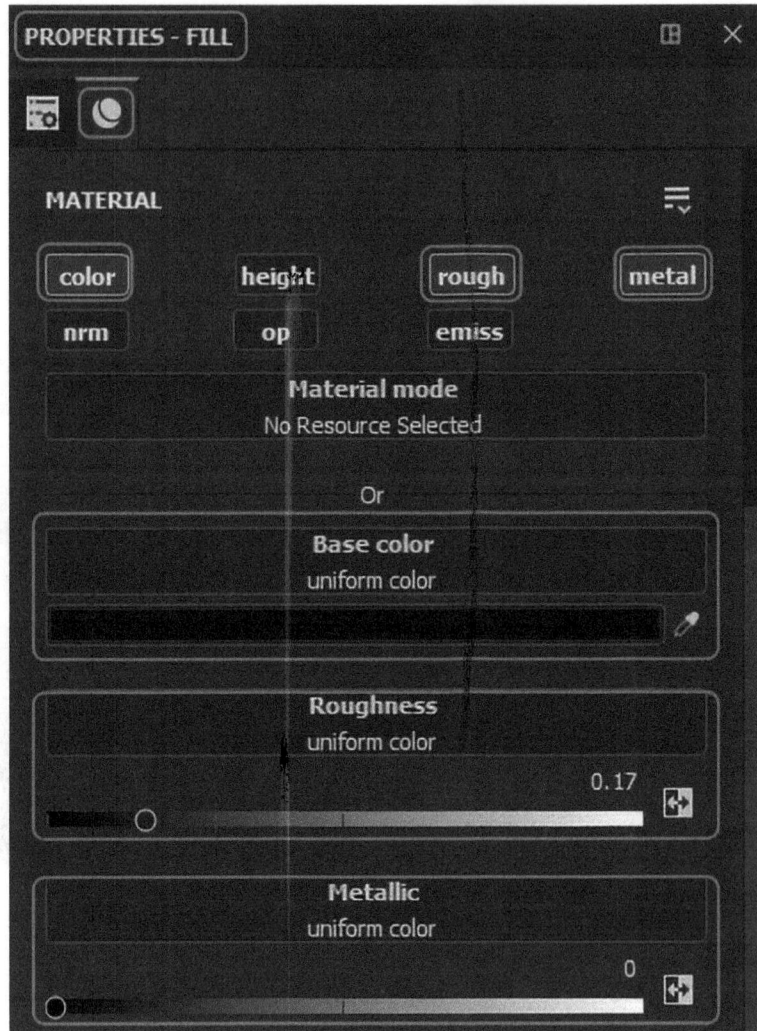

Figure 6.16 – Adjusting the Dark Plastic Parts layer material

15. Collapse the **Dark Plastic Parts** folder, right-click the folder, and select **Create Smart Material**.

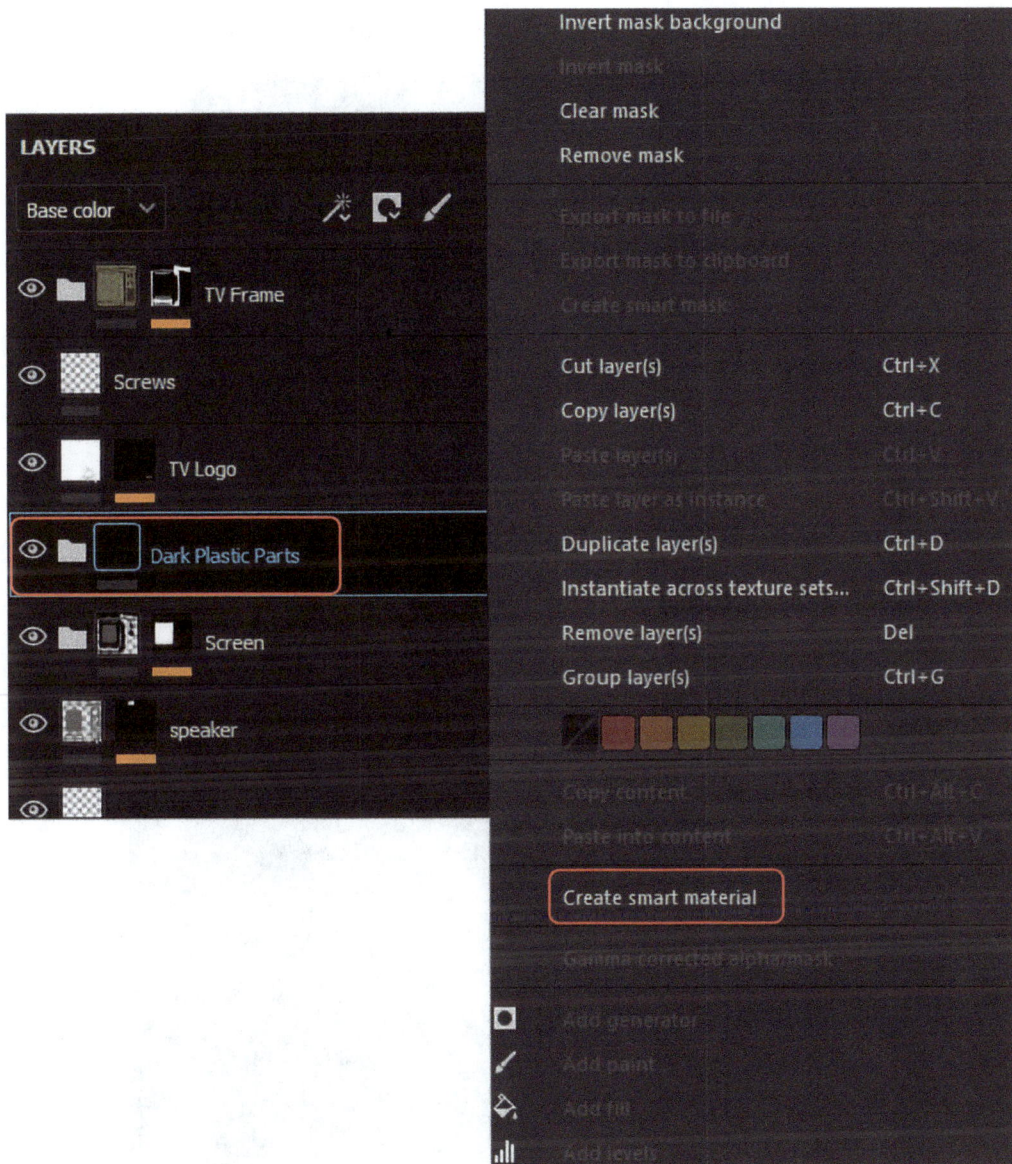

LAYERS

Base color ∨

👁 📁 ⬛ 🖼️ ◻️ TV Frame

👁 ▦ Screws

👁 ◻️ TV Logo

👁 📁 ⬛ Dark Plastic Parts

👁 📁 ▦ ◻️ Screen

👁 ▦ speaker

👁 ▦

Invert mask background	
Invert mask	
Clear mask	
Remove mask	
Export mask to file	
Export mask to clipboard	
Create smart mask	
Cut layer(s)	Ctrl+X
Copy layer(s)	Ctrl+C
Paste layer(s)	Ctrl+V
Paste layer as instance	Ctrl+Shift+V
Duplicate layer(s)	Ctrl+D
Instantiate across texture sets...	Ctrl+Shift+D
Remove layer(s)	Del
Group layer(s)	Ctrl+G
Copy content	Ctrl+Alt+C
Paste into content	Ctrl+Alt+V
Create smart material	
Gamma corrected alpha/mask	
Add generator	
Add paint	
Add fill	
Add levels	

Figure 6.17 – Creating the Dark Plastic Parts Smart Material

16. The **Dark Plastic Parts** Smart Material will be now added to the **ASSETS** library.

Figure 6.18 – The Dark Plastic Parts Smart Material

17. Now, we want the **Dark Plastic Parts Smart Material** only on the front part of the TV and not on the screen. Therefore, select the **Dark Plastic Parts** folder, click on **Add mask**, and choose **Add mask with color selection**.

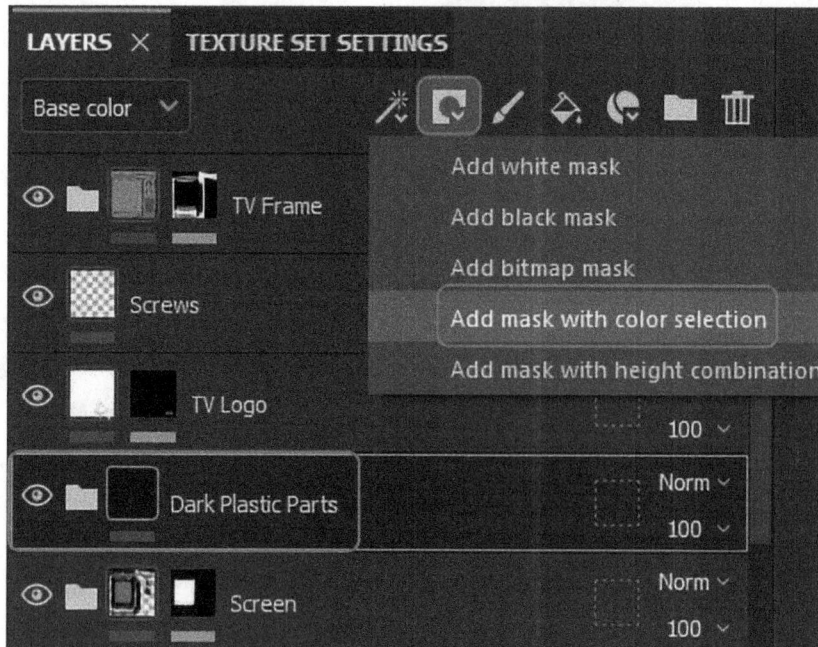

Figure 6.19 – Creating a mask with color selection on the Dark Plastic Parts layer

18. Now, go to the **PROPERTIES - COLOR SELECTION**, select **Pick color** under the **COLOR SELECTION** settings, and click on the front part of the TV, as shown in *Figure 6.20*.

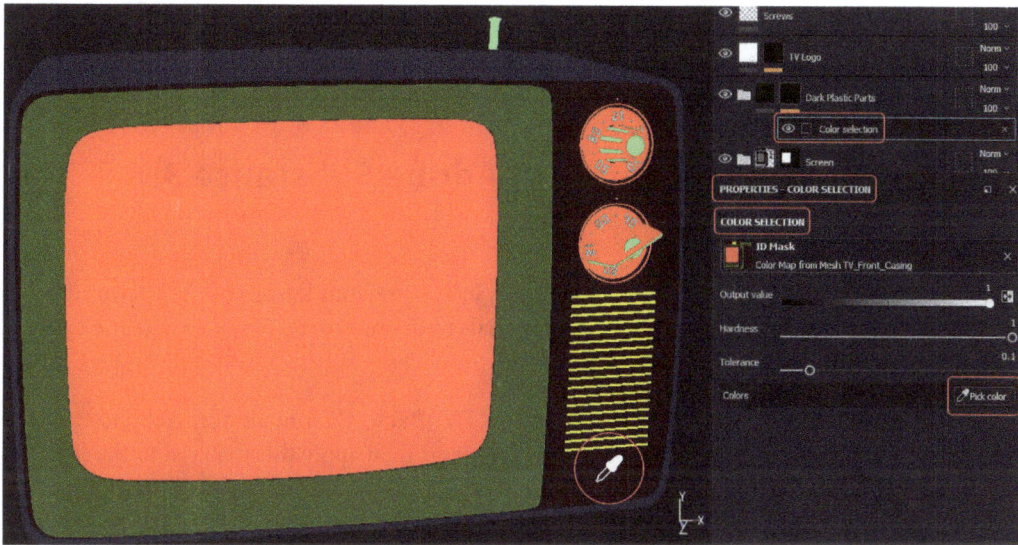

Figure 6.20 – Applying the Dark Plastic Parts material on the front part

19. Now, go to the **TV_Back_Casing**, **TV_Parts**, and **TV_Parts_2** Texture Sets and apply the **Dark Plastic Parts** Smart Material to all the parts with the maroon ID maps in each Texture Set, as shown in *Figure 6.13*, with ID map material applying the technique. To make it clearer, this is done by dragging and dropping in the material and adding a mask with color selection. The final output will look the same as *Figure 6.21*.

Figure 6.21 – The Dark Plastic Parts final output

Hopefully, it has been clear to all of you how we can create Materials or Smart Materials from scratch or any existing Materials or Smart Materials. In the next section, we will add Materials to the rest of the parts, and we will have some fun while applying stickers and decals to the TV.

Applying stickers and decals in Adobe Substance 3D Painter

We have added a lot of Materials and Smart Materials in our previous lessons, and we have added some stickers too. In this section, we will apply Materials to the rest of the TV parts and also apply stickers and decals using different techniques:

1. First, let's complete applying materials to the **TV_Back_Casing** Texture Set. Go to the **ASSETS** panel and search `Plastic Grainy`, and then apply this material to the frame of the **TV_Back_Casing** servlet area while pressing *Ctrl* (Windows) or *Cmd* (Mac) on your keyboard. Once the servlet's frame shows a light green ID map, drop the material onto it, as shown in *Figure 6.22*.

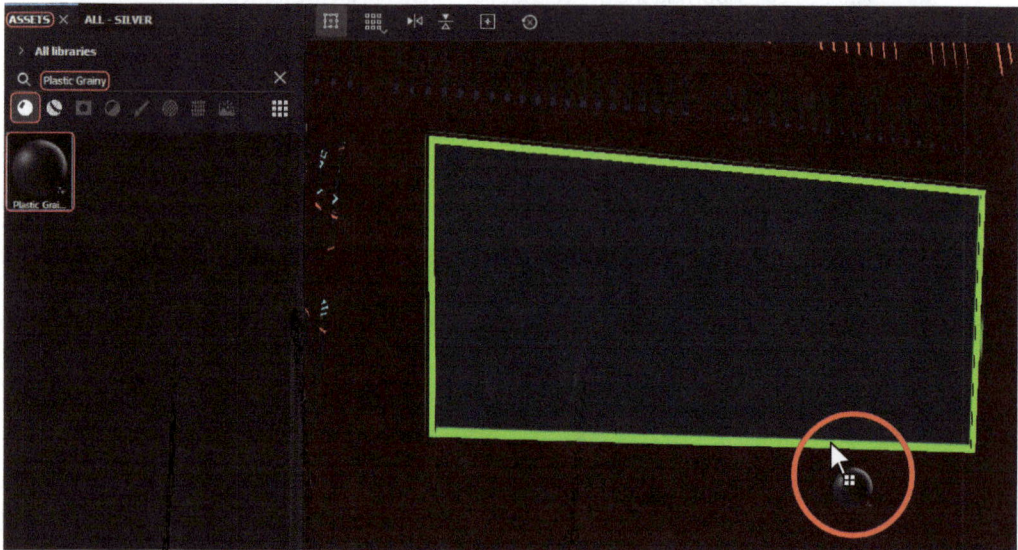

Figure 6.22 – Applying Plastic Grainy to TV_Back_Casing's servlet frame

2. Now, we will apply materials to the **TV_Parts** Texture Set. Go to **ASSETS** and use the same **Plastic Grainy** material used in *step 1*, and then apply this material to the knob numbers while pressing *Ctrl* (Windows) or *Cmd* (Mac) on your keyboard. Once the knob numbers will show a light blue ID map, drop the material onto it, as shown in *Figure 6.23*.

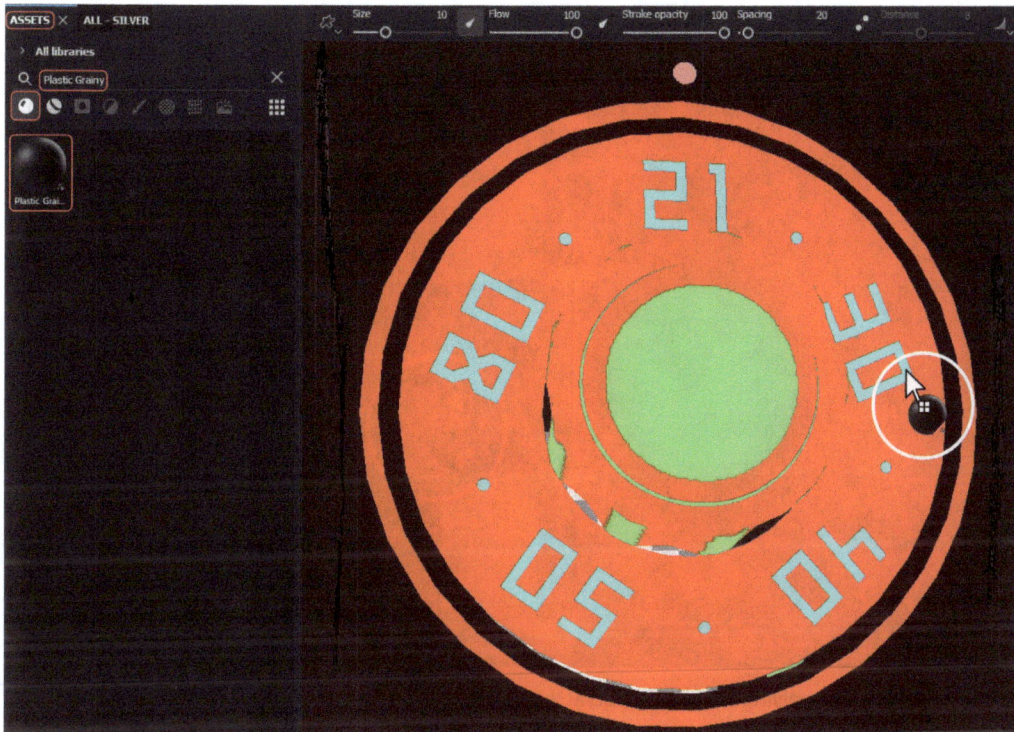

Figure 6.23 – Applying material to knob numbers in the TV_Parts Texture Set

3. Now, let's go to the back side of the **TV_Parts** Texture Set and search for the TV fuse, which is located at the bottom of the back side of the TV. This is the place we will be applying materials. Go to **ASSETS** and search for the `Plastic Dirty Scratched` material, and then apply this material to the TV fuse while pressing *Ctrl* (Windows) or *Cmd* (Mac) on your keyboard. Once the TV fuse shows a dark green ID Map, drop the material onto it, as shown in *Figure 6.24*.

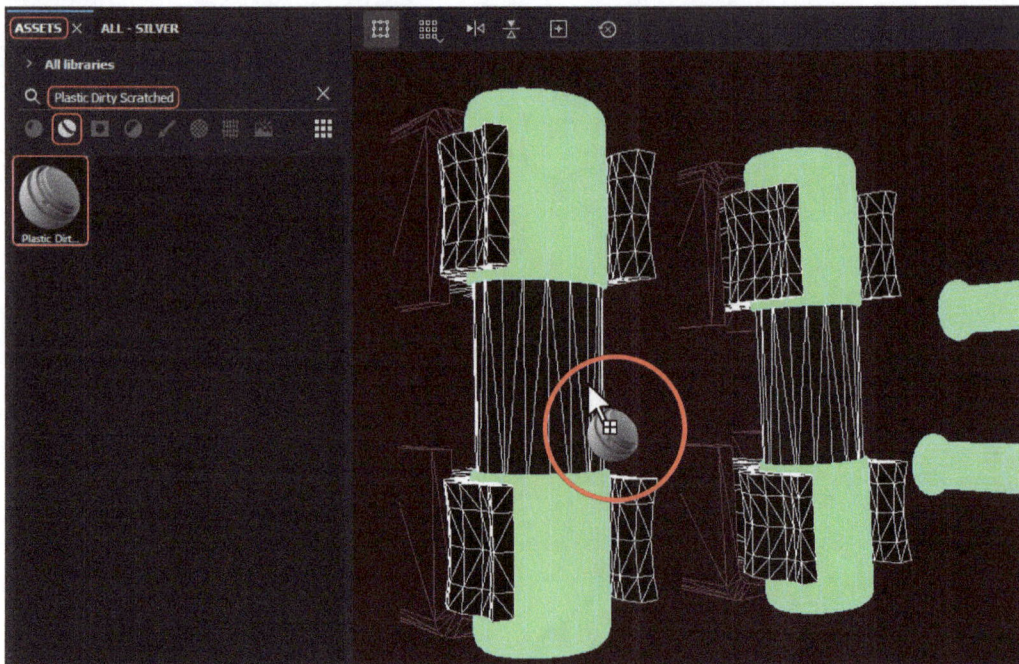

Figure 6.24 – Applying Plastic Dirty Scratched to the TV fuse in the TV_Parts Texture Set

Now that we are done with applying materials to all latent 3D parts, let's now start applying stickers.

4.   First, go to the **ASSETS** panel, select the **Brushes** asset type, and choose the **Basic Hard** brush. Then, go to **All libraries** under the **ASSETS** panel and choose the **Project** library. Then, click on the small arrow next to the **Project** library and select **Retro**. You will see all the stickers that you need to apply.

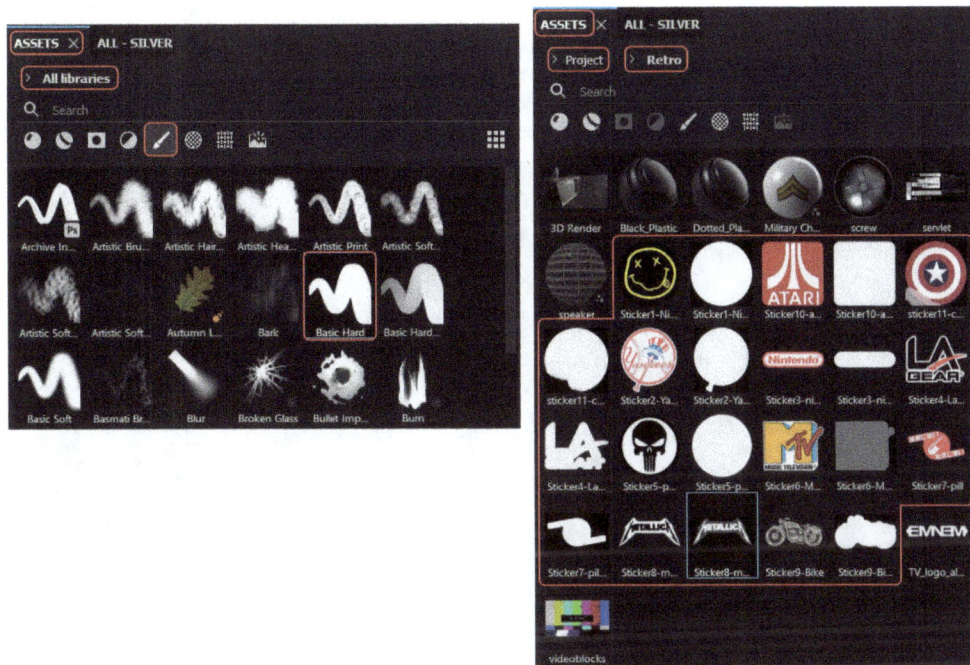

Figure 6.25 – Selecting stickers

5. Let's create a new paintable layer so that we can apply stickers to it. Click on **Add layer** in the **LAYERS** panel to create a new paintable layer, rename it `Stickers`, and keep it selected.

Figure 6.26 – Creating a new paintable Stickers layer

6. Now, let's create a sticker brush so that we can apply stickers to the **TV_Middle_Casing** Texture Set. Go to **PROPERTIES - PAINT** and choose the sticker you want to apply. Drag the base color of that sticker to **Base color** in **PROPERTIES - PAINT**, and the alpha of the sticker to **Alpha** in **PROPERTIES - PAINT**, as shown in *Figure 6.27*. If you do not want to drag images, you can also click on the **Base** color and **Alpha**, and then choose your desired images from the list.

7. For stickers, we need to activate only **color**, **height**, **rough**, and **metal**. Keep **Height** at 0.02; this will give depth to the sticker so that it will have a realistic thickness. Change **Roughness** and **Metallic** to your desired settings; **Roughness** will give you a shine and **Metallic** will give you reflection. I have chosen **Roughness** as 0.1497 and **Metallic** as 0.1463.

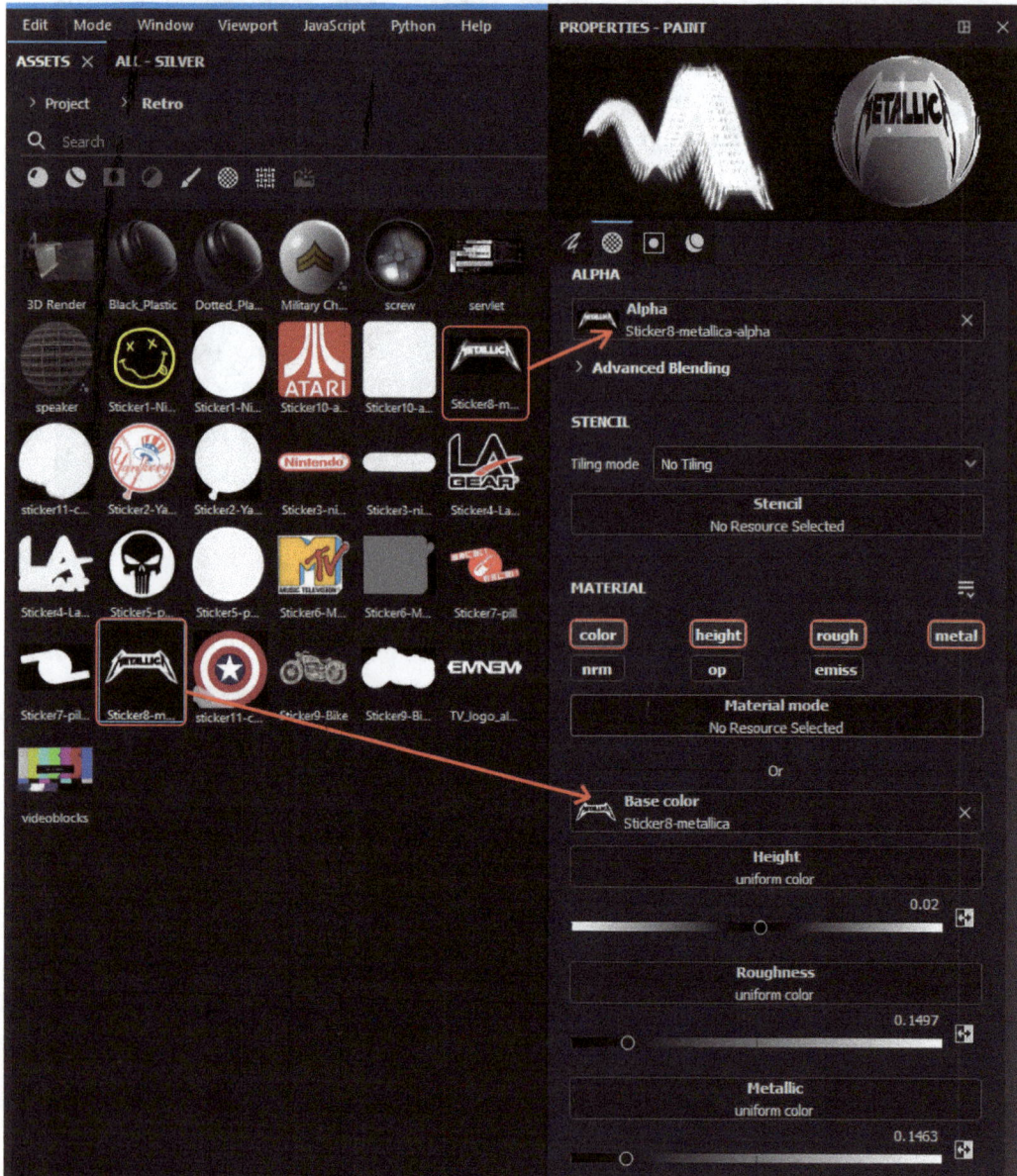

Figure 6.27 – Creating a sticker brush

8. Now, decide on a location in the **TV_Middle_Casing** Texture Set to apply your sticker to. You can rescale the sticker brush size by pressing the *Ctrl* (Windows) key or the *Cmd* (Mac) key on your keyboard and dragging the mouse left or right with the right mouse button pressed.

To change the direction/rotation of your sticker brush, keep the *Ctrl* (Windows) key or *Cmd* (Mac) key on your keyboard pressed and drag the mouse up or down with the mouse left button pressed.

To change the flow of your sticker brush, keep the *Ctrl* (Windows) key or *Cmd* (Mac) key on your keyboard pressed and drag the mouse left or right with the mouse left button pressed. You can change these settings using the **contextual toolbar**, as shown in *Figure 6.28*.

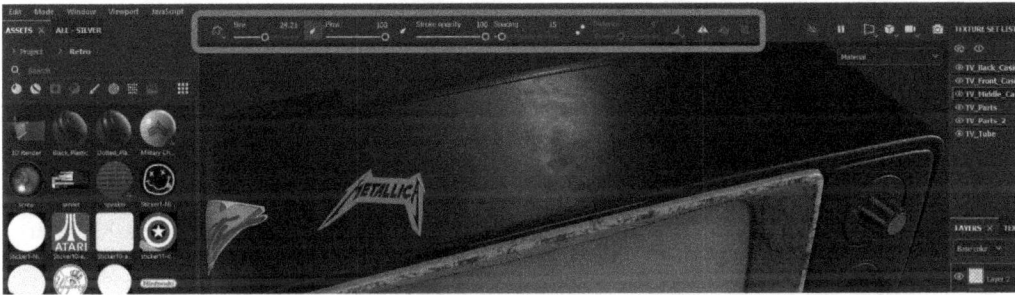

Figure 6.28 – Settings to adjust the brush

9. To apply new stickers, you need to remove the existing **Base color** and **Alpha** settings by pressing the cross next to them and repeating *step 6*. We can also repeat *step 6* without deleting the existing **Base color** and **Alpha** settings. Keep applying the stickers of your choice. However, do not apply a lot of stickers; otherwise, the design will look unprofessional. The final output should look something like *Figure 6.29*.

Figure 6.29 – The final output of the sticker application

Now that we have applied the stickers on the TV, in the next section, we will learn to apply an overall effect that will make the 3D model more look like a one-piece model.

Adding an overall layer effect with the position map in Adobe Substance 3D Painter

Whenever you want to strengthen or double any effect or apply an effect to a whole 3D model as a one-piece model, we have effects such as **Position** and **Add anchor point**. So, let's learn these effects in detail in two separate sections.

Adding an anchor point

Anchor points are used to expose any resource or element and reference it in various parts of the Layer Stack, for various reasons and with various changes.

They bring up a whole new world of possibilities by letting you link layers or masks together. Having a single anchor point influences several areas of your project, thereby turning Substance 3D Painter into a non-linear experience.

Let's double the effect of the **TV Logo** layer inside the **TV_Front_Casing** Texture Set:

1. Go to the **TV_Front_Casing** Texture Set, and then select the **TV Logo** layer mask in the **LAYERS** panel. Then, go to **Add effect** and select **Add filter** to add a filter over the **Fill** layer. Keep the **Filter (empty)** layer selected, go to **PROPERTIES - FILTER**, click on the **Filter** area under **FILTER** settings, and choose the **Blur** filter, as shown in *Figure 6.30*.

Figure 6.30 – Adding the Blur filter to the TV Logo layer mask

2.   Change the **Blur Intensity** parameters to 0.25 under the **FILTER** settings in **PROPERTIES - FILTER.**

Figure 6.31 – Changing the blur settings

3. Now, select the newly added **Levels** layer, which is added by clicking on the add effect button. Then, go to its **PROPERTIES - LEVELS**, and change **Input gamma** to 3 and **Input maximum** to 0.8 under the **LEVELS** settings by double-clicking on the number above the arrows, as shown in *Figure 6.32*.

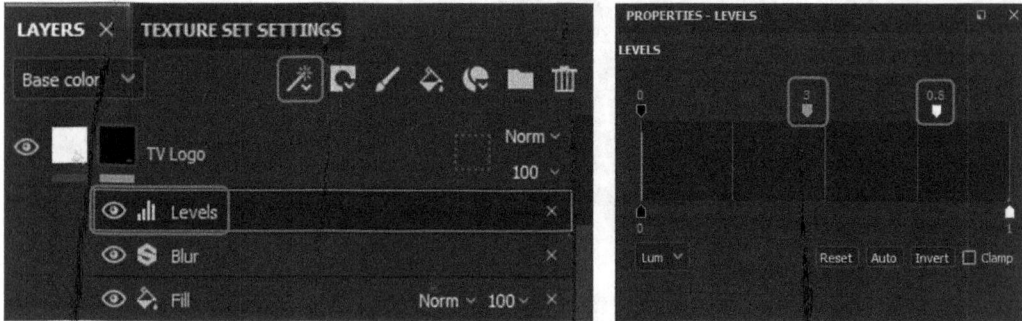

Figure 6.32 – Adjusting the TV Logo levels

4. Now, it's time to add an anchor; therefore, keep the **Levels** layer selected, then click on **Add effect**, and select **Add anchor point**. Once the anchor is created, rename it TV Logo mask.

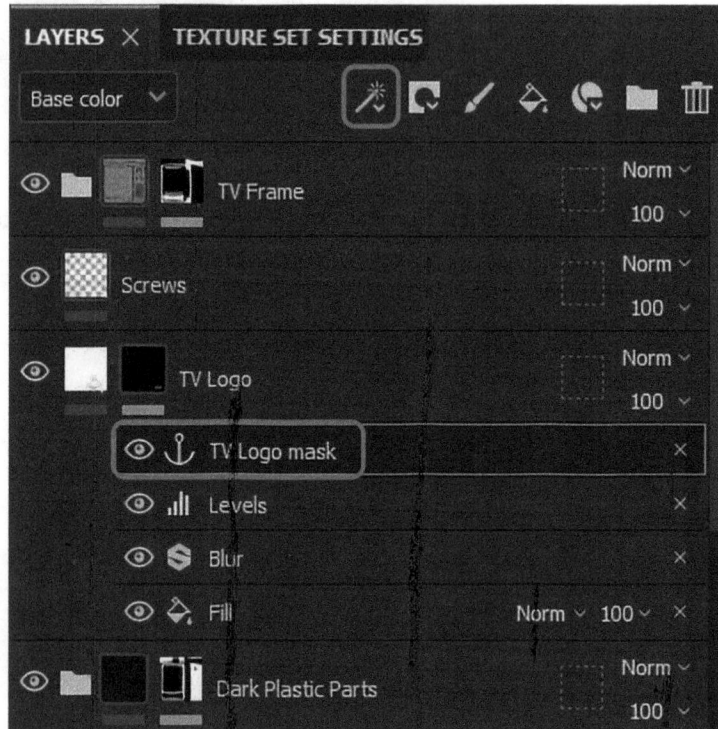

Figure 6.33 – Adding an anchor point

5. We are done with all the effects we need. Now, to double these effects, we have to create another **Fill** layer over **TV Logo** and rename it TV Logo Double Effects. Then, inside **PROPERTIES - FILL**, uncheck all the **MATERIAL** settings except **height**, and change the **Height** value to 1, as shown in *Figure 6.34*.

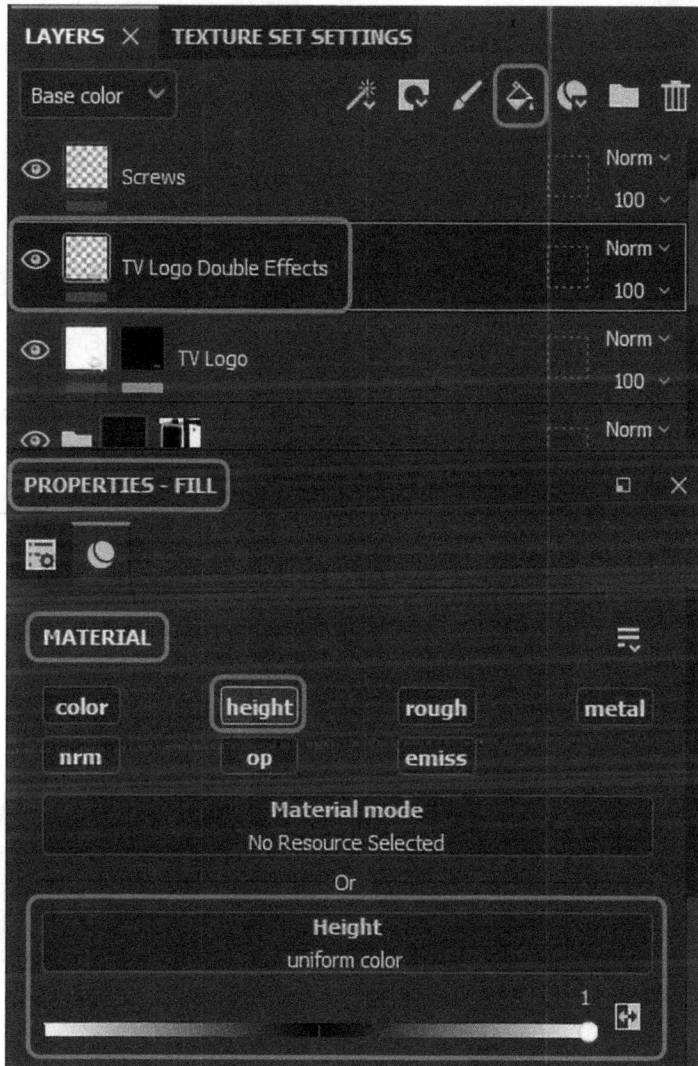

Figure 6.34 – Adding a new TV Logo Double Effects fill Layer

6. Now, add a black mask from the **Add mask** option, then add a new fill layer to that mask from the **Add effect** option, and rename it TV Logo Double Effect Fill, as shown in *Figure 6.35*.

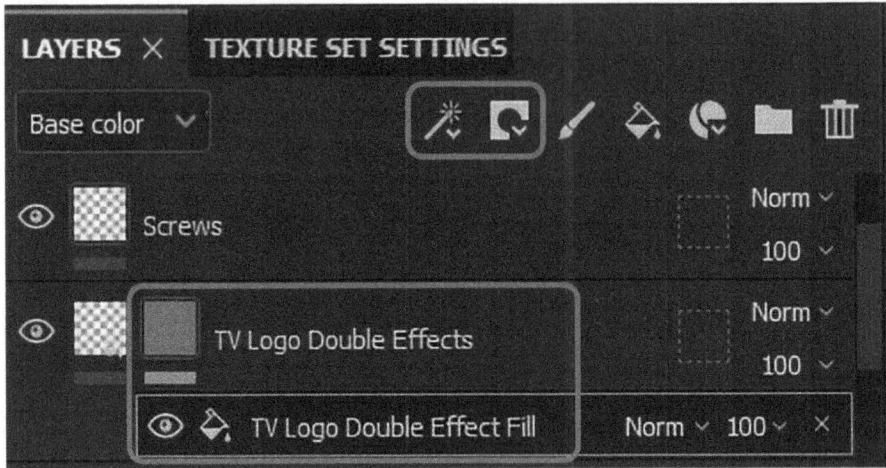

Figure 6.35 – Adding a black mask and Fill layer to the TV Logo Double Effects layer

7. Keep the **TV Logo Double Effect Fill** mask fill layer selected and go to its **PROPERTIES - FILL** window. Under the **GRAYSCALE** settings, click on the **grayscale** option. As soon you will click on it, you will see the **RESOURCES** and **ANCHOR POINTS** tabs; select the **ANCHOR POINTS** tab. Under that tab, you will find the anchor you previously created; click that **TV Logo mask** anchor.

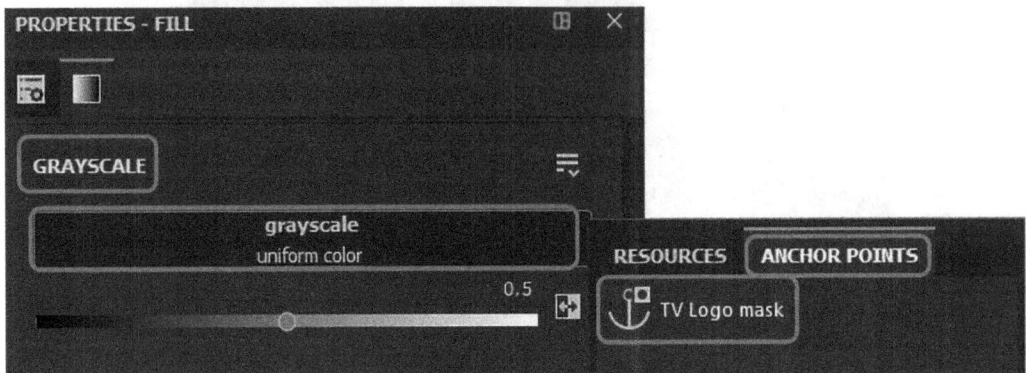

Figure 6.36 – Choosing ANCHOR POINTS

8. Once you will do that, you will notice that the height, blur, and level effects have been doubled. This is how **ANCHOR POINTS** helps you to repeat the effect. Because it is an instance, you can change the original effect settings. Also, wherever these effects are anchored, they will change automatically, as shown in *Figure 6.37*.

Figure 6.37 – The TV Logo Repeated Effects final output

Now that you know the technique to double or repeat any effect, we will move to apply an effect called **Position** using the **Add generator** and **Add effect** options in the next part of this section.

Using Position maps to add an overall dust effect on multi-Texture Set 3D models and 3D meshes

The **Position** map creates a texture with the coordinates of each point on the 3D mesh. To put it another way, this tells Painter where the mesh's top, bottom, front, rear, and left and right sides are. It helps a user to count a 3D model with multiple Texture Sets as a one-piece model.

To work on the **Position** map, you need to make sure you activate it with **Normalization Scale** set to **Full Scene** while baking the textures, as shown in *Figure 6.38*, which we did in *Chapter 1, Getting Started with Adobe Substance 3D Painter*.

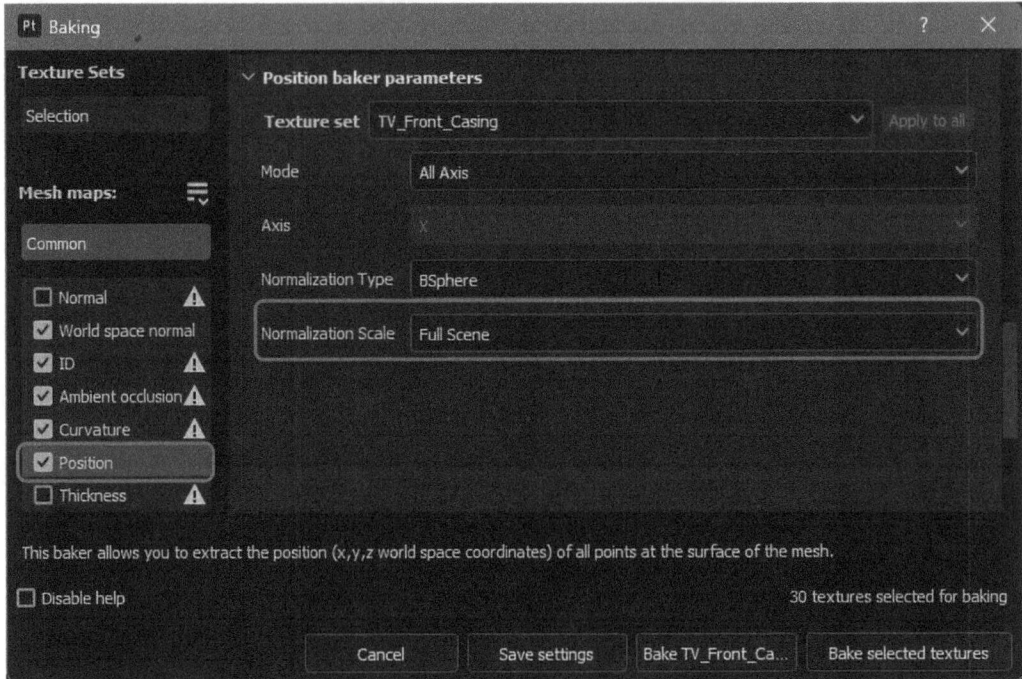

Figure 6.38 – Activating the Position map for baking with a Full Scene normalization scale

9. To add a **Position** map, you can choose any Texture Set; in this project, I'm choosing the
    TV_Middle_Casing Texture Set. Then, create a new fill layer with the help of the **Add fill layer**
    option and rename the new layer Dust Effect. Under **PROPERTIES - FILL**, deselect all
    the **MATERIAL** settings except **color**, **height**, and, **rough**. Now, change **Base color** to C2B280,
    Height to 0.02, and **Roughness** to 1.

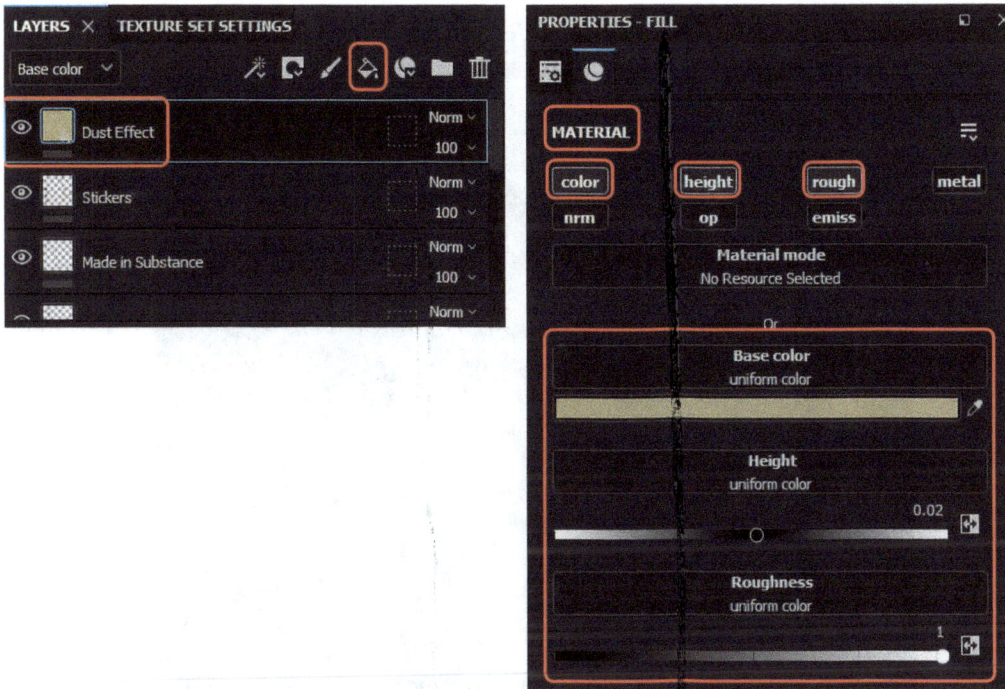

Figure 6.39 – Creating a new Dust Effect fill layer

10. Now, add a black mask to the **Dust Effect** fill layer from the **Add mask** option, and then add a generator to the black mask from the **Add effect** option. Then, select **Generator (empty)**, go to **PROPERTIES - GENERATOR**, and under the **GENERATOR** settings, click on the **Generator** area and choose **Position**.

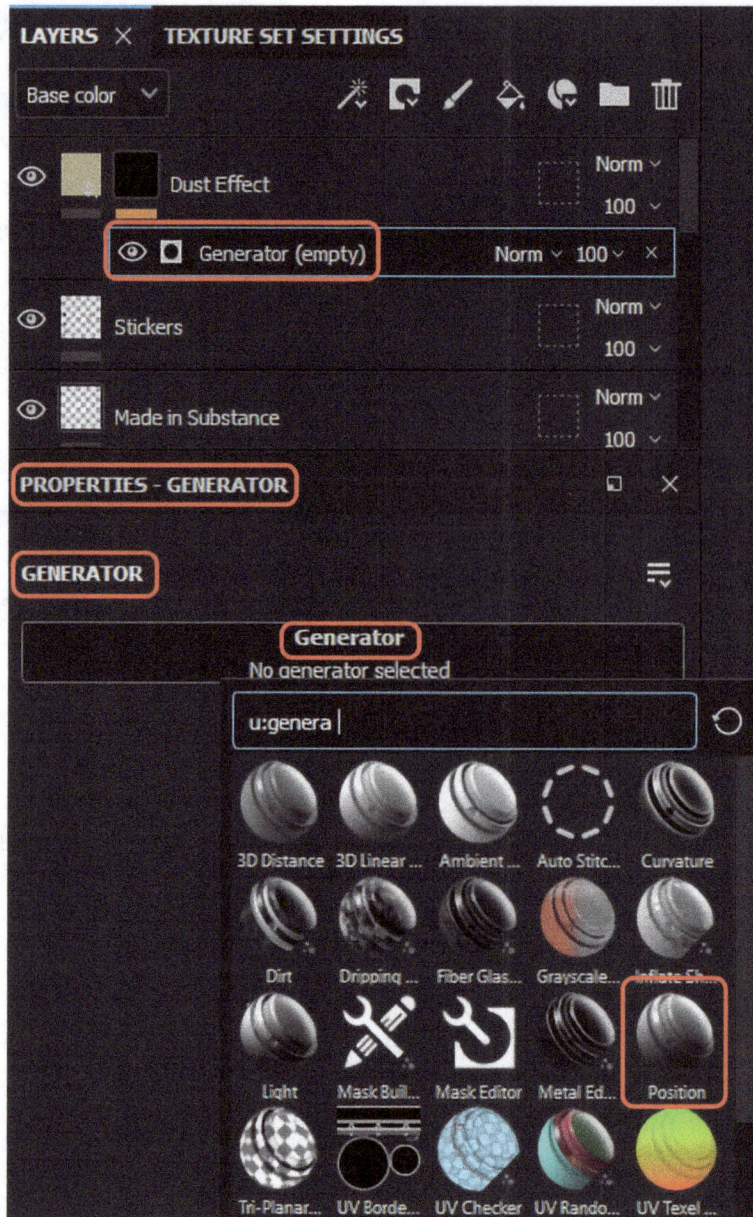

Figure 6.40 – Adding the Position map by adding the Generator effect

Now, the TV with the **Position** map effect will look like *Figure 6.41*.

Figure 6.41 – The Position map effect on TV_Middle_Casing

11. To see the gradient map of the **Position** map effect, press *Alt* (Windows) or *Option* (Mac) on your keyboard and click on **Dust Mask**, as shown in *Figure 6.42*.

The black and white gradient mask will help you identify where the strength of the mask is and where the weakness is. The texture or material will appear with full visibility in the stronger area, which is whiter, and less visible in the weaker area, which is dark. Once you get the idea of the gradient, disable it by simply clicking on the main **Dust Effect** layer.

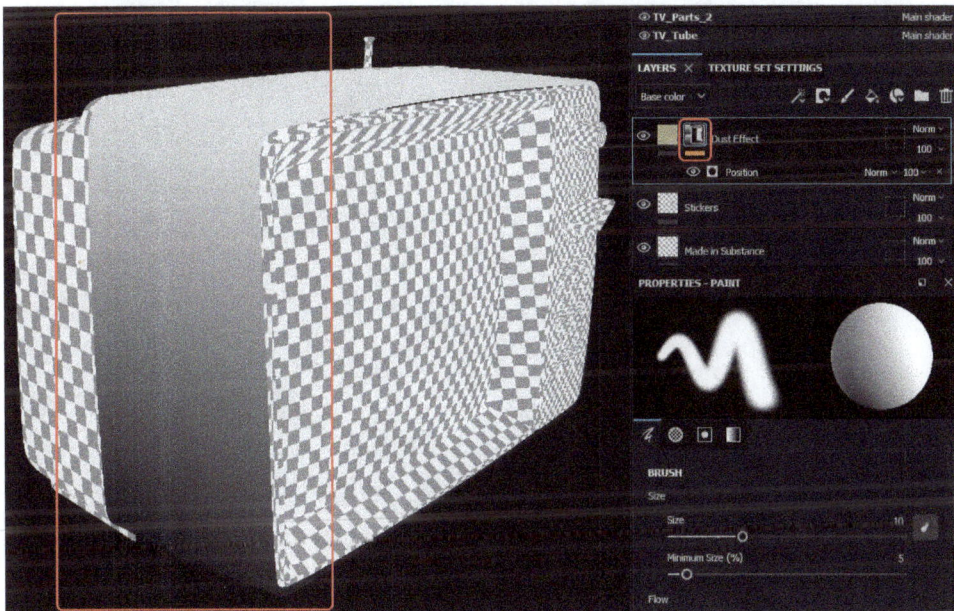

Figure 6.42 – Enabling the gradient map of the Position map effect

12. First, make sure you are still inside the **Dust Effect** mask, then add a new fill layer above the **Position** map effect from the **Add effect** option, rename the new fill layer `Grunge Dust Spread`, keep it selected, and go to **PROPERTIES - FILL**. Under the **GRAYSCALE** setting, click on the **grayscale** area and search for the `Grunge Dust Spread` or `Grunge Dust Small` map, and then select it once it appears.

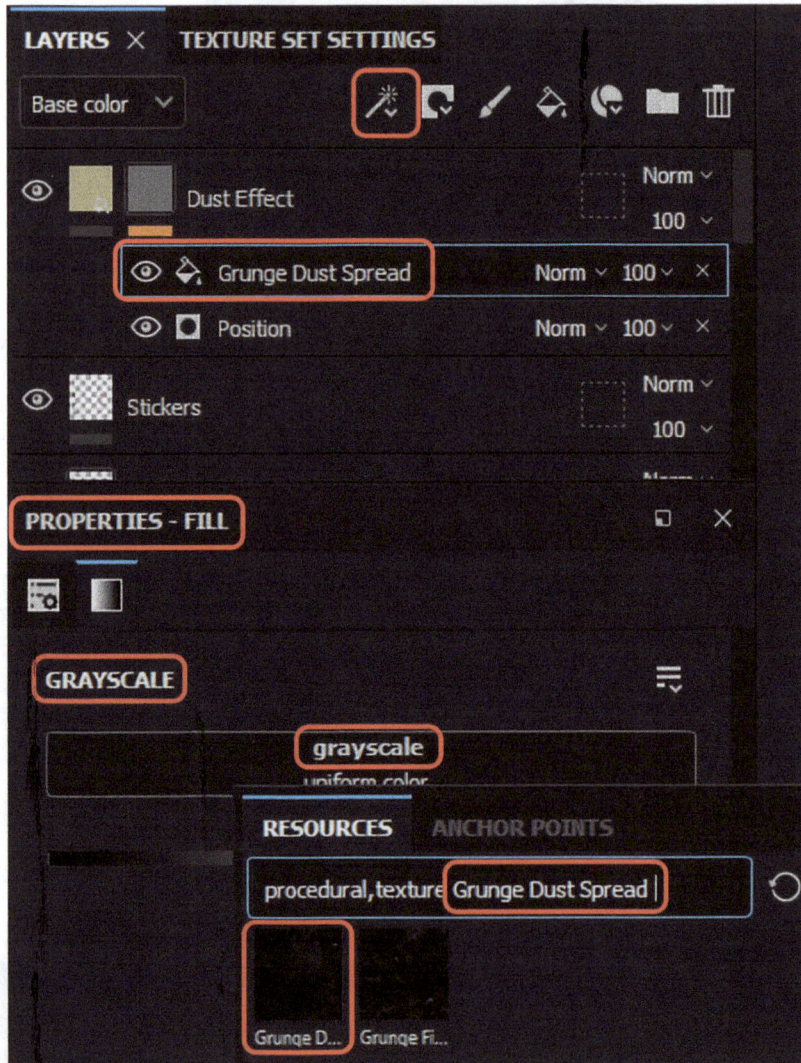

Figure 6.43 – Creating a new fill layer in Dust Effect Mask

13. Now, it's time to apply this effect to all the other Texture Sets. To do that, right-click on the main **Dust Effect** layer and choose **Instantiate across texture sets**.

LAYERS

Base color

Dust Effect

Stickers

Made in Substance

Toggle mask	
Invert mask background	
Invert mask	
Clear mask	
Remove mask	
Export mask to file	
Export mask to clipboard	
Create smart mask	
Cut layer(s)	Ctrl+X
Copy layer(s)	Ctrl+C
Paste layer(s)	Ctrl+V
Paste layer as instance	Ctrl+Shift+V
Duplicate layer(s)	Ctrl+D
Instantiate across texture sets...	Ctrl+Shift+D
Remove layer(s)	Del
Group layer(s)	Ctrl+G
Copy content	Ctrl+Alt+C
Paste into content	Ctrl+Alt+V
Create smart material	

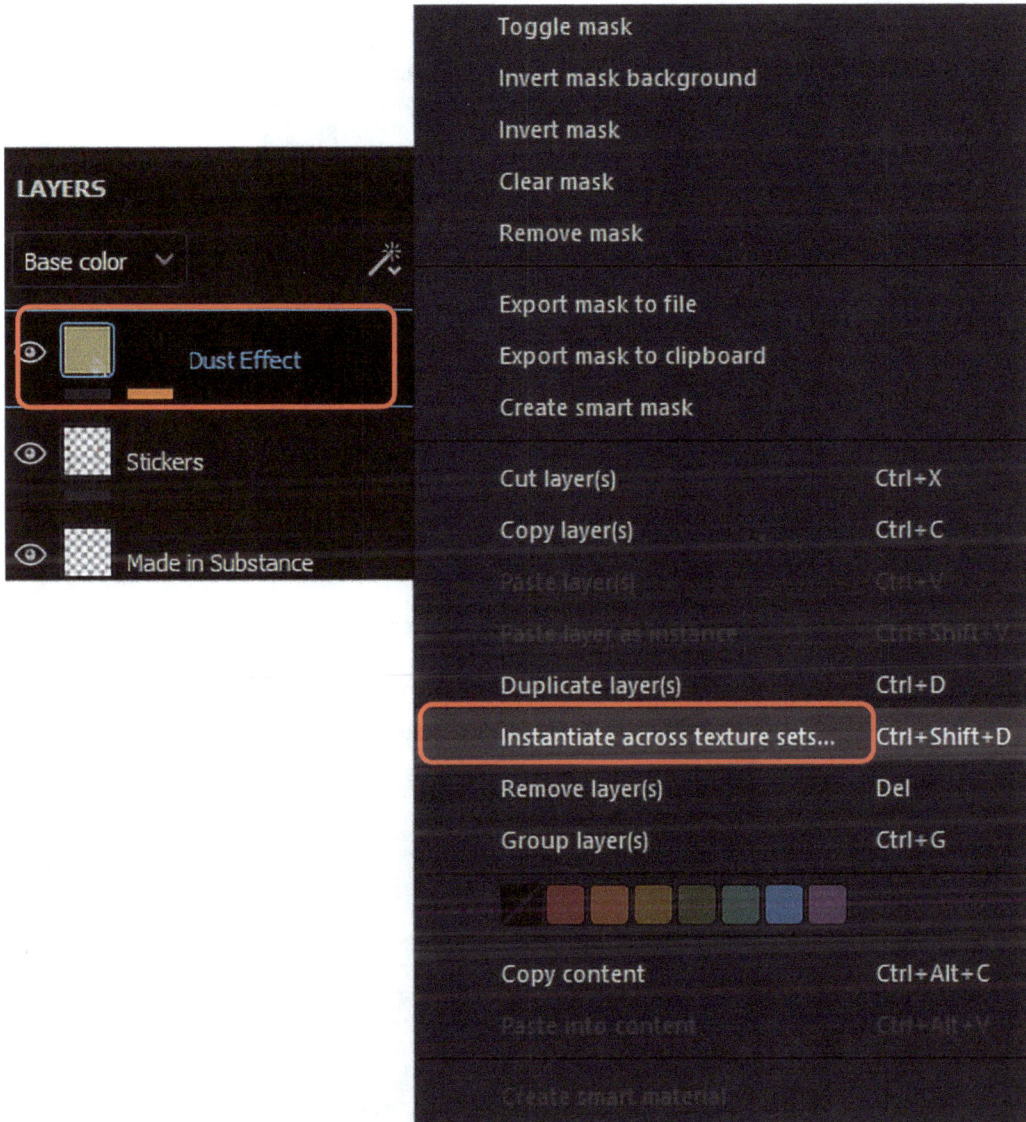

Figure 6.44 – Instantiating Dust Effect on all Texture Sets

14. Once you click the **Instantiate across texture sets** option, the **Instantiate across Texture Sets** window will pop up. Make sure that all the Texture Sets are selected, except for the source Texture Set, which is **TV_Middle_Casing** in our case, and then click **OK**.

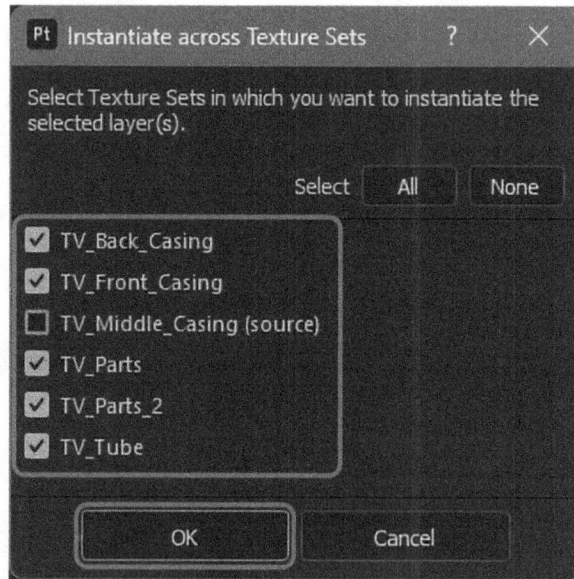

Figure 6.45 – Selecting Texture Sets for instantiation

15. Now, you will notice that the dust effect is all over the 3D model as one object.

Figure 6.46 – The dust effect on the whole 3D model

16. To make the dust effect more realistic, click the black mask of the **Dust Effect** layer and select the **Grunge Dust Spread** fill layer inside the **Dust Effect** layer mask. Then, go to its **PROPERTIES - FILL** window and change the following settings:

- **Projection**: `Tri-planar projection`
- **UV transformations** | **Scale**: `2.4`

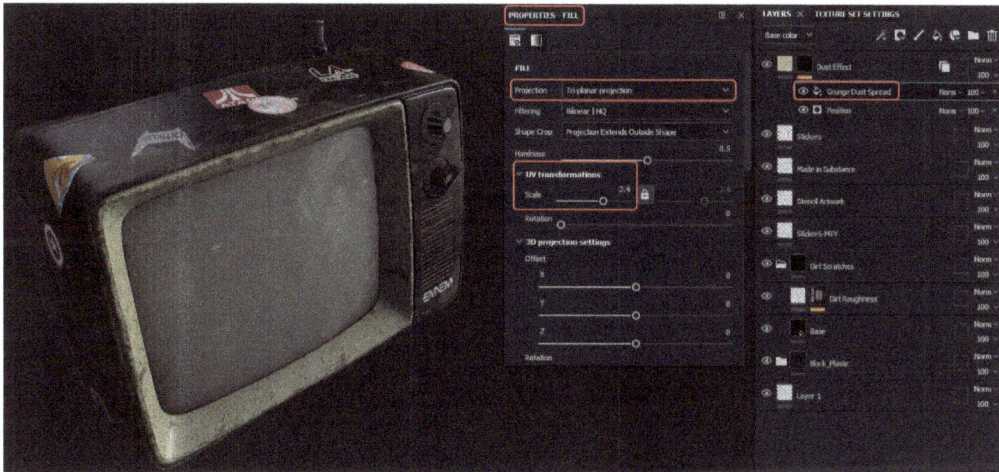

Figure 6.47 – Adjusting the Grunge Dust Spread fill layer

17. The TV model and the texturing are now almost ready, except for the dust effects strength. You will notice that the dust effect is too strong; therefore, you can select the main **Dust Effect** layer and reduce its opacity to around 50, as shown in *Figure 6.48*.

Figure 6.48 – Adjusting Dust Effect's opacity

Hopefully, you now know how to use anchor points and Position maps, which will help you to repeat an effect and apply it to a whole 3D model respectively. Now, as we have completed texturing and painting our 3D model using Adobe Substance 3D Painter, in the next section, we will learn how we can export these textures.

Exporting textures from Adobe Substance 3D Painter

Exporting textures from Painter is the last process when you want to export your completed textures to third-party 3D applications such as Autodesk 3D Studio Max, Maya, or Blender.

So, let's begin exporting them:

1. First, go to the **File** menu and click on it; then, choose the **Export Textures** option.

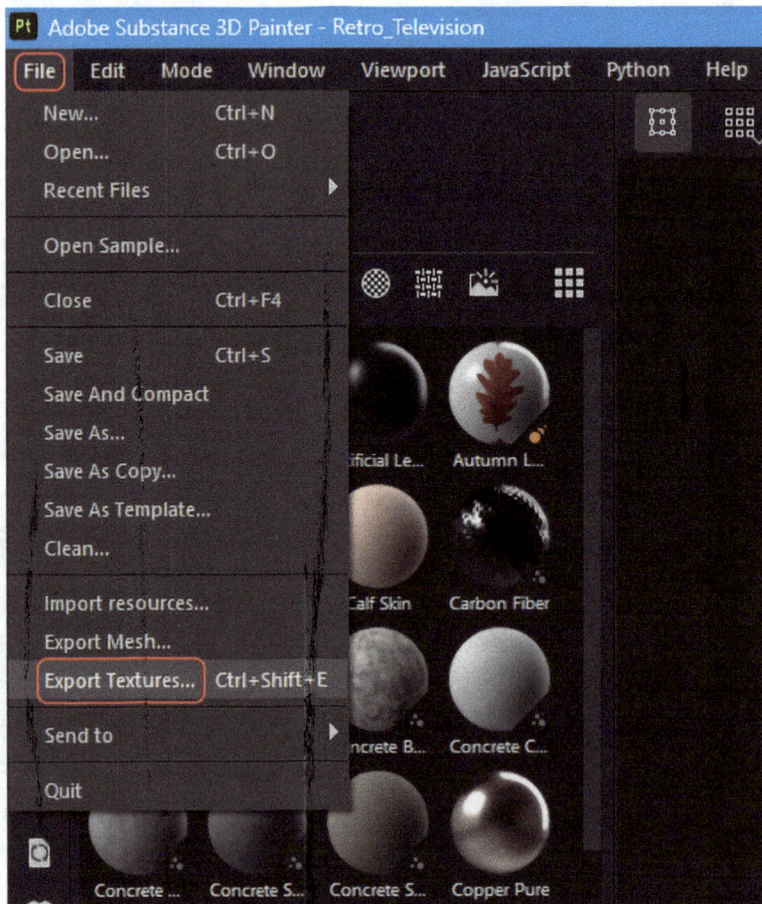

Figure 6.49 – The Exporting Textures option

2. When the **Export textures** window pops up, you need to first adjust **SETTINGS. Global settings** change the **General Export Parameters** of all Texture Sets; however, you can select any specific Texture Set under **Global settings** and change **General Export Parameters**.

3. Under **General Export Parameters**, you can decide where you want to save your texture from the **Output directory** option. You can also select your **Output template** option – for example, if you are working in Unreal Engine, you can choose **Unreal Engine**, or if you are working in V-Ray, you can choose **Vray**.

4. For this project, we will choose **Substance 3D Stager**. For the **Substance 3D Stager** template, **File type** is disabled; however, you can choose different types, such as PNG or JPEG, if you are working on a different output template.

5. You can choose your desired size from the **Size** option; we will choose 4096 to export 4K-quality textures. The last option is **Padding**, which allows you to adjust the border of your textures; the best setting is **Dilation infinite**, because it keeps your texture borders within your UV tiles to avoid showing any visible gaps and borders.

6. Finally, you can also enable **Export shaders parameters** if you want the 3D application, where you are exporting your textures, to adjust its parameters.

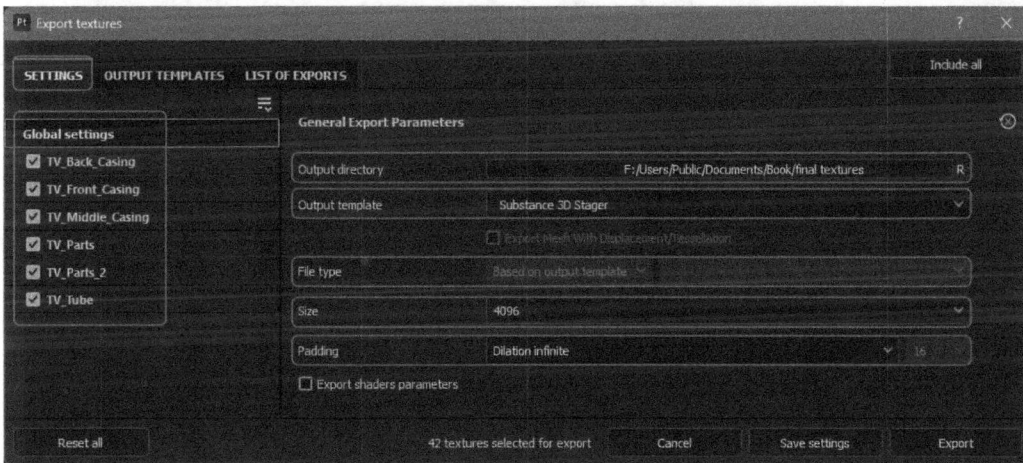

Figure 6.50 – Export textures | SETTINGS

7. Usually, we do not change the settings of any Texture Set and their channels separately. However, if you want to, you can select the desired Texture Set, click on the **Edit** button, as highlighted in *Figure 6.51*, and edit the settings.

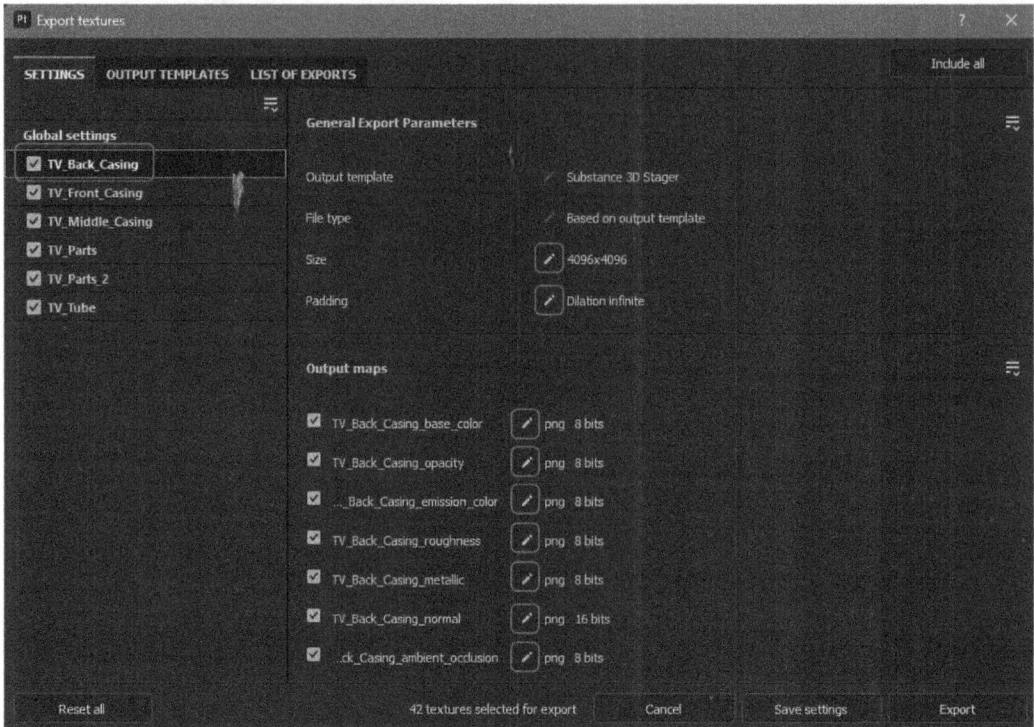

Figure 6.51 – Editing individual Texture Set general export parameters and output maps

8. When you click on the **Edit** button, you will notice that you can change the bits of the roughness texture channel, as shown in *Figure 6.52*; however, you cannot change file types for some output template channels. To do that, you can revisit *step 7*.

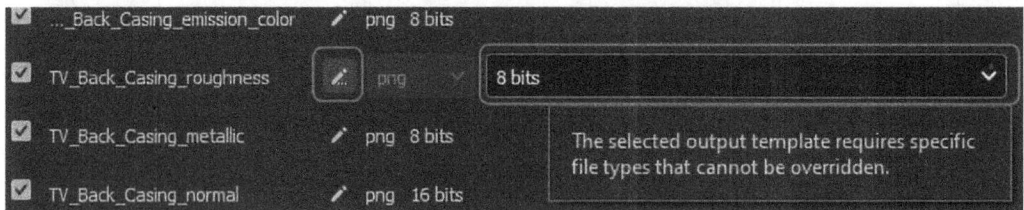

Figure 6.52 – Editing a specific Texture Set

9. When you are unable to change the file type of any specific texture channel or you want to keep a specific texture channel's file type different from other channels, you can go to the **OUTPUT TEMPLATES** tab of the **Export textures** window, select the output template preset, and change the file type or bits of the channel you want.

The other maps in this tab are just to show information. However, we are not going to change anything; this step is only for educational purposes.

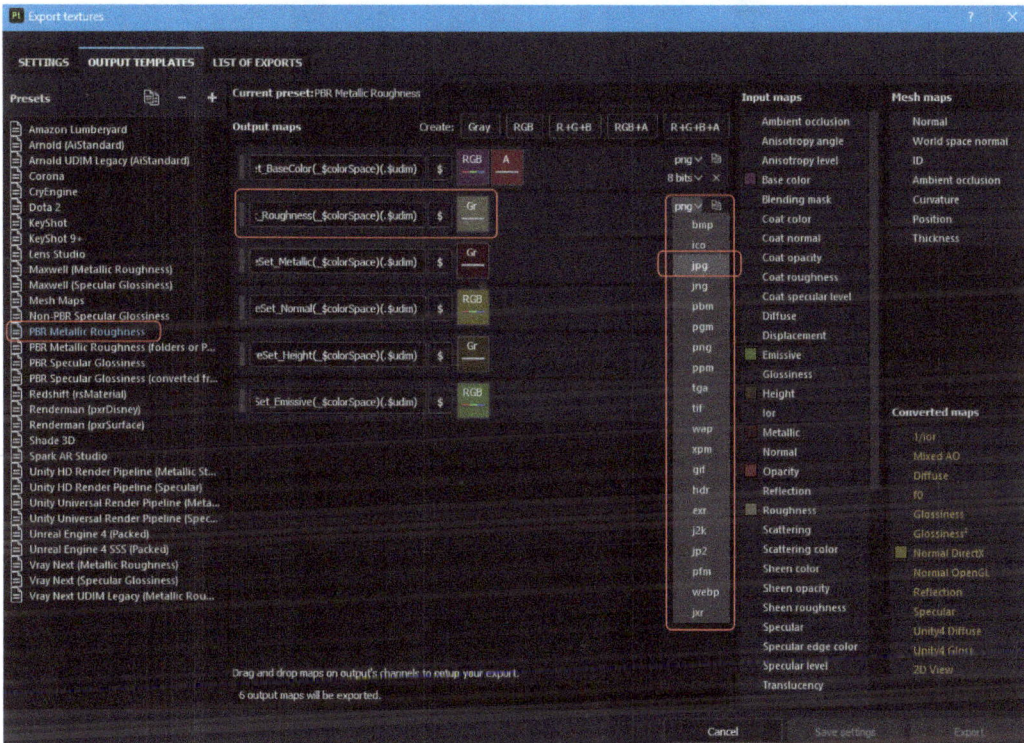

Figure 6.53 – The OUTPUT TEMPLATES settings

10. The last tab in **Export textures** is **LIST OF EXPORTS**, which shows you a list of Texture Sets and their channels that are going to be exported. You need to just click on the **Export** option. It also shows the list of **Texture Sets** and their channels when they are successfully exported in a teal or turquoise color

11. To see your exported textures, click on **Open output directory** in the top-right corner of this tab.

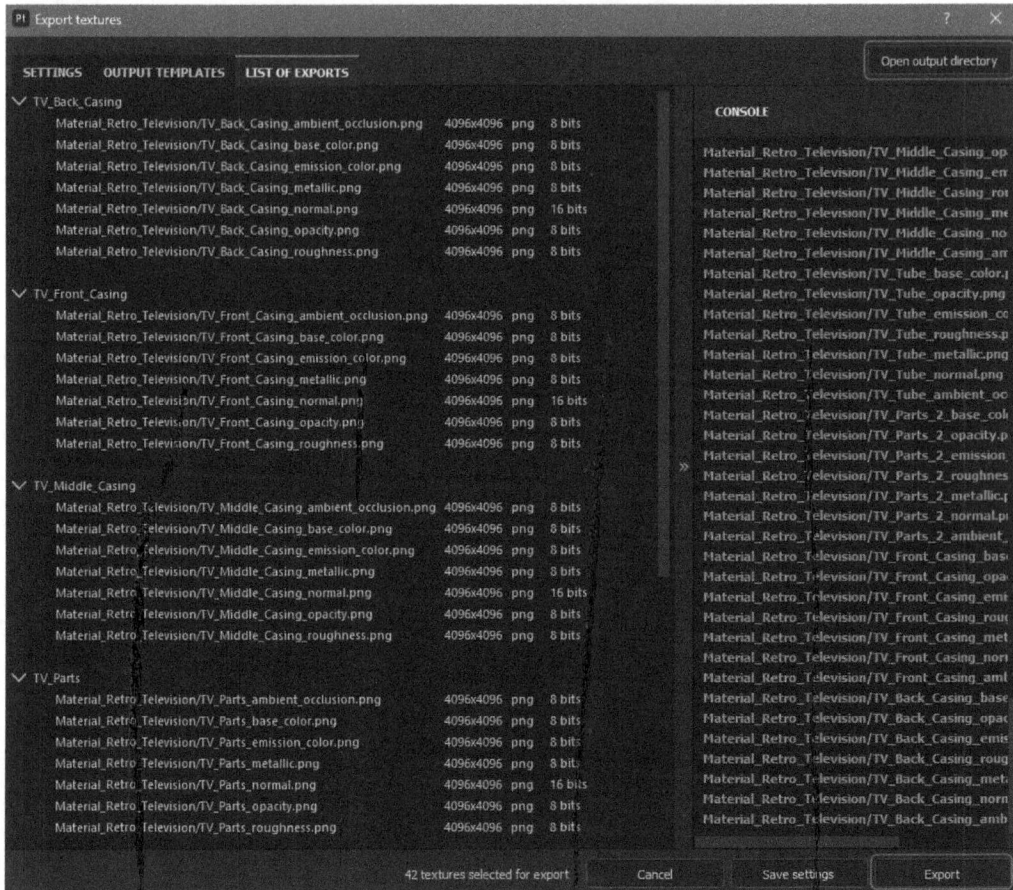

Figure 6.54 – Exporting and opening the export folder from the LIST OF EXPORTS tab

12. Once you click on **Open output directory**, the folder where you exported your textures will open. Now, you can use your desired third-party 3D application and start applying these textures. In our case, we have used **Substance 3D Stager** as our output template, so we will be using these textures again in the *Adobe Substance 3D Sampler* part of this book.

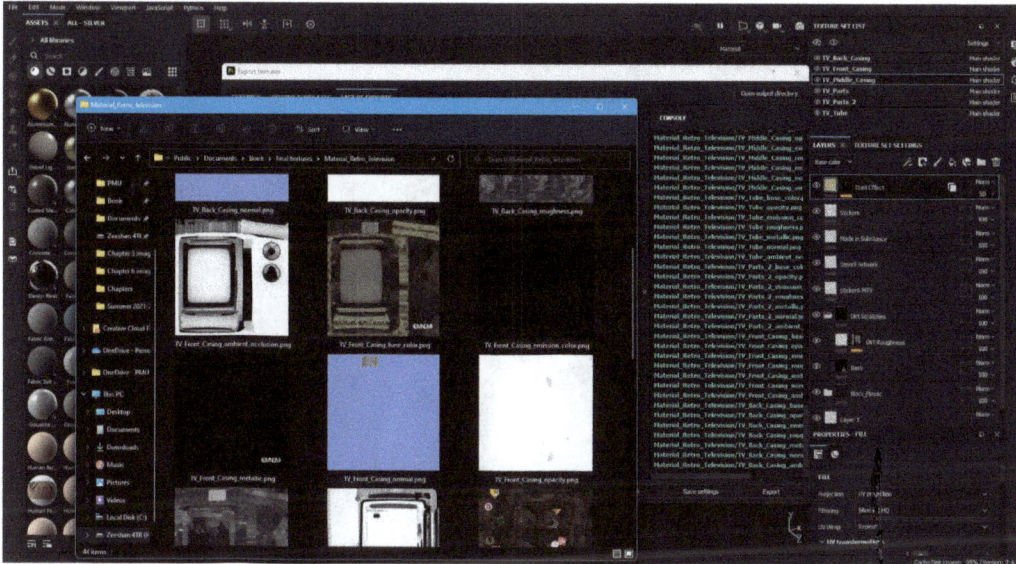

Figure 6.55 – The exported output directory

Now that you have learned one of the most crucial skills of Adobe Substance 3D Painter, which is to export final textures and use them in your desired 3D application, it's finally time to end the *Adobe Substance 3D Painter* part of this book by learning how to render using Adobe Substance 3D Painter's Iray renderer in the next section.

Rendering in Adobe Substance 3D Painter using Iray

After completing a project, users always want to visualize how a model will look in reality; therefore, in this section, you will learn how to render an Adobe Substance 3D designer project:

1. First, you need to go the **Mode** menu and choose **Rendering (Iray)**.

Figure 6.56 – Rendering (Iray) mode

2. Once you select **Rendering (Iray)**, the 3D viewport will switch to rendering mode after a little warm-up. The Iray rendering mode uses real-time rendering, which means you can navigate the viewport and it will keep on rendering every time you adjust it.

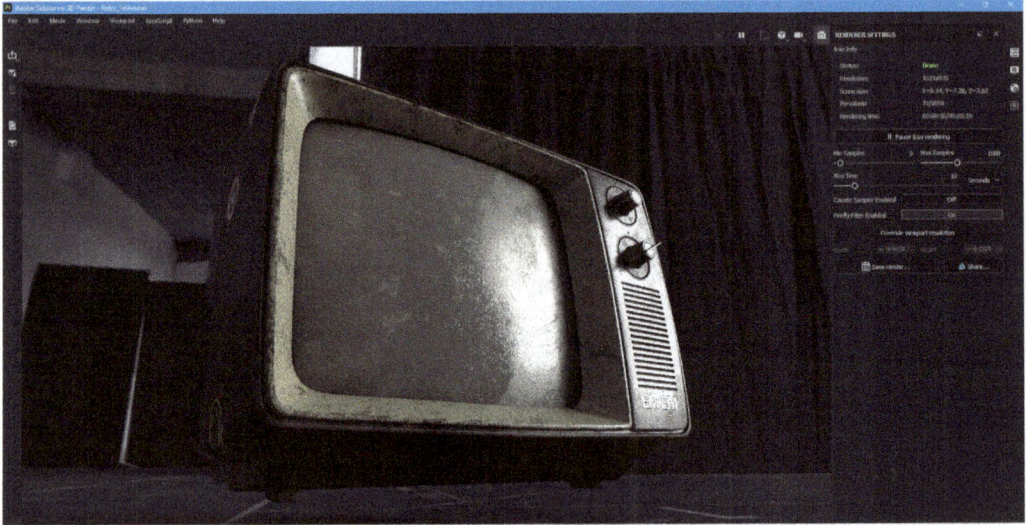

Figure 6.57 – Iray's real-time rendering viewport

3. Note that the TV 3D model is flying in the air, so let's shift the ground up so the TV sits on it, unless you want it to fly in the air on purpose. Go to **Display Settings** and jump to the **Ground** settings. Make sure you tick the **Ground** checkbox to make the ground ivisible, and change its **Y**-axis setting to -3.65.

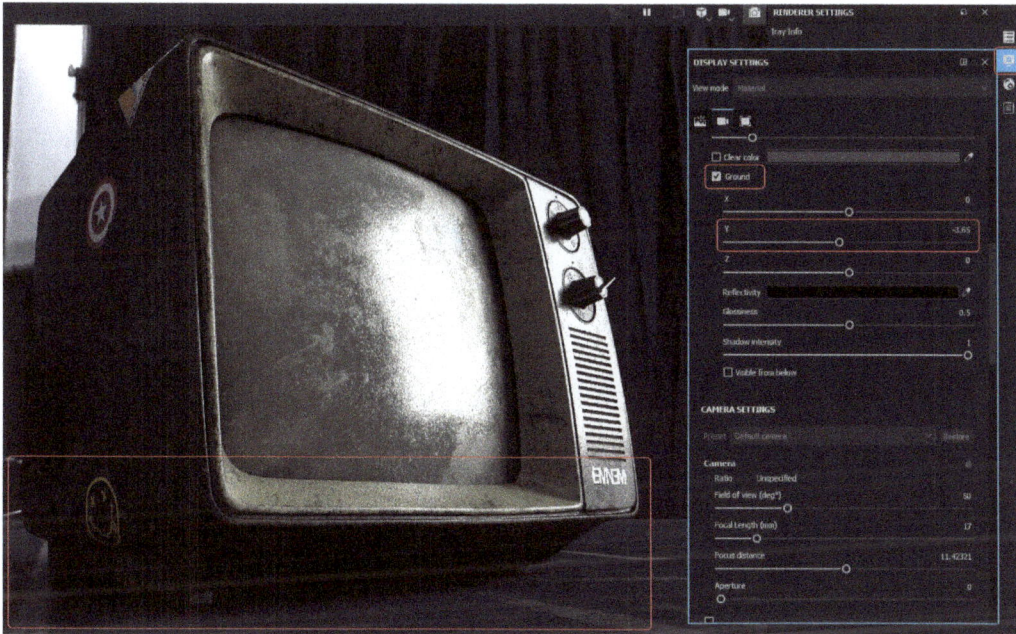

Figure 6.58 – Changing the Ground settings

4. You can also change the light settings by rotating the environment. Either go to **ENVIRONMENT | SETTINGS** and adjust **Environment Rotation**, or do the same by dragging the mouse with the middle mouse button pressed across the viewport, while keeping the *Shift* key on the keyboard pressed. Moreover, in the latest versions, it can be done by holding *Shift* and pressing the right mouse button.

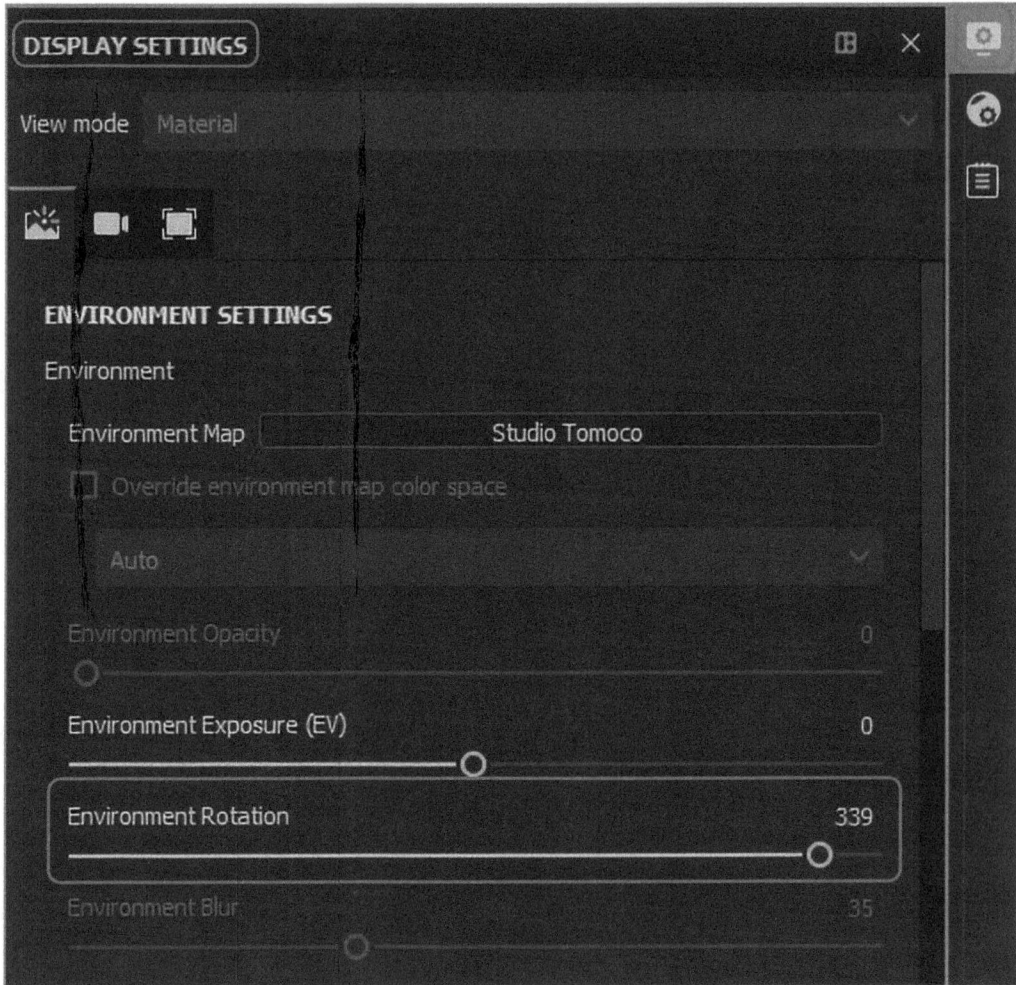

Figure 6.59 – Rotating the environment to change the light settings

5. In **DISPLAY SETTINGS**, you can also change the **Camera** settings and **Activate Post Effects**, so let's try a few effects.

6. First, we will try the **DOF** (**Depth of Field**) effect. For **DOF**, the camera **Aperture** size should be higher; therefore, change it to 0.5. Then, click on the **Activate Post Effects** and **DOF** checkboxes, as shown in *Figure 6.60*. Now, we need a focal point; you can adjust this by changing the **Focus distance** option. You can also press on the 3D model area that you want to focus on with the middle mouse button while keeping the *Ctrl* (Windows) or *Cmd* (Mac) key pressed on your keyboard. Note now that the TV's back part is blurred while the front part is in deep focus.

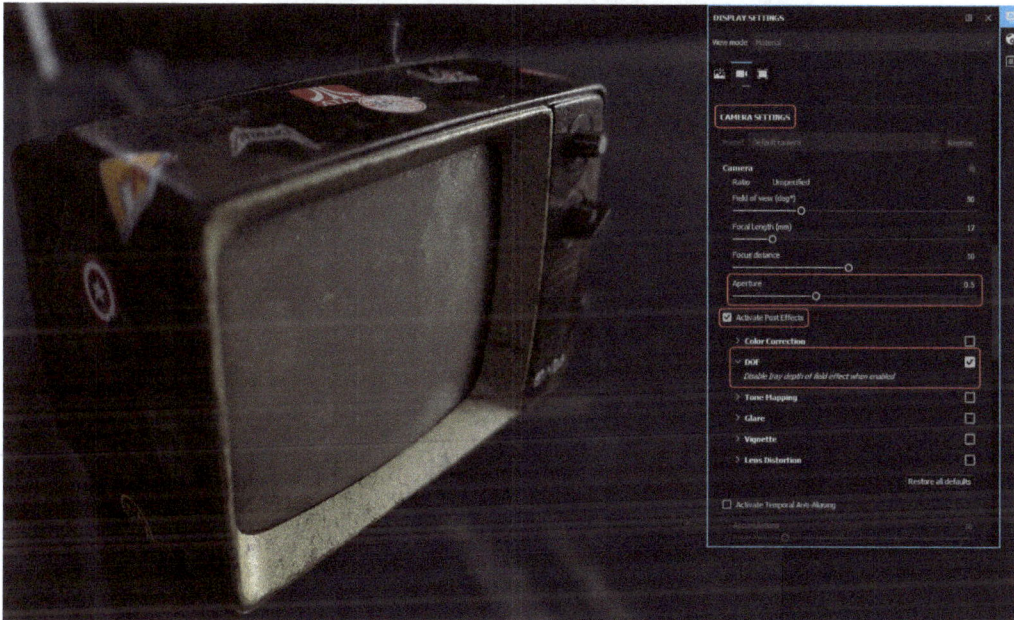

Figure 6.60 – Changing the Camera settings and Activate Post Effects

7. Now, let's do some color correction. First, we need to tick the **Color Correction** checkbox to activate it and then adjust the following settings:

Saturation: 0.75

Contrast: 1.003

Brightness: 1.2

Bias: 0

Sepia Tone Ratio: 0.15

White Balance Temperature (K): 6800

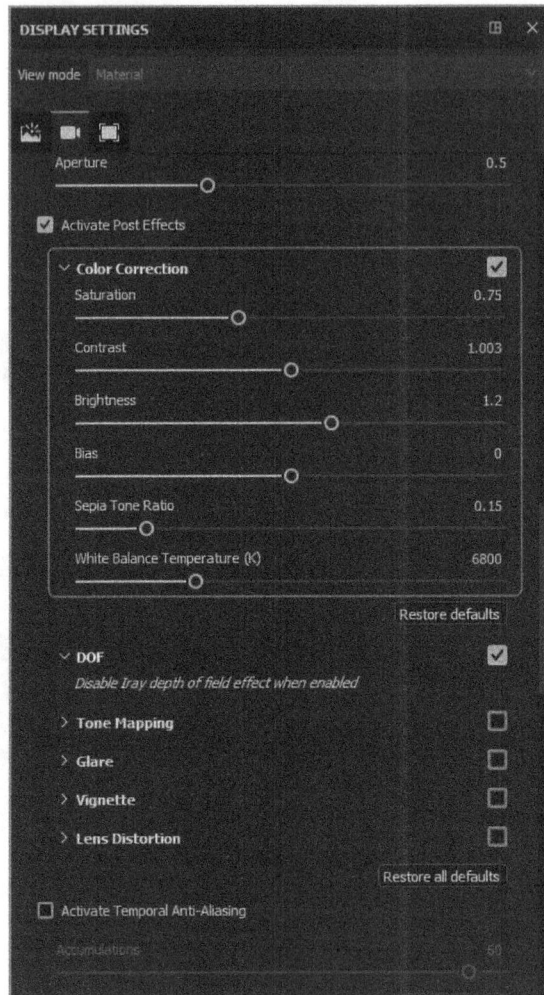

Figure 6.61 – The Color Correction settings

8. The last effects that we will add are **Glare** and **Vignette**. First, you need to tick their checkboxes to activate them, and then adjust the following settings:

    - **Glare**:

        - **Luminance**: 0.6

        - **Threshold**: 0.1

        - **Remap Factor**: 2.75

        - **Shape: Filter Cross Screen**

- **Vignette**:

  - **Strength**: 0 . 3

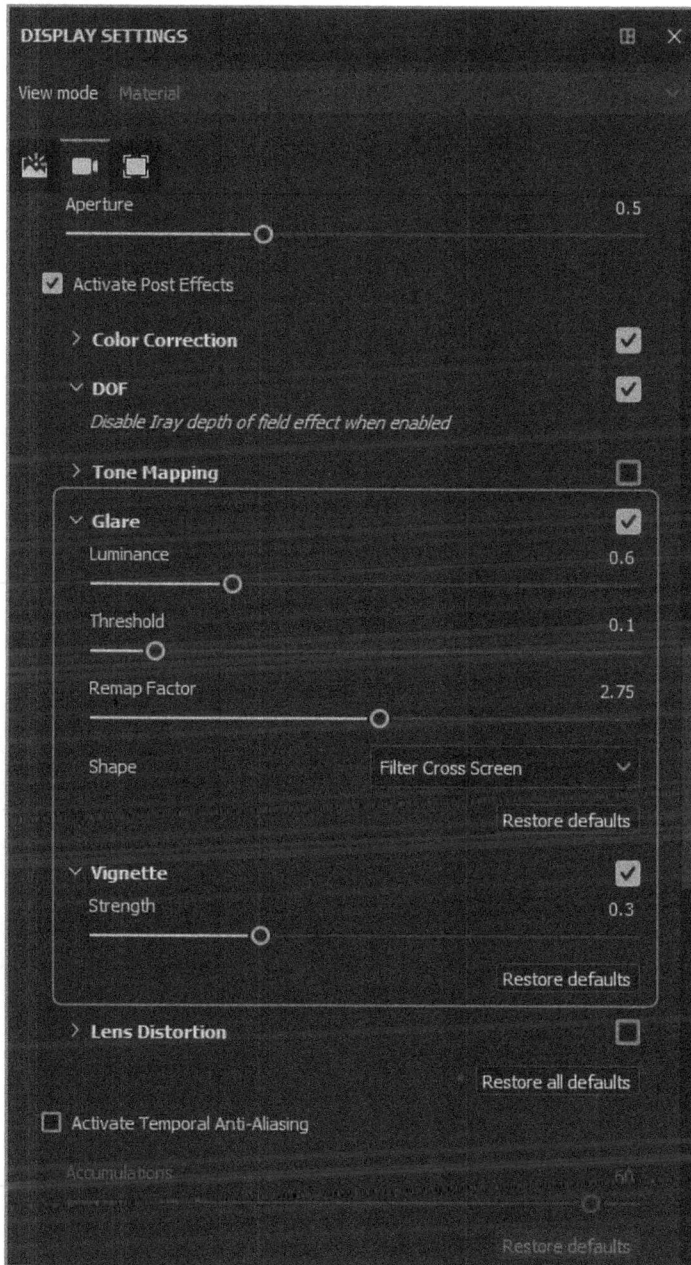

Figure 6.62 – Adjusting the Glare and Vignette Settings

9. There are a couple more settings in **DISPLAY SETTINGS** that you can try on your own. However, the final output after adding the **DOF**, **Glare**, and **Vignette** effects will look the same as shown in *Figure 6.63*.

Figure 6.63 – The final output after applying effects

10. Now, to up the quality of the render, you can either decrease **Min Samples** and increase **Max Samples** or increase **Max Time**. However, you have to compromise with the render time, owing to its higher quality. We will keep the existing samples and time settings; the rendering will stop when it reaches the assigned minimum and maximum samples or maximum time, whichever ends first.

11. The size of the rendering image is the same as your viewport; however, you can change it with the **Override viewport resolution** option. Once you are satisfied with your render, you can save it as an image with your desired file type using the **Save render** option. Moreover, you can also directly upload it to your ArtStation account using the ArtStation **Share** option; however, some versions do not have this option.

Figure 6.64 – Rendering and saving the artwork

Summary

I am glad to announce that we are done with Adobe Substance 3D Painter part of this book, and we have finished learning about Adobe Substance 3D Painter. Now, you can confidently say that you know how to work with Adobe Substance 3D Painter, and after working for a few more weeks with this application, you will surely become a professional.

The Adobe Substance 3D Designer part of this book is about Adobe Substance 3D Designer, which is the industry's go-to program for creating 3D materials. Substance 3D is used by over 95% of AAA gaming projects now in production, as well as the most famous visual effects and animation firms.

Substance 3D Designer is an application intended for creating 2D textures, materials, filters, and 3D models in a node-based interface. It heavily focuses on procedural generation, parametrization, and non-destructive workflows. You will learn Adobe Substance 3D Designer in detail in the next part.

As it is the longest-running application in the Substance 3D ecosystem, the resources made with it are the most versatile and dynamic possible.

7
Getting Started with Adobe Substance 3D Designer

I hope you had fun while learning about Adobe Substance 3D Painter; now it's time to move to our next part and start learning about Adobe Substance 3D Designer, which is one the leading software that is widely used in the gaming industry, film industry, and construction industry, gaming, filmmaking, architecture and so on. Unlike Painter, Adobe Substance 3D Designer relies on nodes rather than layers. Adobe Substance 3D Designer is the industry standard for developing unique materials and allows you total authoring control.

With it, you can create tileable textures and patterns, as well as altering whole texture sets. In this non-destructive, node-based environment, you can use pre-made resources or develop materials from scratch, and make use of Designer's ever-expanding scripting features. Want to tweak or alter the appearance of a material after it's been created? Make modifications whenever you want; you'll never lose any of your work.

Full sets of textures may be edited in real time, and Substance materials can be tiled. For procedural and hybrid processes, you can combine a huge variety of predefined filters and tools. You have complete access to and control over the origins of any resource or filter. So, let us start.

We will cover the following topics in this chapter:

- Substance 3D Designer panels
- The Adobe Substance 3D Designer **EXPLORER** window
- The Adobe Substance 3D Designer **GRAPH** window
- Substance Designer 2D and 3D views
- The Substance Designer **LIBRARY** panel

Substance Designer panels

In our first section, which is about panels, we will briefly go to the different panels of Substance Designer to give you a quick rundown of what each panel does. To work through this chapter, you need to open the Substance_graph.sbs file, which you can download from XXXXXXXXXXXXXXXX. You can see how the Substance Designer user interface looks in *Figure 7.1*.

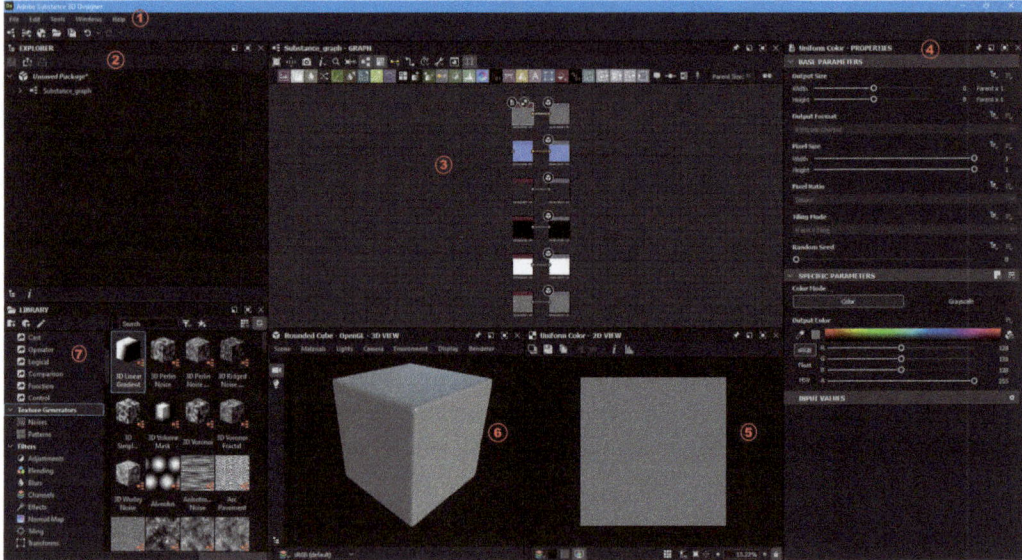

Figure 7.1 – Adobe Substance 3D Designer viewport

The following is a breakdown of the Substance Designer user interface from *Figure 7.1*:

1. **Menu bar**: The menu bar is a general option bar that gives you options such as **File**, **Edit**, **Tools**, **Windows**, and **Help**.

2. **EXPLORER panel**: Let's first jump to the **EXPLORER** panel, which, as you can see, is basically a file management system. This is where you open, close, build, and save your projects and graphs.

3. **GRAPH view/panel**: Then we have the **GRAPH** view. This is where you will actuall be doing most of your work; it is the heart of substance design where all the magic is done.

    This is where people construct their nodes network in order to produce complete materials. That's why we call it the heart of Substance Designer. You can pan around this view by dragging your mouse across the screen with the middle button pressed while the *Alt* (Windows) or *Option* (Mac) key is also pressed on your keyboard. Moreover, in the newer version, you can do it without the *Alt*/*Option* key.

    You can zoom in and out of this view by dragging your mouse up or down with the right button pressed while the *Alt* (Windows) or *Option* (Mac) key is also pressed on your keyboard. Using the hotkey, you may focus on the selected node(s) or the whole graph if none are chosen.

4. **PROPERTIES panel**: Now, here we have properties or parameters. To activate this panel, you need to double-click the desired node to see its property. The **PROPERTIES** panel contains the settings of any node in the graph view, so its contents will vary depending on the type of node that you have chosen.

    Moreover, if you double-click anywhere in the empty area, the **PROPERTIES** panel will simply show you the property of the graph view, instead of the node's property. So, if you want to view a node's properties, just double-click on that node. If you single-click, you just have the properties visible, but in 2D view, the image from the nodes selected before is still visible. If you want to see **Graph view** properties, you just have to double-click on an empty area of the graph view, and it will give you the whole graph property.

5. **2D VIEW**: The 2D view is just a 2D preview of any information that you have just selected inside the graph view. You will notice that the 2D view changes according to whatever node is selected. You can pan around this view by dragging your mouse across the screen with the middle button pressed while the *Alt* (Windows) or *Option* (Mac) key is also pressed on your keyboard. In the newer version, you can do it without the *Alt/Option* key.

    Moreover, you can zoom in and out this view by dragging your mouse up or down with the right button pressed while your *Alt* (Windows) or *Option* (Mac) key is also pressed on your keyboard.

6. **3D VIEW**: The one next to the 2D view is the 3D view. The 3D view is a complete 3D preview of the whole graph. All the nodes in the graph view will be shown inside the 3D view, displaying how the nodes will actually look in the end.

    This is where you get a whole preview of whatever graph you are working on; the complete graph output will be shown in the 3D view. You can pan around this view by dragging your mouse across the screen with the middle button pressed while the *Alt* (Windows) or *Option* (Mac) key is also pressed on your keyboard. In the newer version, you can do it without the *Alt/Option* key.

    You can zoom in and out of this view by dragging your mouse up or down with the right button pressed while the *Alt* (Windows) or *Option* (Mac) key is also pressed on your keyboard, and you can orbit around this view by dragging your mouse across the screen with the left button pressed while the *Alt* (Windows) or *Option* (Mac) key is also pressed on your keyboard. In the newer version, you can do it without the *Alt/Option* key.

7. **LIBRARY**: A library is simply a set of nodes from which you can construct your own node network. It comprises generators, filters, material filters, and so on. You can create your own custom library and also add folders to an existing library.

All the panels inside Adobe Substance 3D Designer are dockable; you can undock them and place them wherever you want, as shown in *Figure 7.2*.

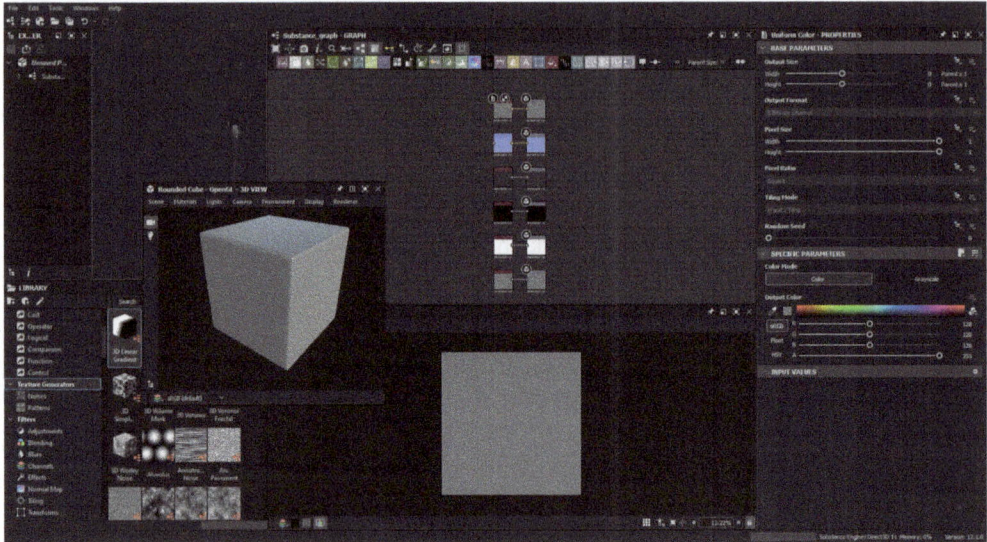

Figure 7.2 – Undocking the panels and docking them to different locations

Also, you can dock one panel into another and work through each panel using tabs, as shown in *Figure 7.3*.

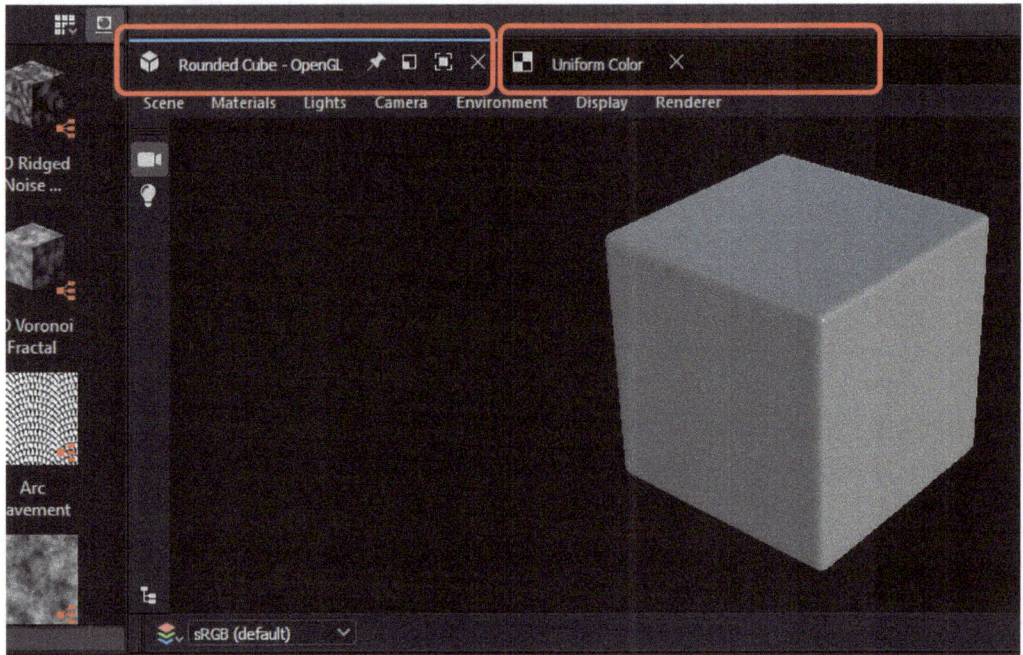

Figure 7.3 – Dropping panels into one another and using tabs to switch

If any panel is closed by mistake, you can reopen it through the **Windows** option located in the top menu bar. You can also restore the default layout from the **Windows** option as well.

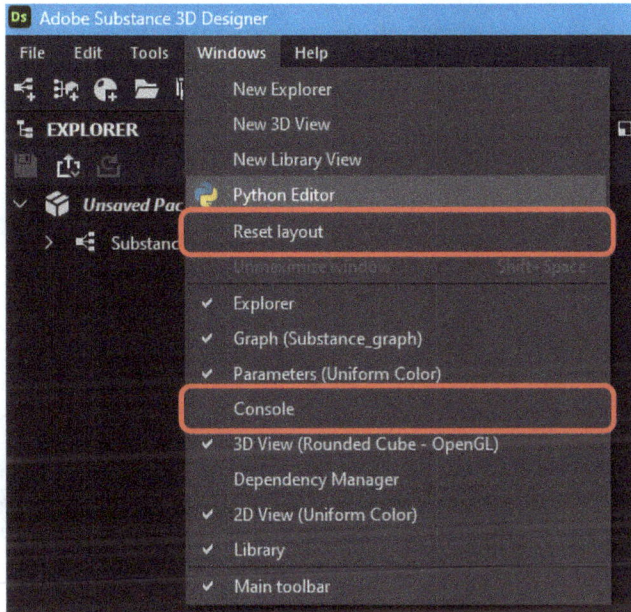

Figure 7.4 – Reseting the layout and reloading any closed panel

You can also pin any window inside Substance Designer; for example, if you want to switch between a 2D normal map view and a 2D rough map view without double-clicking the node each time, you can choose any one of these nodes and pin it, as shown in *Figure 7.5*.

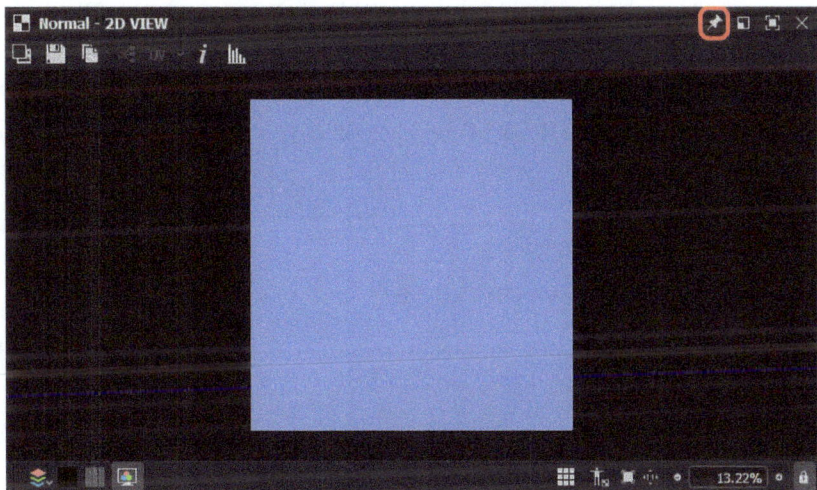

Figure 7.5 – Pinning one of the nodes

Then, you can double-click on another node and pin it as well. This will create tabs for each node, and now you can switch between these tabs very easily, as shown in *Figure 7.6*. You can pin as many nodes as you want. You can also pin graph views the same way if you have multiple graphs.

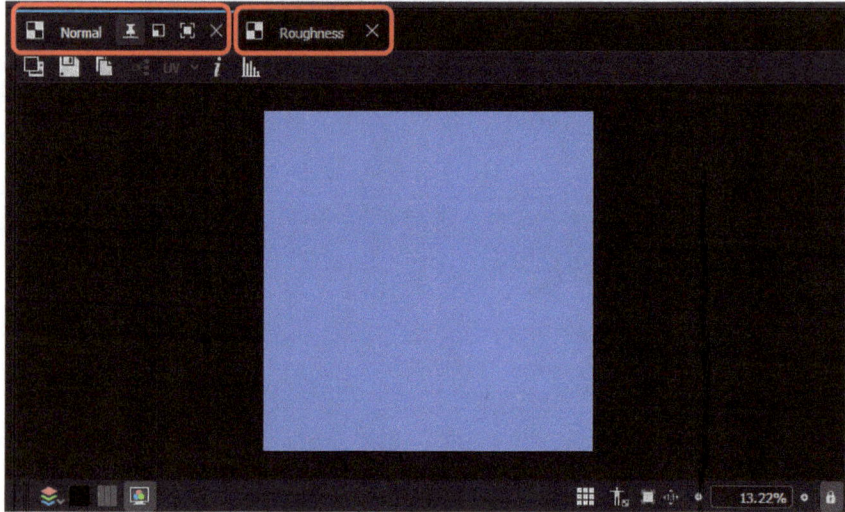

Figure 7.6 – Pinning the nodes in the 2D view and switching between created tabs

I hope these short introductions to each panel are quite helpful to you all; let us dig deep into each panel and learn about them in detail in our next sections.

The Adobe Substance 3D Designer EXPLORER window

You manage your open files and resources in Substance 3D Designer via the **EXPLORER** window or panel. It presents you with a list of all open packages, each of which is extended to reveal its resources as a hierarchy.

You begin and end projects in **EXPLORER** because it enables you to create, store, and export any type of resource.

Let's look at some significant operations you may perform using the **EXPLORER** window.

Creating and closing a new package

Packages are like folders that hold various resources such as graphs, bitmaps, 3D meshes, and so on. Let us see how we can create and close a package:

1. To create a new package, go to the main **File** menu and hover over the **New** option; you will see a bunch of graph options (which we will cover in the next part). From these options, choose **Empty package**, as shown in *Figure 7.7*. You can also create new packages with a right-click in the **EXPLORER** window.

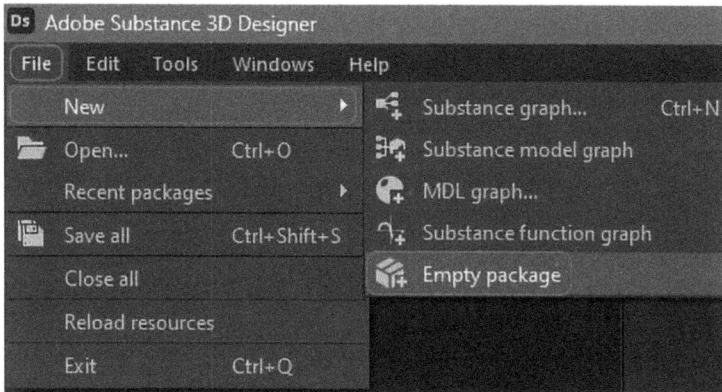

Figure 7.7 – Creating a new package

2. If you want to close the package, you just need to right-click on **Unsaved Package** and choose **Close package(s)**, as shown in *Figure 7.8*.

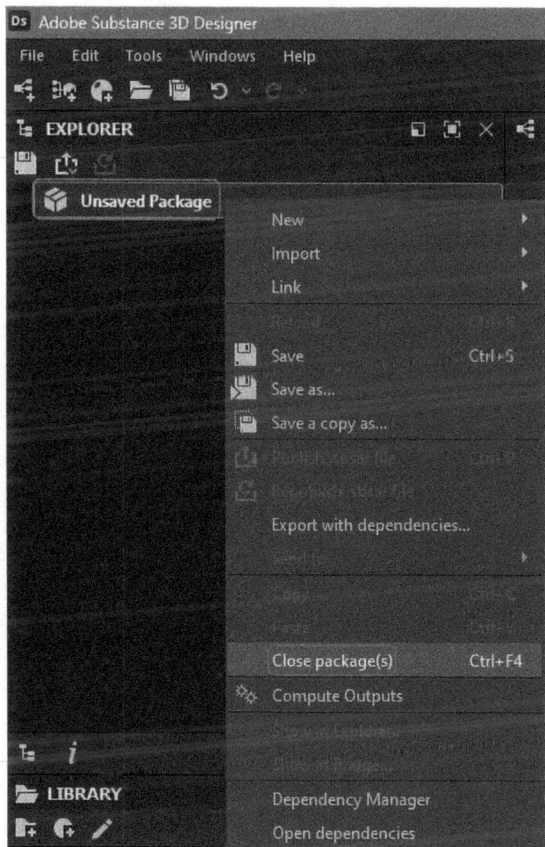

Figure 7.8 – Closing an unsaved package

You can create multiple packages in one Substance Designer file, and each package can hold different resources. Designers usually create multiple packages to avoid working on separate files, which is quite tedious as they have to keep opening files from different folders, saving them, closing them separately, and maintaining them.

As you create more and more resources and add them to your packages, your Substance Designer file size will grow and grow.

Creating and closing Substance graphs

The primary resource type utilized in Designer is the **Substance graph**, often known as a **compositing graph**. You may create, process, and export 2D picture data to one or more texture outputs using a Substance graph. A project will center on one or more Substance graphs in practically every common use case.

Other than Substance graphs, there are **Substance model graphs** to create primitive 3D models, **MDL graphs** to create the **Material Definition Language** for gaming engines, and **Substance function graphs** for when more intricate operations or the fine-tuning of certain behaviors are desired.

However, we will be focusing only on the Substance graph, and you will become familiar with the rest of the graphs after some practice.

To create a new Substance graph, you can execute the following steps:

1.   You don't need to create an empty package to create a Substance graph; you can simply go to the **File** menu, hover over **New**, and choose **Substance graph…**.

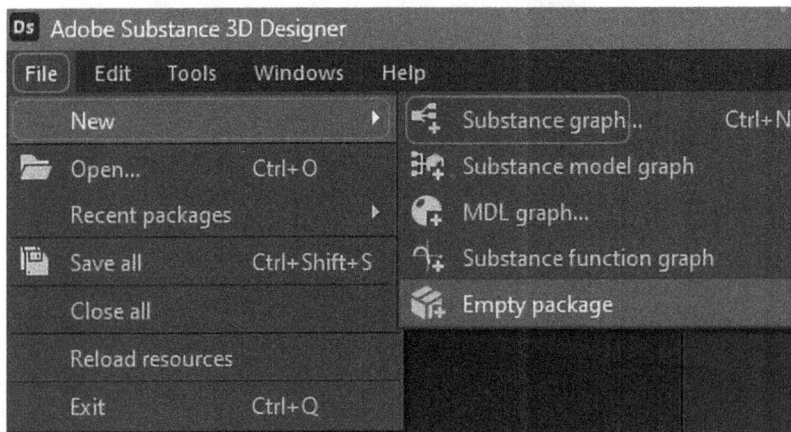

Figure 7.9 – Creating a new Substance graph

2.   After choosing a new Substance graph, you will get a **New Substance Graph** window, and in the **TEMPLATES** section, you have to select your desired template.

3. For this project, we will go for **Metallic Roughness**. In the **PROPERTIES** section, you can rename your graph, select the screen resolution in **Parent size**, and if you are choosing **Output format** as **Relative to parent**, then the Substance graph will take the base parent graph settings as its resolution.

4. Then, choose the package in which you want to create the graph. In our case, we have already closed our empty package earlier; therefore, you will notice it says **New package** under the **Create graph in package** option. Once everything is done as shown in *Figure 7.10*, click **OK**.

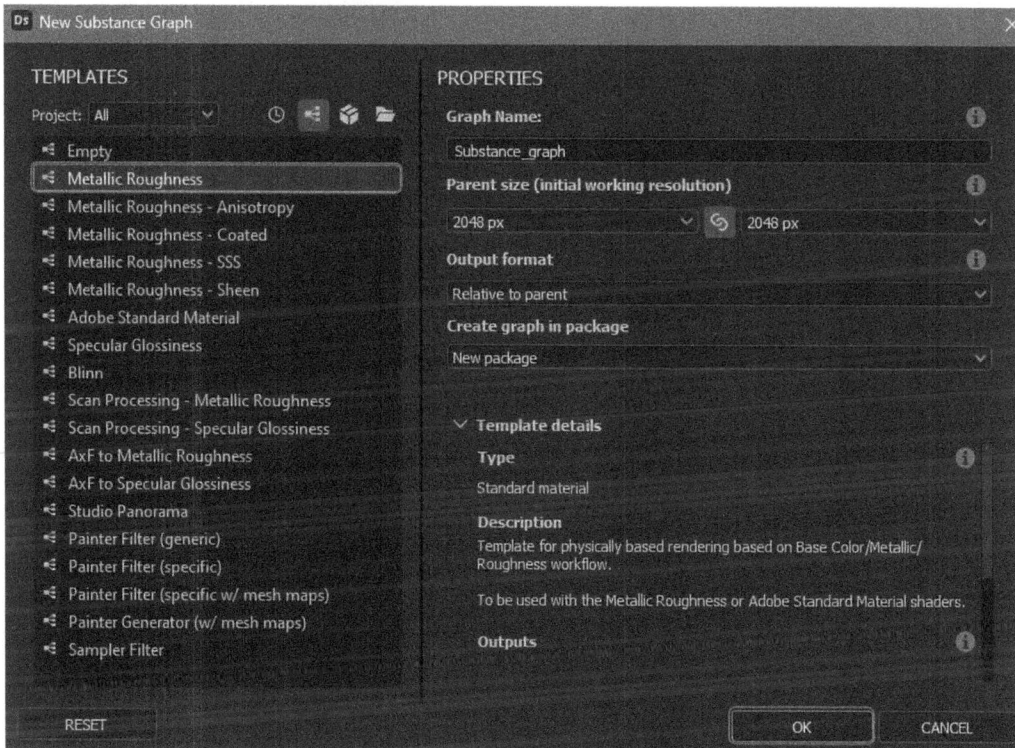

Figure 7.10 – New Substance Graph window

5. Once you press **OK**, you will get **Substance_graph** under **Unsaved Package***. You can also rename **Substance_graph** by right-clicking and choosing the **Rename** option or pressing *F2* (Windows and Mac) on the keyboard, and you can also remove it by right-clicking or pressing *Delete* (Windows and Mac) on the keyboard.

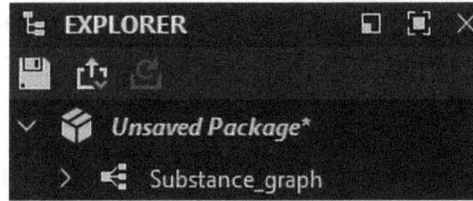

Figure 7.11 – Newly created Substance graph

Now that you are familiar with Substance graph creation, let us learn how to import and link resources in the next section.

Importing and linking resources

To bring any resources into Substance Designer, you can either import or link them. When you import any resources, they are copied into Adobe Substance Designer, but **Link** does not copy the resources into Adobe Substance Designer; instead, it just creates a reference of the file from the original location. However, you can only import **Bitmap** and **Vector graphics**, whereas, using **Link**, you can link **3D Mesh** and **Font** as well.

Let's import and link some resources and observe their differences:

1. Right-click on **Unsaved Package***, go to the **Import** option, and select **Bitmap**.

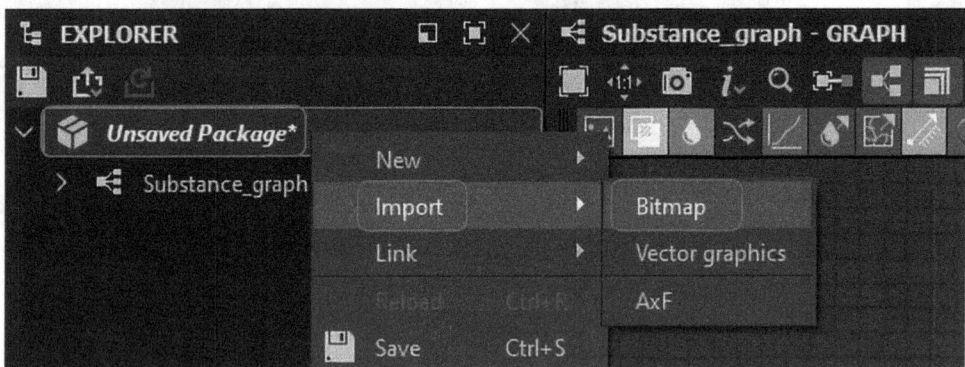

Figure 7.12 – Importing Bitmap

2. Then, go to the `Substance_Designer_Exercise_Files` folder that you downloaded earlier, select `logo.png`, and click **Open**.

3. Once you click **Open**, `logo.png` will be imported inside Substance Designer and will be shown under a newly created folder called **Resources** in the **EXPLORER** window; drag `logo.png` from the **EXPLORER** window to the **GRAPH** window, as shown in *Figure 7.13*.

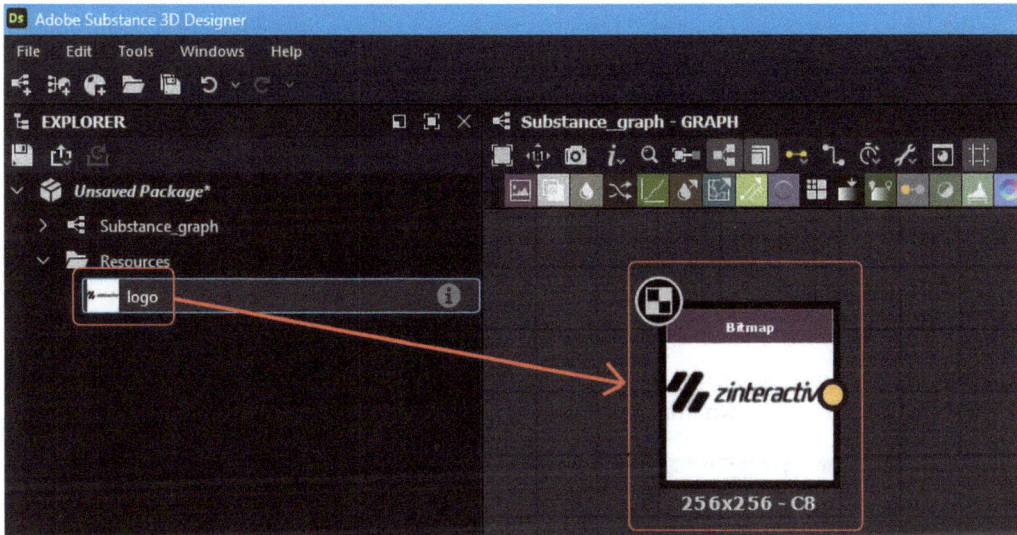

Figure 7.13 – Dragging logo.png to the GRAPH window

4. Now, again, right-click on **Unsaved Package***, go to the **Link** option, and select **Bitmap**. Repeat *Step 2*; this will create `logo_1` under the **Resources** folder. Drag `logo_1` to the **GRAPH** window as well. You can keep `logo_1` under `logo` in the **GRAPH** window for better observation, as shown in *Figure 7.14*.

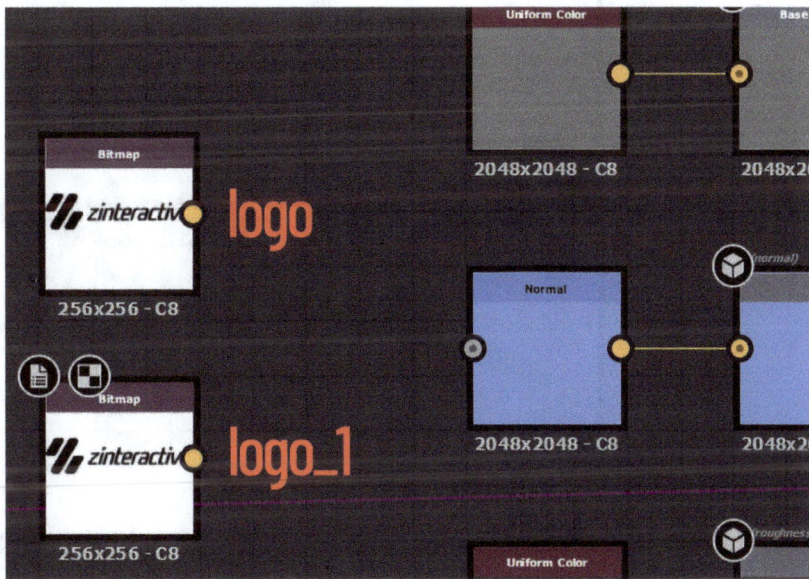

Figure 7.14 – logo and logo_1 in the GRAPH window

5.   The `logo` resource is an imported resource, while `logo_1` is a linked resource, as shown in *Figure 7.15*.

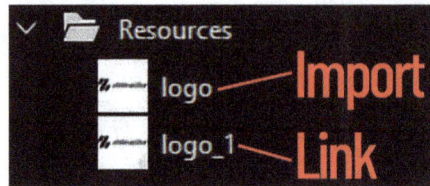

Figure 7.15 – Import versus Link

6.   Now, open the original `logo.png` image in Photoshop, make some changes to it, and save it with the changes, as shown in *Figure 7.16*.

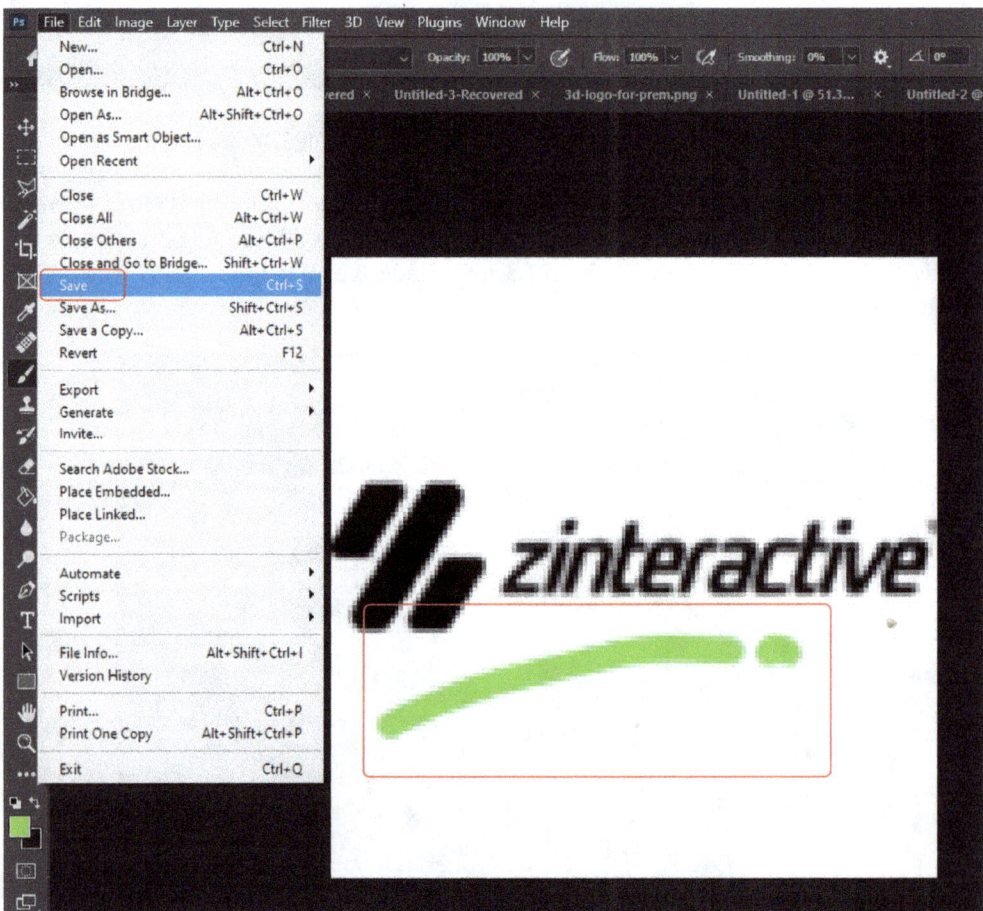

Figure 7.16 – Changing the original logo.png

7. Now go back to Substance Designer; you will notice that the imported `logo` resource is not changed because it is copied inside Substance Designer and not referring to the original file, whereas `logo_1` is changed because it was linked and is referring to the original file instead of being directly copied inside Substance Designer.

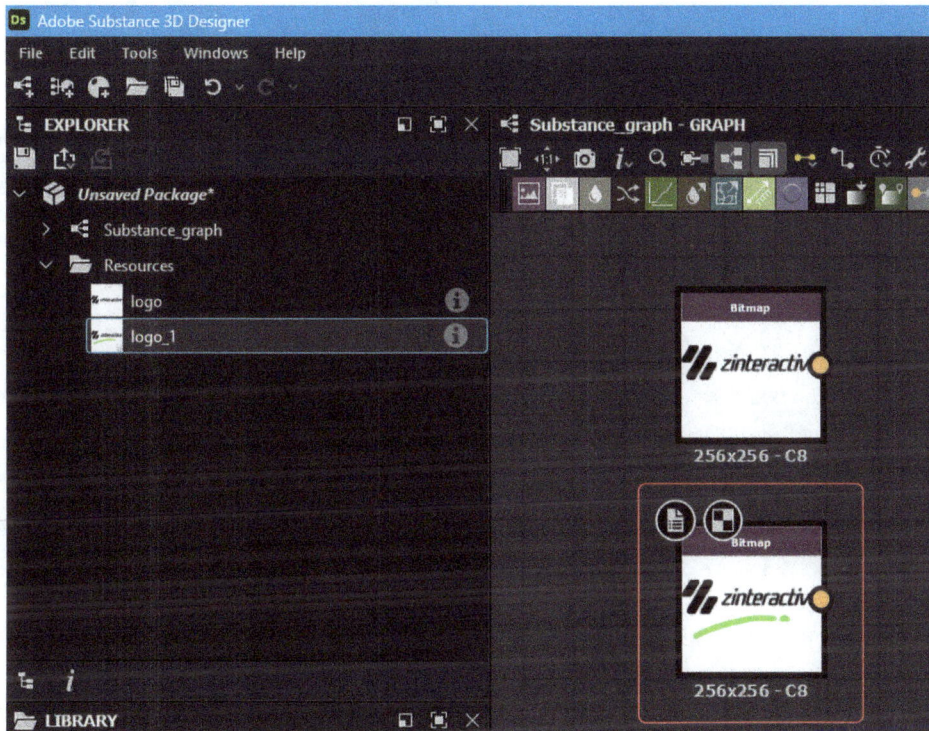

Figure 7.17 – The linked resource changes if the original file changes outside Substance Designer

Deleting the resources

If you have used any resources in the **GRAPH** window and you want to delete them from the **EXPLORER** window, then you must delete them first wherever you have used them, otherwise, you will get an error; therefore, we will execute the following steps to avoid any errors:

1. Let us first delete `logo` directly from the **EXPLORER** window to see what sort of error we get if we don't delete the resource from the **GRAPH** window.

2. Just go to the **EXPLORER** window and press *Delete* (Windows and Mac) on your keyboard, or right-click and select **Remove**.

   Once you get the **Confirm item removal** window, just press **Yes**; as `logo` has been removed from the **EXPLORER** window, the **GRAPH** window cannot locate it; hence, it will give an error, as shown in *Figure 7.18*.

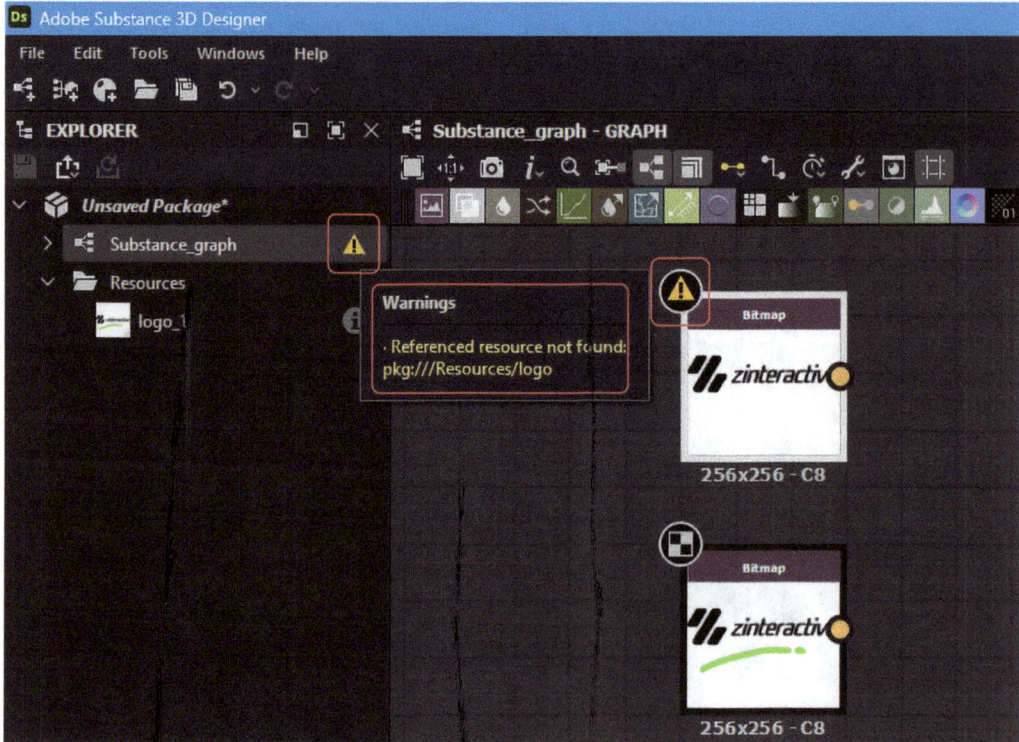

Figure 7.18 – Error after removing the resource from the EXPLORER window directly

3. You can now undo the removal task. To do that, go to the **Edit** menu and select **Undo**, or press *Ctrl + Z* (Windows) or *Command + Z* (Mac).

4. To remove logo correctly, go to the **GRAPH** window, select logo and right-click on it, then choose **Delete Selection** or press *Delete* (Windows and Mac) on your keyboard. Then, go to the **EXPLORER** window and delete logo as we did in *step 1*.

    This way, you will not have any errors because you first removed the usage and then the source.

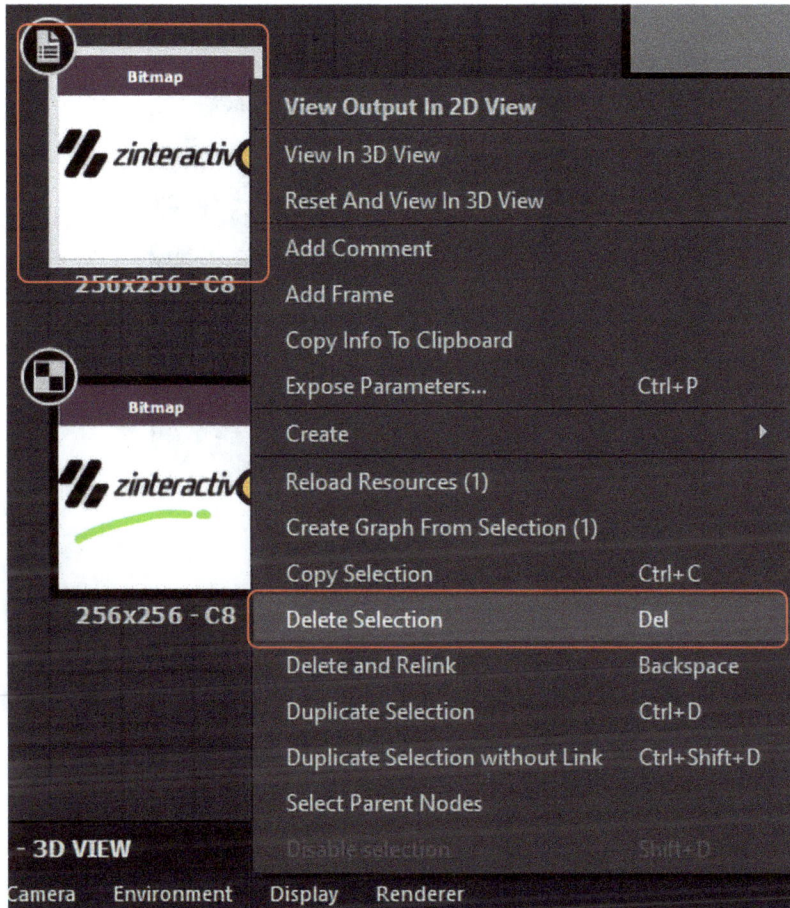

Figure 7.19 – Deleting the resource from the GRAPH window

Saving packages

Saving any package is quite a simple process:

1. You will notice that the package is right now showing as **Unsaved Package*** in the **EXPLORER** window (the * sign shows that the package has not been saved right now).

Figure 7.20 – Unsaved package with asterisk sign

2. Right-click on **Unsaved Package*** and click on **Save as…**.

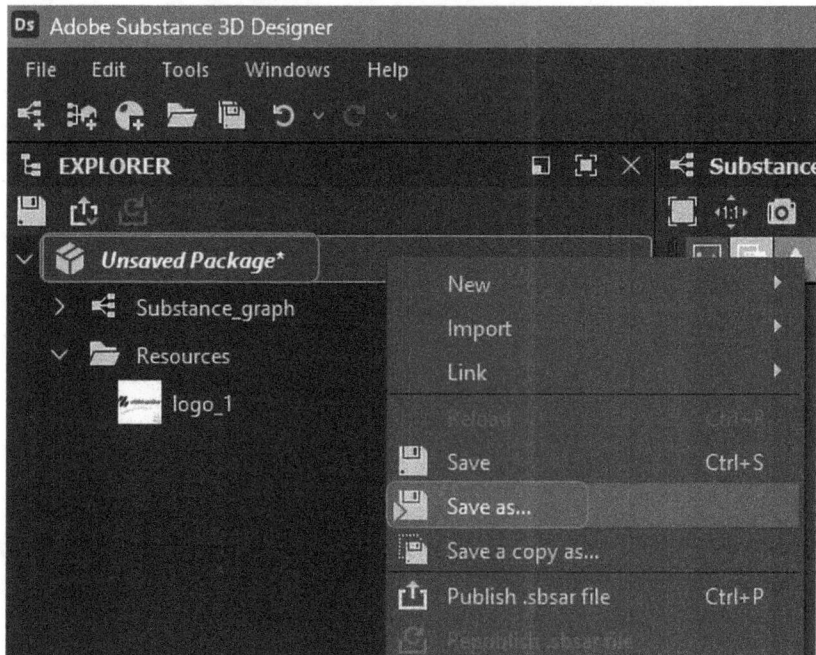

Figure 7.21 – Saving the package

3. Save the package as Test-1. There will be only one **Save as type** option, which is **Substance 3D Designer files (*.sbs)**.

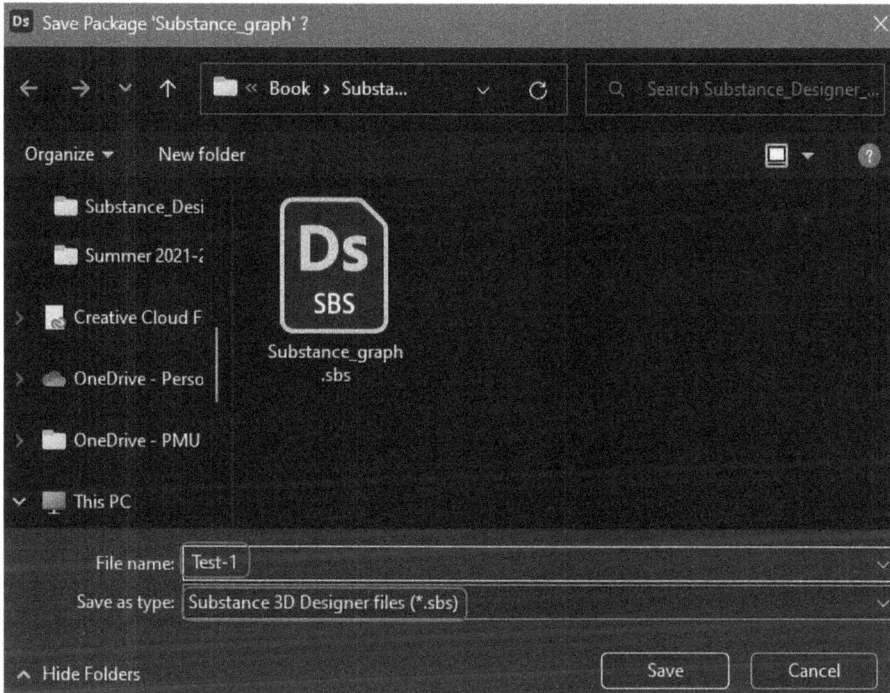

Figure 7.22 – Saving the package as Test-1

4.   As you can see in *Figure 7.23*, the package is no longer showing as **Unsaved Package***; it is now showing as **Test-1.sbs**. Moreover, the package file is always saved as * . sbs, which is the Substance 3D Designer file, also known as the main source file.

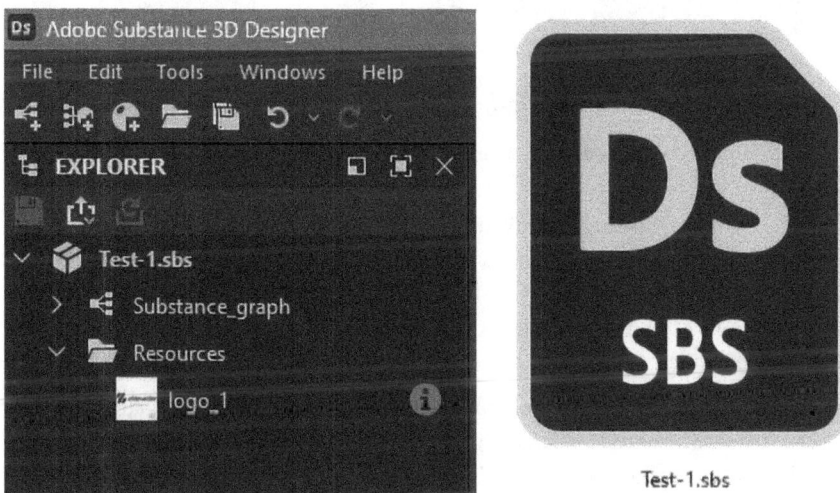

Figure 7.23 – After saving the package as a Substance 3D Designer file

I hope you have learned how to save your packages in Substance Designer. In the next section, we will learn how we can export nodes as separate images in Substance Designer.

Exporting nodes as separate images

When you work on any Substance graph, you can simply export all the output nodes inside that graph as separate image files, and later, you can use them as you want. To export nodes as separate image files, execute the following steps:

1. Go to **Substance_graph** in the **EXPLORER** window, right-click on it, and choose **Export outputs as bitmaps**.

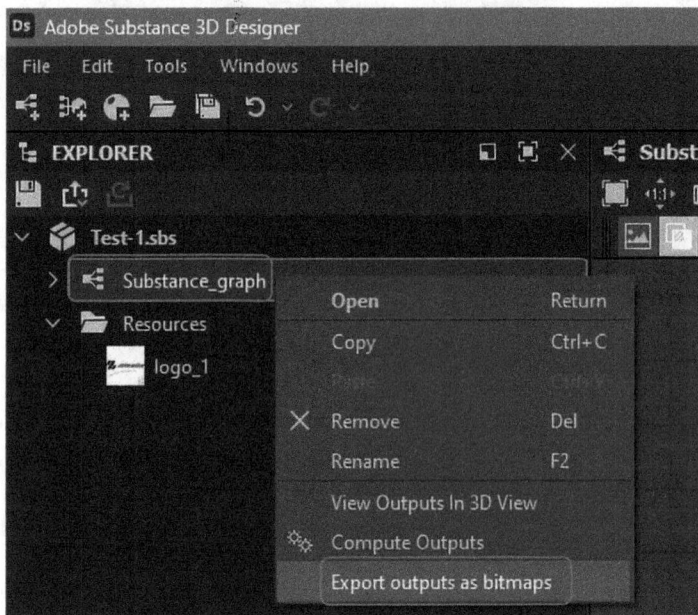

Figure 7.24 – Exporting outputs as bitmaps

2. When you click **Export outputs as bitmaps**, the **Export outputs** window will pop up. As we are exporting the graph nodes, we need to keep the **From graph** option selected, choose an output folder from **Destination**, set up file type from **Format**, choose your desired naming convention from **Pattern**, and choose the desired outputs. Moreover, you can also choose **Exported Color Space**.

3. If you want Substance Designer to export images every time there are changes in the nodes, then check **Automatic export when outputs change**. When everything is set, click on **Export outputs**. The **Batch** export option can be used to export 8K textures, especially if you are using a powerful computer.

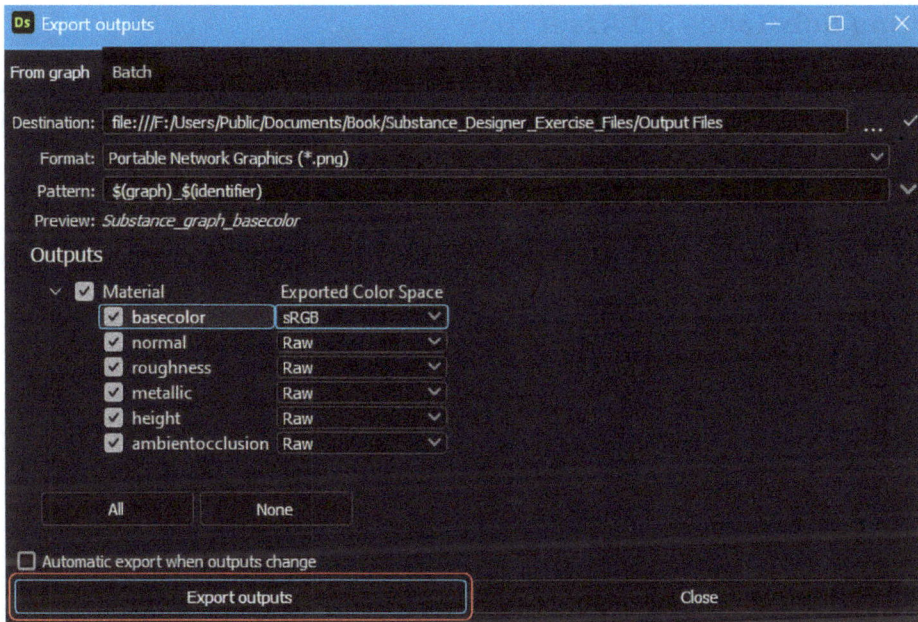

Figure 7.25 – Export outputs

4. Once your export is done, click on the **Close** button. Now you can check the output folder where you have exported your files, as shown in *Figure 7.26*.

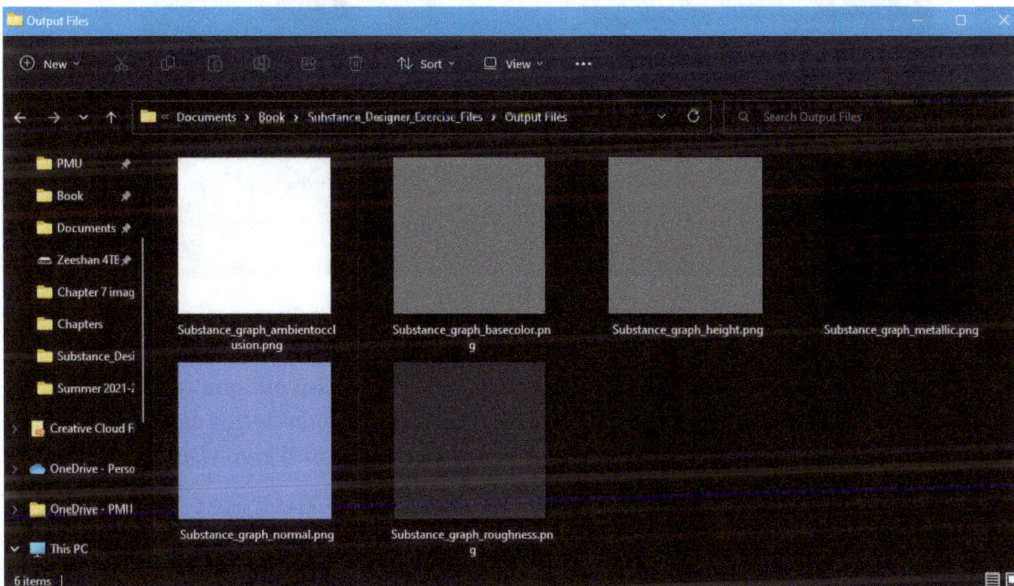

Figure 7.26 – Output files

Exporting Substance 3D Asset files

Adobe Substance 3D Designer can export *.SBSAR files, also known as **Substance 3D Asset files**. The *.SBSAR file is supported by various third-party 3D applications. Therefore, it holds quite high importance.

Let us execute the following steps to export the *.SBSAR file:

1. Go to **Test-1.sbs*** and right-click on it; once the menu appears, choose **Publish .sbsar file**.

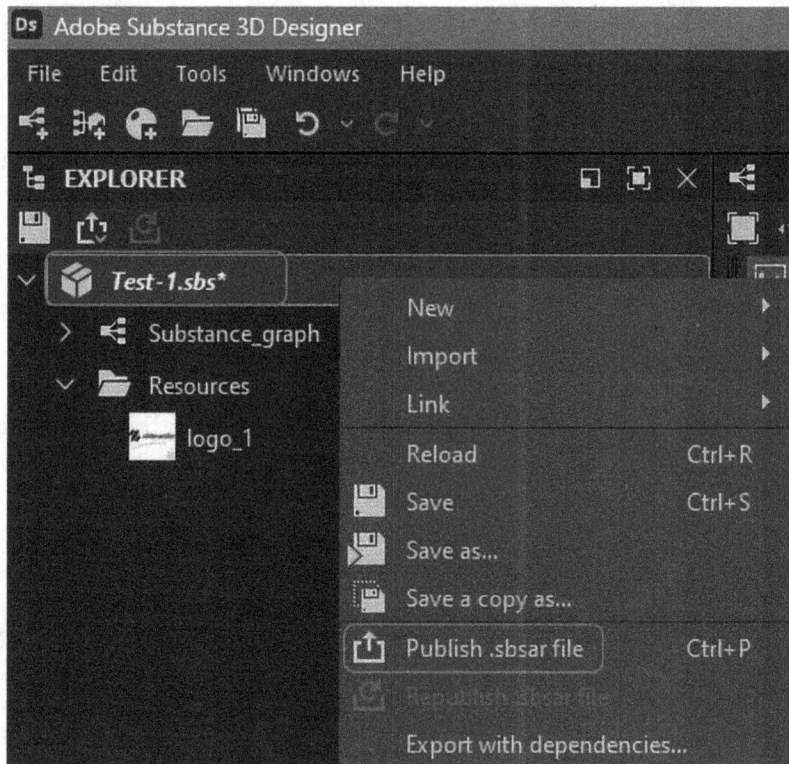

Figure 7.27 – Publish .sbsar file

2. After selecting **Publish .sbsar file**, the **Substance 3D asset publish options** window will pop up. You can choose the output folder from **File path**, keep the compression as **Auto**, and you can generate thumbnails if any icons are missing from the **Exposed graphs** option. Click on **Publish** once all the settings are done.

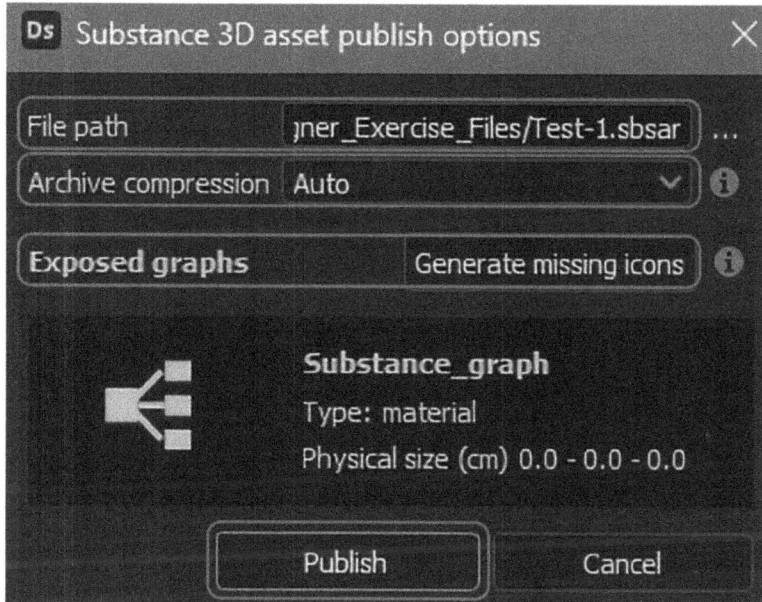

Figure 7.28 – Substance 3D asset publish options window

3. You can open the folder where you exported the *.SBSAR Substance 3D Asset file; the icon will look similar to *Figure 7.29*.

Test-1.sbsar

Figure 7.29 – SBSAR file icon

4. If you want to export *.SBSAR with the same settings, you can select **Republish .sbsar file**, as shown in *Figure 7.30*.

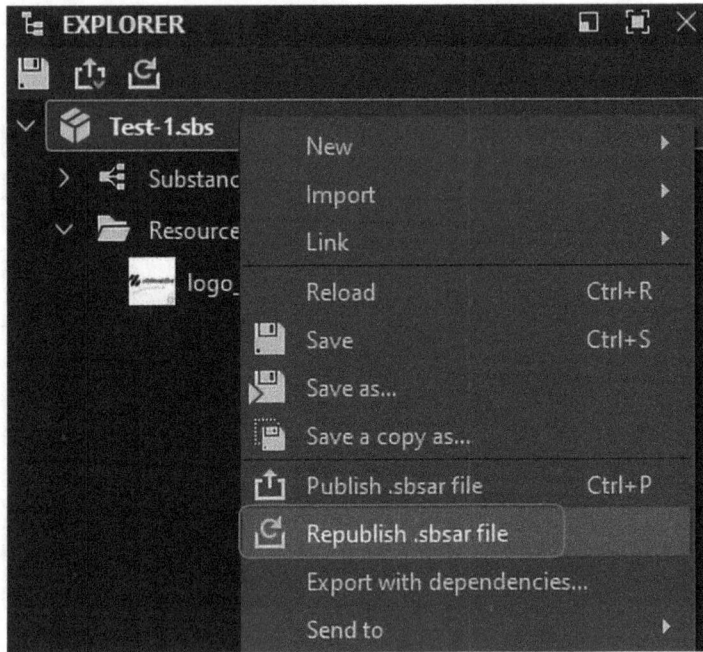

Figure 7.30 – Republish .sbsar file

5. Once all the files are saved and you want to save them again together, go to the **File** menu and choose **Save all**.

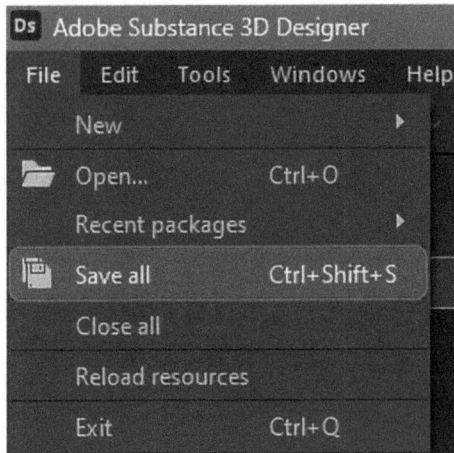

Figure 7.31 – The Save all option

I hope the **EXPLORER** window is all clear to you; in the next section, we will cover the Substance Designer **GRAPH** window.

Adobe Substance 3D Designer GRAPH window

Substance 3D Designer's main window, the **GRAPH** view, is where you create and update your graphs. The toolbar at the top, which offers easy access to some features, and the actual graph area, where nodes are inserted, are the two primary components of the **GRAPH** view.

You can use `Test-1.sbs` to practice on the **GRAPH** window.

Figure 7.32 – Toolbar at the top

The **GRAPH** view is the same for all graph kinds, with the toolbar area of the graph views for **Substance graphs**, **Models**, **Functions**, **MDL Materials**, and **FX-Maps** being the only difference. Let us dissect the different functionalities of the **GRAPH** window in the following subsections.

Atomic nodes

The most fundamental and lowest-level building pieces in Substance 3D Designer are called **atomic nodes**. These nodes can be used to build everything else. Any network would only have these atomic nodes if you reduced it to its simplest components. Moreover, these nodes are hard coded, and you can't open their references.

Figure 7.33 – List of atomic nodes

Placing atomic nodes

There are five possible placements for atomic nodes. They feature icons that remain constant throughout all methods and are constantly color-coded:

- **A node bar**: The atomic nodes are accessible with ease and are color-coded on the node bar, which is located directly above the graph display, as shown in *Figure 7.32*.

- **The Tab key/Spacebar menu**: In the **GRAPH** view, using the *spacebar* or *Tab* key brings up a list that, by default, shows the atomic nodes. It may also be used to look up library nodes in the library. The easiest and most flexible way to place nodes is often through this menu.

- **The Add Node menu**: A similar menu to the *spacebar/Tab* key approach is displayed by right-clicking on the graph and selecting **Add Node**. However, you cannot choose library nodes from this menu.

- **Library**: All atomic nodes are located in a portion of the library that is equally accessible to all other library nodes.

- **Custom node hotkeys**: Although there aren't any hotkeys set up for nodes by default, you may create your own using the **Preferences** pane.

Substance Designer will attempt to automatically link any nodes that are inserted when another node is chosen. The new node is always positioned behind the old one in the flow via this automated connection.

Depending on how you want a lost link to be handled, there are two approaches to removing nodes:

- Pressing *Delete* or selecting **Delete Selection** from the context menu after selecting a node. All connections that are currently in place are severed, perhaps resulting in functional issues, as shown in *Figure 7.34*.

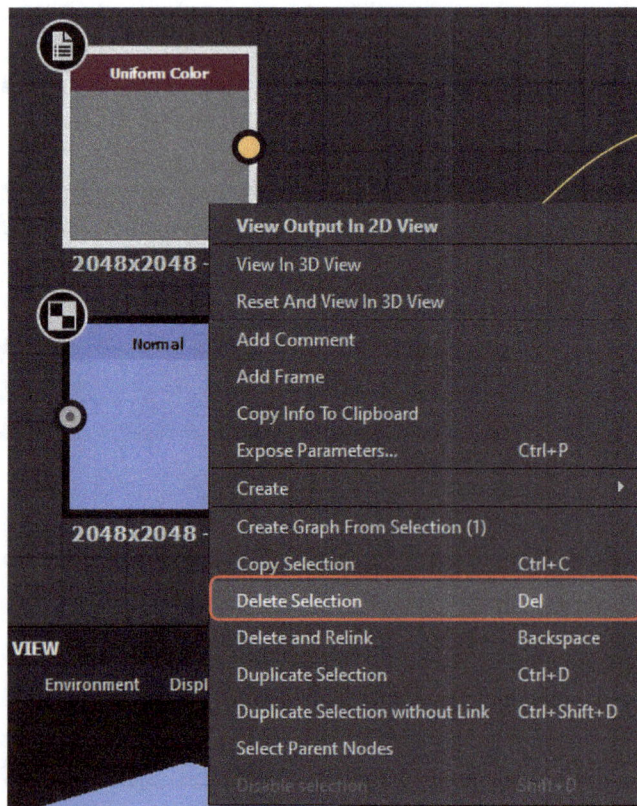

Figure 7.34 – Delete Selection

- Pressing the *Backspace* key after selecting a node or right-clicking the mouse and selecting **Delete** and **Relink** from the menu. When feasible, linkages are kept in order to avoid broken functionality.

Links

In Designer, a link can be manually dragged between two connections or produced when a node is placed, as shown in the following screenshot:

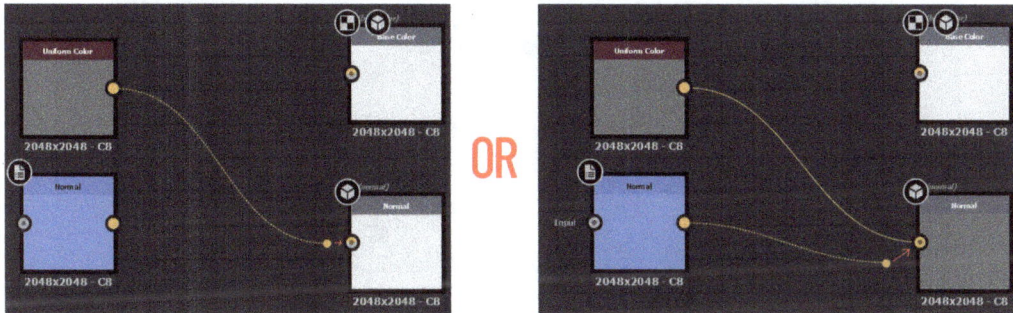

Figure 7.35 – Manually dragging the node links

Selecting the link and hitting *Delete* or pressing *Alt* and clicking on any connection with links will erase them. Delete all links on that connection by using the *Alt* key, as shown in *Figure 7.36*.

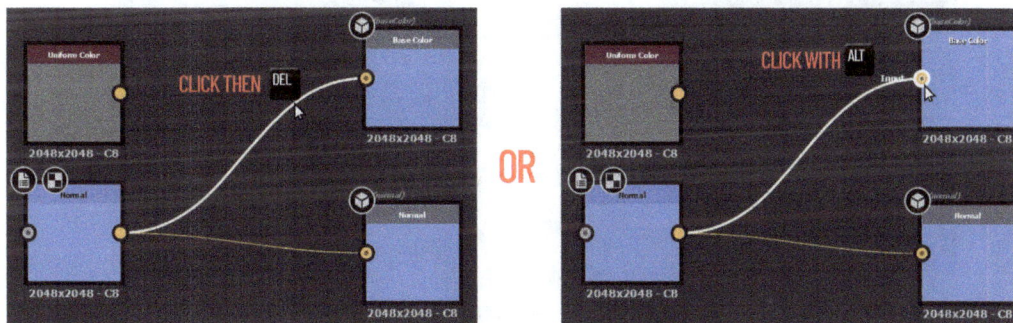

Figure 7.36 – Deleting the node links

By holding *Ctrl* and dragging from a connection, links can be replicated.

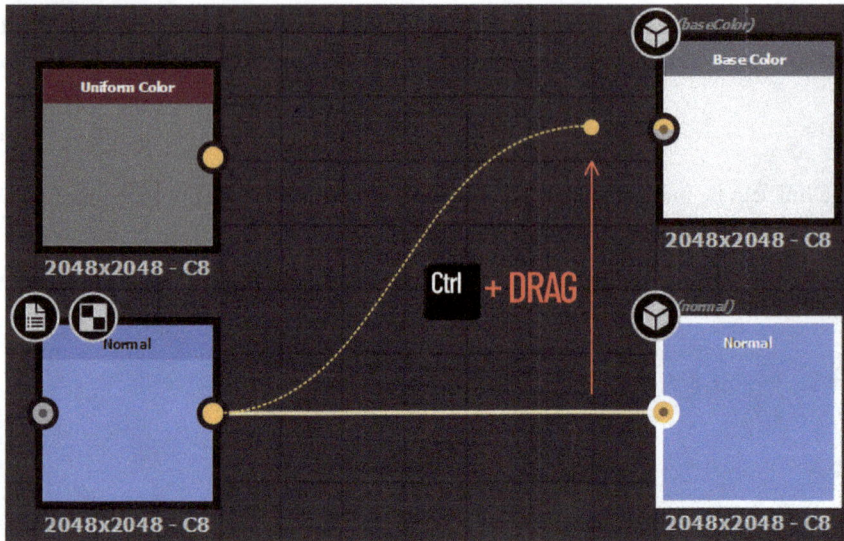

Figure 7.37 – Duplicating the node links

Press *Shift* and drag links between connections, allowing you to pick them up and move them.

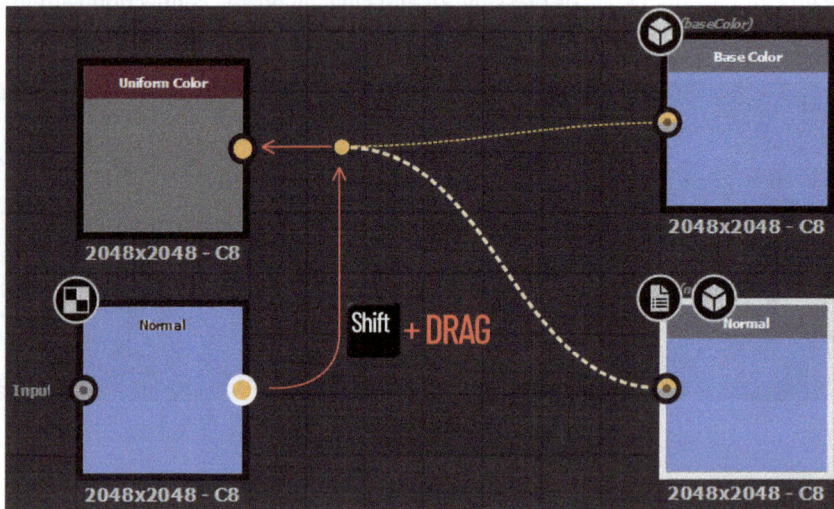

Figure 7.38 – Shifting existing node links to a different node

Disabling nodes

Nodes do not have to be disconnected or removed; instead, they can be deactivated so they have no impact on the graph.

You can disable any node by selecting it, then right-clicking on it and choosing **Disable selection**, or pressing *Shift + D* on your keyboard, as shown in *Figure 7.39*.

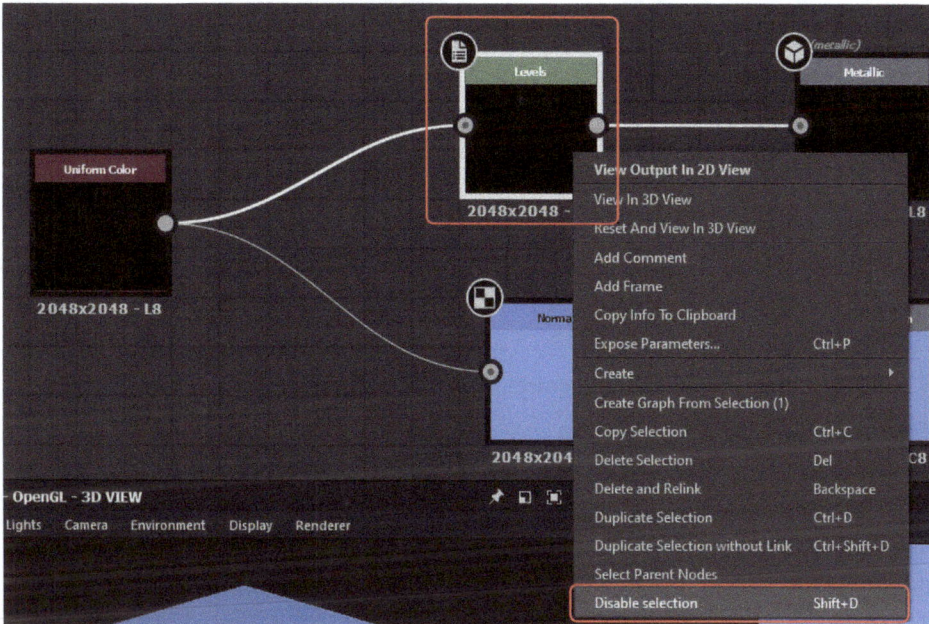

Figure 7.39 – Disabling node by right-clicking

You can also disable multiple nodes by dragging your mouse across them with the left mouse button pressed and then right-clicking on them and choosing **Disable selection** or pressing *Shift + D* on your keyboard.

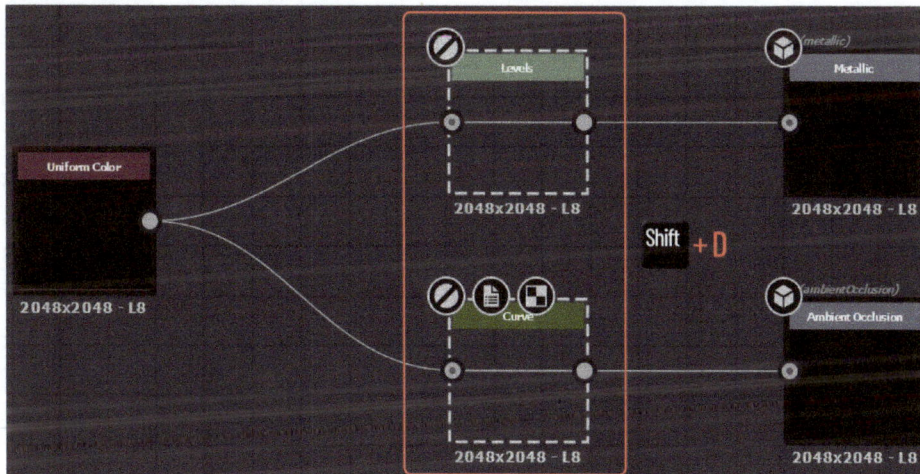

Figure 7.40 – Disabling multiple nodes

Disabled nodes act in the following manner:

- Instead of a thumbnail, they are shown with the disabled badge (⊘), a dashed outline, and an internal rerouting link, as shown in *Figure 7.40*

- The data received in the nodes' primary input will be output

- Disabled nodes can be connected in a chain

- Their relationships and properties are left alone

- They retain and save their disabled state across sessions

- What you see is what you get when publishing to SBSAR since the resultant file considers the nodes' disabled state

---

Only nodes that meet the following requirements may be disabled::

- At least one input exists in the node.

- The node only has a single output.

- It is necessary for the main input and output to be of the same kind, such as grayscale to grayscale, or color to color.

- The same rule for enabling them applies to all of the chosen nodes, which means they must all be in the same condition of being enabled.

Manipulating 2D and 3D views

The 2D and 3D views are meant to be in constant communication with the graph view and its nodes.

Let us see how we can work with the 2D and 3D views in the following steps:

1. Double-click on a node or right-click and choose **View Output in 2D View** to see something in the 2D view.

2. Right-click a node, select **View in 3D View**, then select a channel on the active shader to view something in 3D.

3. Alternatively, you may right-click a blank area of the graph and select **View Outputs in 3D View** provided your outputs are configured properly. You can see the outputs for both 2D and 3D views in *Figure 7.41*.

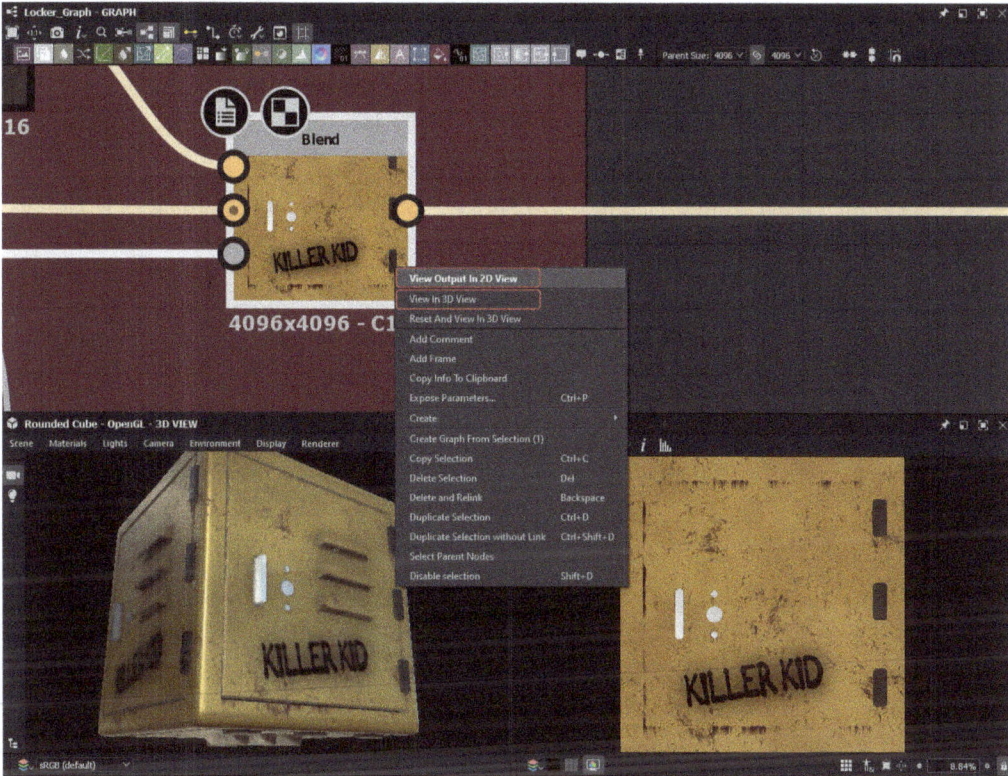

Figure 7.41 – Viewing outputs in 2D and 3D views

Graph view toolbar

The graph view toolbar is divided into different sections, so let us check them out one by one.

Main toolbar

Every type of graph has a primary toolbar that offers general operations and toggles to make the other toolbars visible.

Figure 7.42 – Main toolbar

You can find the following features:

1. **Focus** (hotkey *F*): If the selection is empty, this command will focus on the entire scene.

2. **Reset zoom** (hotkey *Z*): Centers the display in the center of the graph and returns the current zoom level to its default setting, referring to zooming in or out.

3. **Export graph view**: Creates a picture with a 1-to-1 resolution of the whole graph view. Useful for sharing a snapshot of your graph.

4. **Settings for node info**:

   - **Display Connector name**: Toggles the name display of each individual connector on a node.

   - **Display node Size**: Toggles node resolution display.

   - **Display node Badges**: Toggles node badges on all nodes.

   - **Display Timings**: Toggles display of millisecond timings for each node.

   - **Limit text scaling when zooming out**: Maintains the text of graph elements at a consistent screen size after a certain zoom level, preserving the text's legibility while zooming out.

   - **Comments:** This allows you to write small descriptions in the **GRAPH** window.

   - **Frames titles:** This allows you to create a frame around the collection of nodes. You can also give a title to the created frame.

   - **Pins**: This pins the windows and panels, which will turn them into tabs. With the help of these tabs, you can easily switch between them.

5. **Node Finder tool**: Reveals additional toolbar sections that may be used to search for and highlight nodes in your graph.

6. **Highlight Flow**: Any related nodes placed before or after the presently selected node will be highlighted. Suitable for tracking a convoluted network of nodes.

7. **Node toolbar**: Enables or disables the node toolbar.

8. **Parent toolbar**: Displays the **Parent Resolution control** settings as a toggle.

9. **Link Creation mode: Standard** (hotkey *1*), **Material** (hotkey *2*), and **Compact Material** (hotkey *3*) are the available link generation modes.

10. **Rectangle links**: Toggle links between nodes that are round or rectangular in form.

11. **Reset (Timing Control)**: This allows you to reset all nodes and timings.

12. **Tools**: This option contains a set of tools that allow you to perform the following tasks:

   - **Clean**: All nodes that aren't (directly) linked to output are removed.

   - **Export Outputs**: The **Bitmap Export** interface is opened.

   - **Reexport Outputs**: Repeats the last export operation.

   - **PSD Exporter**: Opens the interface for the PSD Exporter.

Node toolbar

The node toolbar shown in *Figure 7.43* contains the most common nodes that are used frequently in Substance Designer.

Figure 7.43 – Node toolbar

Parent toolbar

Only Substance compositing graphs have access to the parent toolbar, which affects the output size of the graph if it utilizes the relative to parent inheritance technique by setting the output size of the graph's parent.

For textures that are not square, the horizontal and vertical sizes can be separated from one another. Default values (256*256) can also be restored.

Figure 7.44 – Parent toolbar

Thumbs toolbar

For Substance compositing graphs, the thumbs toolbar (shown in *Figure 7.45*) is a straightforward toolbar that manages the creation of thumbnails for each node. **All**, **Displayed**, and **None** are the choices. **None** disables all thumbnail computation, **Displayed** just calculates them for the current node in the 2D view, and **All** always calculates all thumbnails. Both **Displayed** and **None** can result in notable performance improvements. In the new version, it is just a box to activate or deactivate, which enables the node image cache.

Figure 7.45 – Thumbs toolbar

Align tools

The align tools are a collection of straightforward aids for organizing and aligning your nodes. All forms of graphs can use them.

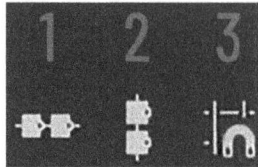

Figure 7.46 – Align tools

The options present in this toolbar are as follows:

1. **Align Horizontally**: Horizontally aligns all of the chosen nodes
2. **Align Vertically**: Vertically aligns all of the chosen nodes
3. **Snap nodes to Grid**: Snaps to the closest grid place all chosen nodes

I hope that the **GRAPH** window is now clear to you. In the next section, we will learn more about the **PROPERTIES** panel.

The Substance Designer PROPERTIES panel

The most technical window is the **PROPERTIES** window. The presentation of sliders, dropdowns, and other components that alter the behavior of a chosen resource or node is always context-sensitive.

The Substance 3D Designer **PROPERTIES** panel's layout contains many rollouts, categories, and parameters. It concentrates on Substance Graph characteristics. The layouts of function graphs, MDL graphs, and FX maps are simpler.

PROPERTIES panel overview

The context of the **PROPERTIES** panel varies depending on the options you choose in the **EXPLORER** window and the **GRAPH** view.

Together with the **GRAPH** view, it is likely the second-most utilized UI panel in Designer since it allows you to modify the characteristics of selected nodes and resources.

Depending on your choice, the **PROPERTIES** panel is divided into a few distinct rollouts (shown in *Figure 7.47*), such as the following:

- **BASE PARAMETERS**
- **SPECIFIC PARAMETERS** (these are node-specific parameters)

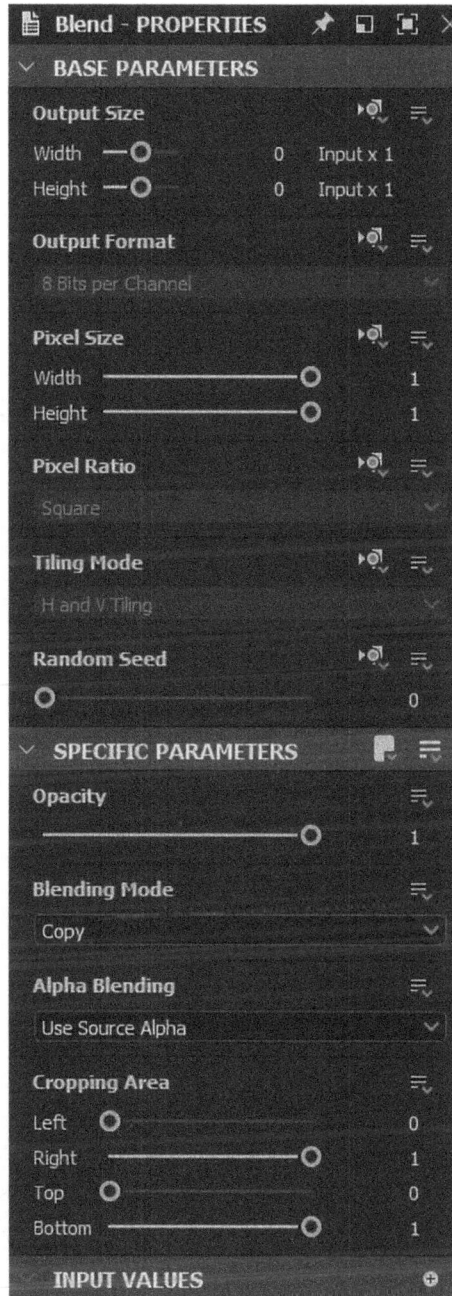

Figure 7.47 – PROPERTIES panel

I hope you have understood the significance of the **PROPERTIES** panel. It's time to move to 2D and 3D views; these views are output views that show you your results.

Substance Designer 2D and 3D views

There are two views inside Adobe Substance 3D Designer, **2D VIEW** and **3D VIEW**. Both views are used to observe and test your graphs, so let us examine them in detail.

2D VIEW

The simplest previewing tool is **2D VIEW**. It strongly integrates with the graph: double-clicking on any node in the **GRAPH** view will show the visual result in **2D VIEW**.

One of the primary panels of Designer's user interface is **2D VIEW**. Its primary goals are as follows:

- Displaying information produced by a certain node or passing through a designated node connection
- Resources for bitmap and vector graphics display
- Displaying further details about the data it presently contains, including color channels or precise color values
- Adjusting GIZMO settings

The 2D view automatically refreshes to reflect changes to a shown picture or value in order to keep up with the most recent condition of the data.

At any given moment, many **2D VIEW** panels can be open and each one can show a distinct set of graphics or values. By using the user interface panel's **Pin** function, you can decide when a new panel should be displayed.

2D VIEW main toolbar

You can interact with your displayed photos in numerous ways thanks to the following capabilities available on the **2D VIEW** panel's main toolbar:

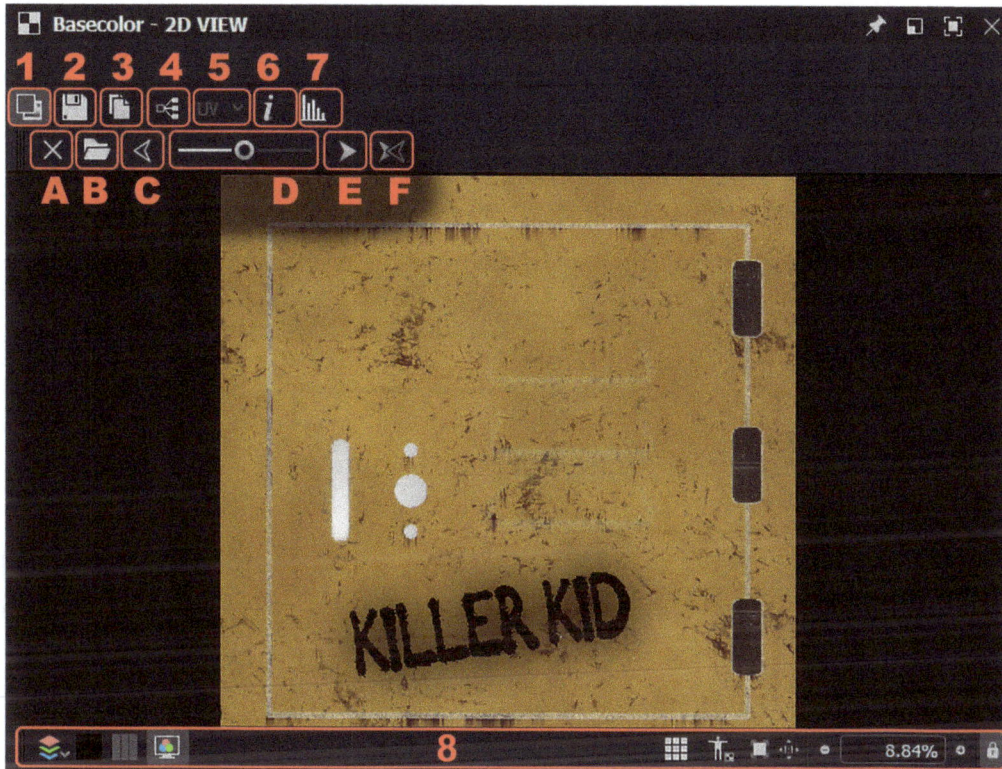

Figure 7.48 – 2D VIEW main toolbar

4. **Background image:** You can place another picture over the one that is now visible. When you click the **Background image** button, a window will appear asking you to choose an image file to serve as an overlay.

A new toolbar with the following controls for the image overlay displays when the file has been chosen:

A. **Close:** Disable the backdrop image overlay and close the overlay controls toolbar

B. **Load Image:** To utilize another picture file as an overlay, choose it

C. **Source Image:** The opacity of the overlay image is set to **0%**

D. **Slider:** The overlay image's opacity may be manually adjusted using a slider

E. **Background Image:** The opacity of the overlay image is set to **100%**

F. **Reset:** The opacity of the overlay image is set to **50%**

5. **Save current image as bitmap:** You can export the presently visible image to an image file. When you click the **Save current image as bitmap** button, a menu asking for the location, name, and file format of the exported file will appear.

6. **Copy view into the clipboard**: You can copy the picture that is now shown to the clipboard. The picture is now ready to be pasted into any other program, such as Adobe Photoshop, when you click the **Copy view into the clipboard** button.

7. **Switch graph outputs**: The **Switch graph outputs** button may be used to swiftly switch to any other graph output if the picture that is presently being displayed is a graph output, as shown in *Figure 7.49*. Other nodes, especially those with many outputs, cannot use this capability.

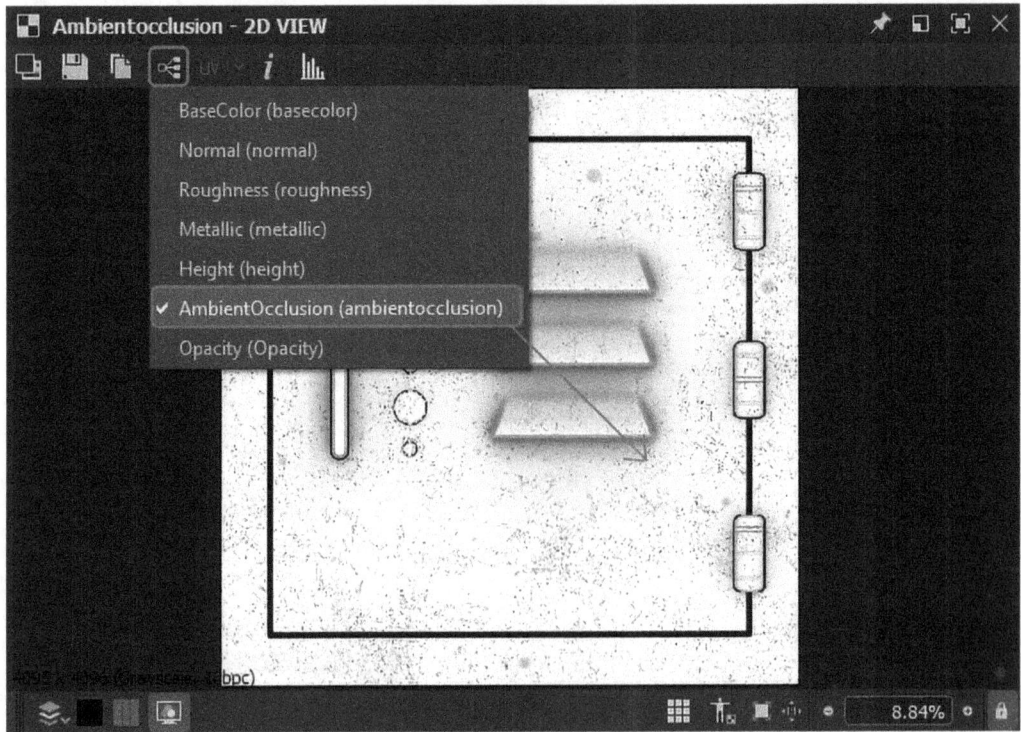

Figure 7.49 – Switch graph outputs

8. **UV overlay**: The UV overlay function is accessible in **2D VIEW** if the **Display UVs in 2D view** option (as shown in *Figure 7.50*) is selected in the **Scene** menu of the **3D VIEW** panel. Utilizing the **UV** button, you can turn it on.

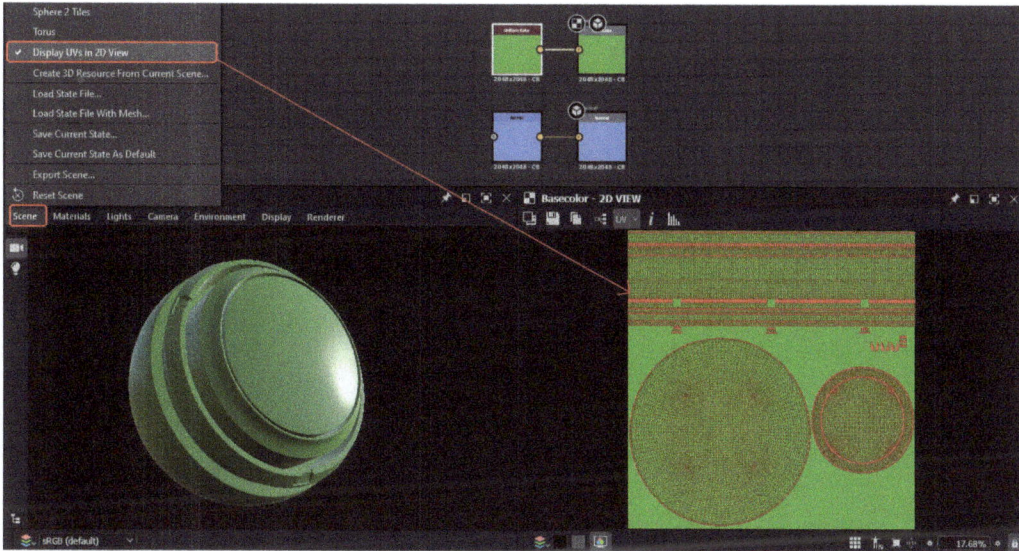

Figure 7.50 – UV overlay

This shows a colored wireframe of the UVs of the current mesh in the **2D VIEW** window. The material color is utilized as the color of the UV overlay if the mesh file has information on the material's color.

If the mesh contains numerous UV sets, the required UVs can be chosen from the drop-down checklist by clicking the arrow next to the UV label in the button, which opens the drop-down checklist.

9. **Image information**: The **Information** panel, which is activated via the **Image information** button, provides the precise pixel values and coordinates of an image. This is extremely useful for reviewing HDR photos, for example, or ensuring that stepping between pixels proceeds as expected.

10. **Display histogram**: The **Histogram** panel, which is activated via the **Display histogram** button, allows you to view the image's histogram. The **Histogram** panel is a floating panel that may be resized, moved, and undocked from the **2D VIEW** panel.

11. The **Display** toolbar: You may adjust how the picture is shown in the viewport using the **Display** toolbar, which, by default, is found at the bottom of the **2D VIEW** panel. The settings for color and transparency are located in the leftmost portion, while the controls for the viewport are located in the rightmost section.

3D VIEW

The most dynamic and sophisticated preview window is the **3D VIEW** window. In contrast to **2D VIEW**, it renders a whole material using a variety of output maps. All channels, including **Base Color**, **Normal**, and **Roughness**, are so visible. The **3D VIEW** tools are shown in *Figure 7.51* and dissected in the following points:

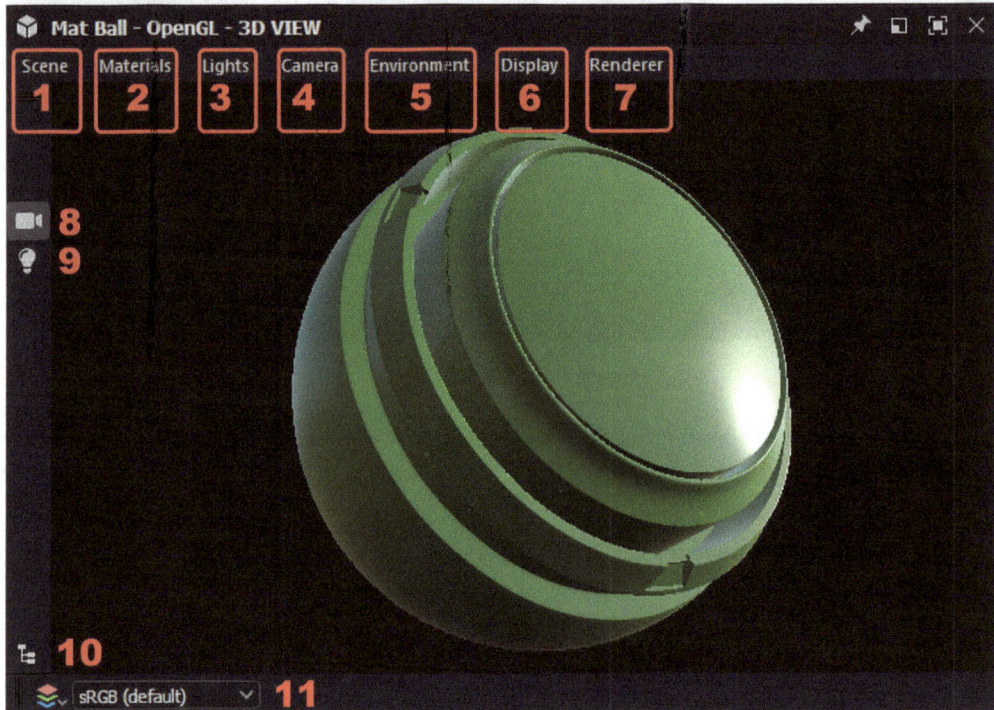

Figure 7.51 – 3D VIEW

1. **Scene**: The geometry (3D resource) shown, and the 3D view states, are dealt with via the **Scene** menu. Only the mesh is a 3D resource; the scene states, which can include lighting, cameras, and associated settings, can also include the mesh.

2. **Materials**: Depending on the loaded 3D model and the renderer being utilized, the **Material** menu varies. In this menu, you can control different types of materials.

3. **Lights**: Only older, legacy, ambient, and point lights are covered in the **Lights** menu. These lights do not produce the same high-quality outcomes as HDR image-based rendering since they are not PBR-compliant.

4. **Camera**: You can adjust camera settings, go to preset angles, and load camera angles from a specific 3D mesh file using the **Camera** menu. You may also access the **Post Processing** options using it.

5. **Environment**: You may change the parameters for the HDRI environment that is used to light PBR-aware components in the **Environment** menu.

6. **Display**: You may toggle a variety of extra helpful features in the **Display** menu.

7. **Renderer**: You may change render settings and switch between renderers using the **Renderer** menu.

8. **Camera**: Activates a 3D camera for control.

9. **Light**: Activates lights for control.

10. **Scene browser**: In **3D VIEW**, a little additional window known as **Scene browser** may be viewed. It shows a hierarchy of every component in your scene. According to the material assignment, your custom 3D mesh will be divided into many pieces. If the mesh is entirely made of one material, it won't split.

    In a 3D application, the material assignment must be done outside; however, once the material has been separated, it can be reallocated.

11. **Color Channels**: With this option, you can switch different color channels in **3D VIEW**.

I hope you are now familiar with the 2D and 3D views and their tools. It's time to work with the **LIBRARY** panel in the next section.

The Substance Designer LIBRARY panel

You can locate and collect all the resources you need to work with in your graph in the **LIBRARY** panel, which is a split-view resource manager.

It keeps track of any networked or local folders that you add to the list of library monitored paths in the **Project** settings. Any additions, deletions, and content updates made in those folders are also made in the **LIBRARY** panel.

All of the default material is listed in the **LIBRARY** panel, as shown in *Figure 7.52*. Make sure to study *Chapter 8, Nodes in Adobe Substance 3D Designer*, to have a better understanding of the distinction between atomic nodes and instance nodes in the library.

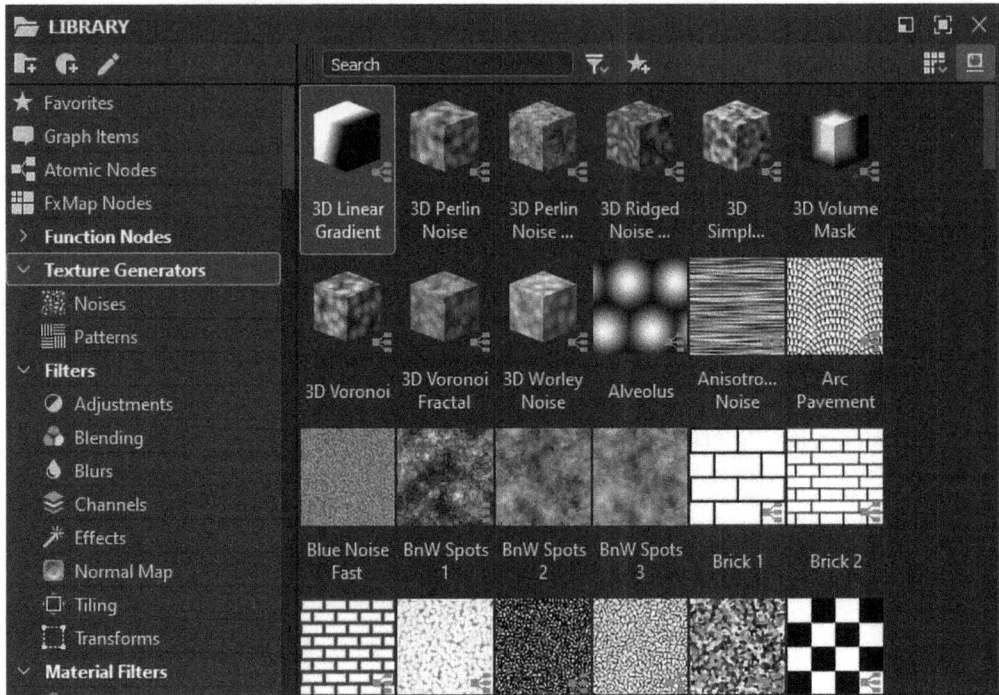

Figure 7.52 – Substance Designer LIBRARY panel

The library can keep track of all supported resources, such as graphs from **Substance packages (SBS)** and **Substance archives (SBSAR)**, bitmap images, vector images, function graphs, MDL graphs, AxF files, fonts, and meshes.

The panel is divided into two major sections:

- The left-hand side's **Categories** section
- The right-hand side's **Content** section

The Categories section

The **Categories** area, which is on the left of the **LIBRARY** panel, has a tree view of all the asset categories (also known as **folders**) and filters. Any item in this tree view may be clicked to reveal both its content and the contents of all of its descendant items.

The Content section

The **LIBRARY** panel's material is presented as labeled thumbnails. Depending on the Substance graphs, bitmaps, **vector graphics (SVG)**, 3D models, function graphs, MDL graphs, typefaces, and AxF, these thumbnails will have a distinct appearance.

Summary

I hope I have walked you through all of the various UI windows and views to get you started with the app. Also, the **EXPLORER** window demonstrates how to build and save. We also learned how to publish packages, and we explained how to create, handle, and export graphs and how to connect or import tools. Moreover, we covered meshes, where we focused on how to begin baking meshes and how to build, save, and publish sets.

Apart from some advanced features of Substance, we have gone through the basic features, where I explained how to create, maintain, and export graphs. Also, I explained basic tools and covered areas such as how to connect or import tools, and how to begin baking meshes.

Hopefully, you have also learned about the **PROPERTIES** window, having explored how to see various properties, some of the most important properties, and all of the different property types.

In addition, now we know how to use the 2D view while designing diagrams and what basic buttons and hotkeys do. We also covered the 3D view, learning how to navigate, how to set up meshes and materials, and how to find our way through the menus.

Last but not least, I hope we are now familiar with the **LIBRARY** panel, its various divisions, how to adjust the thumbnail view, and how to create your own directories and filters.

In the next chapter, we will go through nodes inside Adobe Substance 3D Designer; to create graphs, Substance 3D Designer offers a large number of nodes. Since there are so many nodes, it would be impossible to list them all on a single page.

In the following chapter, we'll go through some of the nodes that are most frequently utilized. In that chapter, we will get a chance to explore these nodes and learn why they are so crucial and what their purpose is.

8

Nodes in Adobe Substance
3D Designer

Nodes are graph inputs and outputs, which are the main tools for creating substances inside Adobe Substance 3D Designer, and we will examine these nodes in this chapter. Adobe Substance 3D Designer provides a huge selection of nodes for the creation of graphs. There are so many nodes that listing them all on a single page would be infeasible.

As a result, we'll review some of the nodes that are used the most. You will get the opportunity to discover more about these nodes in this chapter, as well as why they are so important and what purpose they serve.

We will cover the following topics in this chapter:

- Working with nodes and understanding the basics in Adobe Substance 3D Designer
- Tile **Generator** node
- **Flood Fill** node
- **Quad Transform** node
- **Height Blend** node
- **Curve** node
- **Dirt and Dust** node
- **Shape Mapper** node
- **Shape Splatter** node
- Histogram Scan

Working with nodes and understanding the basics in Adobe Substance 3D Designer

In the first section of this chapter, we will learn how the node system works inside Adobe Substance 3D Designer. As mentioned before, Adobe Substance 3D Designer contains a variety of nodes and the usage of these nodes differs between them, but you do not need to worry about that, as they are pretty much self-explanatory.

However, the first thing we need to know is how we can work with the nodes inside Adobe Substance 3D Design in a very simple way, so let's do that:

1. Once Adobe Substance 3D Designer is launched, a welcome window will pop up, and from there, you can choose **Substance graph** under the **Create** option.

2. As soon you do so, the **New Substance Graph** window will open, and from there, you can choose **Metallic Roughness** from the **TEMPLATES** list because this contains all the useful premade nodes, which you can edit later.

3. Then, change the **Graph Name** details to whatever you want. We will name this project Node Tests. Make sure **Output format** is set to **Relative to parent** so that everything you produce uses the same format as the parent:

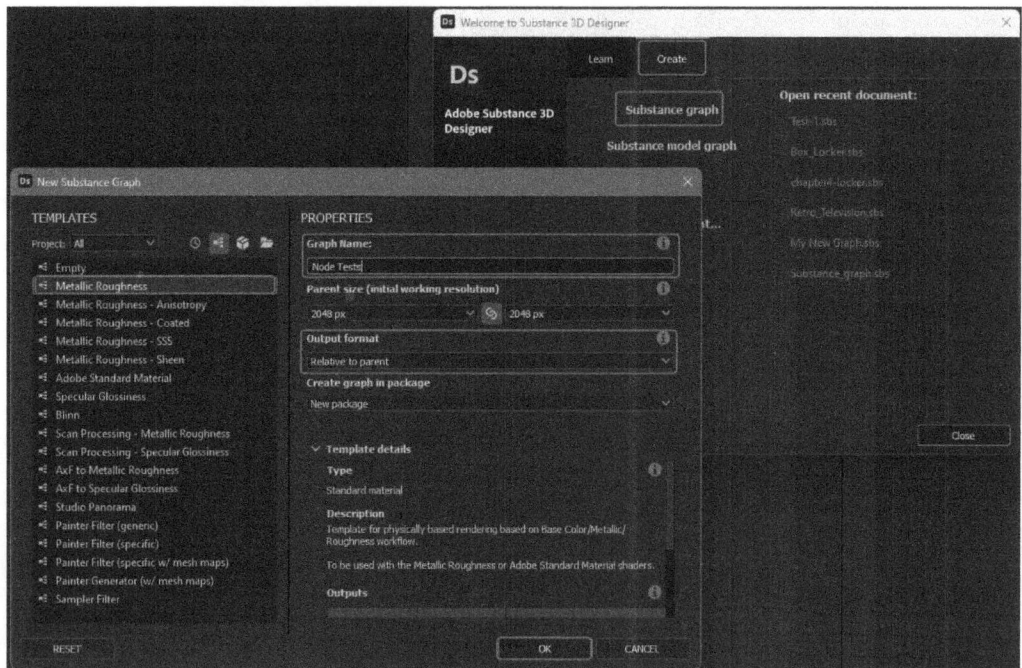

Figure 8.1 – Creating a new substance graph

4. You can also go to the **File** menu, select **New**, then choose **Substance graph**, and repeat *step 2*. The setup of all the basic output nodes will be inside the **GRAPH** window.

5. To test different nodes, let's go to **3D VIEW**, click on the **Scene** menu, and choose **Plain (hi-res)**. We will select this **Scene** mesh because it is simple and easier to observe the results produced by different nodes:

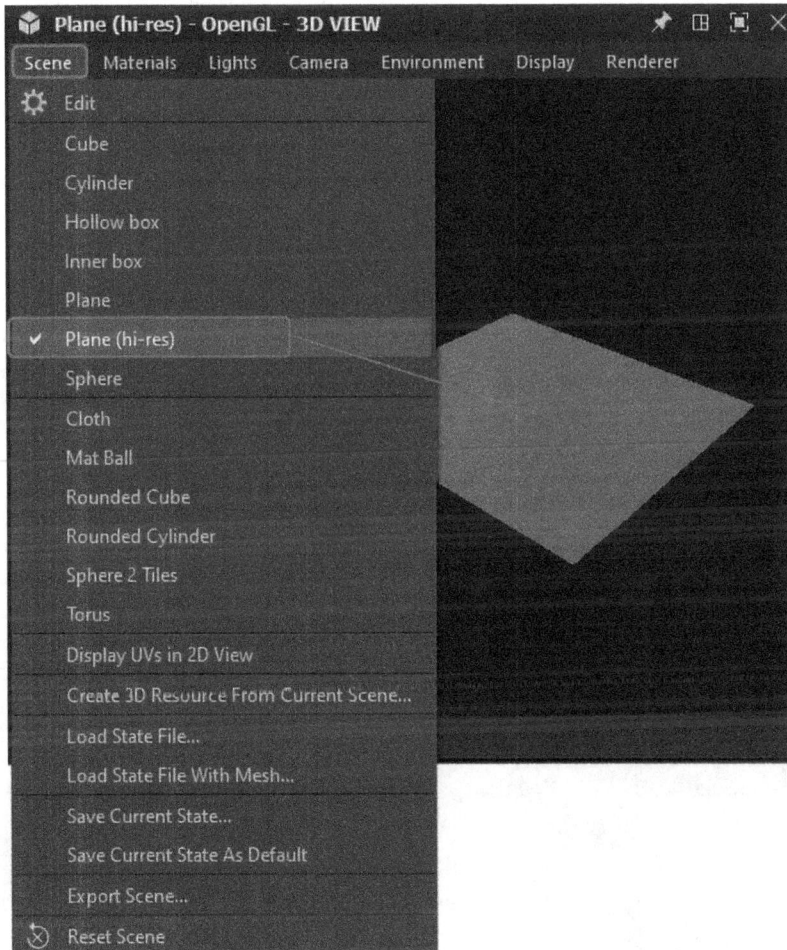

Figure 8.2 – Changing the 3D VIEW scene to a hi-res plane

6. In the **GRAPH** window, you will see six input nodes, which you can replace with your own inputs, and six output nodes, which are the **Metallic Roughness** output nodes:

    • **Base Color**: This node defines the base color of the mesh or substance, often taking **Uniform Color** or **Gradient** as its input.

- **Normal**: A grayscale input picture is interpreted as a height map and used by the **Normal** filter to produce a normal map. Normal maps also fake the height and produce a kind of height illusion, which is often used in games or films instead of **Height** or **Tessellation + Displacement**.

- **Roughness**: This node defines the roughness of the material surface, often taking **Uniform Color** or **Gradient** as its input. If the **Uniform Color** option is dark, the material will become shinier like glossy plastic, glass, or metal, and if the **Uniform Color** option is lighter, the material will become matte like paper.

- **Metallic**: This node defines how metallic the material is, often taking **Uniform Color** or **Gradient** as its input. If the **Uniform Color** option is darker (such as black), the material will not be reflective (like paper), and if the **Uniform Color** option is lighter (such as white), the material will be reflective (like chrome).

- **Height**: This node takes a grayscale input and creates an output based on the blacks and whites – a whiter color will give more height while a blacker color will produce negative height, and a grayish color will keep the height flat. However, to enable it, you need to make a few discreet settings adjustments.

- **Ambient Occlusion**: This node mimics the way light strikes an item. Although it affects all facets of the item, the shadow regions are where it is most obvious, often taking the same map used for the **Height** map as its input.

7. Now, drag **Checker 1** from **LIBRARY** to the **GRAPH** window, as shown in *Figure 8.3*:

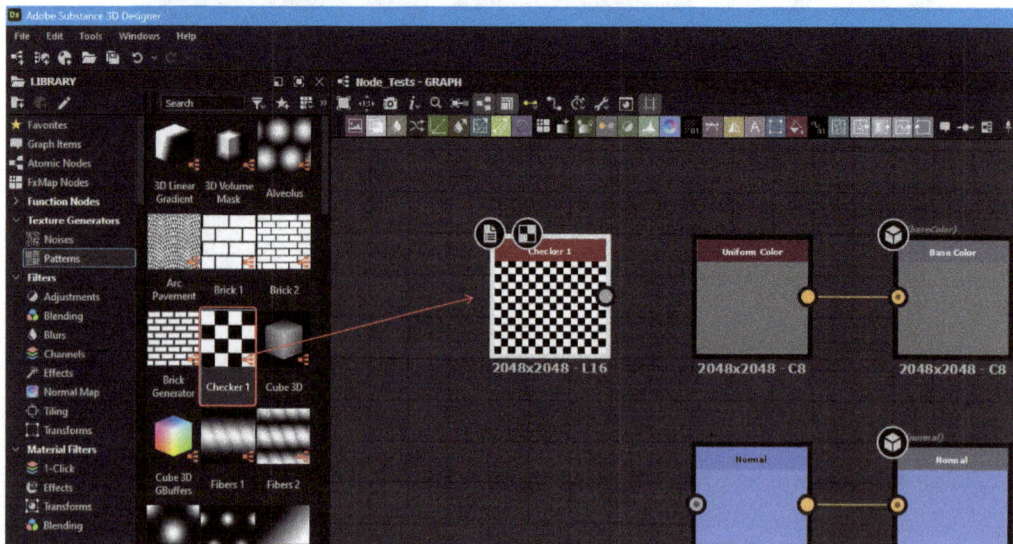

Figure 8.3 – Dragging the Checker 1 pattern to the GRAPH window

8. Now, delete the **Uniform Color** node from the **Ambient Occlusion** and **Height** map nodes so that we can use our own node:

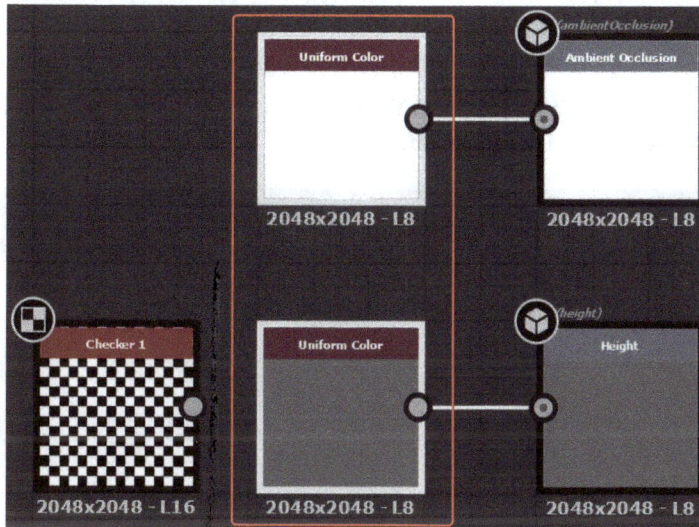

Figure 8.4 – Deleting the existing Uniform Color nodes

9. Now, connect the **Checker 1** node to **Height** so that it can produce the checker pattern height, and connect it to **Ambient Occlusion** too so that it can produce the shadow effects based on the checker pattern. Right-click anywhere in the **GRAPH** window and choose **View Outputs In 3D View**:

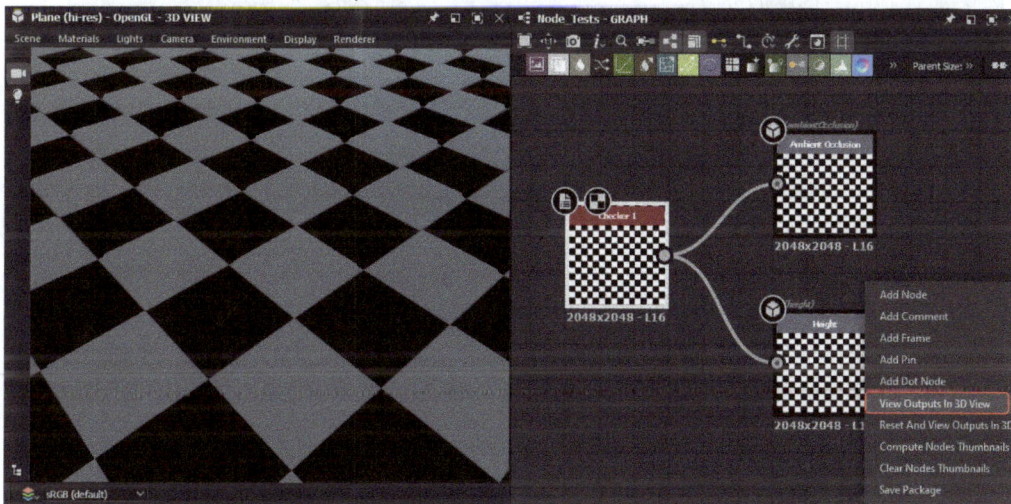

Figure 8.5 – Producing the Height map and Ambient Occlusion

10. Now, let us reveal the secret I spoke about in terms of the **Height** map node before. First of all, we can increase the height amount. However, to increase the height, the **Scene** mesh of our **Plane** needs a higher resolution, which is referred to as **Tesselation** in Adobe Substance 3D Designer. The height displacement here is specifically for helping visualize the material as you make it while it is inside of Adobe Substance 3D Designer.

11. Therefore, go to the **Materials** menu in **3D VIEW**, select **Default**, then select **Metallic Roughness**, and choose **Tesselation + Displacement**:

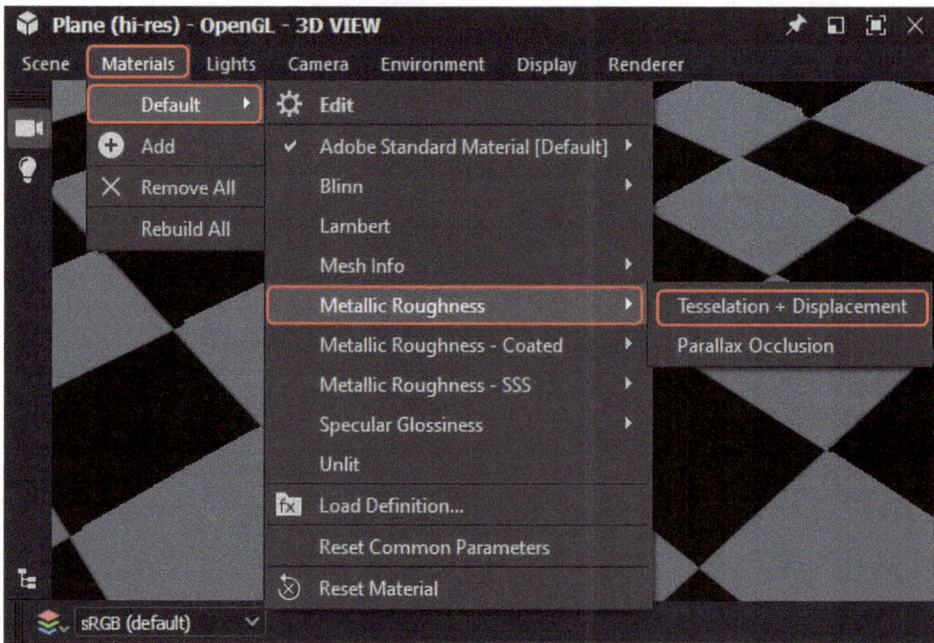

Figure 8.6 – Metallic Roughness material

12. Now that you have the **Height** parameters in the **Properties** window, you can increase the height of the **Height** map node by increasing the **Height Scale** value. Therefore, slide the **Scale** slider to the maximum, **10**.

At first, it seems like you cannot increase the value to more than 10 through the slider. Nevertheless, you can type in a value greater than 10 using your keyboard.

13. To increase the resolution of the height map, you can type a higher value into **Tessellation factor**, so type in `16`. Keep the **Scalar Zero/Height Level** value to `0.5`. This slider actually sets the gray value:

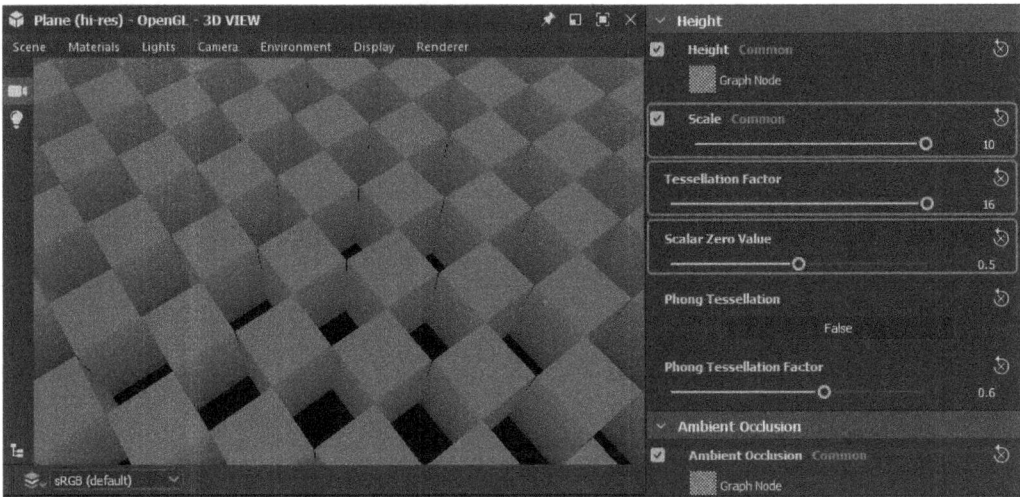

Figure 8.7 – Material "Default" | OpenGL | PROPERTIES

14. To create a **Normal** map height, move the **Normal** map node next to the **Checker 1** pattern node, and connect the **Checker 1** pattern node to the **Normal** map node, as shown in *Figure 8.8*:

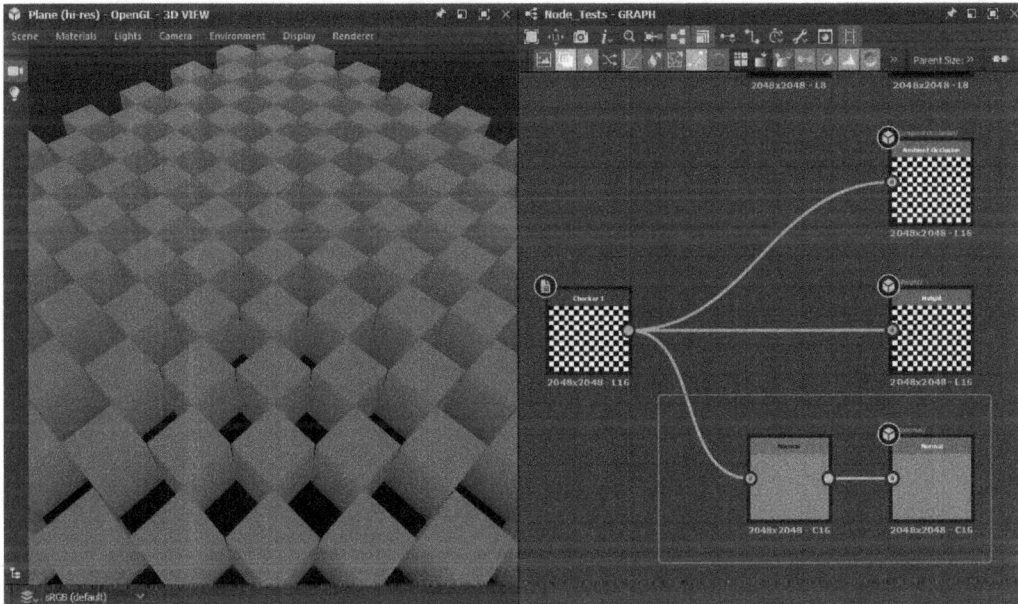

Figure 8.8 – Connecting the Checker 1 pattern node to the Normal map nodé

15. To organize the graph better, move the **Base Color**, **Roughness**, **Metallic**, and all **Uniform Color** nodes connected to them closer to the **Checker 1** pattern node, as shown in *Figure 8.9*:

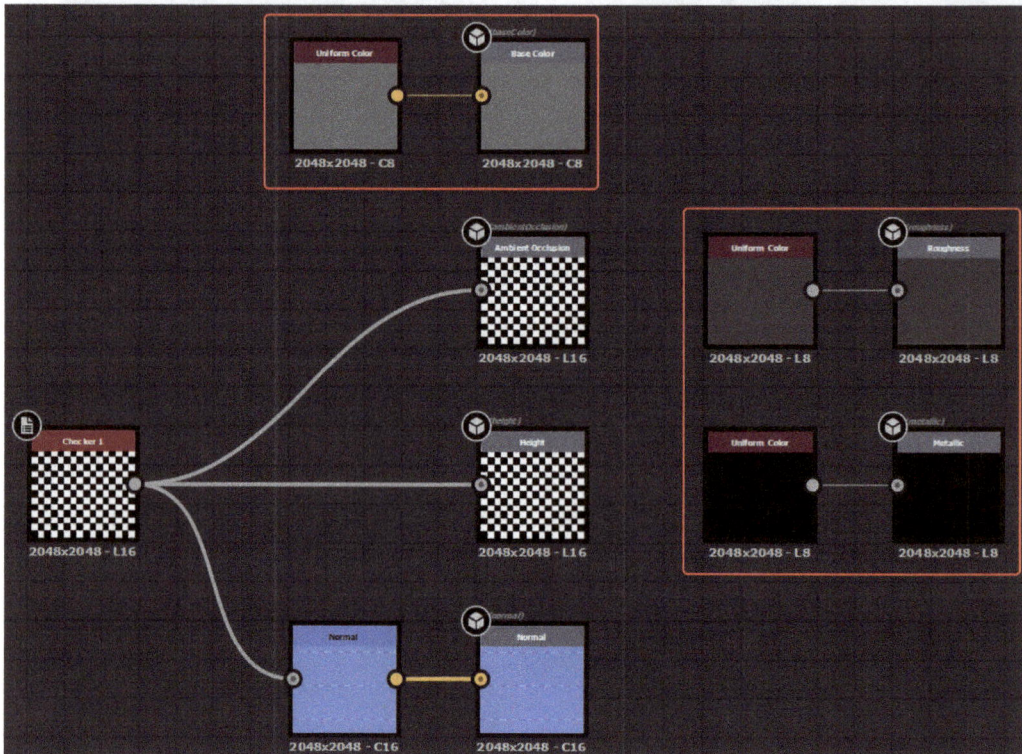

Figure 8.9 – Organizing the nodes

16. Double-click **Base Color's Uniform Color** node so that its properties appear, keep **Color Mode** set to **Color** in the **PROPERTIES** window, and change it to whatever color you want. If they are already in the right color mode, you don't need to apply any changes.

17. To give it a metallic look, change the **Uniform Color** mode of the **Metallic** node to **Grayscale** and change the color to white. To keep the metal shiny, change the **Roughness** node's **Uniform Color** property to black:

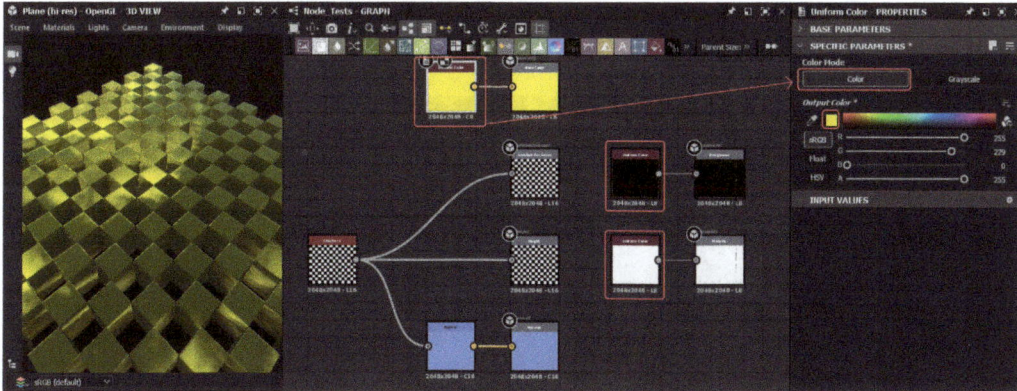

Figure 8.10 – Changing the Base Color, Metallic, and Roughness nodes

Some nodes inside Adobe Substance 3D Designer are **Color Mode**-specific; the *yellow and gray* circle input link means it can take both **Color** and **Grayscale** outputs, and the yellow circle input link means it can only take **Color** outputs, while the gray circle input link means it can only take **Grayscale** outputs:

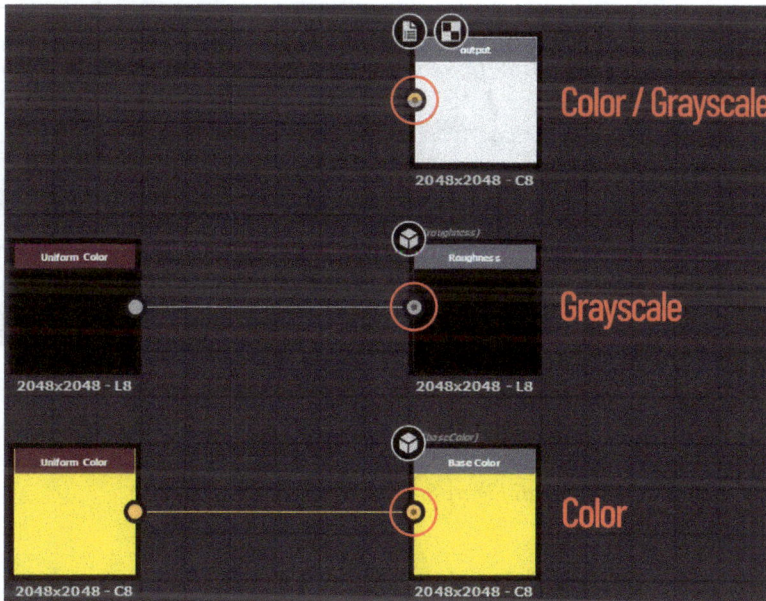

Figure 8.11 – Color Mode-specific links

Therefore, if you want to connect a **Grayscale** node to a **Color** input, you will get a red dotted line error, as shown in *Figure 8.12*:

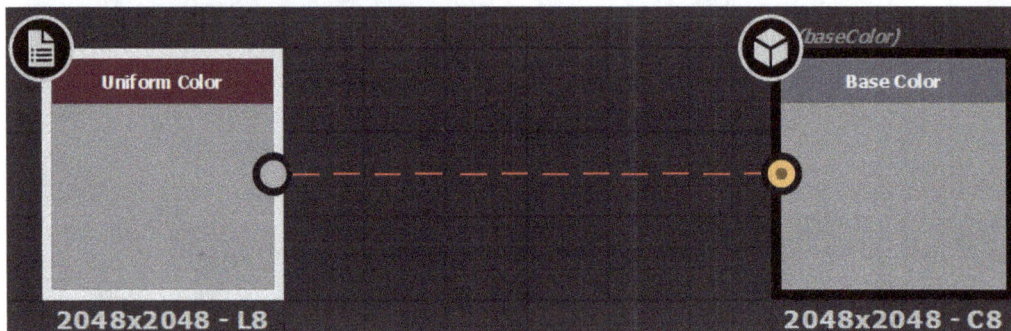

Figure 8.12 – Connecting different mode nodes

18. To avoid this error, you can select the red dotted line link and press the *spacebar* to get the **Nodes** menu. From the menu list, choose **Gradient Map**. This will convert the grayscale to color, as shown in *Figure 8.13*:

Figure 8.13 – Converting Grayscale to Color

19. Similarly, you can choose the **Grayscale Conversion** node if you want to input a **Color** node to a **Grayscale** node, as shown in *Figure 8.13*:

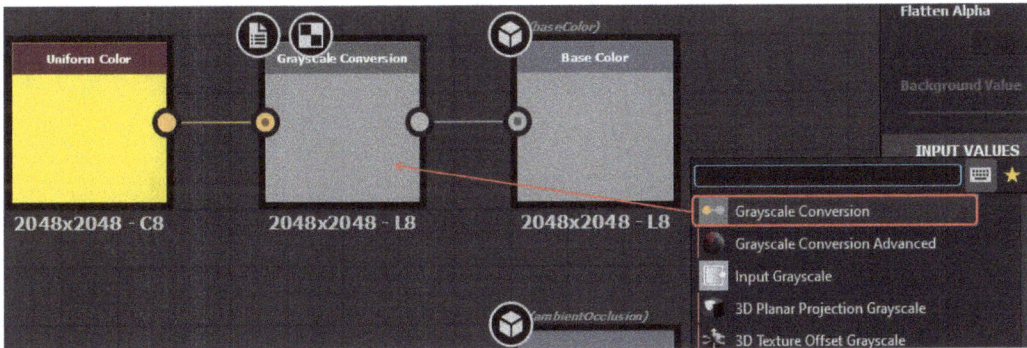

Figure 8.14 – Converting Color to Grayscale

20. Now, the issue is that if you re-open your project, your 3D settings will be reset even if you save it. Therefore, to avoid this, you have to go to **3D VIEW** and click **Scene**.

21. Choose **Save Current State** and save it as `Nodes_Test.sbsscn`. You can also choose **Save Current State As Default**. However, it will become a default scene:

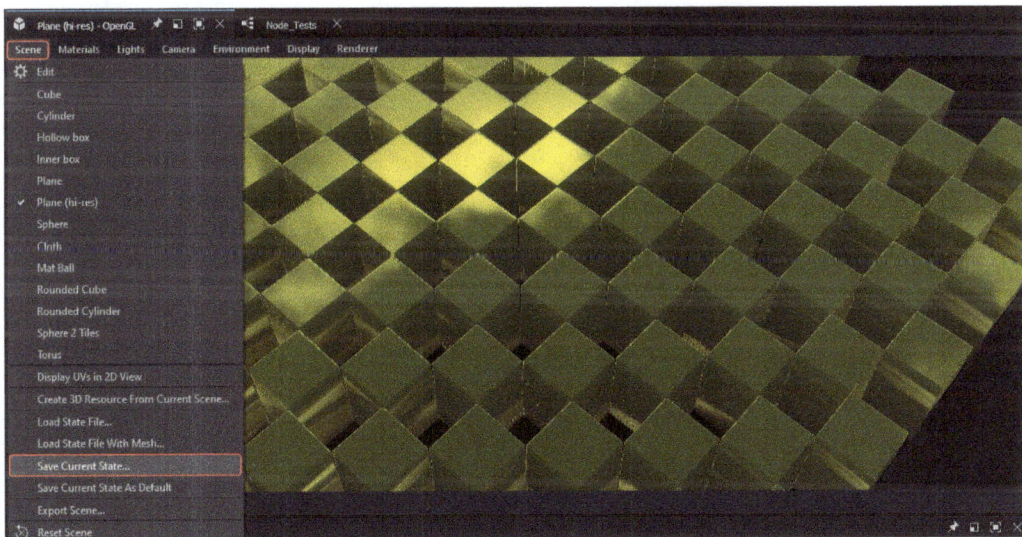

Figure 8.15 – Saving the current state

Save your graph and re-open it – you will notice that your 3D Scene has been reset.

22. To retain the saved state, go to **3D View**, click on **Scene | Load State File**, choose **Nodes_Test. sbsscn**, and select your **Scene Material**, which was **Plane (hi-res)**. This will load your saved state with all the settings:

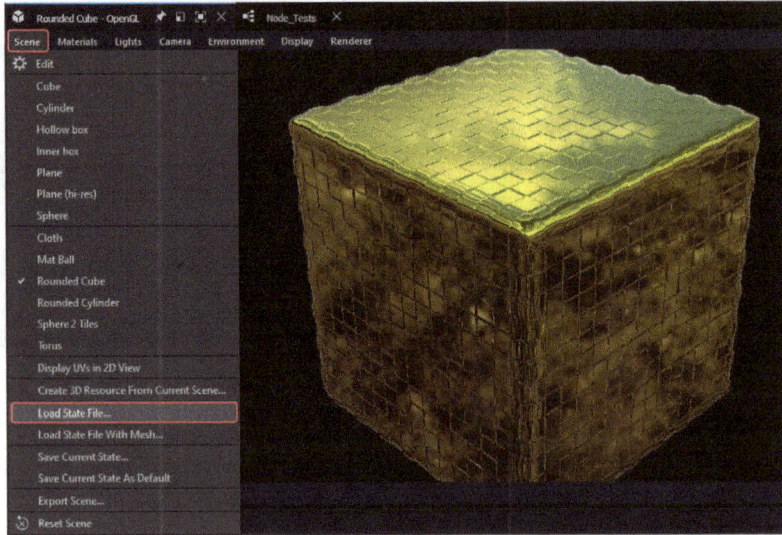

Figure 8.16 – Loading a state file

Hopefully, you may have been familiar with the usage of nodes inside Adobe Substance 3D Designer. Now, let us go through the **Atomic** nodes, which are basically the most frequently used node.

Atomic nodes

A designer's most fundamental building blocks are called **Atomic nodes**. There are 26 **Atomic** nodes in the older versions of Adobe Substance 3D Designer and 28 in the newer version, but we'll just discuss the most commonly used ones here:

Figure 8.17 – Atomic nodes

- **A – Blend**: With a user-defined blend mode and an optional opacity mask, the **Blend** node combines or blends two distinct inputs. It is the most beneficial node in Designer and you will use it in practically every graph you create.

  It's comparable to stacking two layers one on top of the other in Painter or Photoshop and blending them using the top layer's mode. *Chapter 9, Blending Modes in Adobe Substance 3D Designer,* is entirely based on this node.

- **B – Uniform Color**: A consistent, user-defined color or value is returned by the **Uniform Color** node. It is a straightforward node that is frequently used as a foundation for adding color or developing certain values. If you want to produce a shade of gray, convert this node to grayscale. Graphs are quicker to edit and perform better when in the grayscale color mode.

- **C – Bitmap**: The **Bitmap** node can be used to produce a bitmap for use with the 2D painting tools or to import an existing bitmap into your graph. This node should always be used in your graph if a bitmap file is required. Either start from scratch to construct the node or drop a bitmap in a suitable format into the graph view.

- **D – Levels**: You can remap the tones of input by configuring the input and output remap factors in the **Levels** node. The **Levels** node's user interface is comparable to Photoshop's levels tool. A Histogram preview is also available in the **Levels** node's settings. One of the fundamental nodes in Substance Designer, the **Levels** node is frequently used to remap and modify values in a graph since it offers the most precise interface for doing so.

- **E – Output**: A unique node called an **Output** node acts as the finish line for your graphs. The input to **Output** nodes is not processed in any way. Only one input slot, which automatically switches between color and grayscale, is present on the **Output** node. In contrast to **Input** nodes, which offer properties that help you arrange your assets, output nodes lack parameters.

These are a few most commonly used **Atomic** nodes, in the later chapters we will study more about **Atomic** nodes, for the time being, let us study some useful **Library** nodes.

Library nodes

A library of graph instance nodes created by **Atomic** nodes is included with Designer. Some are designed for particular use circumstances, while others are frequently useful.

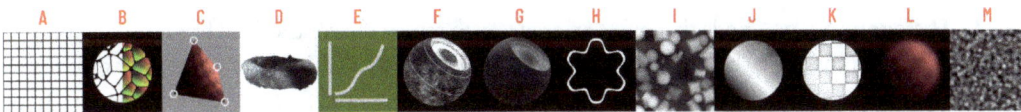

Figure 8.18 – Library nodes

We'll have an overview of a few typical ones:

- **A – Tile Generator**: One of the library's more sophisticated nodes is the **Tile Generator** node. When you master the tile generator, you can use it to make the majority of designs. Almost any graph can utilize the tile generator because it is very adaptable and doesn't require patterns. Use custom shapes and parameter adjustments to generate an infinite number of variants.

- **B – Flood Fill**: **Flood Fill** is one of a sophisticated group of effects that let you give a simple binary tile texture a lot more diversity. It is more of a starting point for other Flood Fill effects and is not intended to be utilized alone.

- **C – Quad Transform**: A special transform node that interacts with the corner points of a quadrilateral shape to transform it. It allows for precise hands-on transformations.

- **D – Height Blend**: This is based on the height information of the two height maps combined. It creates a black and white mask that may be utilized anywhere in addition to a blended height map.

- **E – Curve**: As with other 2D image editing programs, the **Curve** node offers an interface for picture tone remapping. The input, which can either monochrome (grayscale) or colored, can be remapped by the user by placing points and adjusting Bezier curves.

- **F – Dirt**: This node creates a black and white mask using user parameters and baked maps, similar to Painter's Smart Masks. This doesn't need baked **Ambient Occlusion (AO)** – you can also use built-in AO and Curvature. This mask represents dirt on obscured and sunken edges and corners.

- **G – Dust**: This node creates a black and white mask using user parameters and baked maps, similar to Painter's Smart Masks. This mask symbolizes dust that has gathered exclusively in places that face upward and in occluded, lowered locations. It only works with properly baked AO and World Space Normal. This doesn't need baked AO – you can also use built-in AO and Curvature.

- **H – Shape Mapper**: This transforms and accurately maps an input pattern along a polygonal or circular route. It's comparable to Splatter Circular, but with the form distortion which is necessary to precisely follow the course. When you want a height map to be twisted into a circle, for instance, it might be helpful in some very particular situations.

- **I – Shape Splatter**: Its primary function is to enable the insertion of forms on and inside a height map, which is then used to build different maps from the Splatter Data – for instance, arranging pebbles, twigs, and leaves on a landscape that is guided and orientated by different maps. Then, despite using the same common Splatter Data for all channels, different maps may be used for **Height**, **Normal**, **Base Color**, **Roughness**, and any other channel.

- **J – Histogram Scan**: A node that offers an easy way to remap the contrast and brightness of input grayscale photos – very basic yet helpful. It can be used to dynamically *grow* and *shrink* masks.

- **K – Ambient Occlusion (HBAO/RTAO)**: This node converts a height map into an AO map by taking it as input. It employs the sophisticated technique known as **Horizon-Based Ambient Occlusion (HBAO)**. It may be used to convert generative height maps into excellent AO maps. Although more computationally costly, a **Ray Tracing Ambient Occlusion (RTAO)** node is also offered.

- **L – Blur HQ**: A high-quality Gaussian blur is applied to the input by Blur HQ, precisely like the effect in Photoshop. While taking longer to calculate, it has a lot better quality than the typical atomic **Blur** node. The blur's quality level can be altered using a node-specific option.

- **M – Gaussian Noise**: This node creates soft, straightforward random blobs on an adjustable scale, producing one of the most basic yet most useful noise maps. Additional parameters are available to add looping motion to the noise. Gaussian noise is frequently a wise choice if you are unsure of the noise to employ.

I hope the **Atomic** and **Library** nodes are clear to you, so let us use these nodes in practice. We will also go through some more common nodes while we will be working on them practically.

The Tile Generator node

As discussed before, the **Tile Generator** node helps you to create different types of tiles using different shapes. In this section, we will create brick-shaped tiles, and we will also learn how to work with **Gradient Map**.

1. Open Node_Tests.sbs in Adobe Substance 3D Designer, go to **3D View** | **Load State File**, choose Nodes_Test.sbsscn from the **Scene** menu, and choose **Plane (hi-res)**.

   Now, we need to replace the existing **Checker Map** node with **Tile Generator**.

2. To do that, click inside the **GRAPH** window with your left-hand mouse button and press the *spacebar* to load the nodes library. Type Tile Generator into the search field and select the **Tile Generator** node.

3. Remove the **Checker Map** node and connect the **Tile Generator** node with the existing nodes, as shown in *Figure 8.19*:

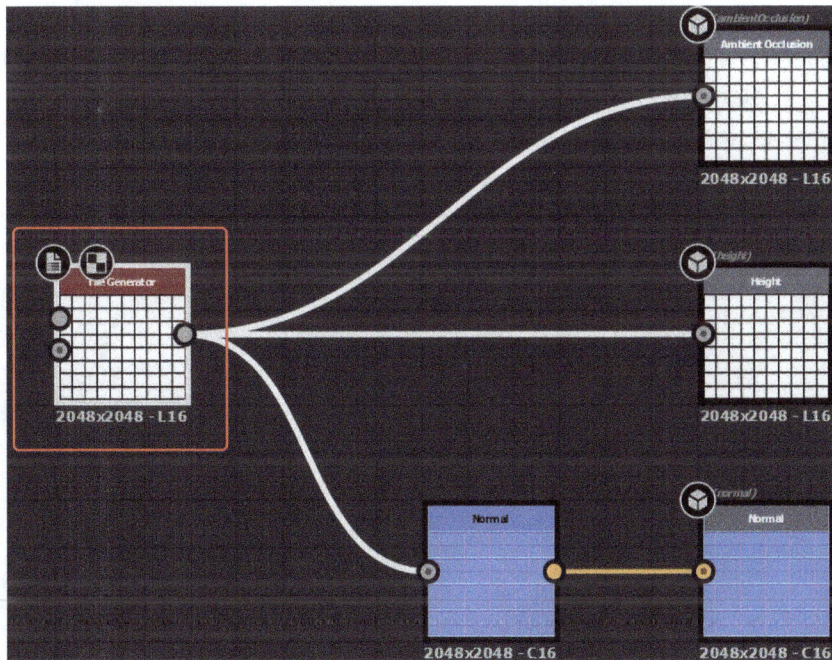

Figure 8.19 – Adding a Tile Generator node

4. Right-click with your mouse inside the **GRAPH** window and choose **View Outputs In 3D View**.

5. Double-click on the **Tile Generator** node and once its properties start showing in the **PROPERTIES** window, adjust the following settings:

    - **INSTANCE PARAMETERS**: **X Amount**: 5, **Y Amount**: 15

    - **Position**: **Offset**: 0.5   **Offset Random**: 0.1

    This will create a brick-shaped tile with a little randomness, as shown in *Figure 8.20*:

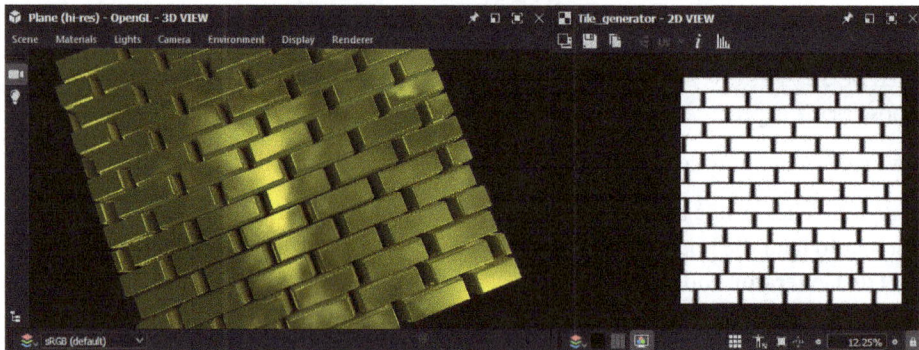

Figure 8.20 – A brick-shaped tile

6. You will notice that the height value is too much for the brick map. Therefore, go to **3D VIEW**, click on the **Materials** menu, select the **Default** option, and choose **Edit**.

7. Once the material parameters have loaded in the **PROPERTIES** window, change the height scale to 1.5, as shown in *Figure 8.21*.

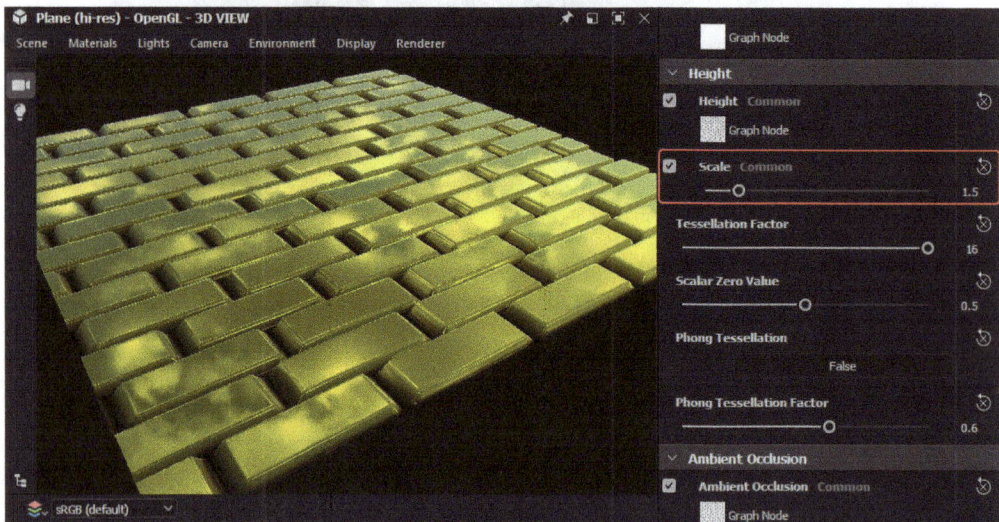

Figure 8.21 – Changing the height

8. Now, let us give the brick tile some realistic color. Therefore, we will not use **Uniform Color** this time; we will use **Gradient Map** with the **Noise** node that matches the brick tile. So, disconnect **Uniform Color** from **Base Color**, remove **Uniform Color**, and add **Gradient Map** instead, as shown in *Figure 8.22*:

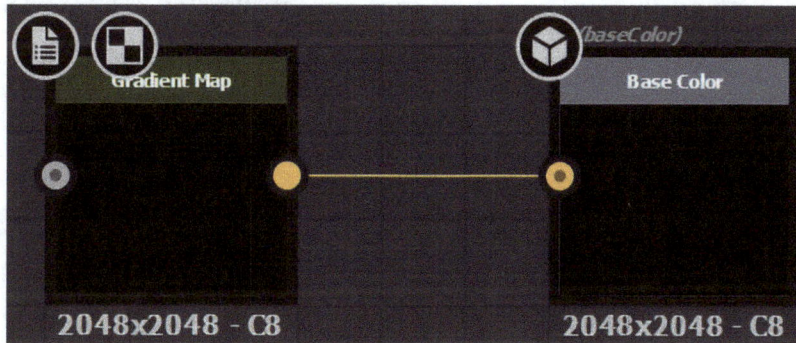

Figure 8.22 – Gradient Map

9. The gradient map needs a node that has variation, such as a **Noise** node, to generate a gradient. Therefore, we will add **Grunge Concrete** and connect it to **Base Color**. After that, double-click **Grunge Concrete** and change its **Balance** to 0 . 55 and **Contrast** to 3 . 5:

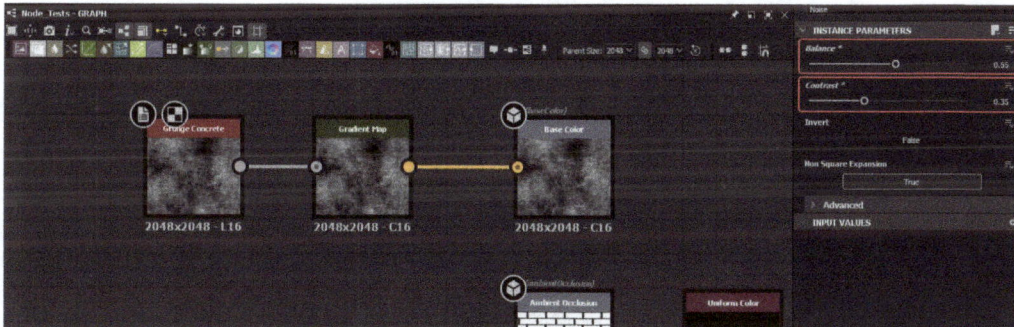

Figure 8.23 – Adding Gradient Concrete

10. Now, go to your internet browser and load any high-resolution red brick image, keep the internet browser and the Adobe Substance 3D Designer window open side by side, and execute the following steps:

  I.   Double-click **Gradient Map** to load its parameters inside the **PROPERTIES** window.

  II. Go to the **Gradient Map**'s properties and click **Gradient Editor**.

  III. Once the **Gradient Editor** window opens, click **Pick Gradient**.

IV.   Once you click **Pick Gradient**, you will load a Gradient Picker Cursor. Therefore, with that cursor, drag across the red brick image with your left-hand mouse button held down, as shown in *Figure 8.24*:

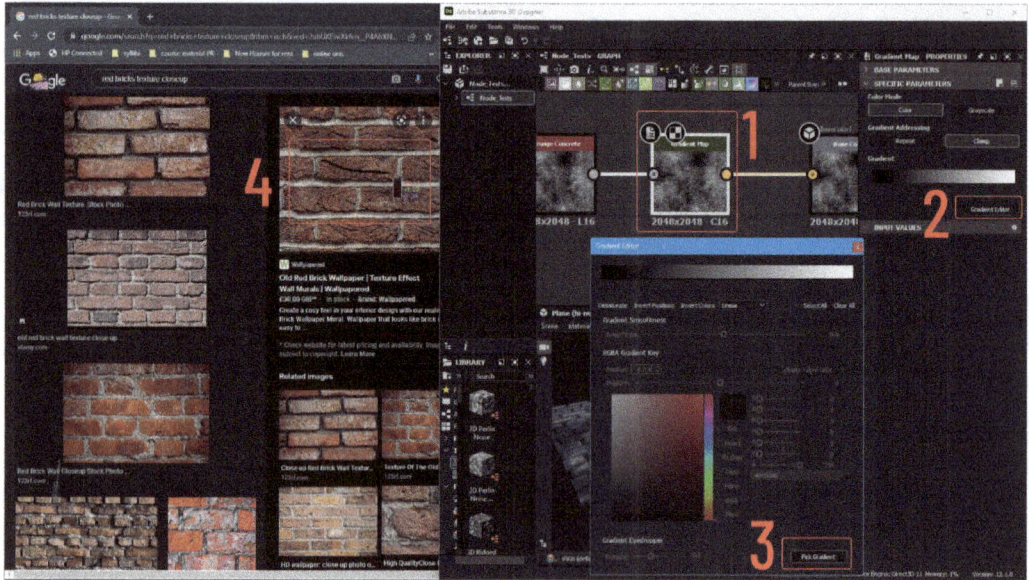

Figure 8.24 – Picking a gradient

11. Now, change the **Roughness** node's **Uniform Color** property to white and the **Metallic** node's **Uniform Color** property to black. These colors will make the brick look rougher, without any reflective quality, like real bricks:

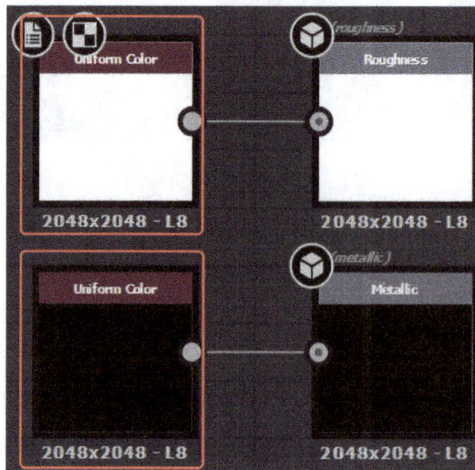

Figure 8.25 – Changing the Color properties for the Roughness and Metallic nodes

Now, you will have a much more realistic-looking red brick wall, as shown in *Figure 8.26*:

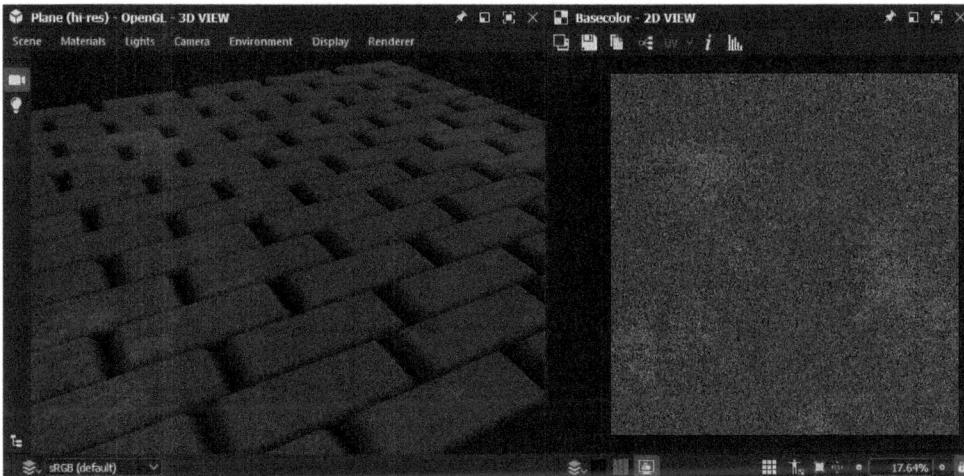

Figure 8.26 – Final output of the red brick tiles

I hope you have learned how the **Tile Generator** node works. **Tile Generator** has many other different parameters that appear in the **PROPERTIES** window. Similarly, there are many other parameters you can set for other different nodes, and studying them all in this book is not possible.

Nevertheless, these parameters are pretty much self-explanatory, which means you can easily learn by yourself – you just need to explore and experiment with them. In the following section, we will learn how to work with the **Flood Fill** node.

The Flood Fill node

Flood Fill is a part of a complex series of effects that enable you to greatly increase the diversity of a straightforward binary tile texture. It is not meant to be used on its own. Rather, it serves as a starting point for other **Flood Fill** effects. This segmented, separate data provides a more adaptable, effective, and harm-free method.

The **Flood Fill** node works with a secondary **Flood Fill** effect node, which includes **Flood Fill to Gradient, Flood Fill to Color or Grayscale, Flood Fill to Random Color, Flood Fill to Random Grayscale, Flood Fill Mapper, Flood Fill to bbox size, Flood Fill to Position, Flood Fill Mapper**, and **Flood Fill to Index**. However, we will only work on **Flood Fill to Random Grayscale, Flood Fill to Position**, and **Flood Fill to Gradient**, because these three are the most frequently used **Flood Fill** nodes.

The input map must be suitable for **Flood Fill** for it to function. An ideal binary map (black and white exclusively, no grayscale) would have a border around each tile that is completely black (0,0,0) for every pixel, separating it from the adjacent lines.

Tile Generator is one instance of software that is ideal for this. Problems occur when grayscale, sloping values are utilized – notably when tiles are not separated by whole black pixels.

This is discernible by a general lack of red values in the outcome and sometimes, by odd artifact lines. In these situations, either change the input map's contrast or remove it. You test whether anything changes and make sure to adjust the **Safety/Speed trade-off** option.

1. Let's try the **Flood Fill to Random Grayscale** node first. So, connect the **Tile Generator** node's output to the **Flood Fill** node's **Mask** input, and then connect the **Flood Fill** node's **Box** output to the **Flood Fill to Random Grayscale** node's **Flood Fill** input. Finally, connect the **Flood Fill to Random Grayscale** node's output to the rest of the inputs, as shown in *Figure 8.27*:

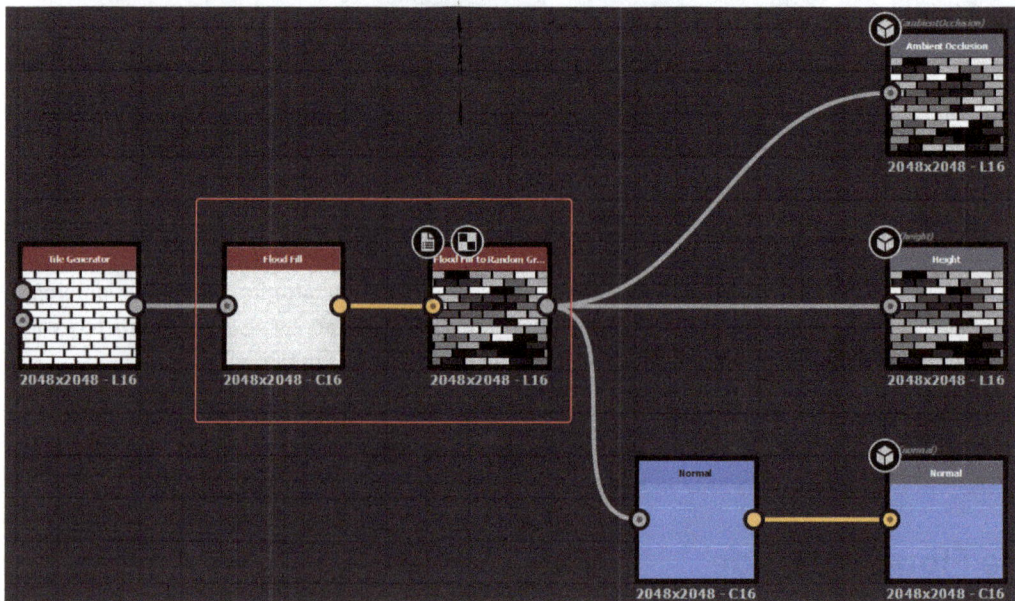

Figure 8.27 – Flood Fill to Random Grayscale

2. **Flood Fill to Random Grayscale** will produce brick walls with random heights because all the heights of the bricks will become randomized, with variations of blacks and whites creating different height maps, normal maps, and AO, as shown in *Figure 8.28*:

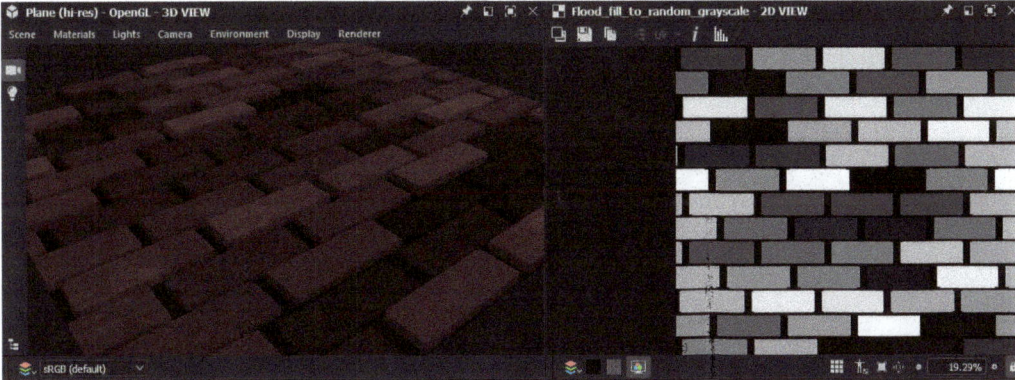

Figure 8.28 – Flood Fill to Random Grayscale output result

3. Now, let us try **Flood Fill to Position**. So, replace **Flood Fill to Random Grayscale** with **Flood Fill to Position**. However, the **Flood Fill to Position** node's output is colored. Therefore, we need to add the **Grayscale Conversion** node to any grayscale node – for example, we have to add the **Grayscale Conversion** node between the **Flood Fill to Position** node and the **Normal** map node, as shown in *Figure 8.29*:

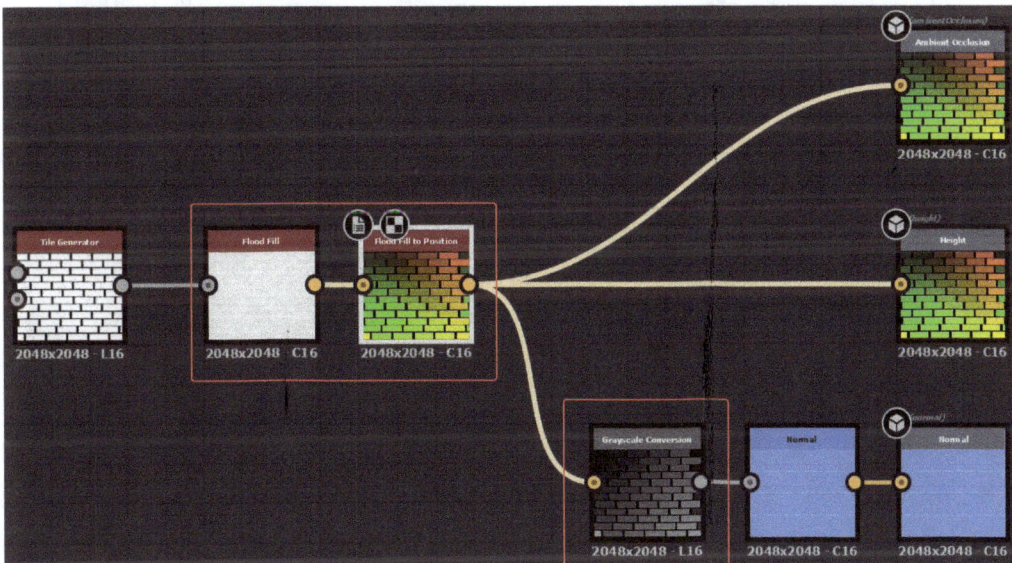

Figure 8.29 – Adding to the Flood Fill to Position node and the Grayscale Conversion node

4. The **Flood Fill to Position** node will create a gradient-like map, which will produce a stairs-like effect if applied to the height map. You can increase the material height if you want to see a more accurate result, as shown in *Figure 8.30*, and later, you can put it back to normal:

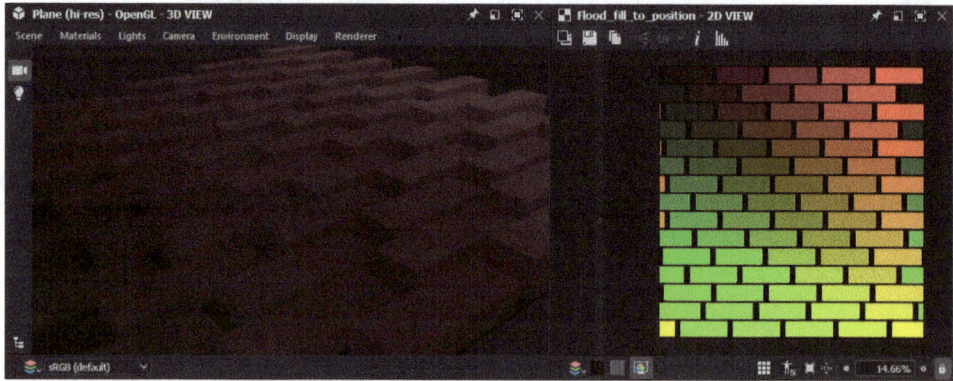

Figure 8.30 – The result of Flood Fill to Position

5. Let us replace **Flood Fill to Position** with **Flood Fill to Gradient and** remove the previously added **Grayscale Conversion** node because the **Flood Fill to Position** node creates a grayscale output:

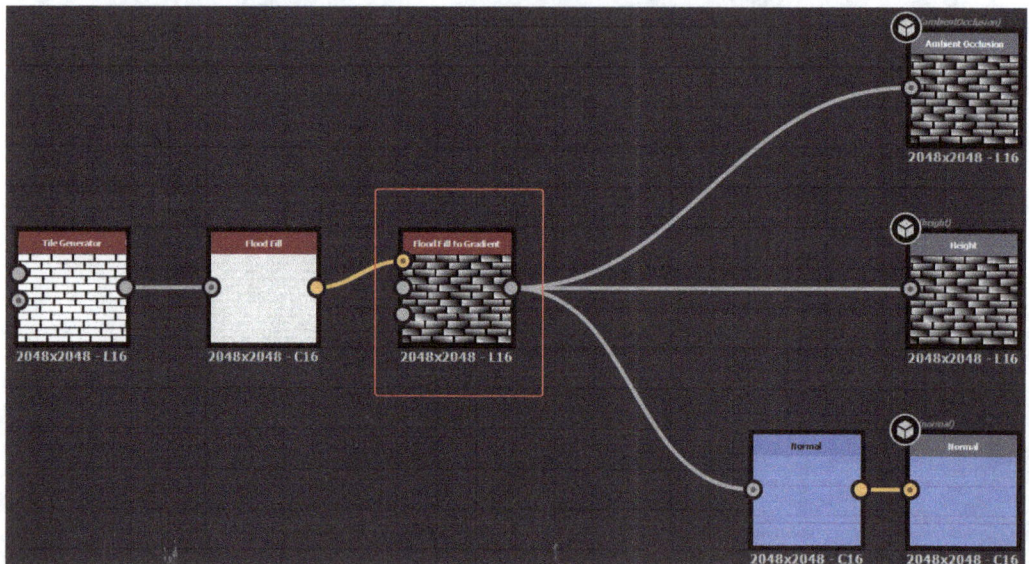

Figure 8.31 – Adding the Flood Fill to Gradient node

You will notice that the **Flood Fill to Gradient** node creates a gradient map on each brick rather than the whole brick wall.

6. The **Flood Fill to Gradient** node also has some useful parameters. Let us change one of the parameters, so double-click on the **Flood Fill to Gradient** node and increase **Angle Variation** up to 1.

This will randomize the direction of the gradient on each brick and it will look something like *Figure 8.32*. The **Flood Fill to Gradient** node is good for creating broken, eroded, or non-uniform textures.

7. You can also add your own variation by adding different noises to the **Gradient** node's **Angle Input** and **Slope Input** settings and changing their parameters from the **Flood Fill to Gradient** node's properties.

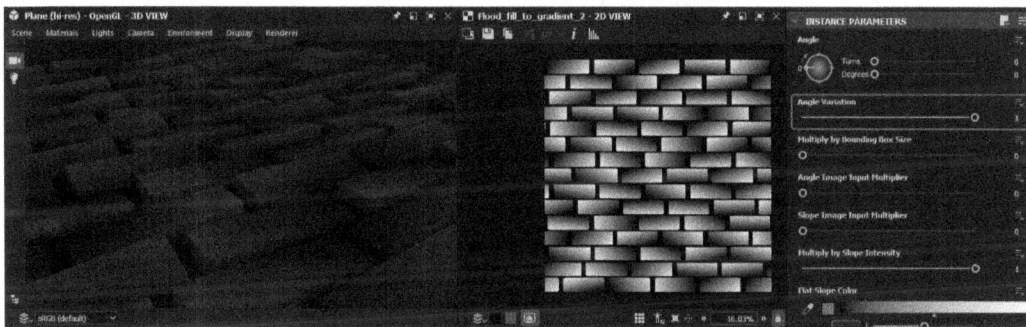

Figure 8.32 – The result of the Flood Fill to Gradient node

I hope these three useful **Flood Fill** effect nodes are now clear to you, and you have understood how to use them with the main **Flood Fill** node. In the next section will learn how the **Quad Transform** node works.

The Quad Transform node

The **Quad Transform** node allows you to distort the connected node either using the mouse or changing the parameters in the **PROPERTIES** window. Let us try to add a shape over the brick wall and distort it using the **Quad Transform** node and blend it using the **Blend** node:

1. In the **GRAPH** window, add the **Star** and **Quad Transform Grayscale** nodes – we will use the grayscale version of **Quad Transform** because the **Star** node is grayscale per se. Then, connect the **Star** node's output to **Quad Transform Grayscale** node's input, as shown in *Figure 8.33*:

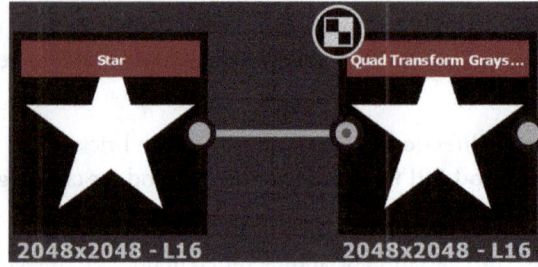

Figure 8.33 – Adding the Star and Quad Transform Grayscale nodes

2. Now, we have to blend the **Star** node using a secondary color with the brick wall. Therefore, add a **Blend** node and a different **Uniform Color** (as shown in *Figure 8.34*), so that we can blend the **Star**, existing **Gradient Map** color, and a different **Uniform Color**:

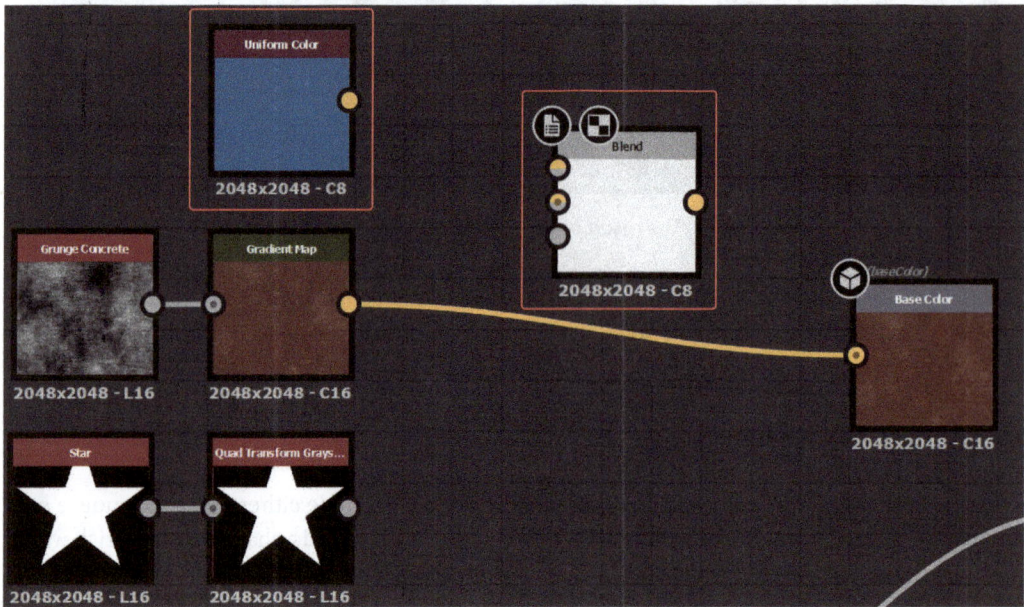

Figure 8.34 – Adding the Uniform Color and Blend nodes

3. Now, connect **Uniform Color**'s output to the **Blend** node's foreground, connect **Gradient Map**'s output to the **Blend** node's background, and **Quad Transform Grayscale**'s output to **Blend** node's **Opacity** input. Then, connect the **Blend** node's output to the **Base Color** node's input. Your connection should look the same as in *Figure 8.35*:

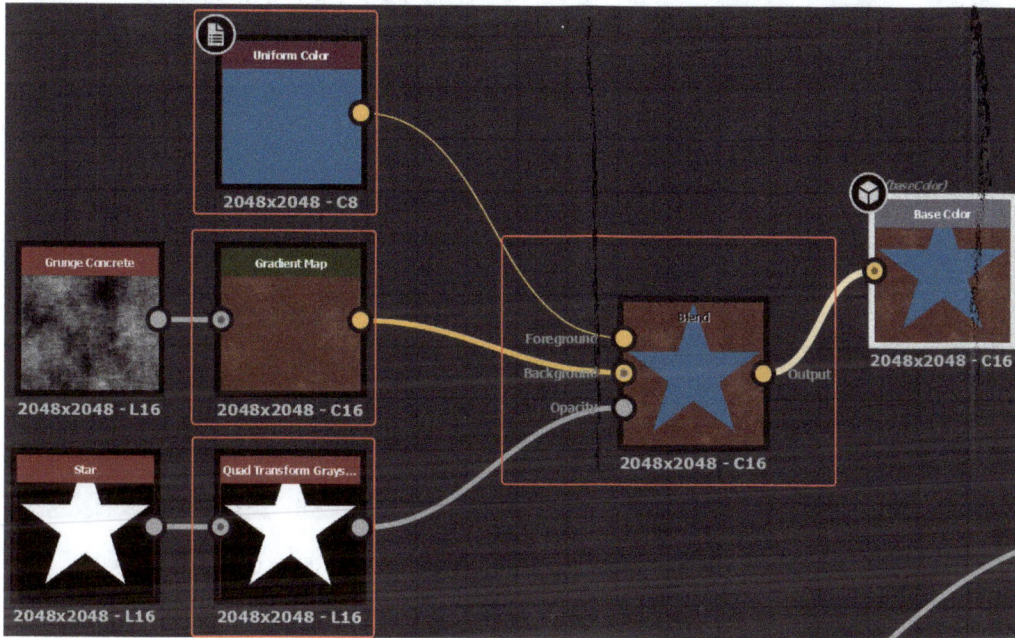

Figure 8.35 – Connecting Uniform Color, Gradient Map, and Quad Transform Grayscale to Blend

4. Now, let us distort the **Star** shape node using the **Quad Transform Grayscale** node. Double-click on **Quad Transform Grayscale** with your mouse. The **Star** shape node shape will load in **2D VIEW** with four corner points that you can move around with the mouse – or you can change its parameters under the **p00**, **p01**, **p02**, and **p03** settings, as shown in *Figure 8.36*:

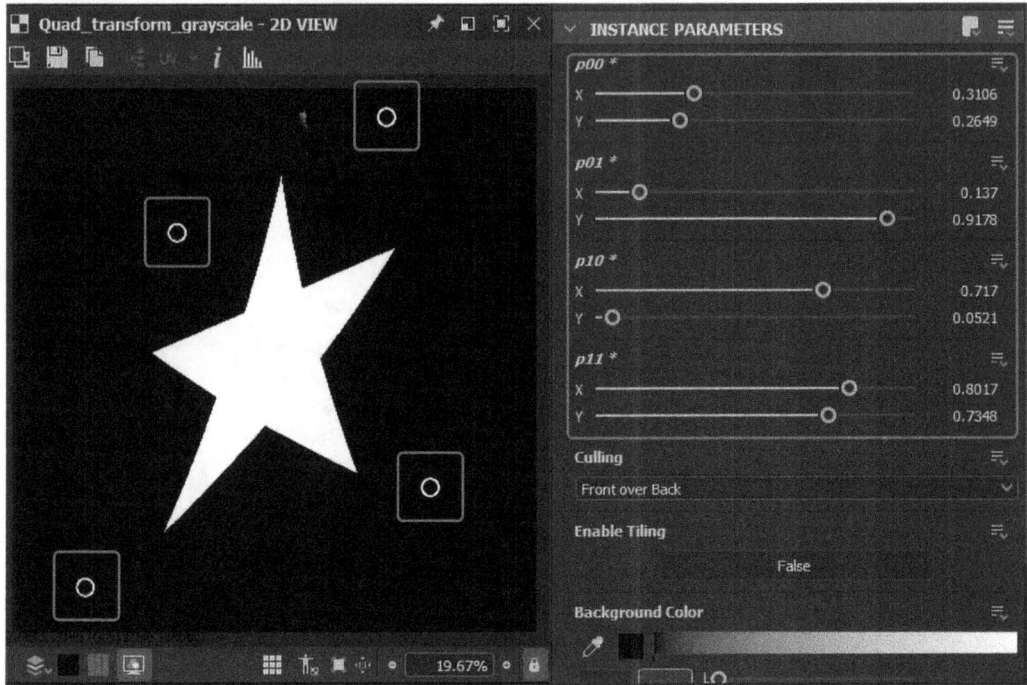

Figure 8.36 – Distorting the star-shaped node using the Quad Transform Grayscale node

5. The **Quad Transform Grayscale** node will distort the **Star** shape node and the **Blend** node will blend two nodes based on the mask. You can also double-click on **Base Color** with the mouse to produce the result, as shown in *Figure 8.37*.

    If you don't see the result in **3D VIEW**, then right-click in the **GRAPH** window and choose **View Outputs In 3D View**:

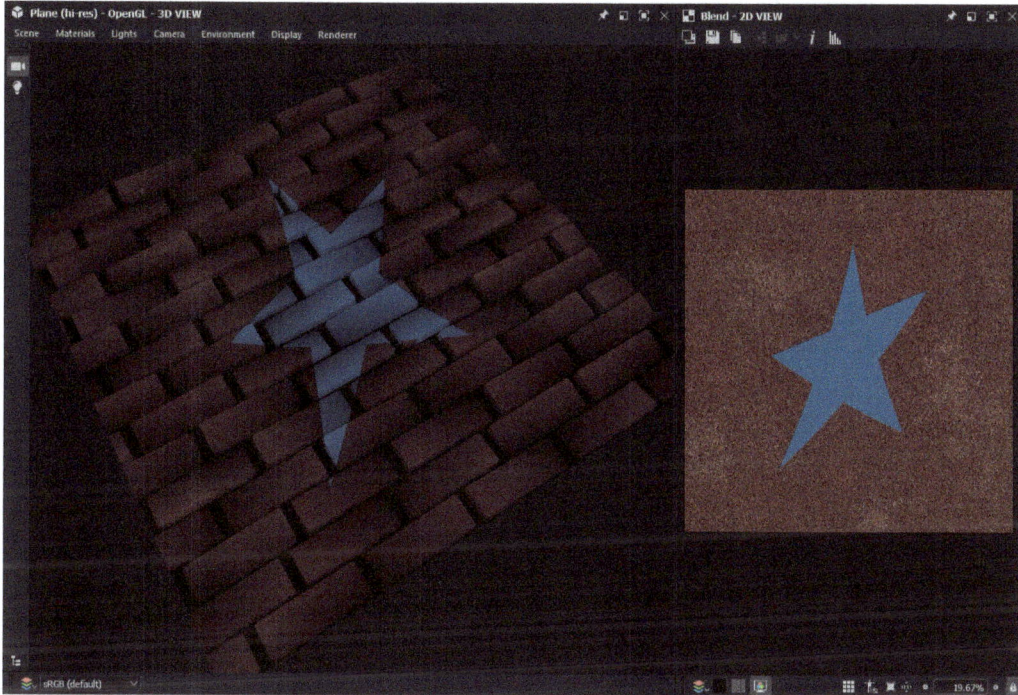

Figure 8.37 – The final result of the Quad Transform Grayscale and Blend nodes

I hope it's clear now to you how to work with the **Quad Transform** node and the purpose of the **Blend** node. In the following section, we will learn about the **Height Blend** node.

The Height Blend node

The height data from the two combined height maps forms the foundation of the **Height Blend** node. It produces a blended height map as well as a black and white mask that may be used everywhere.

So, let us make two height maps and combine them using the **Height Blend** node.

1. Select the existing **Flood Fill to Gradient** node and using the keyboard, press *Ctrl* + *C* (on Windows) or *Command* + *C* (on Mac), and then press *Ctrl* + *V* (on Windows) or *Command* + *V* (on Mac) to copy and paste it. You can also use *Ctrl* + *D* (on Windows) or *Command* + *D* (on Mac) to duplicate the nodes instead of copying and pasting them. Then, change the **Angle Variation** setting of the copied **Flood Fill to Gradient** node to 0.5:

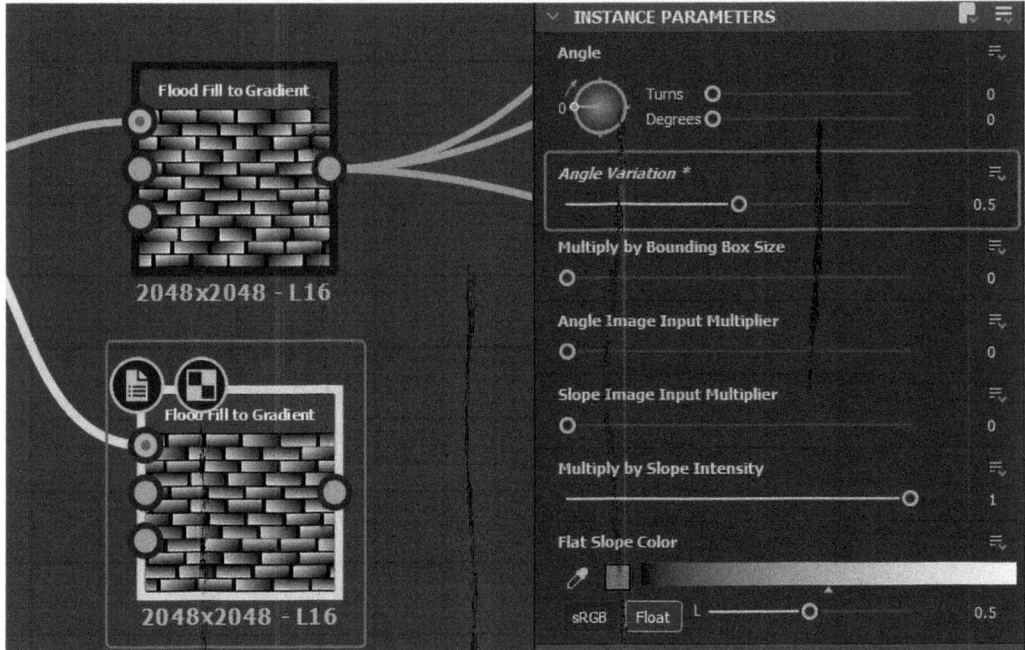

Figure 8.38 – Copying and pasting Flood Fill to Gradient and changing its angle variation

2.   Now, add the **Height Blend** node to the **GRAPH** window and connect the original **Flood Fill to Gradient**'s output to the **Height Blend** node's **Height Top** and the copied **Flood Fill to Gradient**'s output to the **Height Blend** node's **Height Bottom** and connect the **Height Blend** node's **Blended Height** to the inputs of the **Ambient Occlusion**, **Height**, and **Normal** nodes, as shown in *Figure 8.39*:

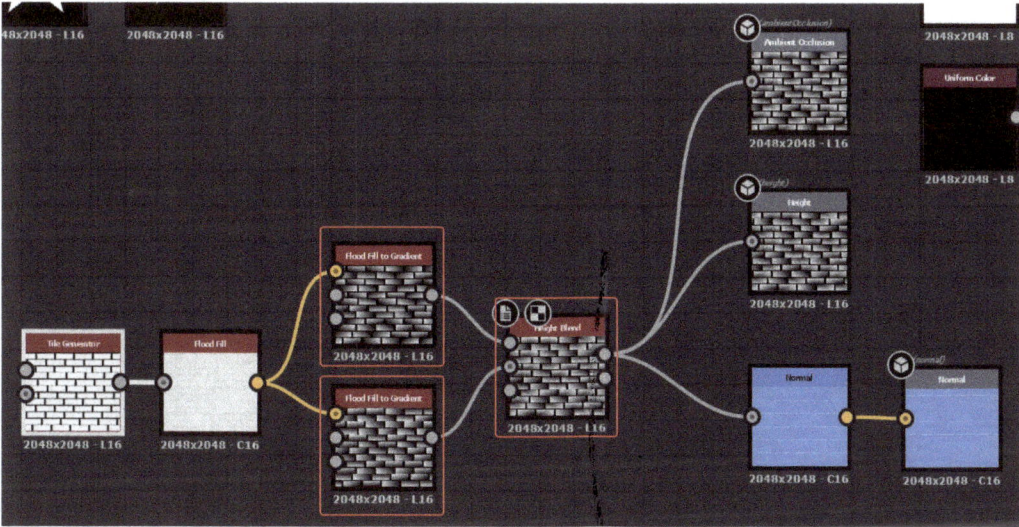

Figure 8.39 – Adding the Height Blend node

3. The **Height Blend** node will combine the gradients of the original and copied **Flood Fill to Gradient** nodes and will produce an interesting result, as shown in *Figure 8.40*:

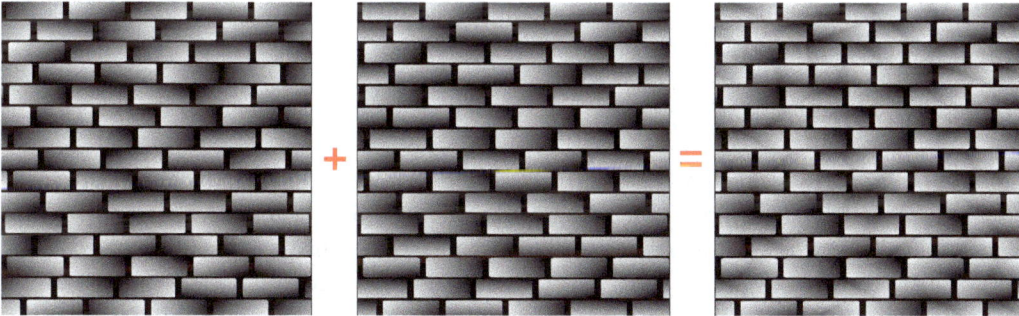

Figure 8.40 – The result of the Height Blend node

4. If you want, you can change the parameters of the **Height Blend** node to get the desired result. If you don't see the result in **3D VIEW**, then right-click in the **GRAPH** window and choose **View Outputs In 3D View**:

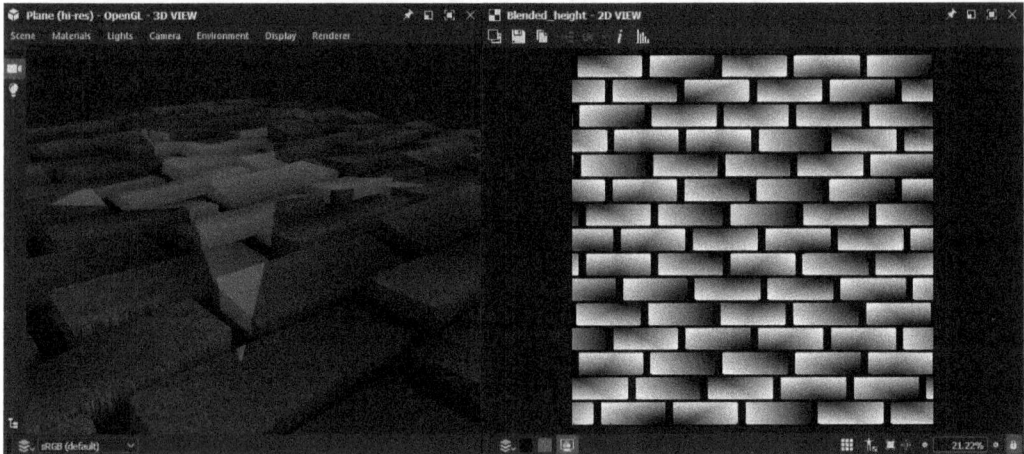

Figure 8.41 – The final output of the Height Blend node

It is used to blend two height maps based on a mask. Additionally, it generates a mask and has controls for the contrast, because you can use it to blend all kinds of height maps. In the following section, we will learn about the **Curve** node.

The Curve node

The **Curve** node provides a picture tone remapping interface, similar to other 2D image-altering tools. The user can remap the input, which can be either monochrome or colored, by setting points and modifying Bezier curves. It is extremely helpful when used to remap gradient transitions to a particular height profile.

This enables incredibly accurate modeling of bevel profiles and other similar shapes. The **Curve** node provides a full-featured curve editor instead of the conventional basic UI with sliders and settings that most other nodes have.

So, let us apply **Curve** to some existing nodes:

1. In *Figure 8.42*, you will notice that the **Tile Generator** brick map is blurry at the edges. We need to remove this blurriness:

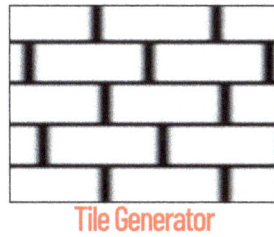

Figure 8.42 – Tile Generator blurry edges

2. Add the **Curve** node between **Tile Generator** and **Flood Fill**, as shown in *Figure 8.43*, and then double-click **Curve** to activate its parameters inside the **PROPERTIES** window.

3. Once you see the diagonal curve graph, double-click in the middle of the line, as shown in *Figure 8.43*, to add a **Curve** point. This will also add points at both ends of the curve:

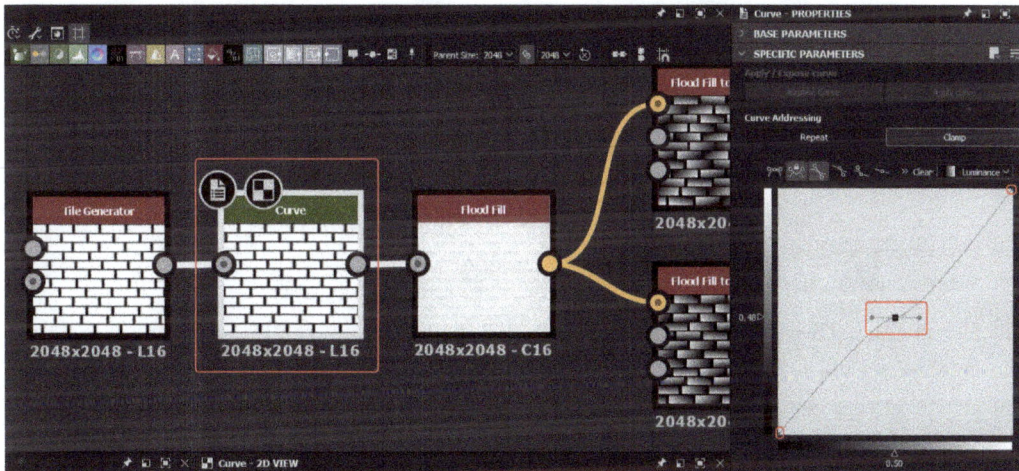

Figure 8.43 – Adding points to the curve

Now, we only need the points at both ends of the curve. Therefore, we will delete the middle point because we only added it to activate the endpoints.

4. After deleting the middle point, move the endpoint at the bottom left all the way to the right, which will make the **Tile Generator** bricks sharp and clear, as shown in *Figure 8.44*:

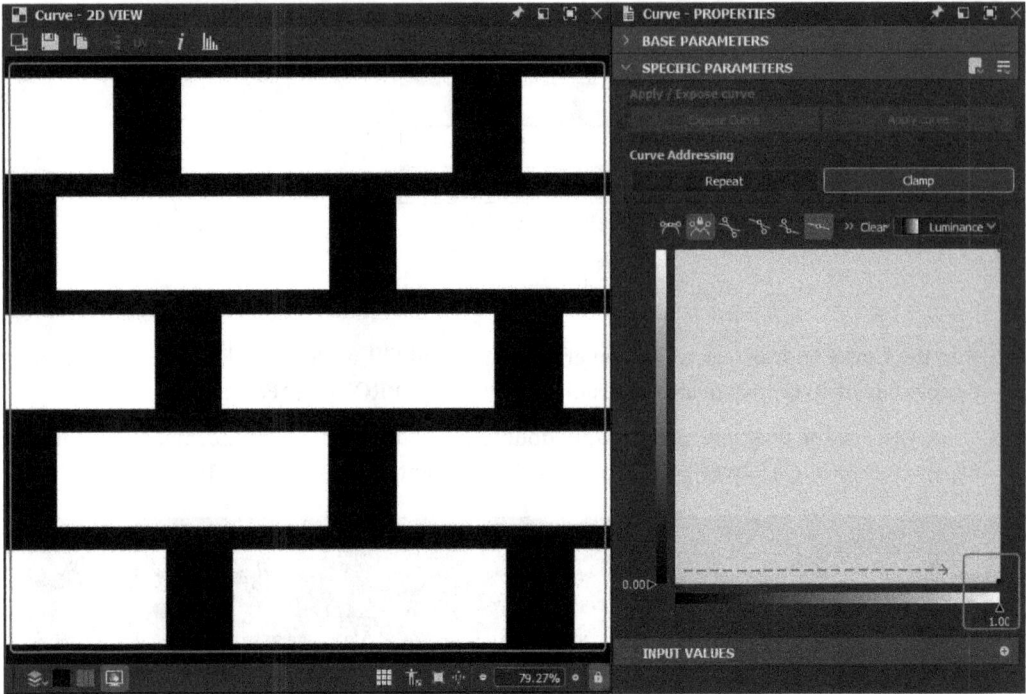

Figure 8.44 – Moving the bottom left endpoint to the right

You will now notice that the **Tile Generator** bricks now have more gaps between them.

5.   Left-click on **Tile Generator** (don't double click – otherwise, **2D View** will change to **Tile Generator**) and change **Pattern Specific** to 0.03, as shown in *Figure 8.45*:

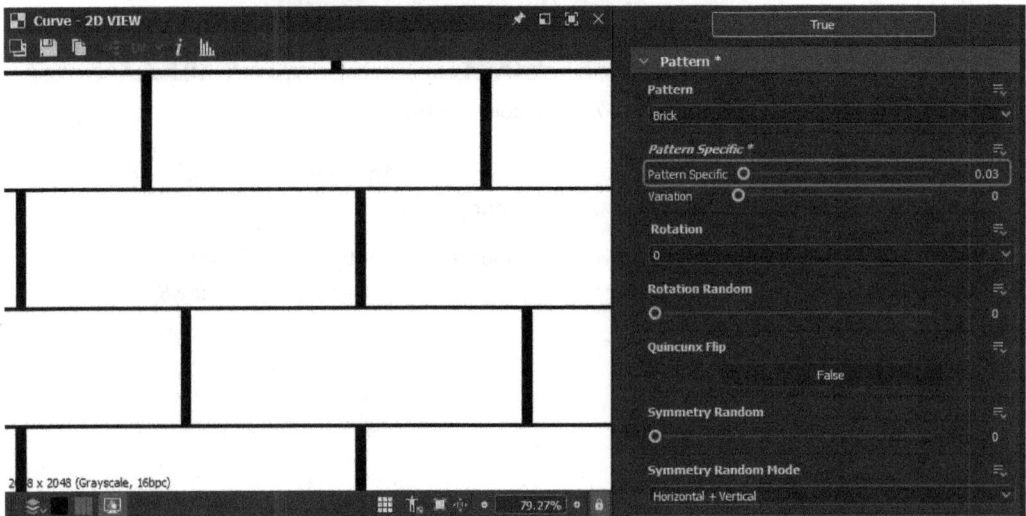

Figure 8.45 – Changing the Pattern Specific value of Tile Generator

You can also add more points to the **Curve** node – move the Beziers around it and reshape the curve as required. I hope the purpose of this node is clear to you. We will work on the **Curve** node more in *Chapter 10, Creating Television Shelf in Adobe Substance 3D Designer*. In the following section, we will add some **Dirt** and **Dust** to the existing brick wall.

The Dirt and Dust nodes

The **Dirt** and **Dust** nodes work with inputs from the **Ambient Occlusion, Curvature, Position, World Space Normal**, and **Mask (optional)** nodes.

The **Dirt** node produces a dirt map effect using all the aforementioned nodes, while the **Dust** node produces a dust map effect using only **Ambient Occlusion, World Space Normal**, and **Mask (optional)**.

So, let us apply the **Dirt** and **Dust** effects to our project:

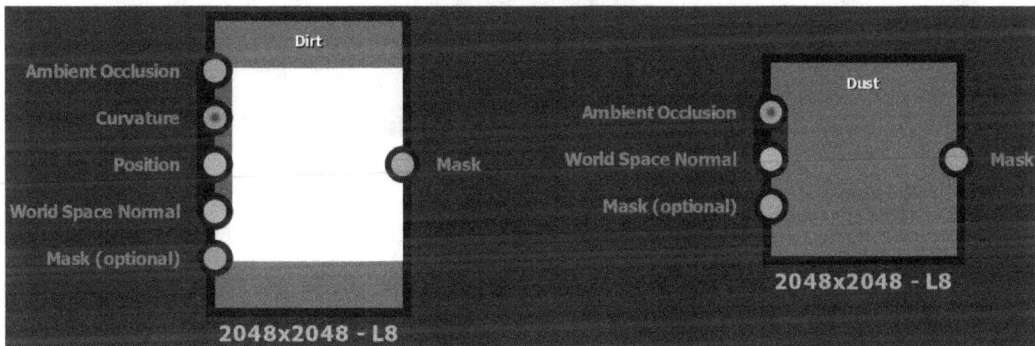

Figure 8.46 – The Dirt and Dust nodes

The **Ambient Occlusion** node that we have is merely an **Output** node as shown in *Figure 8.47*:

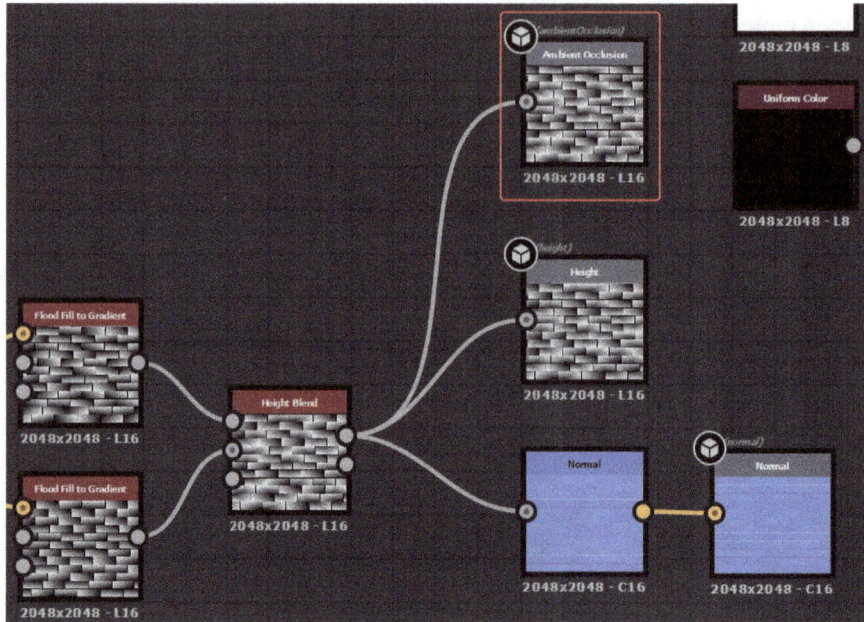

Figure 8.47 – A single Ambient Occlusion input node

Therefore, we need to add **Ambient Occlusion (HBAO)** between the **Height Blend** and **Ambient Occlusion** inputs, as shown in *Figure 8.48*.

You can study more about **Ambient Occlusion (HBAO)** and **Ambient Occlusion (HTAO)** in the *Atomic nodes* section.

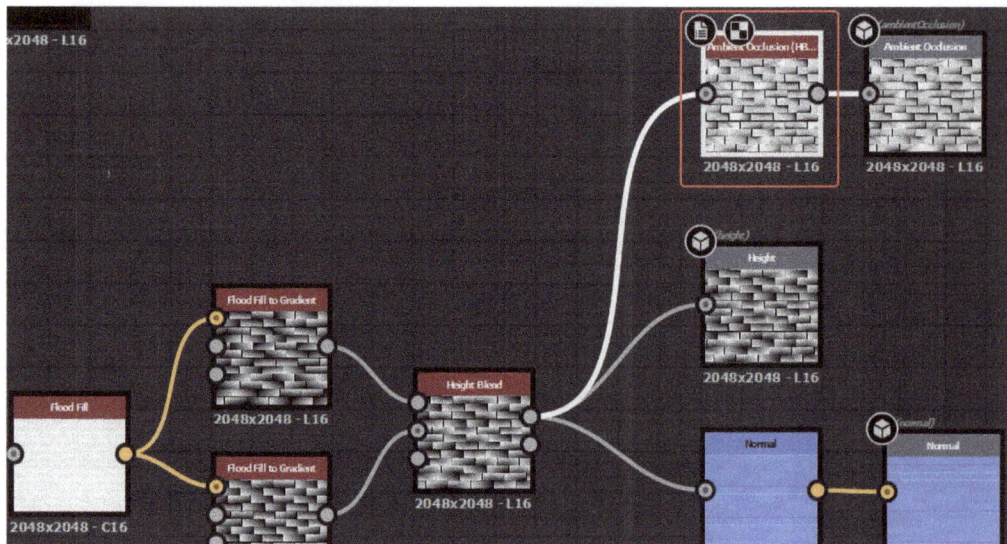

Figure 8.48 – Adding Ambient Occlusion (HBAO)

6. Now, add the **Curvature Smooth** node and connect the **Normal** map node's output to the **Curvature Smooth** node's **Normal** input, as shown in *Figure 8.49*:

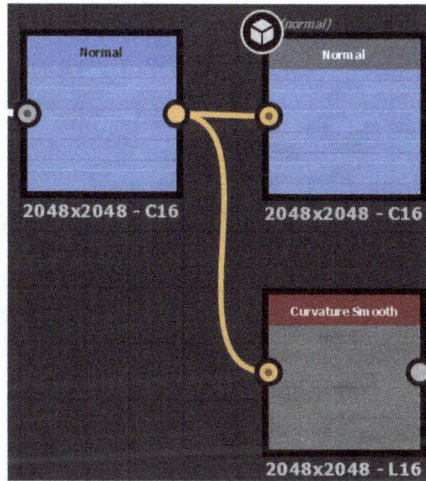

Figure 8.49 – Adding the Curvature Smooth node

7. Once all the nodes are ready, it is time to connect them. Therefore, connect **Ambient Occlusion (HBAO)** to the **Dirt** node's **Ambient Occlusion** input, **Curvature Smooth** to the **Dirt** node's **Curvature** input, and **Normal** to the **Dirt** node's **World Space Normal** input, as shown in *Figure 8.50*:

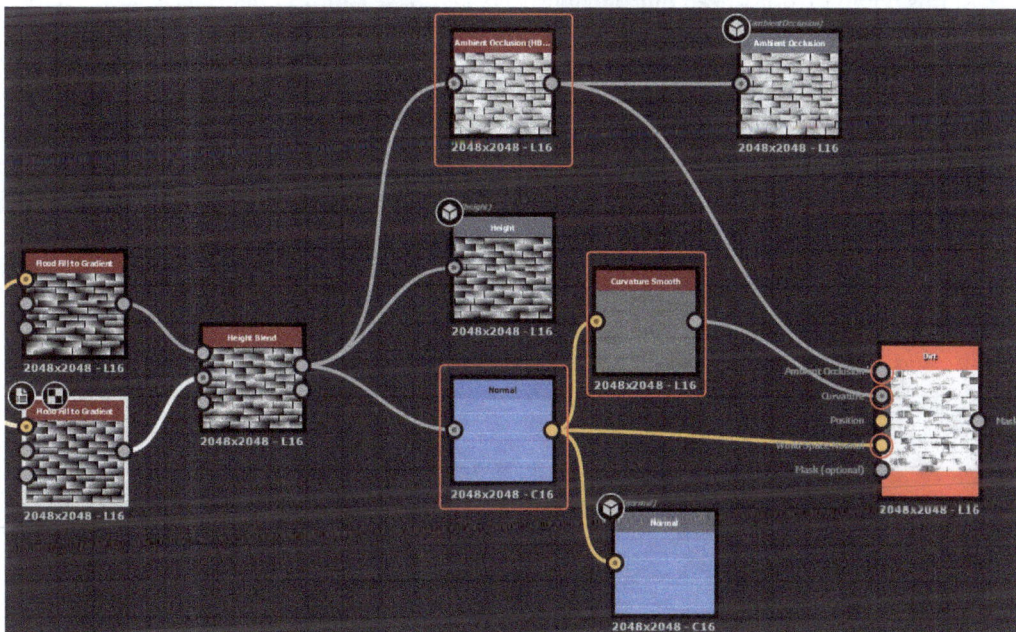

Figure 8.50 – Adding nodes to the Dirt node's inputs

8. Now, we need to blend our **Dirt** map with the dirty-looking color. Therefore, add a dark brownish **Uniform Color** node and a **Blend** node, and connect the dark brownish **Uniform Color** node's output to the newly created **Blend** node's **Foreground** input and the **Dirt** node's **Mask** output to the newly created **Blend** node's **Opacity** input, as shown in *Figure 8.51*:

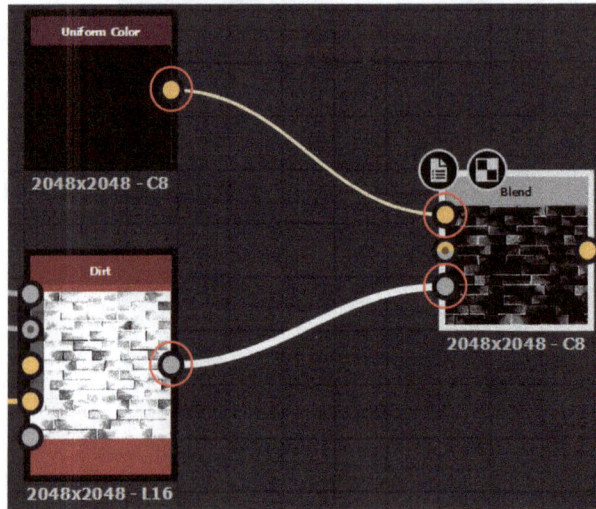

Figure 8.51 – Connecting the Uniform Color and Dirt nodes to the Blend node

At this point, we don't need our previously created **Base Color** node. Hence, we need to disconnect it from its **Blend** node in order to create a new **Base Color** node with the **Dirt** effect. However, you should only disconnect it and not delete it:

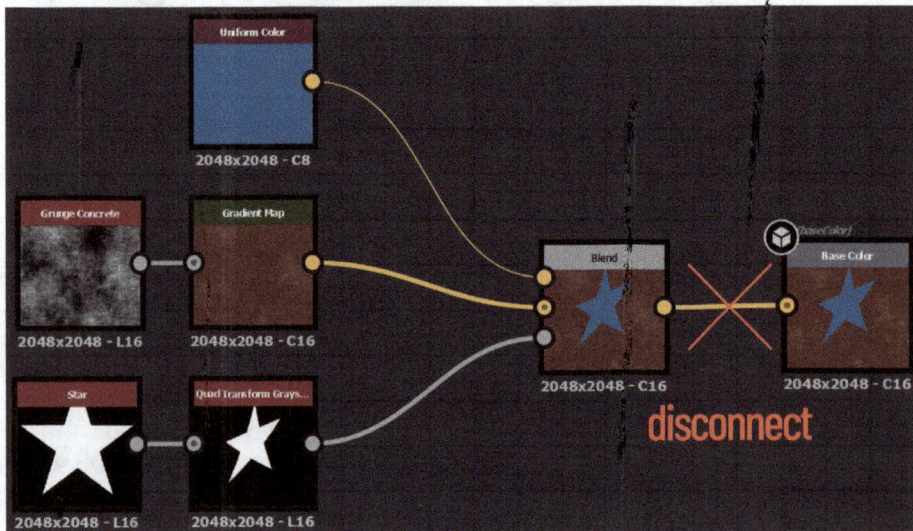

Figure 8.52 – Disconnecting the Base Color node

9. Now, move all the **Base Color**-related nodes to the bottom of the **Height Blend**-related nodes. Connect the old **Blend** node's output to the newly created **Blend** node's **Background** input and the newly created **Blend** node's output to the **Base Color** node's input:

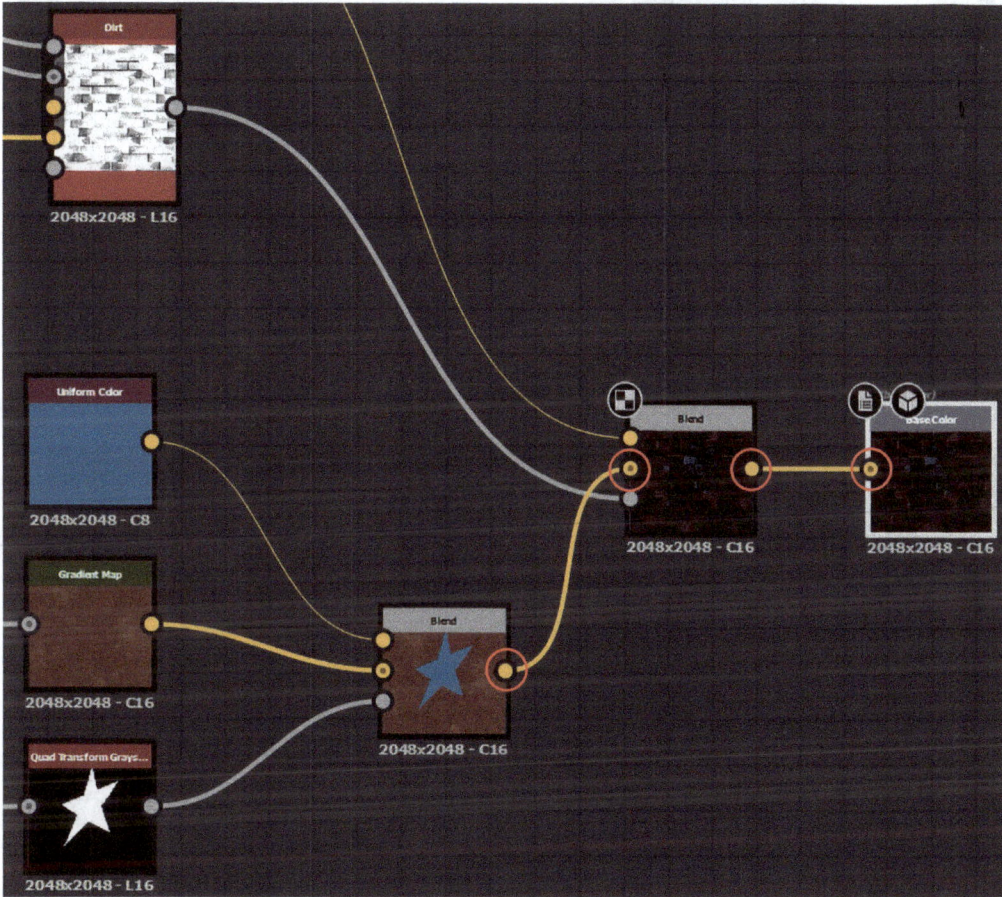

Figure 8.53 – Connecting the old and new Blend nodes to Base Color

10. Double-click the **Dirt** node and change the following settings under **INSTANCE PARAMETERS** (which are specifically related to the selected node):

*   **Dirt Level**: 0 . 7, **Dirt Contrast**: 0 . 2, **Grunge Amount**: 0 . 4

*   **Edges Masking**: 0 . 3, **Grunge Scale**: 1

These parameters will produce a realistic **Dirt** effect on the brick wall, as shown in *Figure 8.54*.

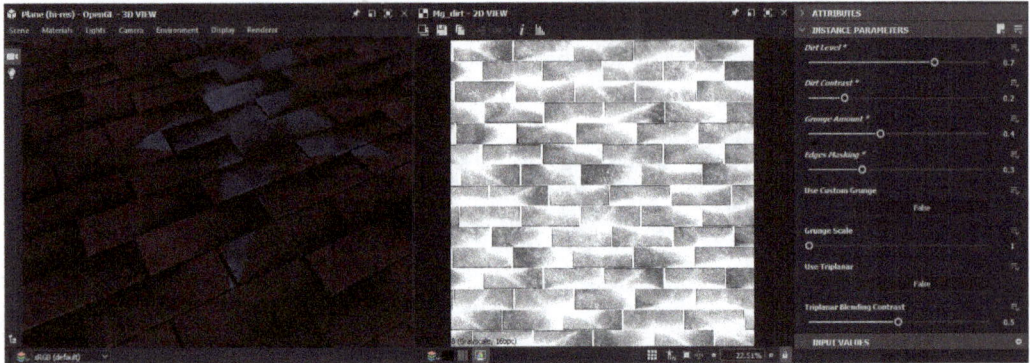

Figure 8.54 – The Dirt node effect

11. Similarly, you can add the **Dust** node, and connect the existing **Ambient Occlusion (HBAO)** node's output to the **Dust** node's **Ambient Occlusion** input, and the **Normal** map node's output to the **Dust** node's **World Space Normal** input:

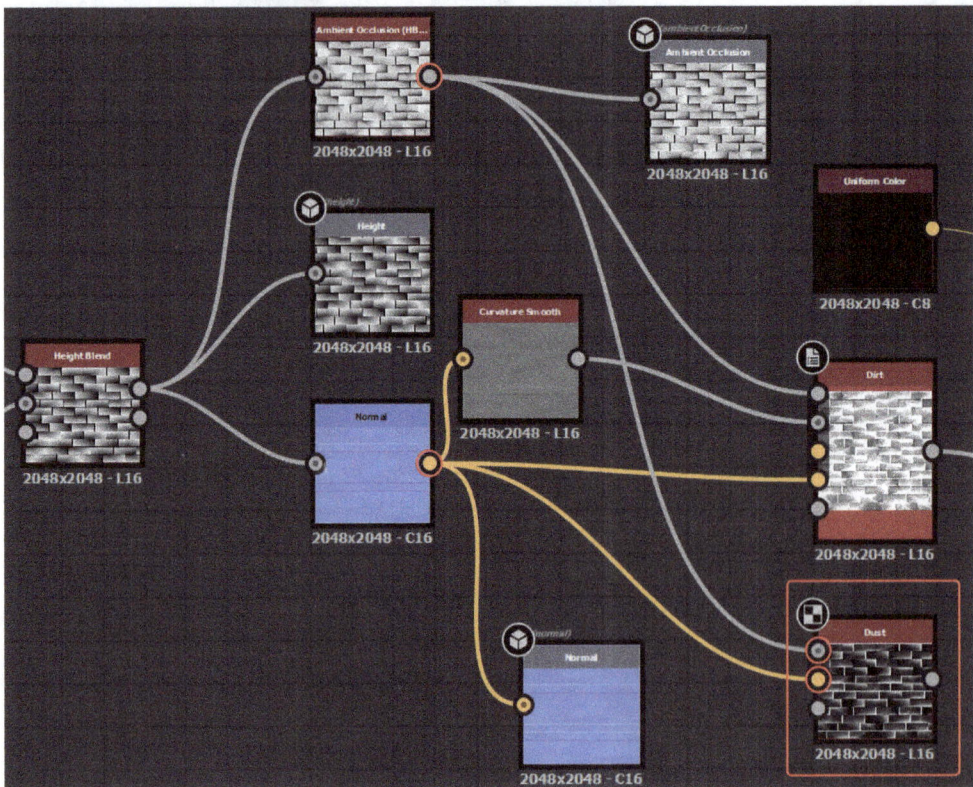

Figure 8.55 – Adding the Dust node

12. Now, add a new **Blend** node for the **Dust** effect. We will name this **Blend** node `Dust Blend` and the older one `Dirt Blend` to avoid confusion.

13. Add a new **Uniform Color** node that resembles sand. Then, connect the sandy **Uniform Color** output to the **Dust Blend** node's **Foreground** input, and the **Dust** node's **Mask** output to the **Dust Blend** node's **Opacity** input, as shown in *Figure 8.56*:

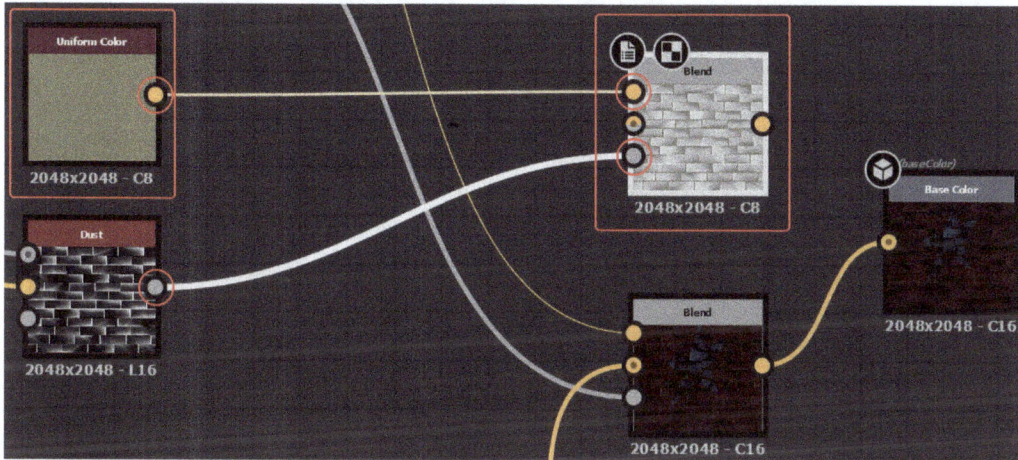

Figure 8.56 – Connecting the new sandy Uniform Color node and the Dust node to the new Blend node

14. Disconnect the **Dirt Blend** node and the **Base Color** node in order to create a new node system. Now, connect the **Dirt Blend** node's output to the **Dust Blend** node's **Background** input and connect the **Dust Blend** node's output to the **Base Color** node's input, as shown in *Figure 8.57*:

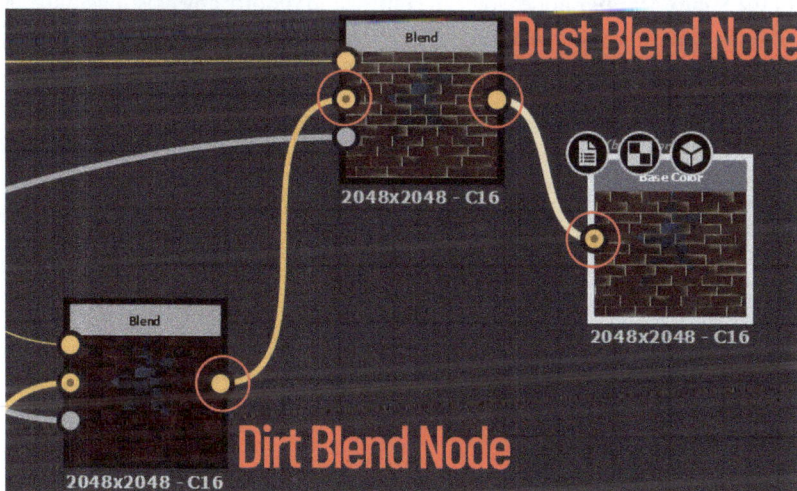

Figure 8.57 – Adding the Dust effect to Base Color

15. You will now notice a visible dust effect on the brick wall. If you want, you can also change the **INSTANCE PARAMETERS** settings of the **Dust** node:

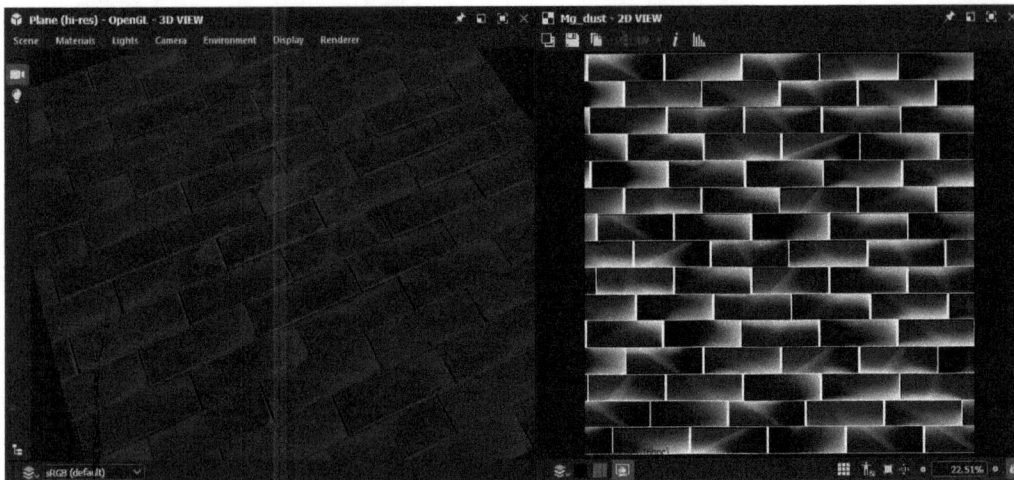

Figure 8.58 – The Dust node's final output

You can also use the dirt effect or the dust effect by themselves. However, we blended both of these nodes so that you could also learn more about the **Blend** node. In the following section, we will learn about the **Shape Mapper** node.

The Shape Mapper node

With the help of the **Shape Mapper** node, you can control and modify any existing **Shape** node, **Star** node, or other similar nodes. The **Shape Mapper** node manipulates your existing shape into a **Circle** or **Polygon** form using other **INSTANCE PARAMETERS** settings.

Let us do some experiments on our existing **Star** node:

1. Add the **Shape Mapper** node in the **GRAPH** window, and then connect the **Quad Transform Grayscale** node's output to the **Shape Mapper** node's input and the **Shape Mapper** node's output to the **Blend** node's **Opacity** input, as shown in *Figure 8.59*:

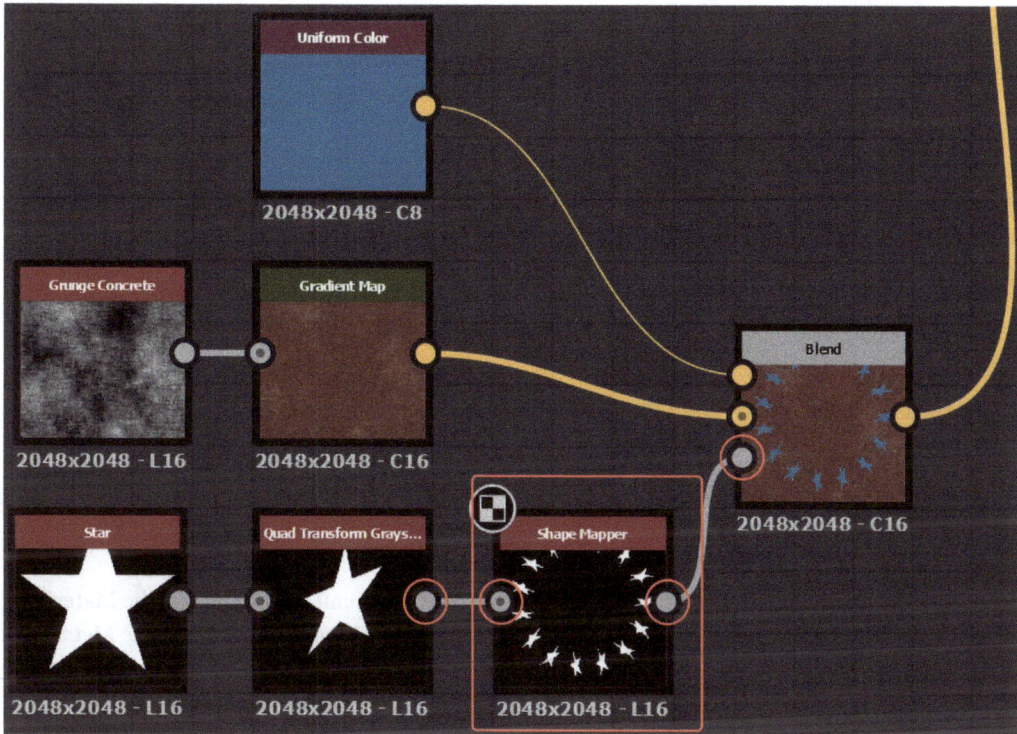

Figure 8.59 – Adding the Shape Mapper node

2. The final output of the **Shape Mapper** node will look similar to *Figure 8.60*. However, you can change the **Shape Mapper** node's **INSTANCE PARAMETERS** settings if you want a different result:

Figure 8.60 – The Shape Mapper node's final output

I hope it is now clear to you the purpose of the **Shape Mapper** node and how it works. In the following section, we will learn about how we can work with the **Shape Splatter** node.

The Shape Splatter node

This is an extremely complicated node that should be utilized with **Shape Splatter Data Extract**, **Shape Splatter to Mask**, and **Shape Splatter Blend**. Nevertheless, it is optional to use these conjunction nodes.

It is similar to **Tile Sampler** or **Tile Generator** in that it is used to produce shapes, but it uses a dynamic, non-destructive method that gives the user control over each stage through a multi-level system.

Shape Splatter is a more sophisticated variant of **Flood Fill**, which creates the map and associated data in a single phase, as opposed to **Flood Fill**, which requires a basic input map from an external source.

However, don't worry, I will make it easier for you to comprehend this node. We need to start by opening a new **Metallic Roughness** substance graph.

1. Create a new **Metallic Roughness** substance graph and name it `Shape Splatter Test`.

2. Go to **3D VIEW** and choose **Plane (hi-res)** from the **Scene** menu. Then, click on the **Material** menu, select the **Default** menu, and choose **Tesselation + Displacement** from the **Metallic Roughness** menu.

3. Once the **Height** parameter has loaded, change **Scale** to 5 and **Tessellation Factor** to `16`:

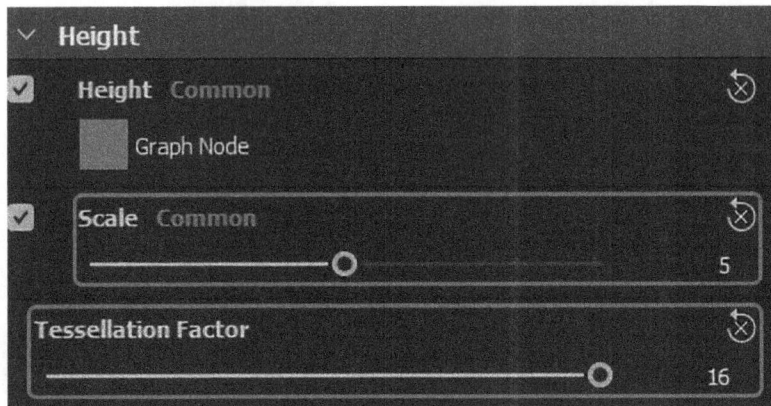

Figure 8.61 – Adjusting the material height settings from the PROPERTIES window

4. Now, add the **Shape Splatter** node into the **GRAPH** window. However, we need a **Background Height** node and two **Pattern** nodes to connect to the **Shape Splatter** node. Therefore, we will add a **Perlin Noise** node, a **Shape** node, and a **Star** node, as shown in *Figure 8.62*:

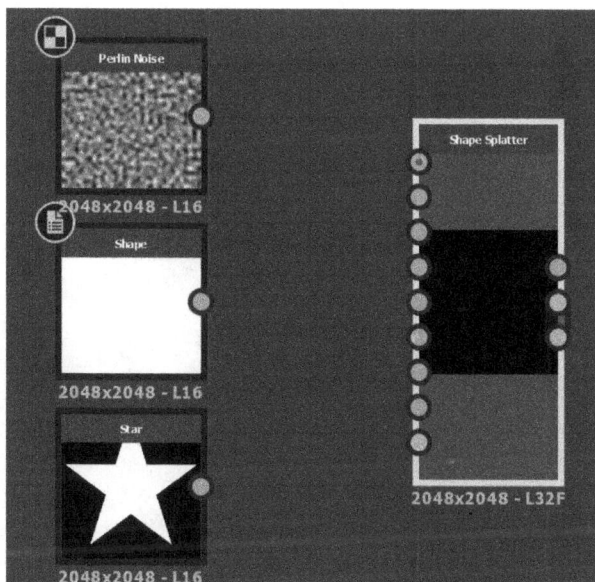

Figure 8.62 – The Shape Splatter, Perlin Noise, Shape, and Star nodes

5.   We need to change the shape inside the **Shape** node, so double-click the **Shape** node, and from the **PROPERTIES** window, change **Pattern** to **Disc** and **Scale** to 0.7 under the **INSTANCE PARAMETERS** settings:

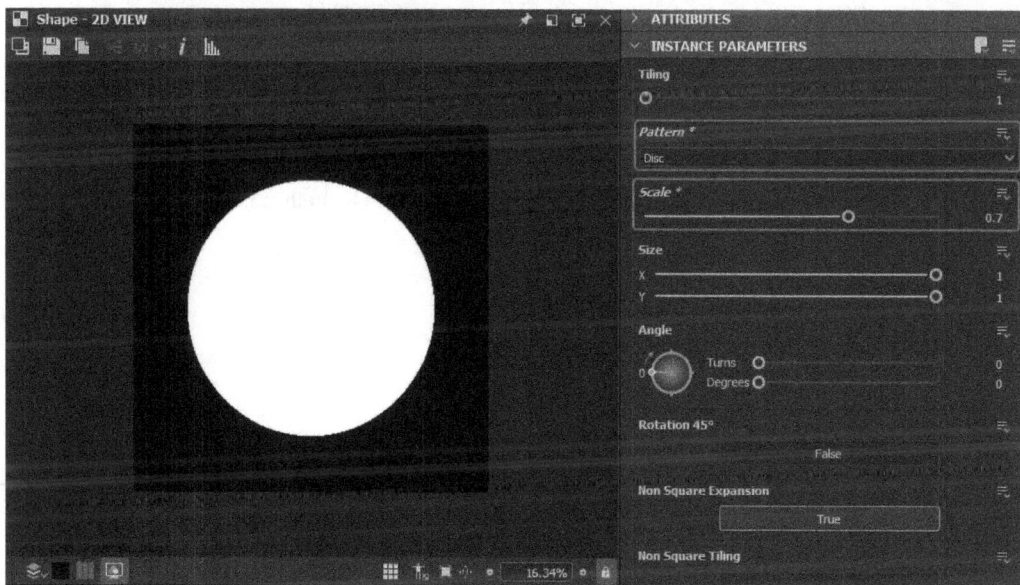

Figure 8.63 – Adjusting the Shape node

6. Now, connect **Perlin Noise** to the **Shape Splatter** node's **Background Height**. Then, we have to connect the **Shape** and **Star** nodes to two **Pattern** nodes. However, we only have one – therefore, double-click the **Shape Splatter** node, and under the **INSTANCE PARAMETERS** settings, change **Pattern Input Number** to 2, and then connect the **Shape** node to the **Shape Splatter** node's **Pattern 1** input and the **Star** node to the **Pattern 2** input, as shown in *Figure 8.64*:

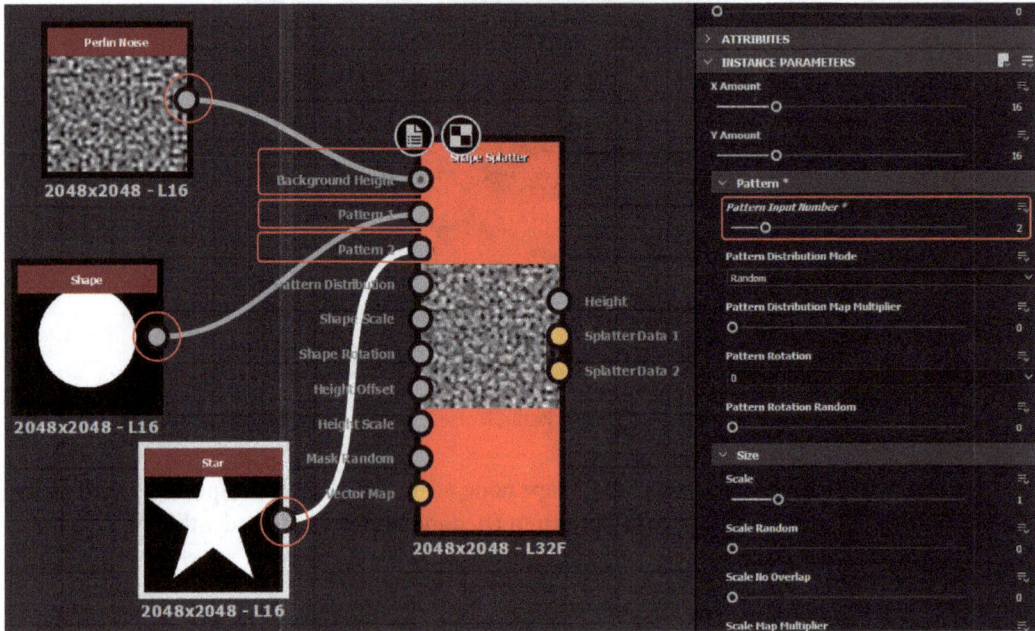

Figure 8.64 – Connecting the nodes to Shape Splatter

7. Now, connect the **Shape Splatter** node's **Height** output to the **Normal** map node, the **Height** map node, and the **Ambient Occlusion** node via **Ambient Occlusion (HBAO)**, as shown in *Figure 8.65*:

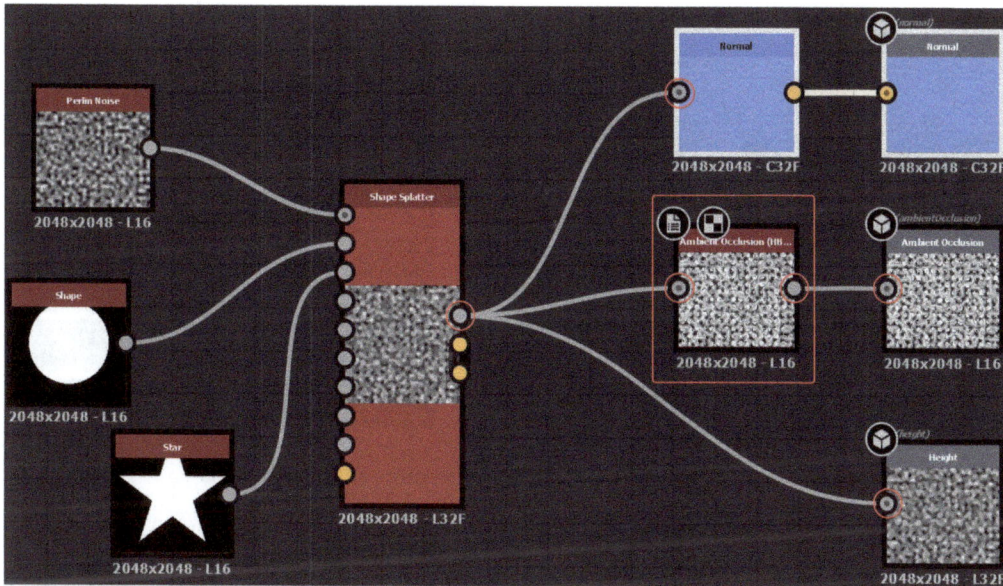

Figure 8.65 – Connecting the Shape Splatter node to Normal, Height, and Ambient Occlusion

8. Double-click on the **Perlin Noise** node and change its **Scale** to 12 under **INSTANCE PARAMETERS** in the **PROPERTIES** window.

9. Now, double-click on the **Shape Splatter** node to change the following parameters in the **PROPERTIES** window:

   - **INSTANCE PARAMETERS: X Amount** = 8, **Y Amount** = 8 (to reduce the number of patterns)
   - **Size: Scale** = 0.6 (to reduce the overall size of the patterns)
   - **Position: Position Random** = 0.5 (to randomly place the patterns)
   - **Rotation: Rotation Random** = 0.17 (to randomize the rotation angle of each pattern)
   - **Height: Height Offset** = 1 (to increase the height distance of patterns)
   - **Height: Height Scale** = 0.97 (to increase the height of displacement)
   - **Height: Conform to Background** = 1 (this will allow all the patterns to conform with **Background Height** and follow its height peaks)

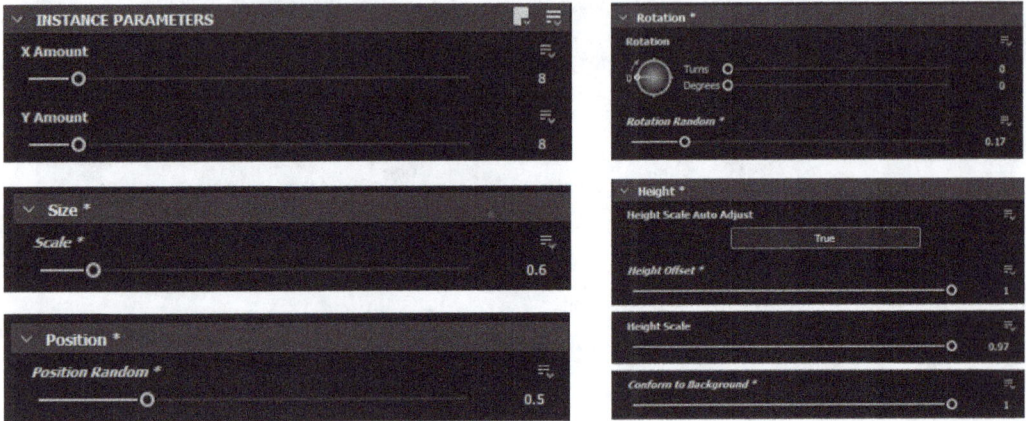

Figure 8.66 – Applying the changes to the Shape Splatter parameters

Let us add some color to the **Shape Splatter** node using the **Shape Splatter Blend Color** node.

10. Add the **Shape Splatter Blend Color** node and increase its **Pattern** input to 2, and connect three different **Uniform Color** nodes to the **Shape Splatter Blend Color** node's **Background**, **Pattern 1**, and **Pattern 2** inputs.

11. Then, connect the **Shape Splatter** node's **Splatter Data 1** and **Splatter Data 2** outputs to the **Shape Splatter Blend Color** node's **Splatter Data 1** and **Splatter Data 2** inputs, and connect the **Shape Splatter Blend Color** node's output to the **Base Color**'s input, as shown in *Figure 8.67*:

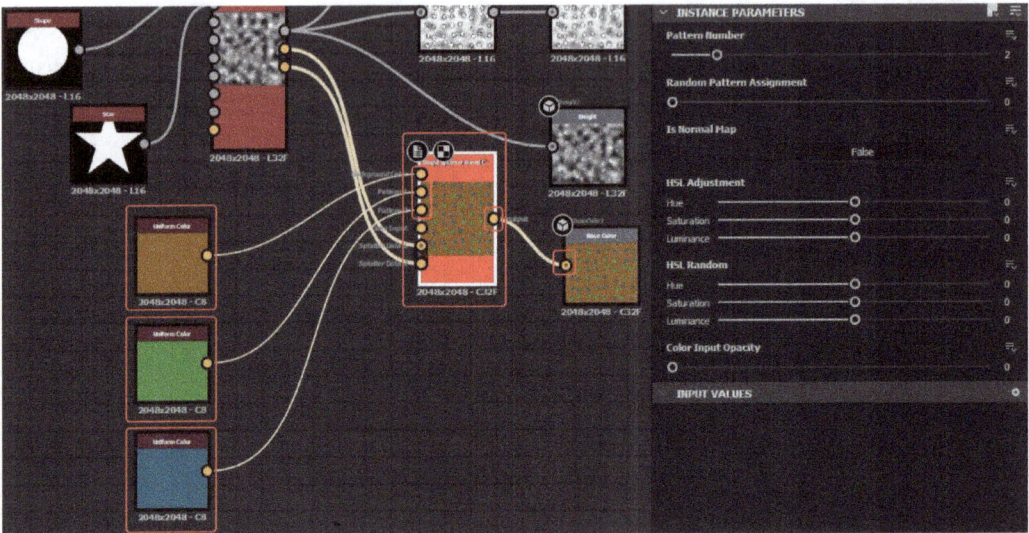

Figure 8.67 – Connecting the Shape Splatter node to the Shape Splatter Blend Color node

The **Shape Splatter** and **Shape Splatter Blend Color** nodes will produce a combination of different patterns conforming to the height peaks of the background in different colors, as shown in *Figure 8.68*:

Figure 8.68 – The final result of the Shape Splatter node and the Shape Splatter Blend Color node

Hopefully, you got the idea behind the **Shape Splatter** node. You can use this node in combination with the **Shape Splatter Blend Color** node and create interesting effects, such as adjusting the **Height** map of coins on the ground or the **Base Color** of stickers placed on the wall. In the following section, we will learn about the **Histogram Scan** node.

Histogram Scan

Histogram Scan is a node that offers an easy way to remap the contrast and brightness of input grayscale maps. This node is very basic yet helpful and can be used to dynamically *grow* and *shrink* masks.

So, let us work with this node:

1. In *Figure 8.68*, you will notice that the **Perlin Noise** node's height is quite smooth – what if we need **Perlin Noise** to have flat height peaks like the Grand Canyon? This is where **Histogram Scan** comes in handy. Therefore, add the **Histogram Scan** node into the **GRAPH** window.

2. Connect the **Perlin Noise** node's output to the **Histogram Scan** node's input and connect the **Histogram Scan** node's output to the **Shape Splatter** node's **Background Height** input.

3. Then, change the **Histogram Scan** node's **Position** parameter to 0.43 and the **Contrast** parameter to 0.69 in the **PROPERTIES** window under the **INSTANCE PARAMETERS** settings, as shown in *Figure 8.69*:

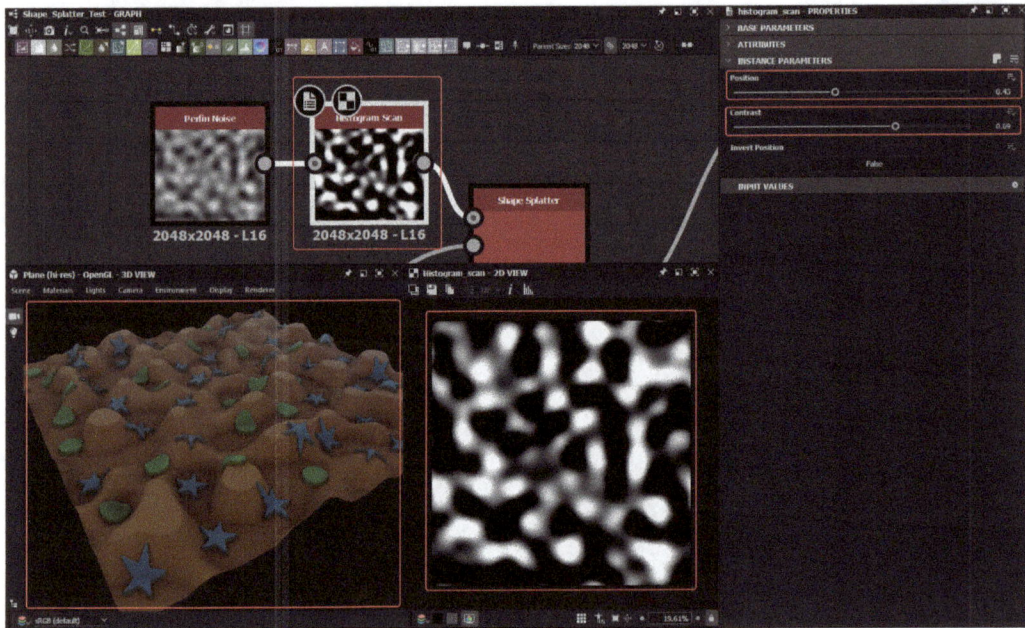

Figure 8.69 – The final output of the Histogram Scan node

I hope the **Histogram Scan** node's purpose is clear to you. With this node, you can easily turn fuzzy and blurry shapes sharper, and this node can be also used instead of the **Curve** node.

Summary

As we are finally done with this chapter, I hope you comprehended all the topics and gained the skills to work with the nodes explained in this chapter. With the knowledge and expertise, you have gained in this chapter, you can now create materials such as brick walls, cracked walls, chipped paint, and other damage effects.

In the next chapter, you will get a chance to explore the Blending Modes in Adobe Substance 3D Designer. This feature is crucial owing to its ability to blend different nodes using procedural method to create interesting results. Every blending mode in the next chapter will be explained in detail, so get ready for it!

9
Blending Modes in Adobe Substance 3D Designer

In digital image editing and computer graphics, blend modes (also known as **blending modes** or **mixing modes**) are used to control how two layers are merged with one another. In most applications, the bottom layer (**Background Input** in Substance 3D Designer) is simply covered by whatever is in the upper layer (**Foreground Input** in Substance 3D Designer) as the usual blend mode. Nevertheless, there are many alternative methods to mixing two layers because each pixel has a numerical value.

On a per-pixel level, the mix operation in Substance 3D Designer is similar to any other math operation. It takes one pixel each from two photos and performs a mathematical operation on them to produce an output. If a pixel is given a value of 0.5, adding it would produce $0.5 + x$. The **Blend** node only produces values between 0 and 1.

While doing calculations, the pixel processor can employ values above and below 1; however, the output will clamp min/max values that are between 0 and 1.

In this chapter, you will get a chance to explore the blending modes in Adobe Substance 3D Designer in a much simpler and easier way. You will also learn why this feature is so crucial and what its purpose is. Every blending mode in this chapter will be explained in detail with a step-by-step guide. But first, you must learn about a few more nodes that are important to work in combination with the **Blend** node.

We will cover the following topics in this chapter:

- Understanding the **Transformation 2D** node, the **Levels** node, and the Grayscale Height value
- Copy blending mode
- Add and Subtract blending mode
- Min (Darken) and Max (Lighten) blending mode
- Multiply and Divide blending mode
- Screen and Soft Light blending mode

- Add Sub and Overlay blending mode
- Switch blending mode

Understanding the Transformation 2D node, the Levels node, and the grayscale height values

There are a couple of nodes we need to know about before learning about blending modes because these nodes are frequently used in combination with **Blend** nodes and other **Library** nodes that we studied in *Chapter 8, Nodes in Adobe Substance 3D Designer*.

Transformation 2D

The first node I would like to explain is called the **Transformation 2D** atomic node. The **Transformation 2D** node enables the scaling, rotation, tiling, and alteration of input proportions. It is comparable to utilizing Photoshop's transform tool (*Ctrl + T/Command + T*).

This node, which may be stretched or squashed, can increase tiling, remove tiling, position an image in a precise location, and do other helpful operations. However, it might not be a great fit for all applications, therefore the **Safe Transform**, **Non-Square Transform**, **Quad Transform**, and **Trapezoid Transform** nodes might be of interest.

Let us see how we can use Transformation 2D in Designer:

1. Open `Transformation_2D.sbs` from the given Substance 3D Designer exercise files, then go to **3D VIEW**, select the **Scene** menu, and click on **Load State File**. Once the **Load state file** window pops up, choose `Transformation_2D.sbsscn`.

2. After that, return to the **Scene** menu and choose **Plane (hi-res)**.

3. Right-click with the mouse in the **GRAPH** window and choose **View Outputs In 3D View**.

4. You will see a **GRAPH** setup, where the **Star** node is connected to a bunch of other nodes. We need to apply the **Transformation 2D** node between the **Star** node and the other connected nodes. Therefore, select the **Star** node and click on the **Transformation 2D** node icon from the atomic nodes bar.

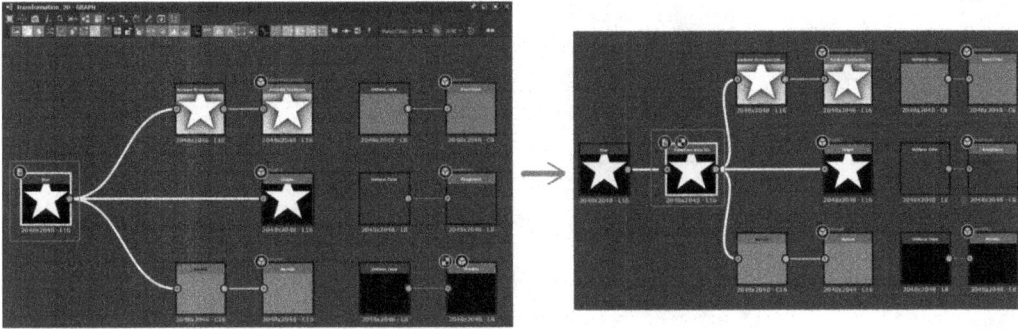

Figure 9.1 – Adding the Transformation 2D node

5. You can scale, rotate, move, and offset any connected node either by using the mouse or by adjusting the parameters in the **PROPERTIES** window as shown in *Figure 9.2*.

Figure 9.2 – Altering connected nodes with the help of the Transformation 2D node

6. You can also toggle the **Tiling** mode of this node from **Horizontal + Vertical** to **No Tiling**, by enabling it from the **PROPERTIES** window under the **Select method of inheritance** icon and choosing **Absolute**, and then choosing your **Tiling Mode** as shown in *Figure 9.3*.

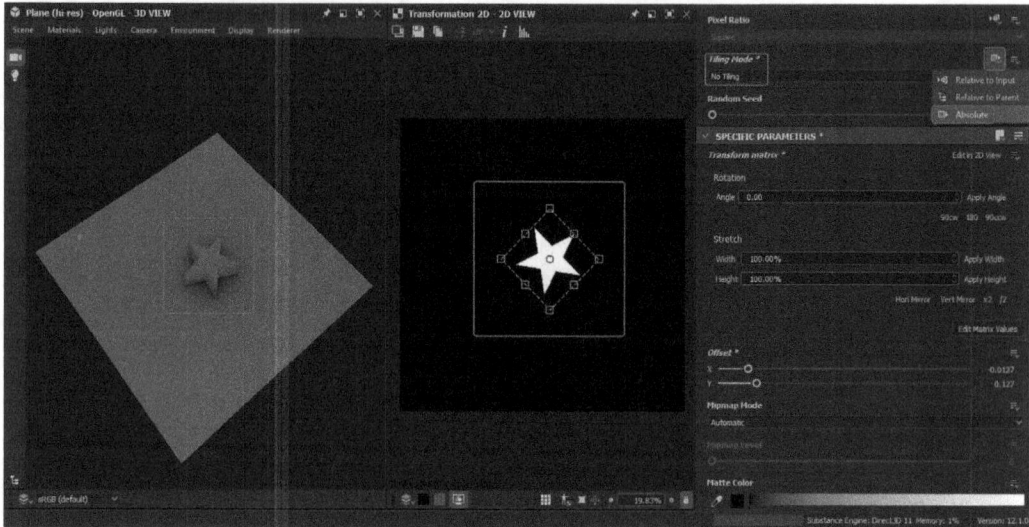

Figure 9.3 – Choosing Tiling Mode for the Transformation 2D node

Hopefully, you are familiar with the **Transformation 2D** node now and will have fun using it in your future projects. In the next section, we will learn about the **Levels** node.

The Levels node

Levels are another useful **Atomic** node. By adjusting input and output remap factors in the **Levels** node's interface, which is reminiscent of other 2D picture editors, you can remap the tones of an input. A **Histogram preview** option is also included in the settings.

As it offers the most exact and accurate interface for modifying values, it is one of the most important and practical nodes in Substance 3D Designer and is frequently used to remap and adjust values in a graph.

Let us work on it and see why this node is so vital:

1. Open `Grayscale.sbs` from the given Substance 3D Designer exercise files, then go to **3D VIEW**, select the **Scene** menu, and click on **Load State File**. Once the **Load state file** window pops up, choose `Grayscale.sbsscn`.

2. After that, return to the **Scene** menu and choose **Plane (hi-res)**.

3. Right-click with the mouse in the **GRAPH** window and choose **View Outputs In 3D View**.

4. As mentioned earlier, the **Levels** node is used for various purposes; however, we will use it for adjusting the heights of different maps. Once `Grayscale.sbs` and its state file are loaded, you will notice that it consists of a **Star** and **Shape** nodes combination.

The **Height** maps of both maps are similar, but what if we want to decrease the height of one of the maps, for example, the **Shape** node? To do that we can add the **Levels** node to the **Shape** node.

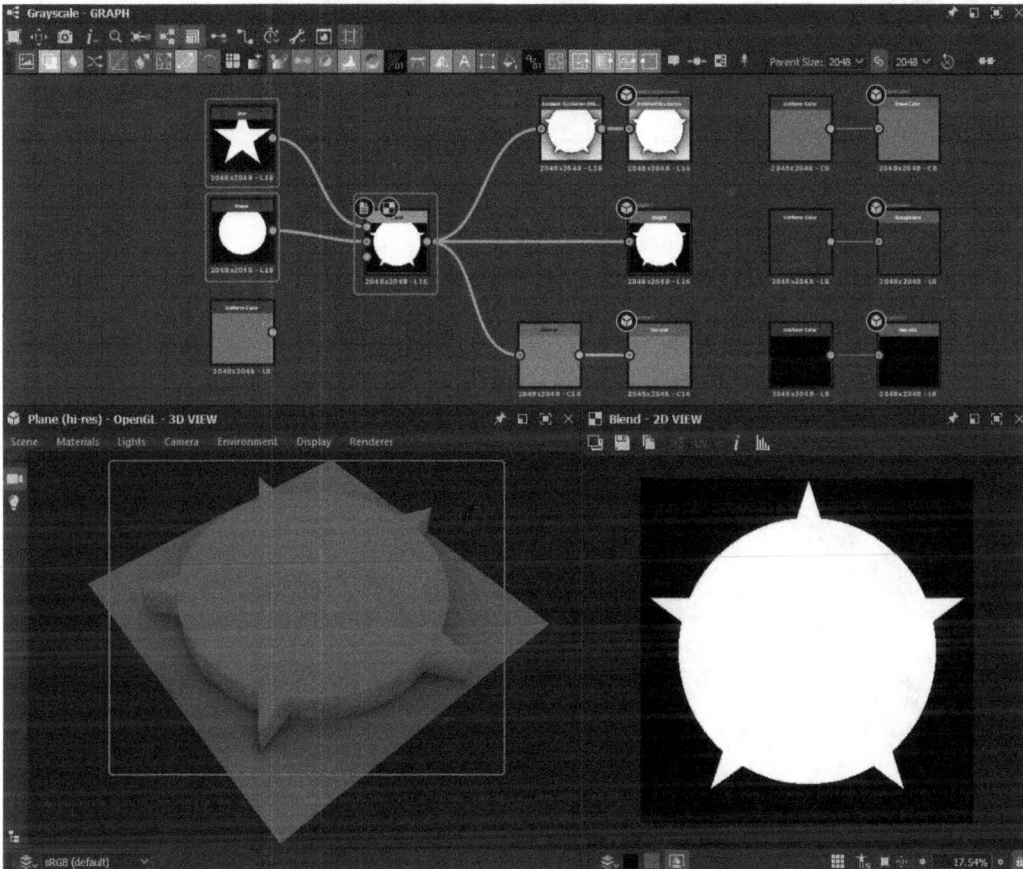

Figure 9.4 – Grayscale.sbs

5. Now select the connection between the **Shape** node and the **Blend** node and click on the **Levels** node from the **Atomic** node bar in the **GRAPH** window.

    This will create a **Levels** node between the **Shape** node and the **Blend** node. You will also notice that there is a **Histogram** in the **Levels** node's properties with a black and white point value slider at the bottom.

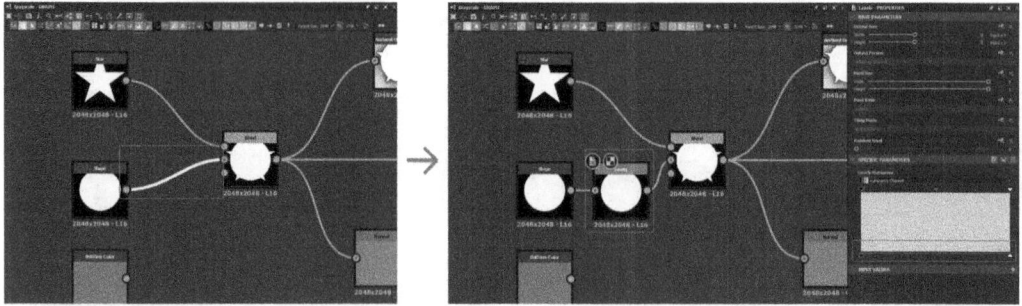

Figure 9.5 – Adding the Levels node

6. Let us slide the white point value to the left as shown in *Figure 9.6*. This will reduce the strength of the **Shape** node's brightness, which will result in its height reduction.

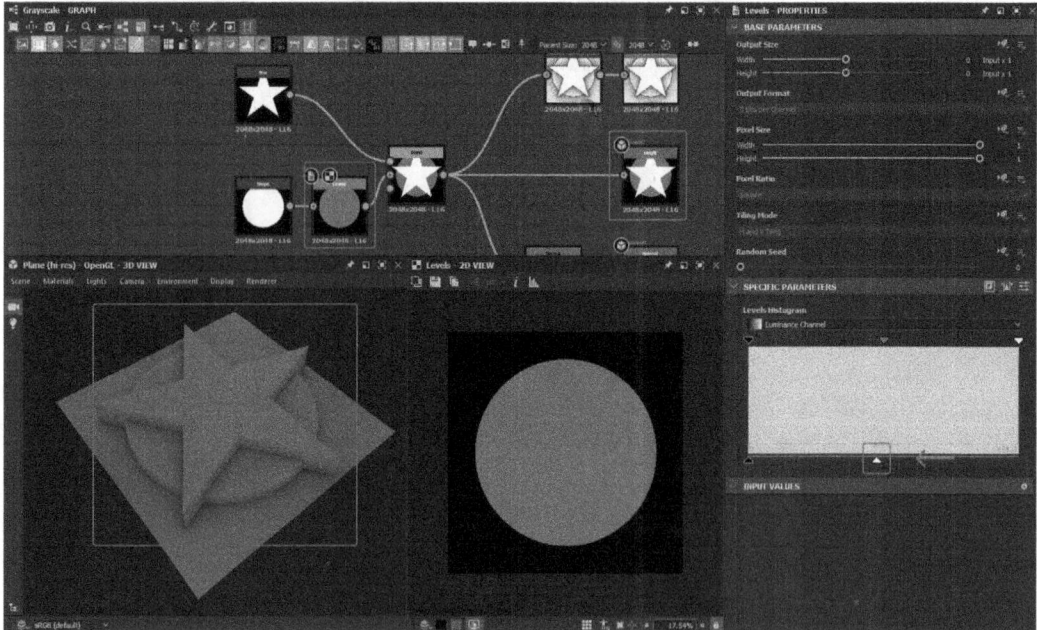

Figure 9.6 – Decreasing the height of the Shape node

You can also control the height of the **Star** node by adding levels to it, and a couple of other nodes in the same way if they are connected altogether.

When you want to control the height of any node, you cannot do it through the **Tesselation** + **Displacement** property because it will affect the whole **GRAPH** rather than the selected node.

This is where the **Levels** node comes in. You can define the height of the overall **GRAPH** through the **Tesselation + Displacement** property and then control the heights of individual nodes through the **Levels** node.

Hopefully, the use and working of the **Levels** node are clear to you. In the next section, we will learn about grayscale height values.

Grayscale height values

It is quite simple to add height to any node inside Substance 3D Designer; however, the problem occurs when you want to subtract height. This problem can be easily resolved by adding a grayscale height value.

So let us explore the concept of grayscale height value:

1. Open `Grayscale_Height_Value.sbs` from the given Substance 3D Designer exercise files, then go to **3D VIEW**, select the **Scene** menu, and click on **Load State File**. Once the **Load state file** window pops up, choose `Grayscale Height Value.sbsscn`.

2. After that, return to the **Scene** menu and choose **Plane (hi-res)**.

3. Right-click with the mouse in the **GRAPH** window and choose **View Outputs In 3D View**. You will get a simple **GRAPH** setup as shown in *Figure 9.7*.

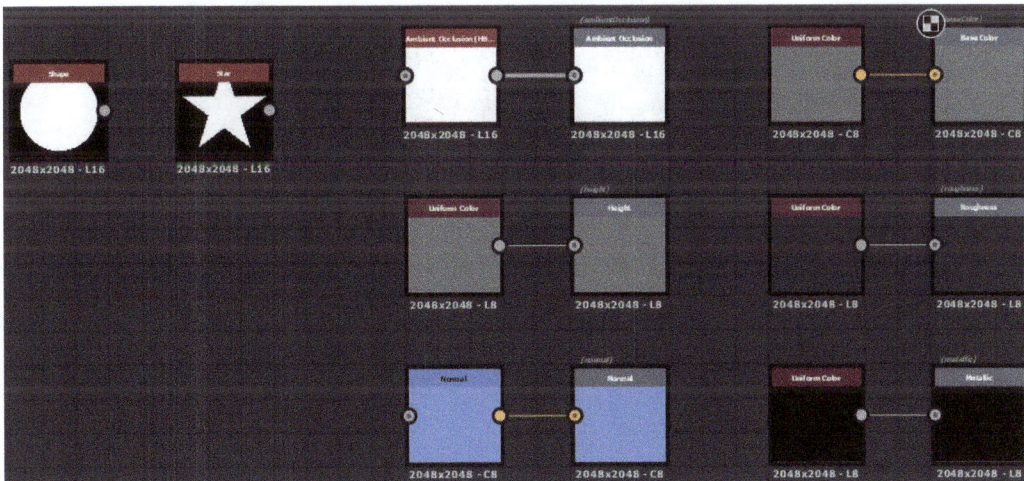

Figure 9.7 – Grayscale_Height_Value.sbs

4. Let us disconnect the grayscale **Uniform Color** from the **Height** map, add the **Shape** node to it, and also add it to **Ambient Occlusion (HBAO)** and the**Normal** map as shown in *Figure 9.8*.

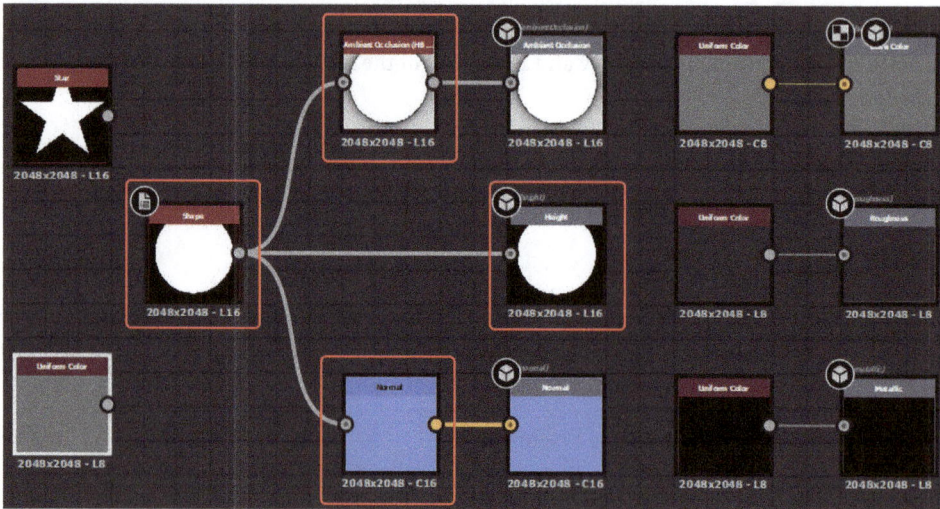

Figure 9.8 – Connecting the Shape node to Height, Ambient Occlusion, and Normal

5.   If you use the **Levels** node, you can only adjust the height of **Shape**; you cannot subtract its height.

6.   To subtract the height, add a **Blend** node to the **GRAPH** window, connect the **Shape** node to the **Blend** node's **Foreground** input, and grayscale **Uniform Color** to the **Blend** node's **Background** input, and the rest of the other nodes to the **Blend** node's **Output** as shown in *Figure 9.9*.

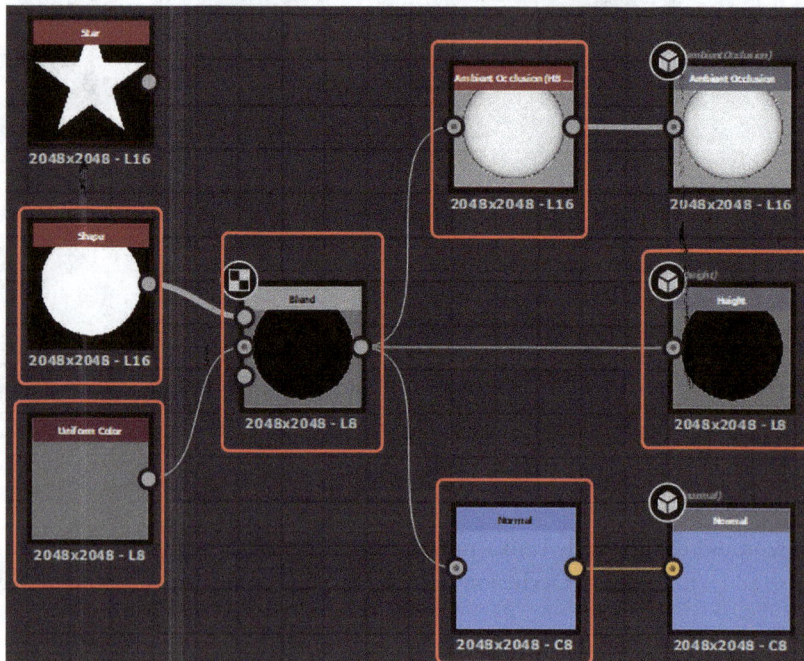

Figure 9.9 – Adding the Blend node

7. Now double-click the **Blend** node and change **Blending Mode** to **Subtract** from the **PROPERTIES** window. This will give a negative height to the **Shape** node, which wasn't possible without a grayscale **Uniform Color**.

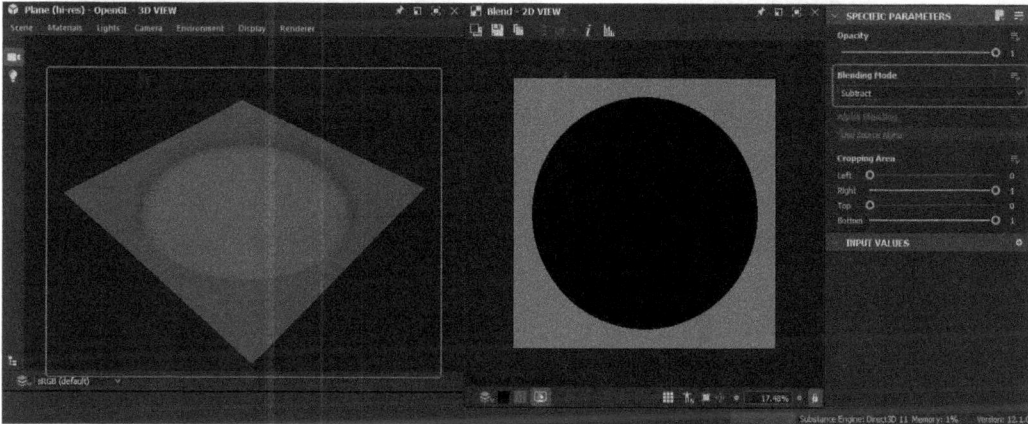

Figure 9.10 – Subtracting the Shape node from grayscale Uniform Color

Do not worry about **Blending Mode** right now because we will learn about it in detail in this chapter. Now **Grayscale** has become part of your **GRAPH**. It is always ideal to add a gray **Uniform Color** to the **Blend** node's **Background** input at the start, even if you don't want to subtract it in the first place. This will give you a middle color value so you can add other nodes to it and subtract from it.

To illustrate this, let us do an experiment in the next steps:

1. Change the existing **Blend** node's **Blending Mode** to **Add (Linea Dodge)**. This will create a normal-looking **Shape** node with a height that we usually see.

2. Now select the existing **Blend** node and click on the **Blend** node on the **Atomic** nodes bar as shown in *Figure 9.11*. This will add a new **Blend** node to your **GRAPH** window.

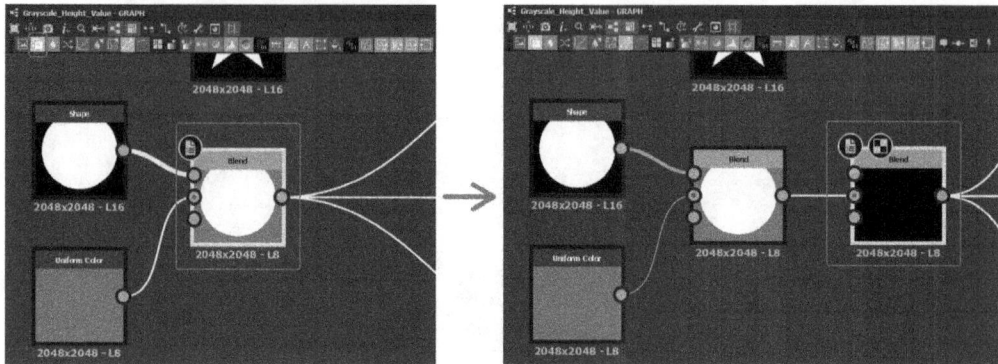

Figure 9.11 – Adding a new Blend node

3. Now connect the **Star** node that is already given in the **GRAPH** window to the **Foreground** input of the new **Blend** node.

4. Then double-click the **Star** node with the mouse and change **Scale** to 0 . 5 in the **PROPERTIES** window under **INSTANCE PARAMETERS** as shown in *Figure 9.12*.

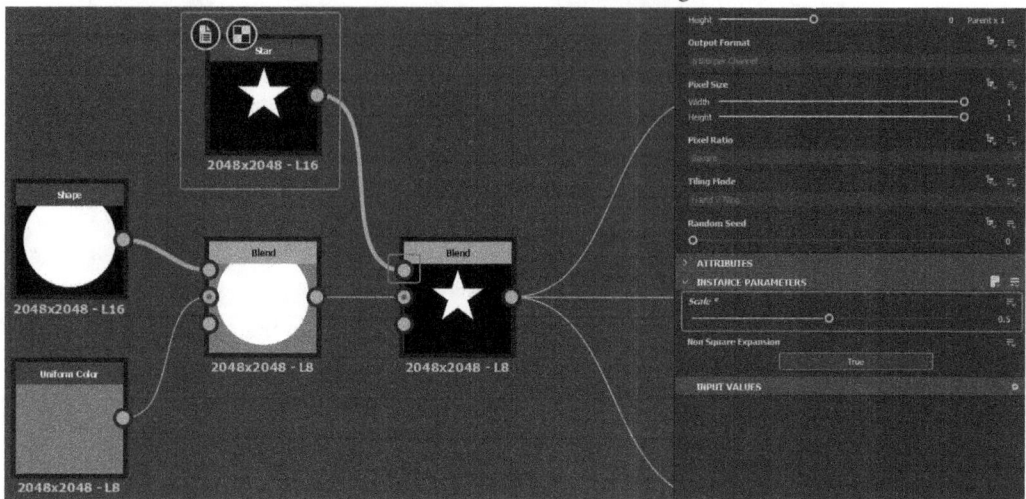

Figure 9.12 – Connecting the Star node to the new Blend node and scaling it up

5. Now double-click the new **Blend** node with the mouse and change **Blending Mode** to **Subtract** from the **PROPERTIES** window, and you will notice that the **Star** node has been subtracted from the **Shape** node as well as the **Plane (hi-res)** material.

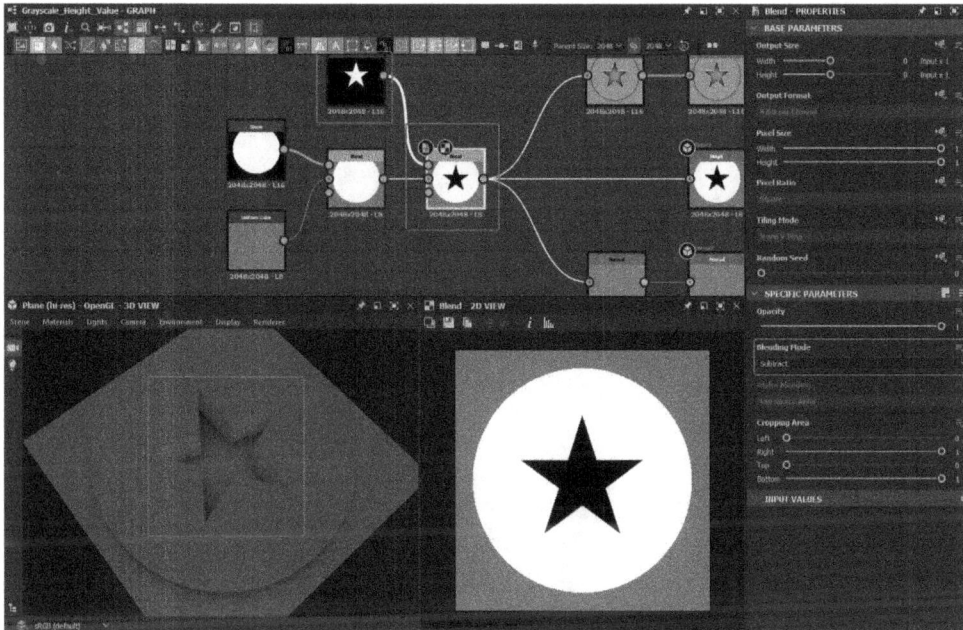

Figure 9.13 – Subtracting the Star node from the Shape node

If there was no grayscale **Uniform Color** applied in the first place, the **Star** node would be subtracted without any depth, which would cause unwanted results.

Hopefully, all these pre-requisite nodes are clear to you. Now, finally, it is time to learn about the blending modes in our next section.

The Copy blending mode

In the **Copy** blending mode, the foreground layer is stacked on top of the background. It is equivalent to the normal overlay mode of Adobe Photoshop.

It especially helps you while utilizing an opacity mask or when the foreground input is alpha.

Let us see how it works:

1. Open the `Copy_Blend_Node.sbs` file from the given Substance 3D Designer exercise files. Load the `Copy_Blend_Node.sbsscn state file`, then choose **Plane (hi-res)** from the **Scene** menu. Right-click with the mouse in the **GRAPH** window and choose **View Outputs In 3D View**.

2. You will notice the **Copy** blending mode's result looks the same as shown in *Figure 9.14*, which is basically a copy of **Foreground** over **Background**.

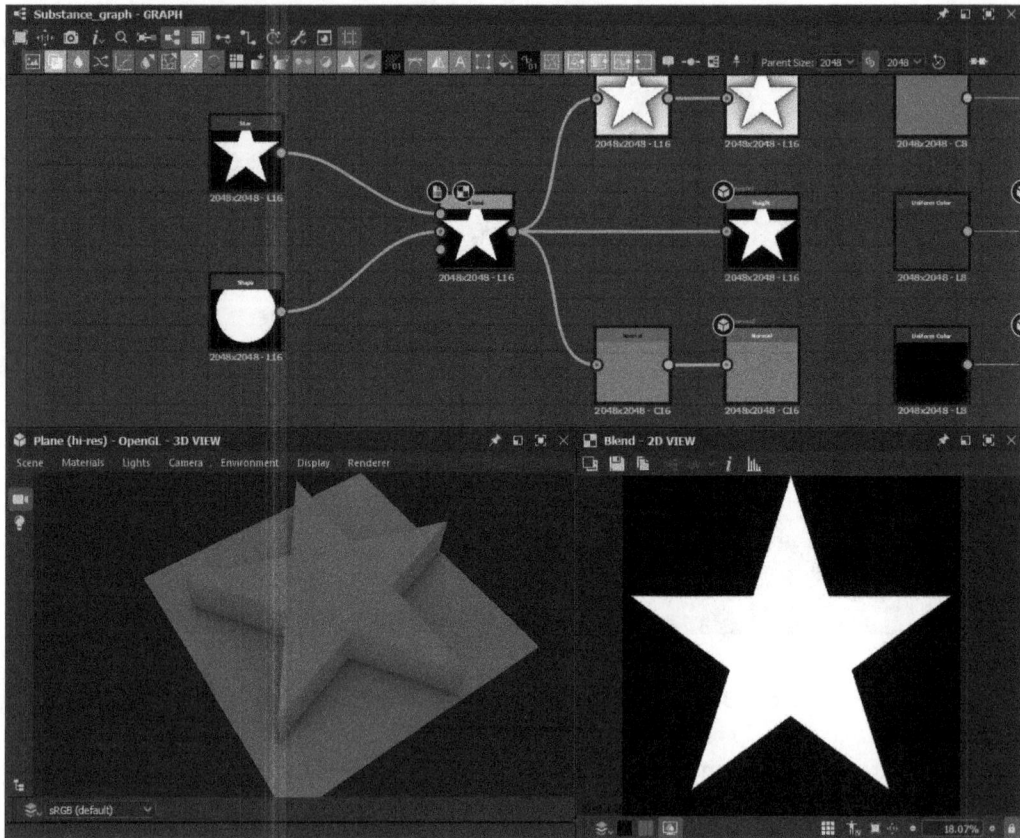

Figure 9.14 – Copy blending mode

3. The **Copy** blending mode is non-commutative, which means when you switch the **Foreground** input and **Background** input, the result changes.

4. So let us try doing that. To switch the **Foreground** input and **Background** inputs, select their connection lines and press *X* on your keyboard. This will switch the two inputs and you will get a different result. You can switch back by pressing *X* again if you are not satisfied with the result.

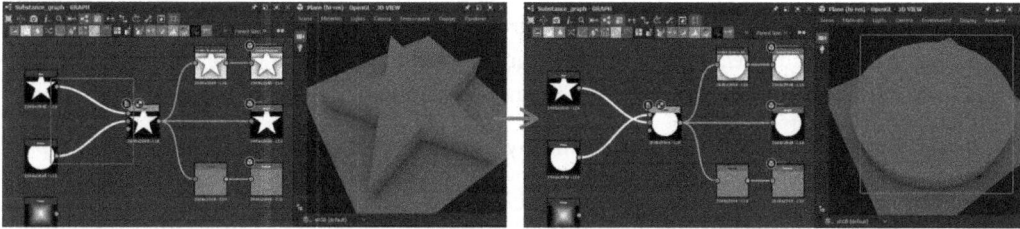

Figure 9.15 – The Copy blending mode's non-commutative nature

5. Now connect the given **Pyramid Shape** node to the **Opacity** input of the **Blend** node. By doing this, the **Pyramid** shape will become the mask for the **Blend** node, and **Foreground** and **Background** will only be affected in this **Opacity** area.

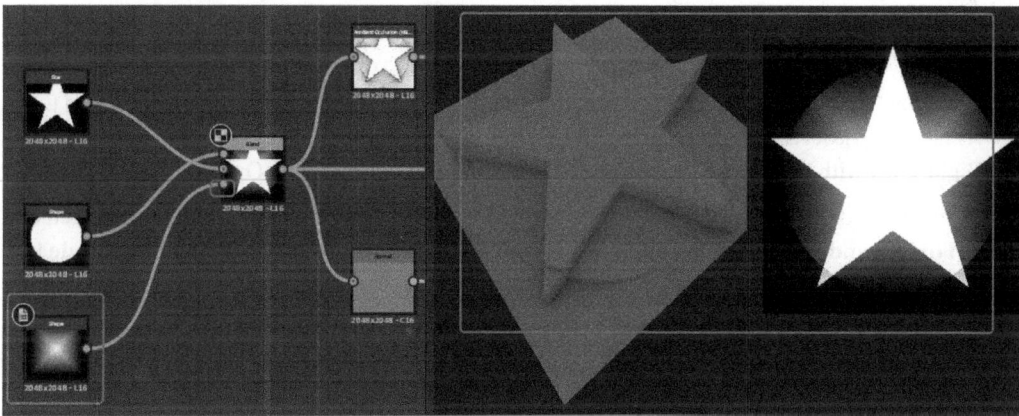

Figure 9.16 – Opacity map in Blending mode

Hopefully, you are now familiar with the first blending mode. In the next section, we'll learn about **Add** and **Subtract**.

The Add and Subtract blending modes

In the **Add** blending mode, the foreground is added to each corresponding background pixel. This mode is commutative, and the node's input order doesn't matter. Values over 1 are trimmed. It adds white values from the foreground and background together.

Whereas in the **Subtract** blending mode, the **Foreground** input value will be subtracted from each corresponding background pixel. If the outcome of the subtraction is less than 0, the value is restricted to 0 and turns the result completely black. This is a non-commutative blending mode.

In this mode, the white portion of the foreground layer is subtracted from the background.

You can return to the *Grayscale height value* section in this chapter to see how these modes work in practice. This is a commutative blending mode.

The Min (darkened) and Max (lightened) blending modes

The **Min** blending mode picks the lower values from either the foreground or background.

The **Max** blending mode picks the brighter values from either the foreground or background. The **Min** (darkened) and **Max** (lightened) nodes are commutative nodes.

Let us see how to work with these nodes in the following steps:

1. Open the Min_and_Max.sbs file from the given Substance 3D Designer exercise files, load the Min_and_Max.sbsscn state file, then choose **Plane (hi-res)** from the **Scene** menu. Right-click with the mouse in the **GRAPH** window and choose **View Outputs In 3D View**.

2. In the Min_and_Max.sbs Substance graph file, the **Gradient Linear 1** node is connected as **Foreground** while the **Clouds 3** node is connected as **Background**, and **Blending Mode** is set to **Min (Darken)**.

    You can also switch **Foreground** and **Background** without affecting the results because the **Min** and **Max** blending modes are commutative.

3. You will notice that when you blend **Gradient Linear 1** and **Clouds 3** using the **Min (Darken)** blending mode, the area that is affected is the solid and bright area, producing a darker blending result, as shown in *Figure 9.17*.

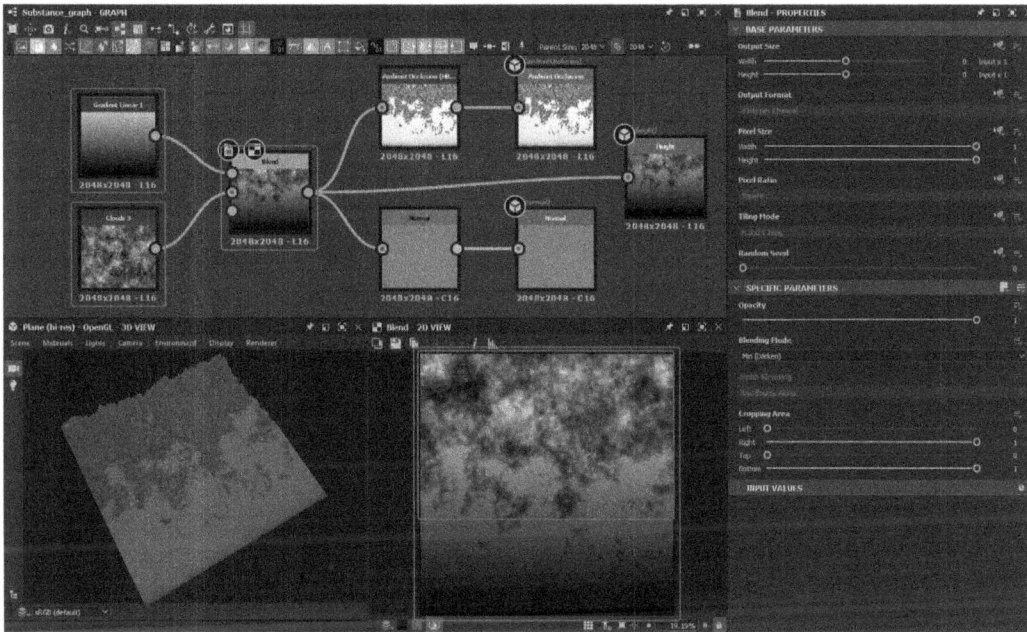

Figure 9.17 – Min (Darken) blending mode result

4. Now change **Blending Mode** to **Max (Lighten)**. With this blending mode, you will notice that the area that is affected is the solid and dark area, producing a brighter blending result, as shown in *Figure 9.18*.

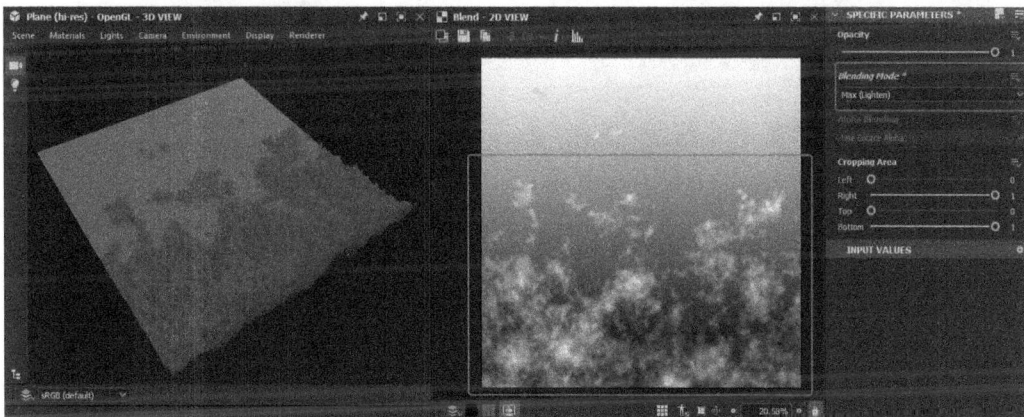

Figure 9.18 – Max (Lighten) blending mode result

Hopefully, this blending mode is clear to you now. In the next section, we will study the **Multiply** and **Divide** blending modes.

The Multiply and Divide blending modes

In the **Multiply** blending mode, the **Background** input value is multiplied by each **Foreground** pixel it corresponds to. Since each pixel's value falls between 0 and 1, the outcome is always lower (darker) than (or equal to) the value of the original pixel.

The **Multiply** blending mode is commutative.

In contrast, the **Divide** blending mode divides the input value of each background pixel by each matching foreground pixel. The **Divide** blending mode is non-commutative.

Let us see how to work with these nodes in the following steps:

1. Open the Multiply_and_Divide.sbs file from the given Substance 3D Designer exercise files, load the Multiply_and_Divide.sbs.sbsscn state file, then choose **Plane (hi-res)** from the **Scene** menu. Right-click with the mouse in the **GRAPH** window and choose **View Outputs In 3D View**.

2. In the Multiply_and_Divide.sbs.sbs Substance graph file (which is provided in the downloadable Substance_Designer_Exercise_Files folder), the **Shape** node is divided into two separate **Transformation 2D** nodes (keep its **tiling** mode to **No Tiling** so that it doesn't repeat). One **Transformation 2D** node is pushing the disc shape to the right side, while the other **Transformation 2D** node is pushing the disc shape to the left side.

   Both **Transformation 2D** nodes are connected to different inputs of the **Blend** node. This is how you can output several nodes from a single node.

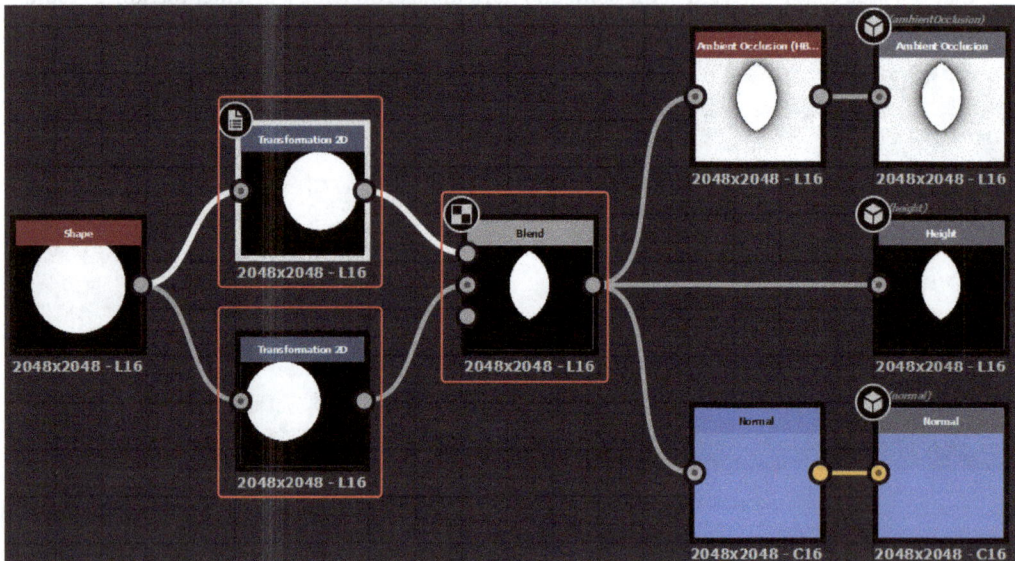

Figure 9.19 – Multiply_and_Divide.sbs.sbs

3. You will notice that when you blend the two **Transformation 2D** nodes using the **Multiply** blending mode, the overlapping area remains while the non-overlapping area is clipped, as shown in *Figure 9.20*.

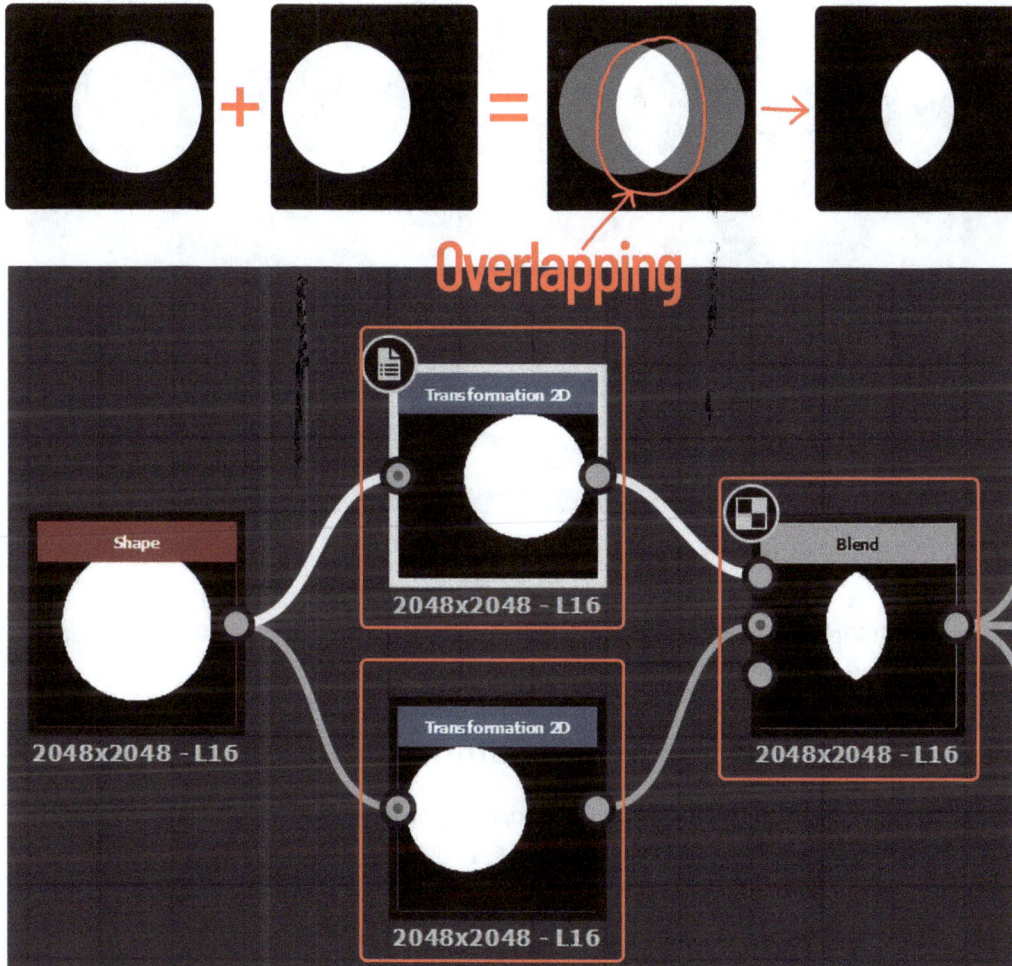

Figure 9.20 – Multiply blending mode technique

4. Now change **Blending Mode** to **Divide**. With this blending mode, you will notice that the non-overlapping area of the **Foreground** input remains while the other area is clipped, as shown in *Figure 9.21*.

    This blending mode is non-commutative, therefore the result changes when the **Foreground** and **Background** inputs are switched.

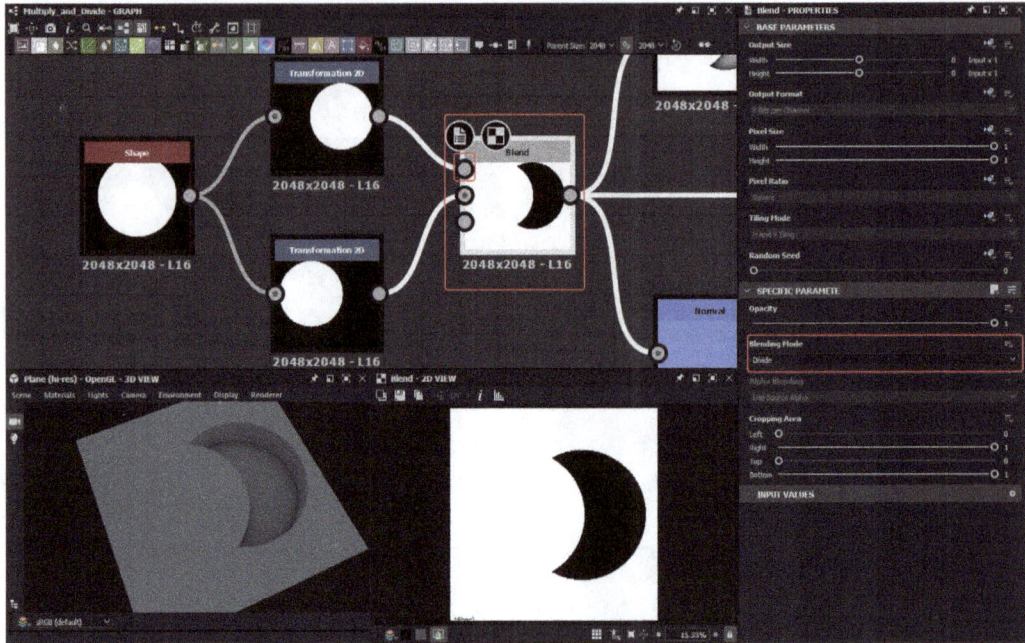

Figure 9.21 – Divide blending mode technique

Hopefully, the **Multiply** and **Divide** blending modes are clear to you. In the next section, we will study the **Screen** and **Soft Light** blending modes.

The Screen and Soft Light blending modes

The values of the pixels in the two inputs are inverted, multiplied, and then inverted once again when using the **Screen** blending mode. The outcome has the opposite effect of multiplying and is consistently equal to or greater (brighter) than the original. This is a commutative blending mode.

In contrast, the **Soft Light** blending mode alters the foreground color's brightness to produce a subtle brighter or darker effect. Blending colors that are brighter than 50% will make the background pixels appear lighter while blending colors that are less brilliant than 50% will make them appear darker. This is a non-commutative mode.

Let us see how to work with these blending modes in the following steps:

1. Open Screen_and_Soft_Light.sbs from the given Substance 3D Designer exercise files. Load the Screen_and_Soft_Light.sbsscn state file, then choose **Plane (hi-res)** from the **Scene** menu. Right-click with the mouse in the **GRAPH** window and choose **View Outputs In 3D View**.

2. In the `Screen_and_Soft_Light.sbs` Substance graph file, the **Gradient Axial** node is connected as **Foreground** while the **Clouds 3** node is connected as **Background**, and **Blending Mode** is set to **Screen**. You can also switch **Foreground** and **Background** without affecting the results because **Screen** is commutative.

3. You will notice that when you blend the **Gradient Axial** node and the **Clouds 3** node using the **Screen** blending mode, the area that will be affected is the darkest area. Whether it is a solid black area or a dark gradient area, the blending will show in the darkest area as shown in *Figure 9.22*.

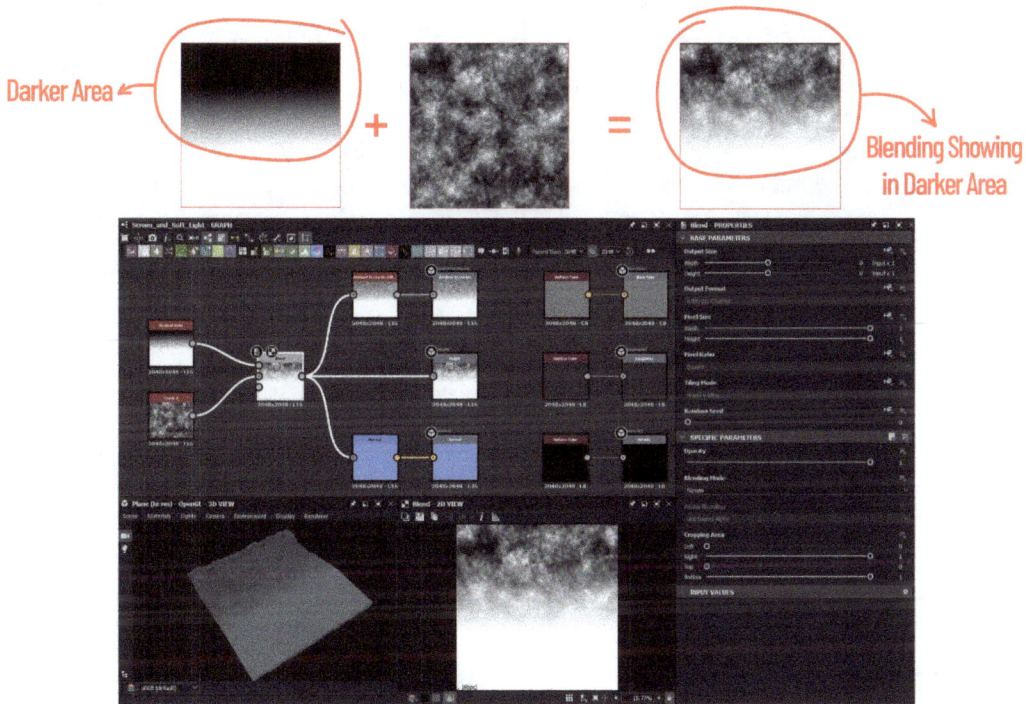

Figure 9.22 – Screen blending mode technique

4. Now change **Blending Mode** to **Soft Light**. With this blending mode, you will notice that the dark part of **Gradient Axial** will make the **Cloud 3** node's blending darker, and the bright part of **Gradient Axial** will make the **Cloud 3** node's blending brighter. This blending mode is non-commutative, therefore the result changes when the **Foreground** and **Background** inputs are switched.

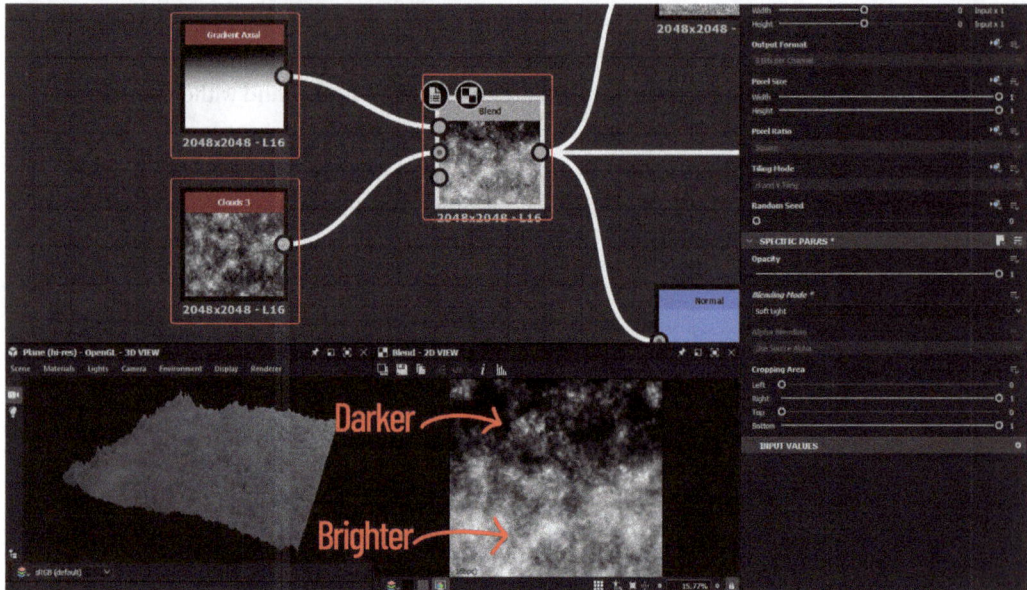

Figure 9.23 – Soft Light blending mode technique

The **Gradient Axial** node used in these steps allows you to modify the gradient with the help of a mouse in 2D View or **INSTANCE PARAMETERS** in the **PROPERTIES** window. Similarly, **Gradient Axial Reflected**, **Gradient Radial**, and **Gradient Circular** can be manipulated with the mouse in 2D View or **INSTANCE PARAMETERS** in the **PROPERTIES** window.

Hopefully, the **Screen** and **Soft Light** blending modes are clear to you. In the next section, we will cover the **Add Sub** and **Overlay** blending modes.

The Add Sub and Overlay blending modes

The following is how the **Add Sub** blending mode operates:

- Any **Foreground** pixels that have a value greater than 0.5 are combined with the corresponding **Background** pixels
- Pixels in **Foreground** that have a value less than 0.5 are deducted from the corresponding pixels in **Background**

The **Overlay** blending mode, which is a non-commutative mode, in contrast, combines the **Multiply** and **Screen** blending mode and operates as follows:

- A blend of the **Multiply** type is used if the lower layer's pixel value is less than 0.5
- **Screen** type blending is used if the lower layer's pixel value is greater than 0.5

Let us work on these two nodes in practice in the following steps:

1. Open `AddSub_and_Overlay.sbs` from the given Substance 3D Designer exercise files, load the `AddSub_and_Overlay.sbsscn state file`, then choose **Plane (hi-res)** from the **Scene** menu. Right-click with the mouse in the **GRAPH** window and choose **View Outputs In 3D View**.

2. In the `AddSub_and_Overlay.sbs` Substance graph file, the **Gradient Axial** node is connected as **Foreground** while the **Fractal Sum 1** node is connected as **Background**, and **Blending Mode** is set to **Add Sub**.

    Switching **Foreground** and **Background** will affect the results because the **Add Sub** blending mode is non-commutative.

3. You will notice that when you blend **Gradient Axial** and **Fractal Sum 1** using the **Add Sub** blending mode, the area that will be affected is the transition area, and the area that is either solid black or solid white will remain unaffected, as shown in *Figure 9.24*.

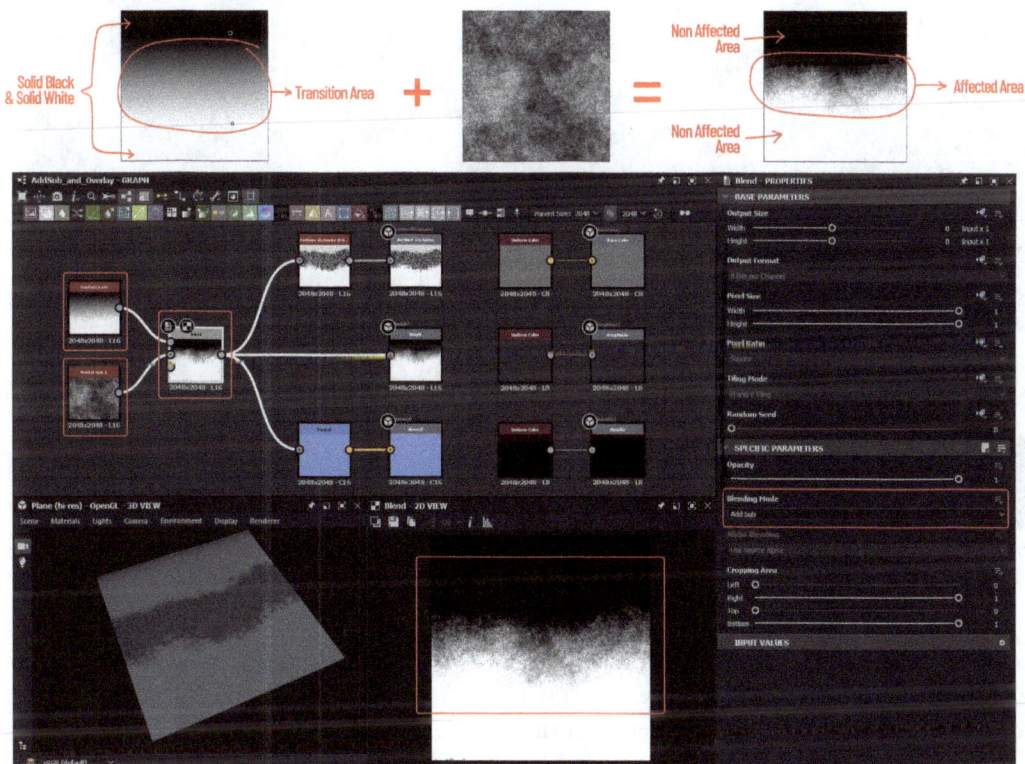

Figure 9.24 – Add Sub blending mode technique

4. Now change **Blending Mode** to **Overlay**. With this blending mode, you will notice that the dark part of **Gradient Axial** will blend with the **Background** input, like for the **Multiply** blending mode, while the bright part will blend like for the **Screen** blending mode.

The **Overlay** blending mode is **non-commutative**, which means switching **Foreground** and **Background** will affect the result.

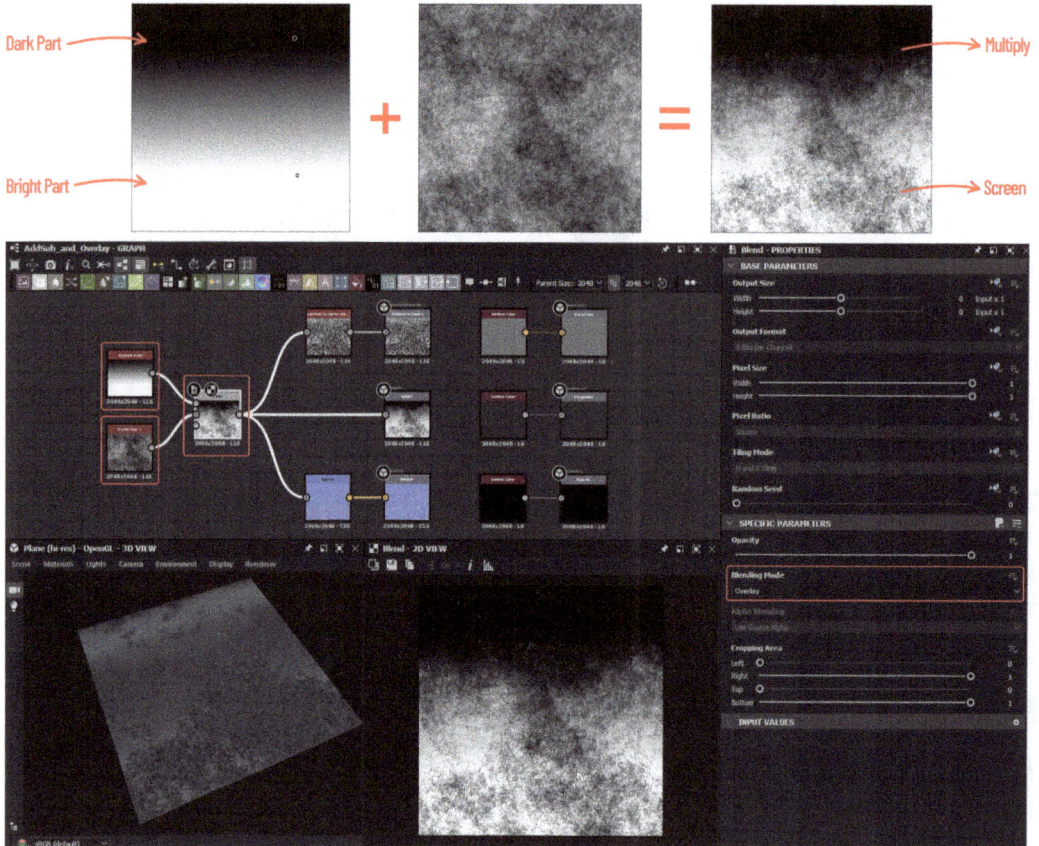

Figure 9.25 – Overlay blending mode technique

Hopefully, you have comprehended the **Add Sub** and **Overlay** blending modes. In the next section, we will study the **Switch** blending mode, which is quite different from all the other blending modes because of its capability of using the **Opacity** attribute to blend.

The Switch blending mode

With the **Switch** blending mode, which is a commutative blending mode, the foreground and background are combined based on opacity, and operate as follows:

- The background will become more apparent as we move nearer to 0
- The foreground will become more obvious as we approach 1

Let us work with the **Switch** blending mode in practice to see how it works:

1. Open `Switch.sbs` from the given Substance 3D Designer exercise files, load the `Switch.sbsscn state file`, then choose **Plane (hi-res)** from the **Scene** menu. Right-click with the mouse in the **GRAPH** window and choose **View Outputs In 3D View**.

2. In the `Switch.sbs` Substance graph file, the **Disc Shape** node is connected as **Foreground** while the **Square Shape** node is connected as **Background**, and **Blending Mode** is set to **Switch**.

3. Switching **Foreground** and **Background** will not affect the results because the **Switch** blending mode is commutative. You can adjust this blending mode with the **Opacity** and **Cropping Area** parameters in the **PROPERTIES** window.

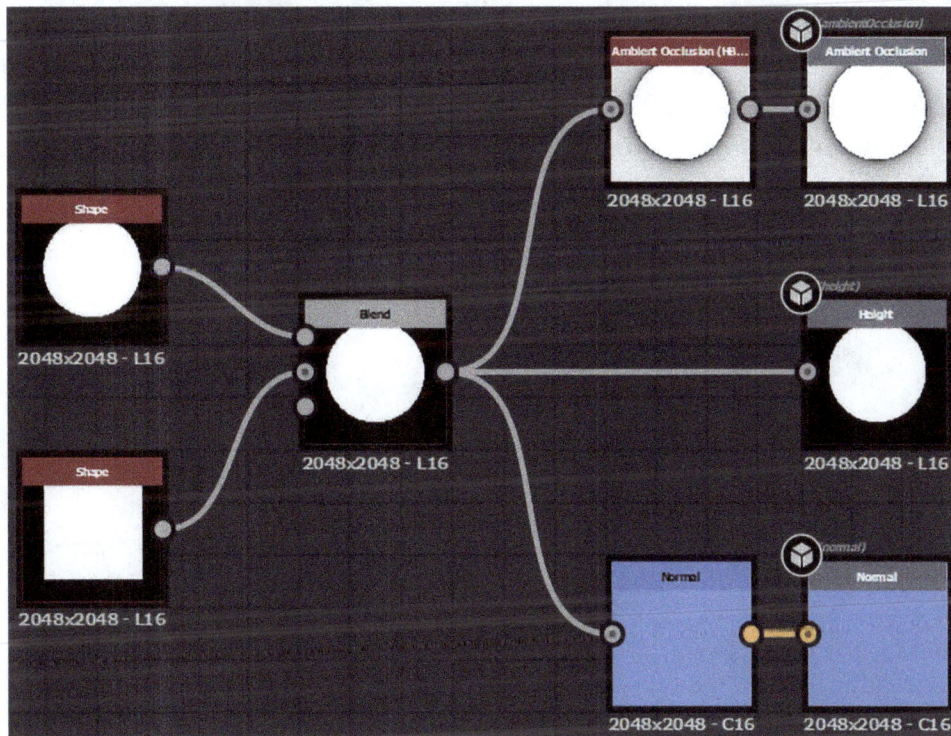

Figure 9.26 – Switch.sbs

4. When you double-click the **Blend** node with the mouse in the **GRAPH** window, its properties will load in the **PROPERTIES** window, and you will also find the **Opacity** and **Cropping Area** parameters there.

5. Just notice that the **Disc Shape** node is visible owing to its input in **Foreground**, and the **Square Shape** node is hidden beneath it because it's in the **Background** input. Therefore, when you reduce the **Opacity** parameter of the **Blend** node's **Switch** blending mode, the **Square Shape** node will start to be revealed, as shown in *Figure 9.27*.

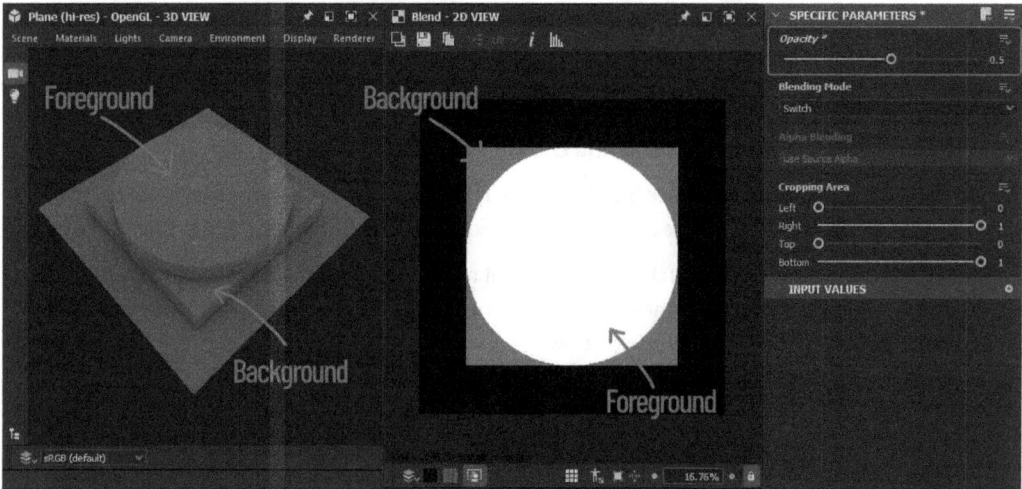

Figure 9.27 – Adjusting the Opacity parameter of the Blend node's Switch blending mode

6. You can also use **Cropping Area** parameters to create interesting shapes. The **Cropping Area** parameters crop the left, right, top, and bottom parts of the **Foreground** input, revealing the **Background** input, as shown in *Figure 9.28*.

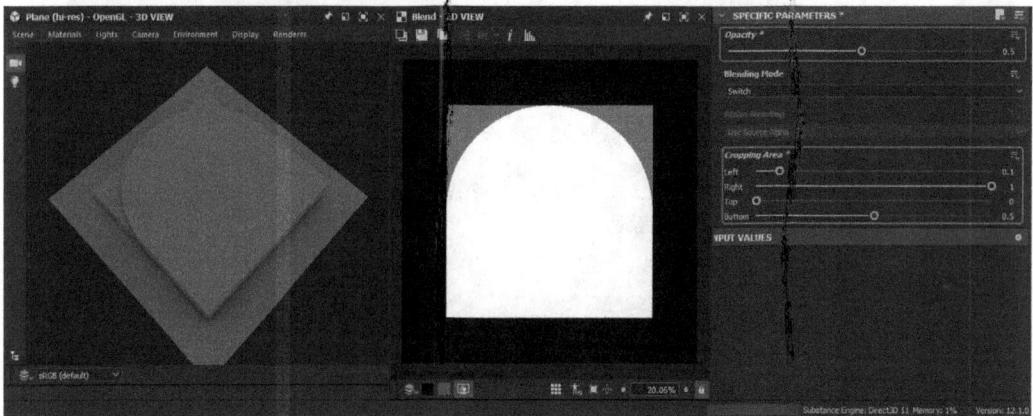

Figure 9.28 – Adjusting the Cropping Area parameter of the Blend node's Switch blending mode

While working on a Substance 3D Designer project, you will notice that there are some icons that appear on the nodes in the **GRAPH** window, as shown in *Figure 9.29*. Let us see what these icons are all about:

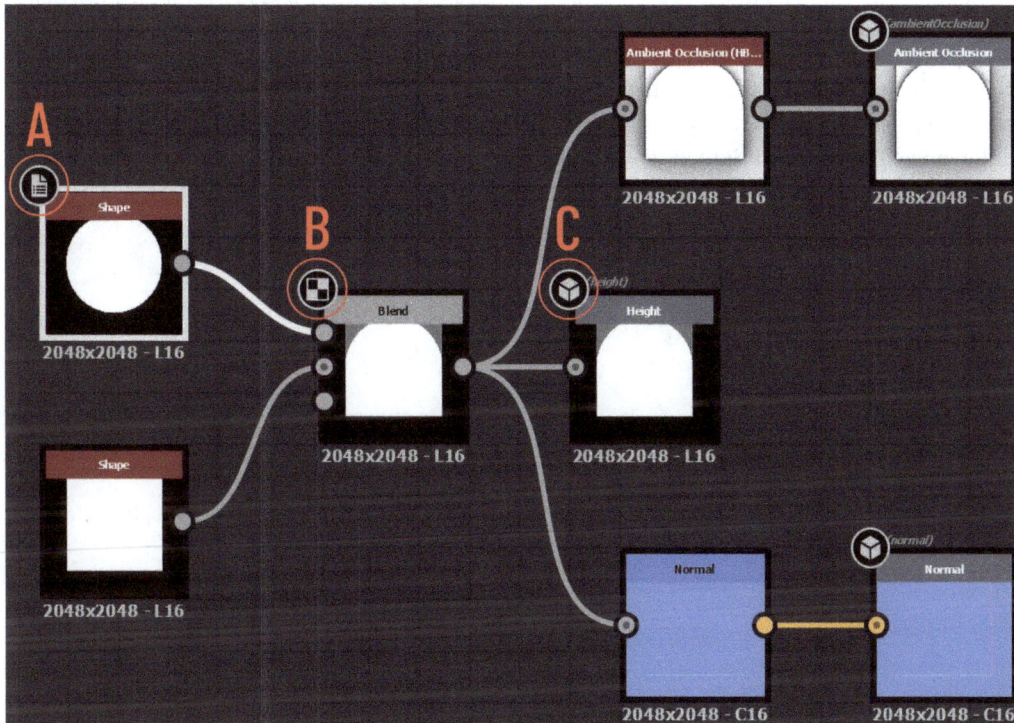

Figure 9.29 – Node icons

- **A – Displayed in Properties**: When you click on a node with the mouse's left button, its properties will appear in the **PROPERTIES** window

- **B – Displayed in 2D View**: When you double-click on a node with the mouse's left button, it appears in **2D VIEW**

- **C – Affected on usage "Node" of material "default" in 3D View**: This icon represents all the nodes that are the final output and appear in **3D VIEW** when you choose **View Outputs In 3D View**

By now, you have hopefully comprehended the importance of blending modes in Adobe Substance 3D Designer. I am sure you will be able to create both basic and complex maps with the skills you have learned in this chapter.

Summary

Finally, we have completed all the blending modes and I hope you have learned how to use them, when to use them, and why to use them. Blending modes are a Substance 3D Designer feature that you will use a lot in your projects.

In the next chapter, you will learn how to create a television shelf using Adobe Substance 3D Designer. Working on projects is the best way to gain practical experience and exposure to any tool; therefore, the following chapter will enlighten you with extensive practical knowledge of Adobe Substance 3D Designer.

10

Creating a Television Shelf in Adobe Substance 3D Designer

In this chapter, we will be looking at combining two of Adobe Substance 3D Designer's most powerful elements, Texture Graph and Model Graph, which are two entirely different things. Moreover, we will go through how these two graphs can work together to create different models and apply maps to them without using any third-party software. You likely won't yet know enough about them to differentiate them. Some insight early on as to how these two types of graphs can be integrated with one another will help.

The best approach to getting hands-on experience with and exposure to any program is to work on projects, thus this chapter will arm you with an in-depth practical understanding of Adobe Substance 3D Designer.

So, let us go through a step-by-step guide to creating a television shelf in an easier way.

We will cover the following topics in this chapter:

- Creating the top and bottom profiles of the television shelf
- Creating the side profiles of the television shelf
- Creating the front doors of the television shelf
- Creating the door vents and knobs of the television shelf
- Creating Albedo Maps, Roughness Maps, and other textures for the television shelf

Creating the top and bottom profiles of the television shelf

First of all, we need a basic 3D mesh for the television shelf. Once we get this primitive, we can enhance it by creating a Substance material for it. In this section, we will first create a primitive, and then we will add a nice-looking profile to its top and bottom of it:

1. After launching Adobe Substance 3D Designer, create a new Substance model graph. Once it's created, right-click on it with your mouse and rename it `TV_Shelf_Model`.

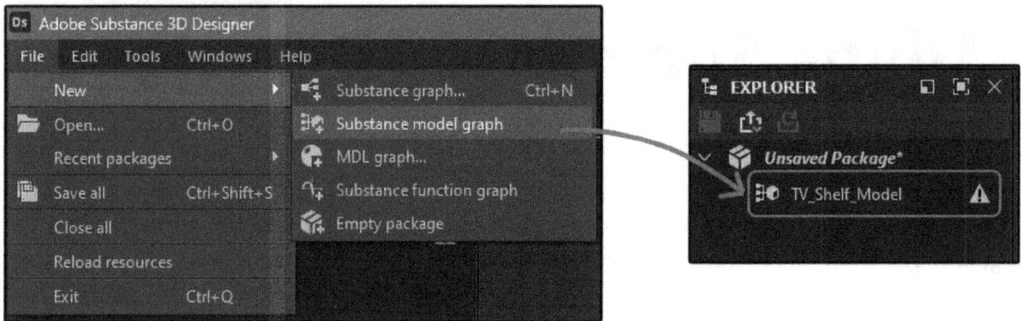

Figure 10.1 – Creating a new Substance model graph

You will notice that there is a yellow alert sign next to the Substance model graph. This is because the graph is empty.

2. Go to the **SUBSTANCE MODEL GRAPH** window, press the spacebar, and type `primitive 3d` in the search field. Once **Primitive 3D** appears in the library, select it.

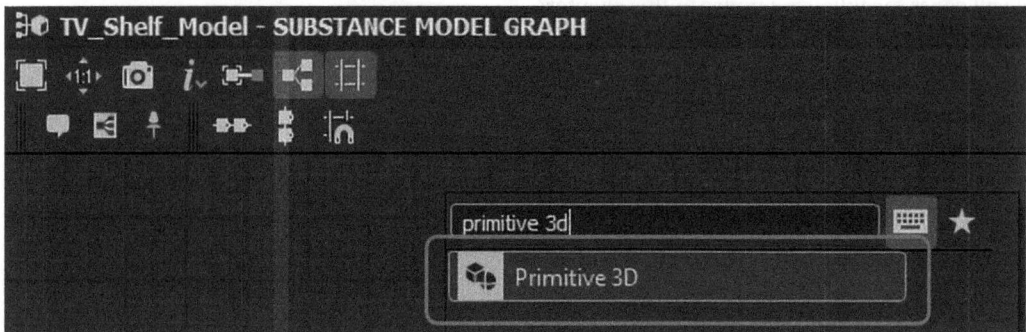

Figure 10.2 – Adding Primitive 3D in the SUBSTANCE MODEL GRAPH window

3. Now double-click on **Primitive 3D** with the mouse left button. This will load it in **3D VIEW** and its parameters will appear in the **PROPERTIES** window.

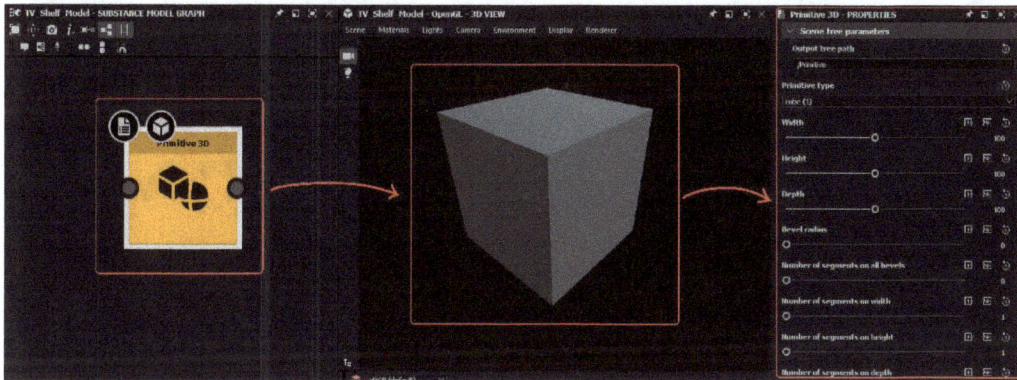

Figure 10.3 – Launching Primitive 3D in 3D VIEW and loading its parameters in PROPERTIES

4. We need a wide and tall TV shelf that can hold a CRT TV. Therefore, apply the following changes:

   - Rename **Output tree path** to /TV_Shelf (this is your root path of the mesh)

   - Set **Primitive type** to **cube (1)**, **Width** to 45, and **Height** to 22

   - Set **Depth** to 21 and **Bevel radius** to 0.2

5. Now, adjust the segments as follows; however, keep in mind that we are increasing the segments to a higher value. Hence, if you do not have a high-end computer, then keep these values low so your computer can handle it:

   - **Number of segments on all bevels**: 10

   - **Number of segments on width**: 180

   - **Number of segments on height**: 100

   - **Number of segments on depth**: 100

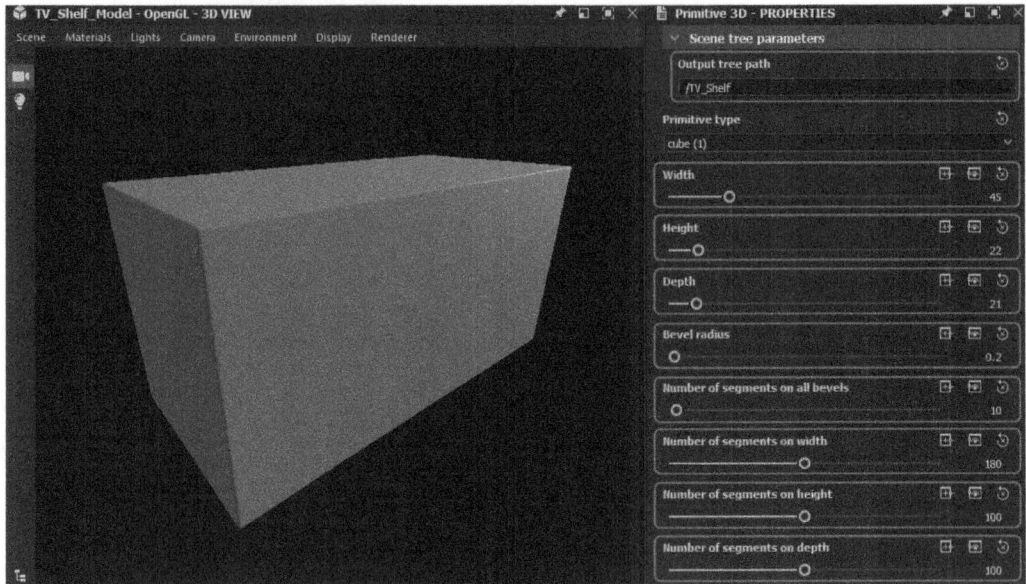

Figure 10.4 – Adjusting the primitive's size

> **Units of Size**
>
> The units of size inside Substance Designer are *centimeters*.

6. Now, we can further enhance the quality of the primitive. Go to the **Materials** menu in **3D VIEW**, hover over **Default** and then **Metallic Roughness**, and select **Tesselation + Displacement**. This will change the material type and load its parameters in the **PROPERTIES** window.

7. In the **PROPERTIES** window, go to the **Height** parameters and increase **Tessellation Factor** to 16. However, if your computer is slow, then do not change it.

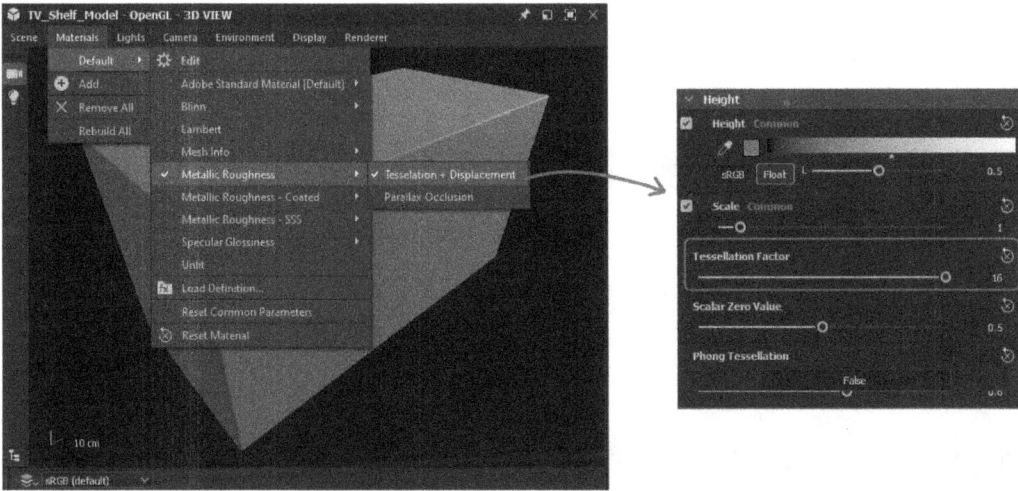

Figure 10.5 – Changing the primitive's material type and Tessellation Factor

8. Now, let us create a new Substance graph. Go to the **File** menu, hover over **New**, and then select **Substance graph…**. Choose **Metallic Roughness** under TEMPLATES and change **Graph Name** to TV_Shelf_Substance in the **New Substance Graph** window.

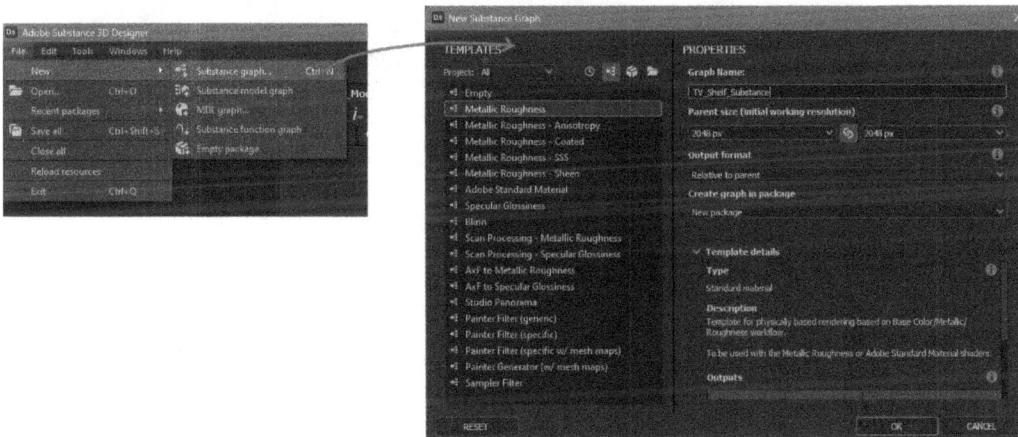

Figure 10.6 – Creating a new Substance graph

9. Right-click with the mouse inside the **GRAPH** window and choose **View Outputs In 3D View**, then put **Ambient Occlusion**, **Height**, and **Normal** on the left side and the rest on the right side, as shown in *Figure 10.7*. We can change the location of these nodes whenever we want to:

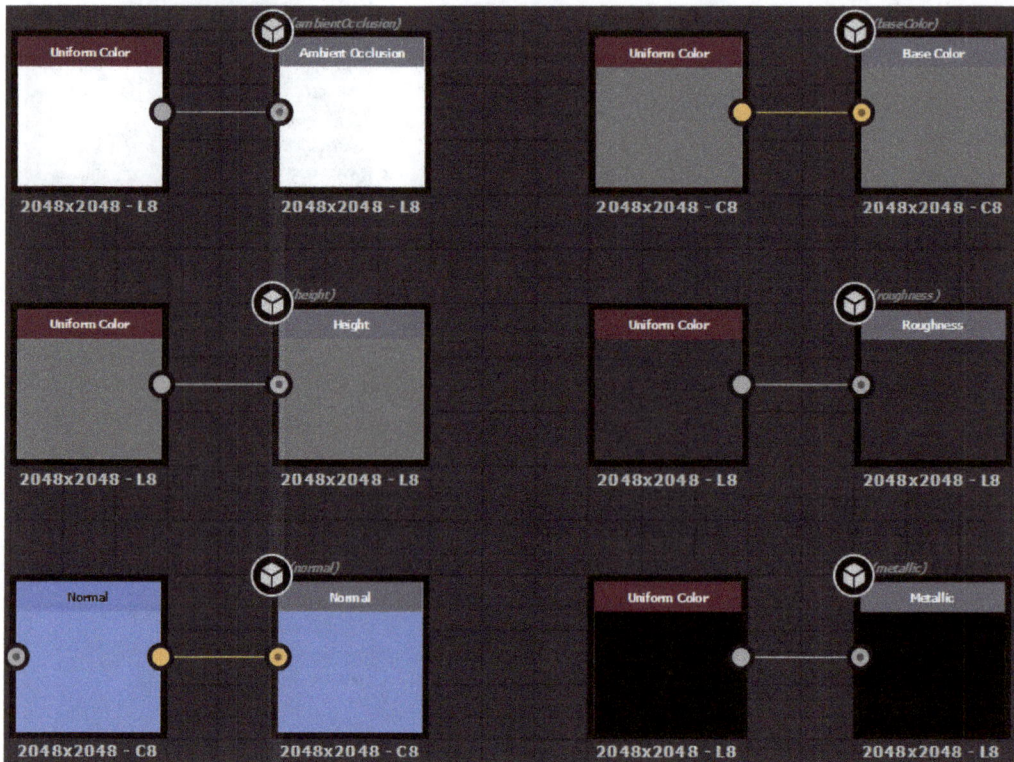

Figure 10.7 – Location of the nodes

10. Now, remove **Uniform Color** from **Ambient Occlusion** and connect **Ambient Occlusion (HBAO)** to it.

11. Add a **Blend** node, click on it with the left mouse button, and then right-click on it with the mouse and choose **Add Comment**. Rename the comment Final Blend, which will be the new name for the **Blend** node. You can also make a hotkey for comments (by pressing *C*). You just need to select the node and press *C* and a comment is automatically attached.

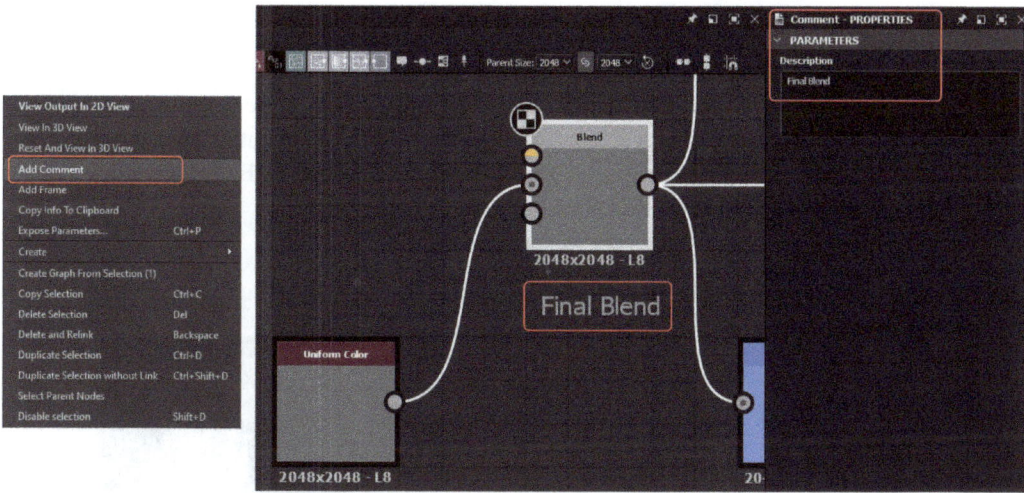

Figure 10.8 – Adding a comment to the Blend node

12. Disconnect **Uniform Color** from the **Height** node and connect it to the **Background** input of the **Blend** node, then connect the **Final Blend** output to **Ambient Occlusion**, **Height**, and **Normal**.

13. Change the **Blending Mode** setting of **Final Blend** to **Add**, as shown in *Figure 10.9*:

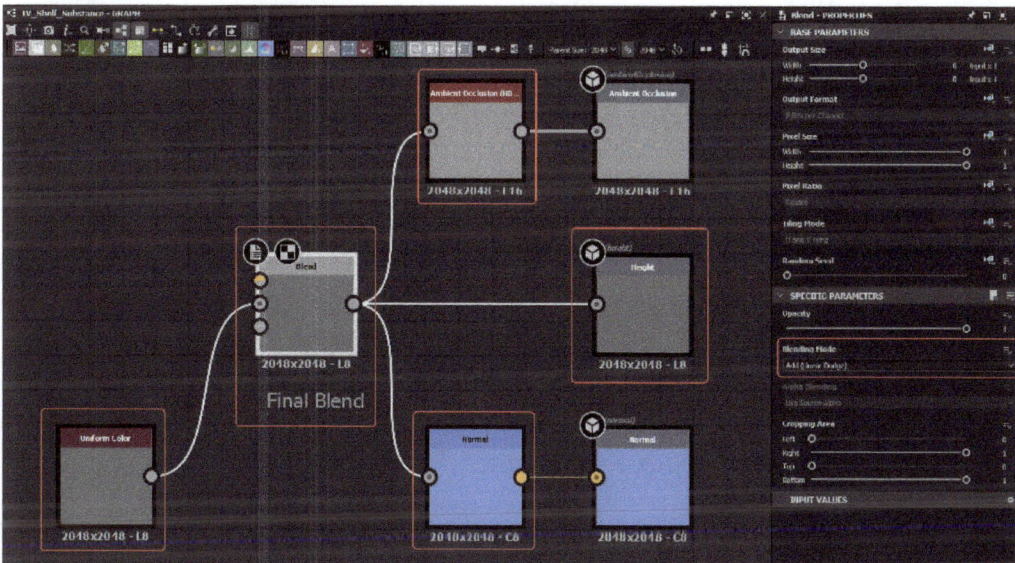

Figure 10.9 – Creating the Final Blend graph setup

Now, **Final Blend** has become our final output, and all the new **Blend** nodes will be connected to it.

14. You have to make sure that the **Output Format** setting of each node is set to **16 Bits per Channel**, so you get a high-quality result. Otherwise, you will have an output showing artifacts and blocks. You can set this by selecting the node, going to its **PROPERTIES** window, and changing **Output Format** to **16 Bits per Channel**.

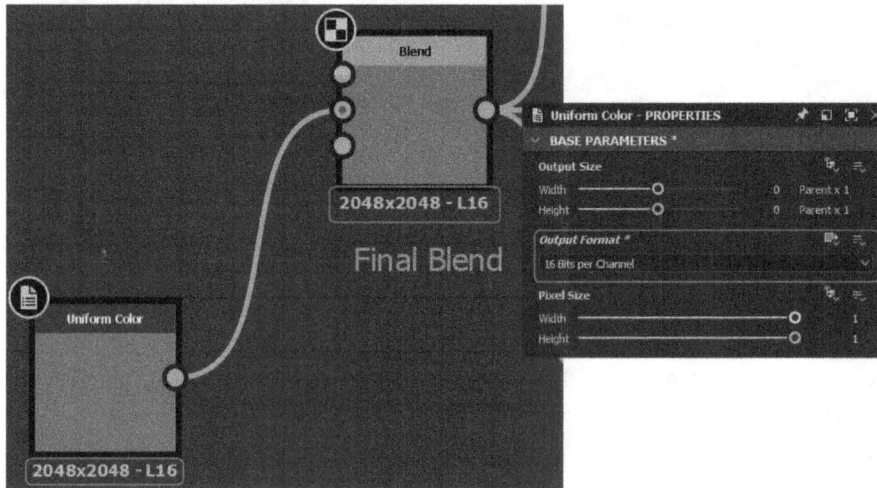

Figure 10.10 – Changing Output Format to 16 Bits per Channel

15. Go to the **Scene** menu in **3D VIEW** and choose **Display UVs in 2D View**, then go to **2D VIEW** and make sure the **UV** option is active.

You will see the mesh's texture UV unfolded, as shown in *Figure 10.11*.

Here **A** is the top of the shelf, **B** is the left of the shelf, **C** is the front of the shelf, **D** is the right of the shelf, **E** is the back of the shelf, and **F** is the bottom of the shelf.

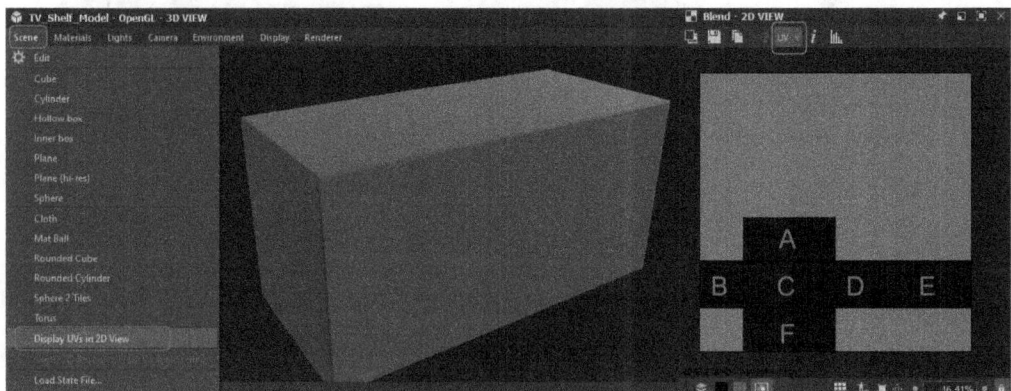

Figure 10.11 – Display UVs in 2D View

Now, let us create a profile at the top of the shelf, as in *Figure 10.12*:

Figure 10.12 – Targeted profile at the top of the shelf

Right Node to Create Profile

You will notice that the profile in *Figure 10.12* has slopes and curves that give it a changing depth. By now, you know that the white map gives the height and raises the mesh outward while black pulls it inward. However, it is not possible to create a profile with solid black or white maps, as shown in *Figure 10.12*, because there are slopes and curves in it. Hence, it's better to create this profile with the help of a **Gradient Linear** node and modify it with a **Curve** node.

16. Add the **Gradient Linear 1** node and connect it to the **Transformation 2D** node, then connect the **Transformation 2D Output** node to the **Final Blend** node's **Foreground** input.

    We will create the profile on the sides and front of the television shelf, so let's move the **Gradient Linear 1** node on top of sides and front of the television shelf with the help of the **Transformation 2D** node.

17. Double-click on the **Transformation 2D** node and change its **Tiling Mode** setting to **No Tiling**, **Stretch Width** to 150%, **Stretch Height** to 2%, **Offset X** to -0.0349, and **Offset Y** to -9.3197.

    Make sure you only set the **Stretch Width** and **Stretch Height** values once, because every time you enter a value, it either increases or decreases the node.

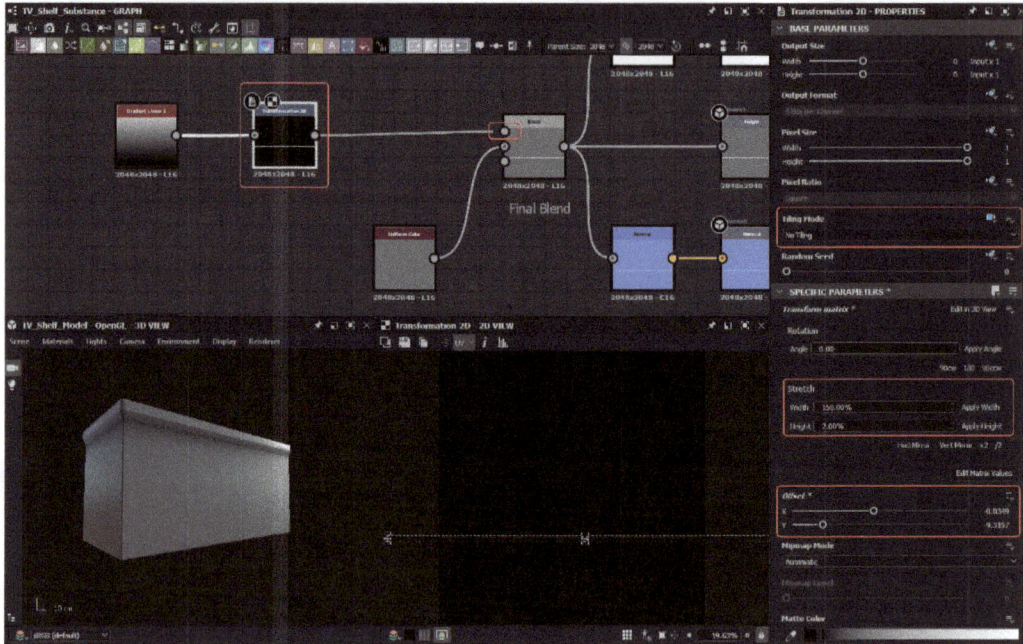

Figure 10.13 – Transforming Gradient Linear 1

Now you will notice that we have a profile but it's kind of flat. To make it look as it does in *Figure 10.12*, we must manipulate it.

18. To manipulate it, add the **Curve** node between the **Transformation 2D** and **Blend** nodes' **Foreground** input, as shown in *Figure 10.14*:

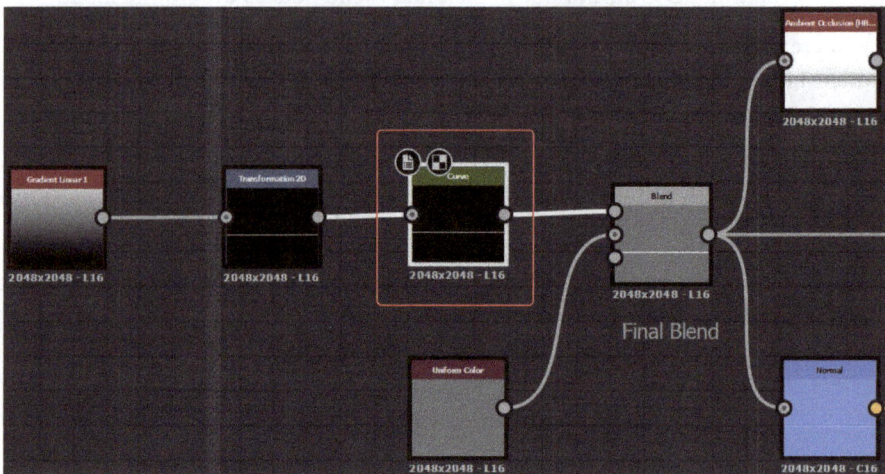

Figure 10.14 – Adding the Curve node to Transformation 2D

19. As we know, the **Curve** node has a **Curve Addressing** graph in the **PROPERTIES** panel, which represents the curve design. In *Figure 10.15*, you will notice that the **Curve Addressing** graph is flat; therefore, the profile in **3D VIEW** is also flat.

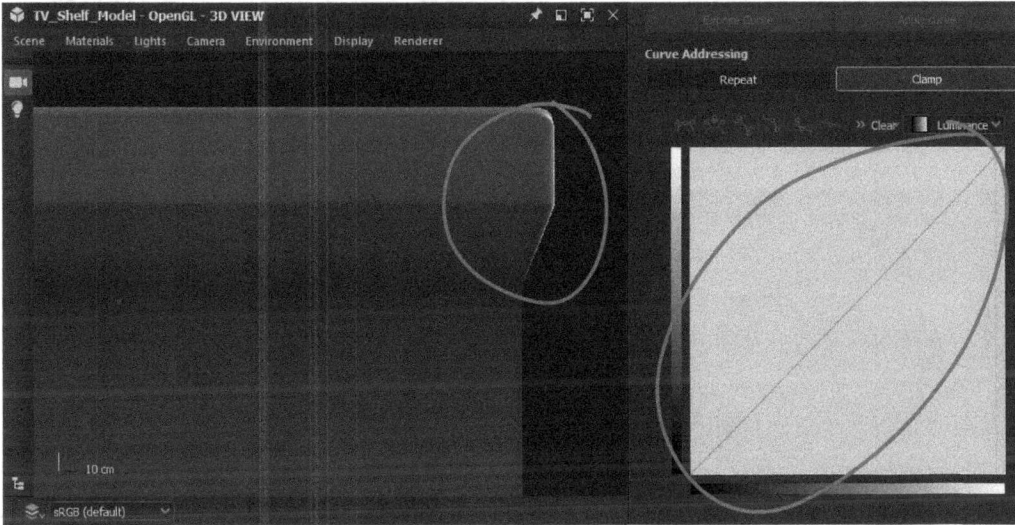

Figure 10.15 – Flat curve profile

20. To make the profile look as in *Figure 10.12*, we must add points to the graphs, as shown in *Figure 10.16*. You can also create your own profile by readjusting the points as you want:

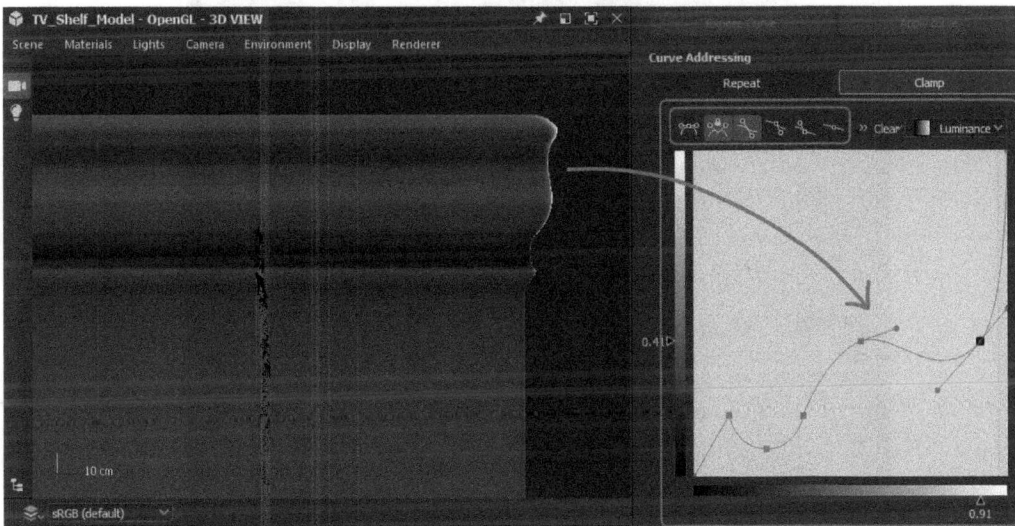

Figure 10.16 – Creating a profile with the Curve node

Now, let us add the same profile to the bottom of the shelf. However, we need to flip it vertically so that it is symmetrical to the top profile.

21. To do that, we need a **Mirror Grayscale** node. Connect **Mirror Grayscale** between the **Curve** and **Blend** nodes, as shown in *Figure 10.17*, then go to the **Mirror Grayscale** node's **INSTANCE PARAMETERS** tab in the **PROPERTIES** window and change the following settings:

- **Mode**: **Mirror Axis Y**

- **Axis Y Offset**: 0.241

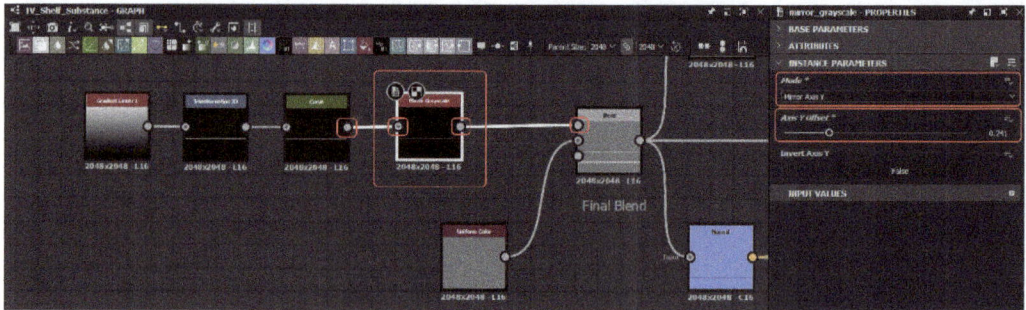

Figure 10.17 – Mirroring the top profile on the bottom

You will notice that an exact mirrored copy of the top profile is now copied to the bottom as well.

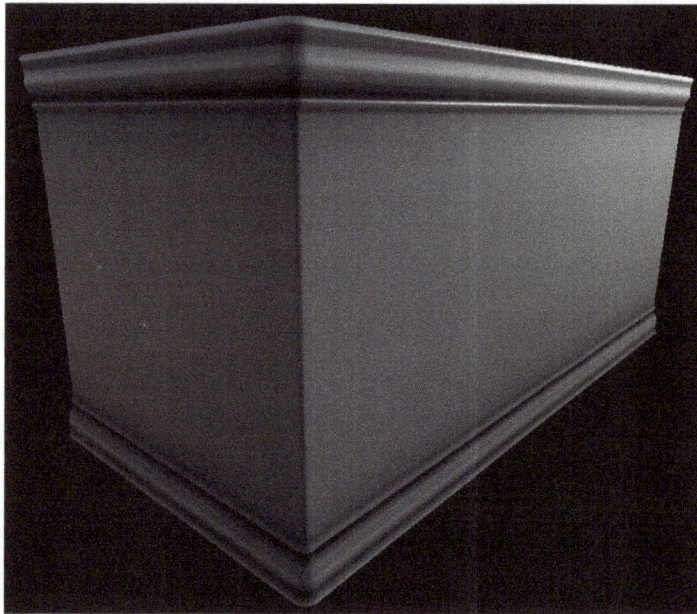

Figure 10.18 – Final output of the profile

You will be creating a lot of nodes while working in Substance Designer. Therefore, it will be necessary to organize them as you work.

22. One way of organizing the nodes is to create a frame around them, select all the top and bottom profile nodes, right-click on their selection with the mouse, and choose **Add Frame**. You can also create a custom hotkey; for example, press *Shift + F* to quickly add frames.

23. Once the frame is added, you can rename it and recolor it from the **PROPERTIES** window. Go to the **PROPERTIES** window of the frame and rename it Top and Bottom Profiles under the **Title** parameter.

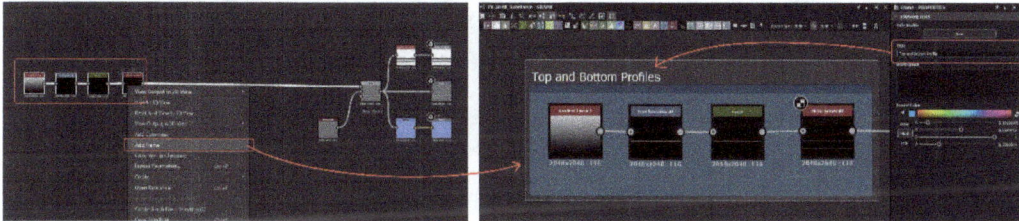

Figure 10.19 – Adding and renaming a frame to organize the nodes

You have now learned how to create profile-type substance materials, using **Curve** and **Mirror** nodes. In the next section, we will create the side profiles.

Creating the side profiles of the television shelf

In this section, we will learn how we can create frame-looking profiles inside Substance Designer. Frames can be created in myriad different ways; however, we will look at the easiest way of doing it.

So, let us start creating frame-looking side profiles with the following steps:

1. First, we need some space between the **Top and Bottom Profile** and **Final Blend** nodes so that we can add new nodes between them.

   Pull the **Top and Bottom Profile** frame toward the left to create space, as shown in *Figure 10.20*. We can create more space later as needed.

Figure 10.20 – Creating space for new nodes

2. To add new nodes to the Substance graph, we need an open node in between the **Top and Bottom Profile** and **Final Blend** nodes. Therefore, select the link connection between these two nodes and add the **Blend** node.

Figure 10.21 – Adding a Blend node

3. Now, similar to how we added the **Blend** node in between, we can add nodes to the Blend node's **Foreground** input; but before that, let us change the **Blending Mode** setting of the newly added **Blend** node to **Add**.

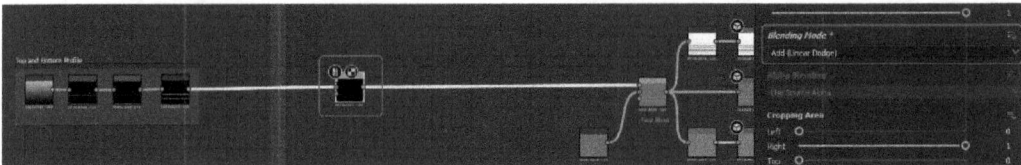

Figure 10.22 – Changing Blending Mode of the newly created Blend node to Add

4. Add the **Shape** and **Transformation 2D** nodes to the **GRAPH** window and connect them.

5. Double-click on the **Transformation 2D** node with the mouse and change the following parameters from the **PROPERTIES** window:

- **Tiling Mode: No Tiling**
- **Stretch Width:** 13%; **Stretch Height:** 10%
- **Offset X:** 3.2441; **Offset Y:** -2.5927

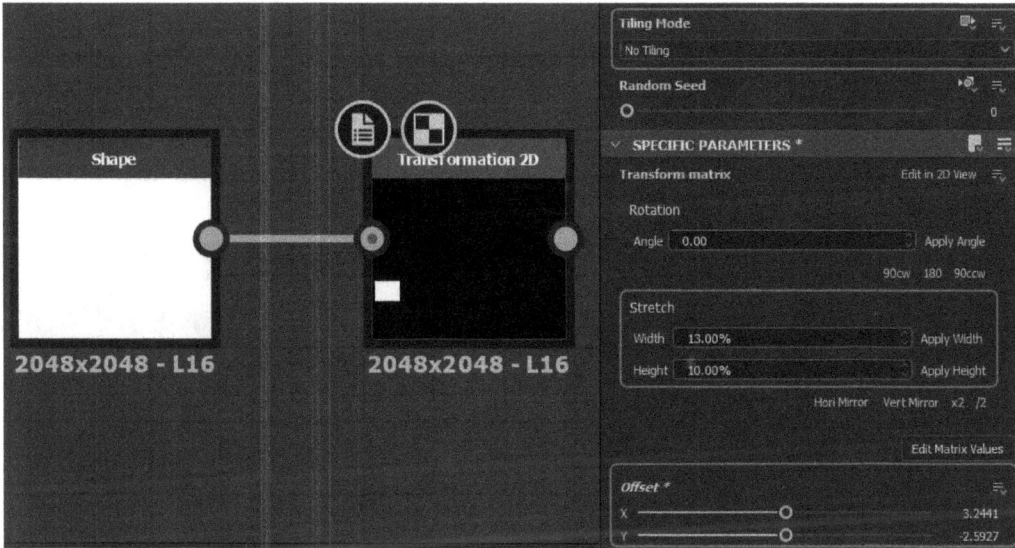

Figure 10.23 – Adding the Shape and Transformation 2D nodes

6. Now, select the **Transformation 2D** node and copy it with the *Ctrl + C* (Windows) or *Command + C* (Mac) hotkey, then paste it with the *Ctrl + V* (Windows) or *Command + V* (Mac) hotkey. Alternatively, you can simply use *Ctrl + D* (Windows) or *Command + D* (Mac) to duplicate the node.

7. Double-click with the mouse on the copied **Transformation 2D** node and change the following settings:

-   **Stretch Width**: 95%; **Stretch Height**: 95%

-   **Offset X**: 3.411; **Offset Y**: -2.7359

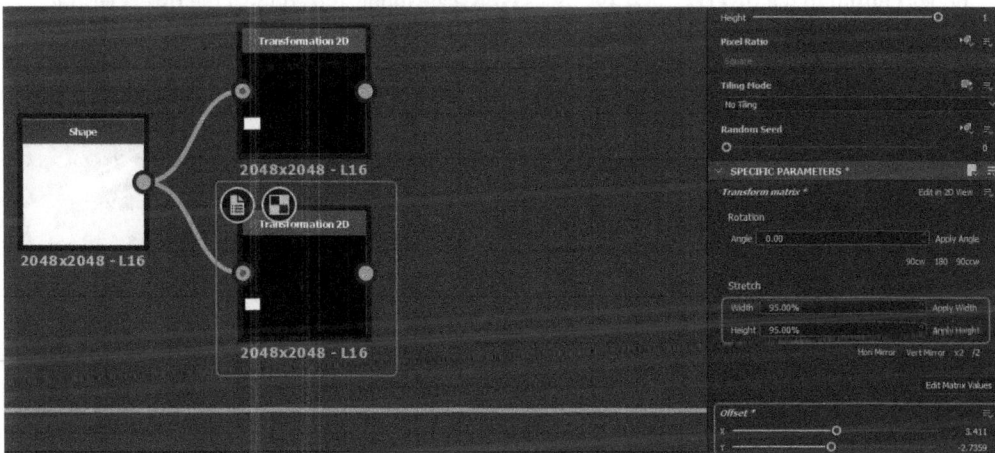

Figure 10.24 – Copying and changing Transformation 2D

8. Now, we need to subtract the smaller box from the bigger box. To do that, we need to create a new **Blend** node and connect the first **Transformation 2D** node to the **Background** input of the **Blend** node.

9. Connect the second **Transformation 2D** node to the **Foreground** input of the **Blend** node. Then, double-click the **Blend** node to view it in **2D VIEW** and change **Blending Mode** to **Divide** so that we have a frame.

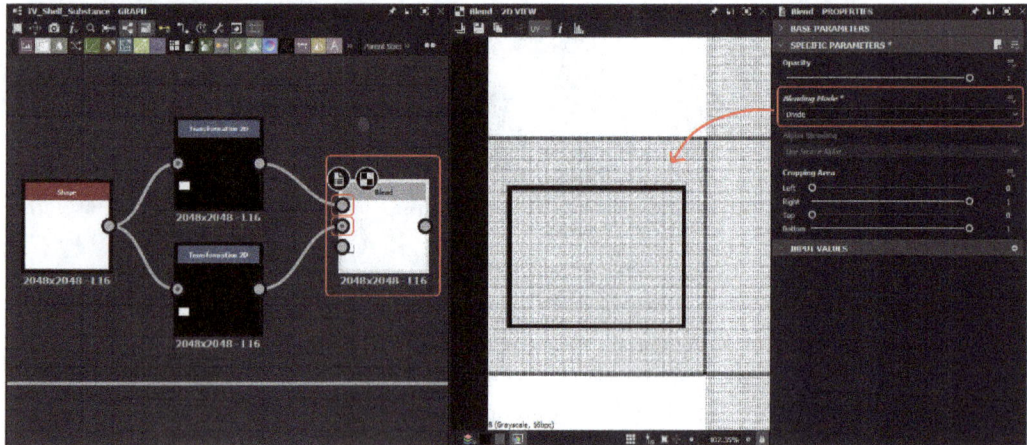

Figure 10.25 – Adding a Blend node

You will notice that the **Divide** blending mode has done the job; however, the frame is black, and everything surrounding it is white, whereas we need it to be the opposite so that we have an extruding frame.

10. Therefore, to invert the colors, add an **Invert Grayscale** node, then connect the new **Blend** node's **Output** to the **Invert Grayscale** node's **Grayscale** input, and connect the **Invert Grayscale** node's **Output** to the old **Blend** node's **Foreground** input, as shown in *Figure 10.26*:

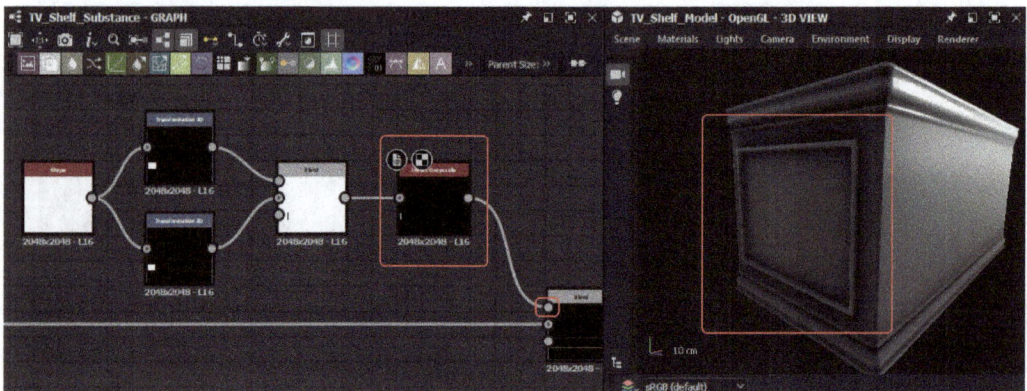

Figure 10.26 – Side profile first version

11. If you want to decrease the depth of the side profile, add the **Levels** node between **Invert Grayscale** and the old **Blend** node, then double-click the **Levels** node with the mouse, go to its **PROPERTIES** window, and for adjustment, toggle from **Histogram** setting to **Values**, then change **Level Out High** to 0.2, as shown in *Figure 10.27*.

This will reduce the brightness of your node, resulting in a shorter height.

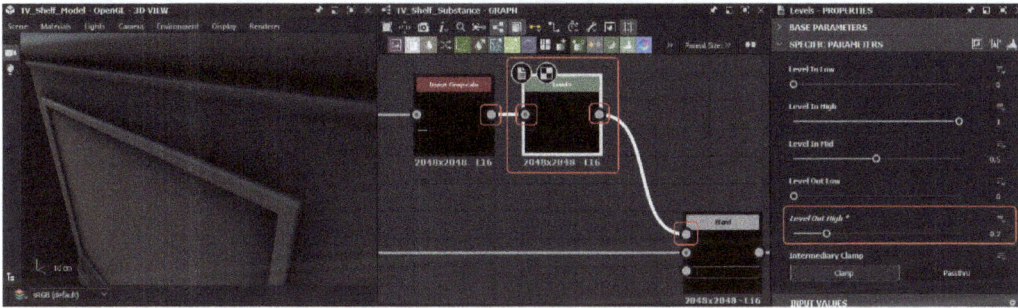

Figure 10.27 – Adjusting the side profile depth with the Levels node

12. Now, we need the same side profile on the other side. Add a **Mirror Grayscale** node between **Levels** and the old **Blend** node. Then, double-click on **Mirror Grayscale** with the mouse, and in the **PROPERTIES** window, change **Mode** to **Mirror Axis X** and **Axis X Offset** to 0.328.

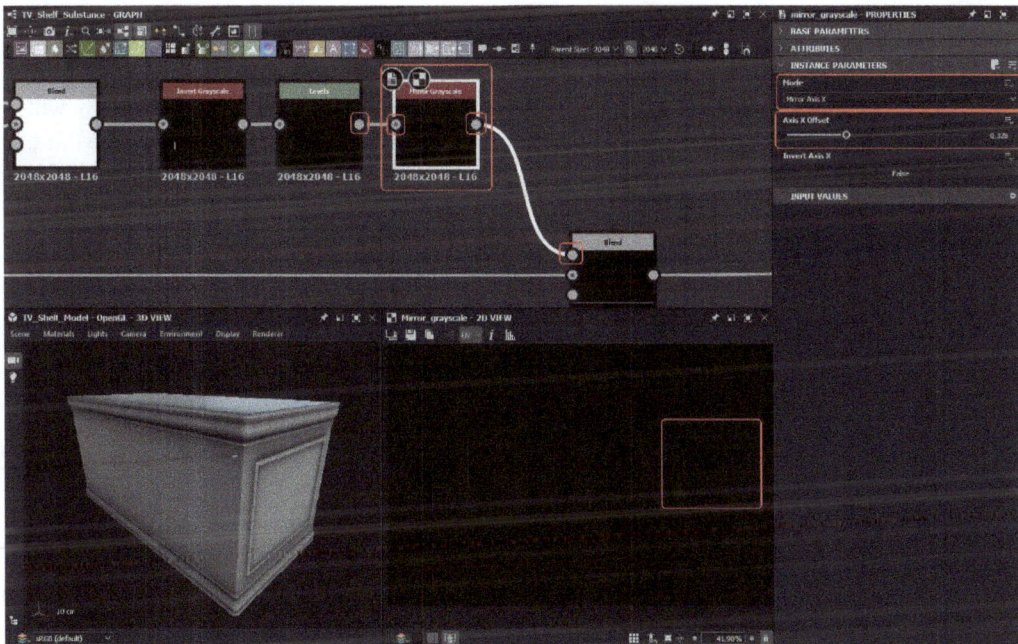

Figure 10.28 – Final version of side profiles

13. Now, select all the nodes related to side profiles and add a frame to it as we did for the top and bottom profiles. Rename the frame title `Side Profiles`, as shown in *Figure 10.29*:

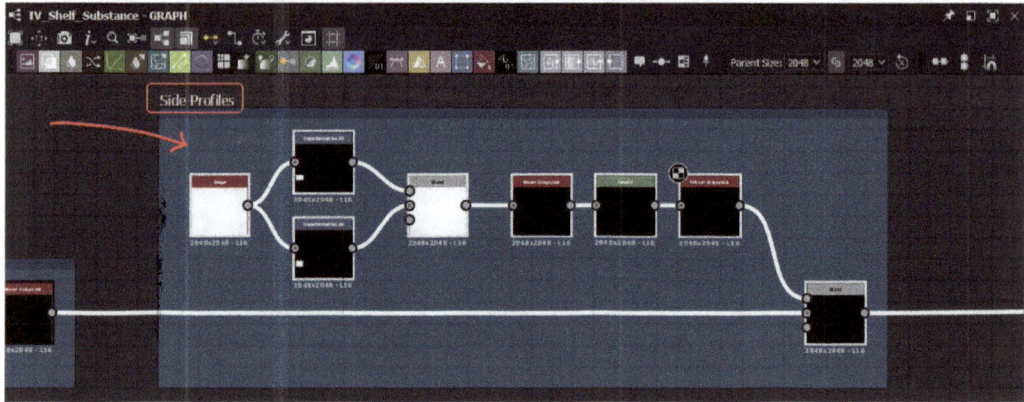

Figure 10.29 – Adding a frame to the side profiles

I hope you have understood the technique and method to create frame-type profiles. With this method, you can also create insets and panel lines, and so on. In the next section, we will learn how to create the television shelf's front doors.

Creating the front doors of the television shelf

To create some substance materials, you need to use the layers of the node. In this section, we will create the doors, then we will add some frame details on the second node and knob details on the third.

So, let us create our first node for the television shelf front doors:

1. Add some space between the **Side Profiles** and **Final Blend** nodes, then select the link connection between them and add the **Blend** node to it. Change its **Blending Mode** setting to **Add (Linear Dodge)**, as shown in *Figure 10.30*:

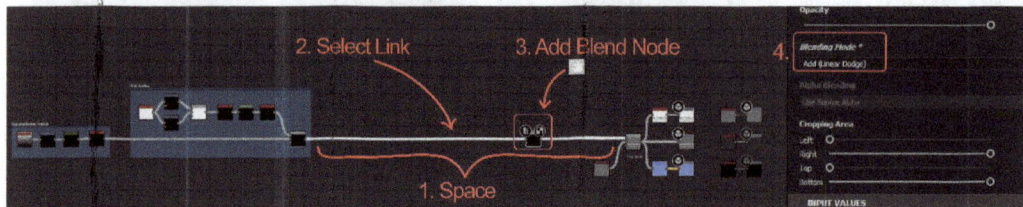

Figure 10.30 – Adding space and a new Blend node with Add (Linear Dodge) Blending Mode

2. Now, add the **Shape** node and connect the **Transformation 2D** node to it, then change the following parameters of the **Transformation 2D** node and connect it to the **Foreground** input of the **Blend** node that you created in step 1, as shown in *Figure 10.31*:

    - **Output Format: 16 Bits per Channel**

    - **Tiling Mode: No Tiling**

    - **Stretch Width**: 15%; **Stretch Height**: 10%

    - **Offset X**: 1.6735; **Offset Y**: -2.5927

    You can also transform **Transformation 2D** with the mouse; however, you need to see UVs in **2D VIEW**.

3. If you cannot see the UVs because of the black background, double-click the **Final Blend** node with the mouse so that the UVs are displayed in **2D VIEW**.

4. Click on the **Transformation 2D** node with the mouse so that the transformation manipulator appears in **2D VIEW** and its parameters appear in the **PROPERTIES** window.

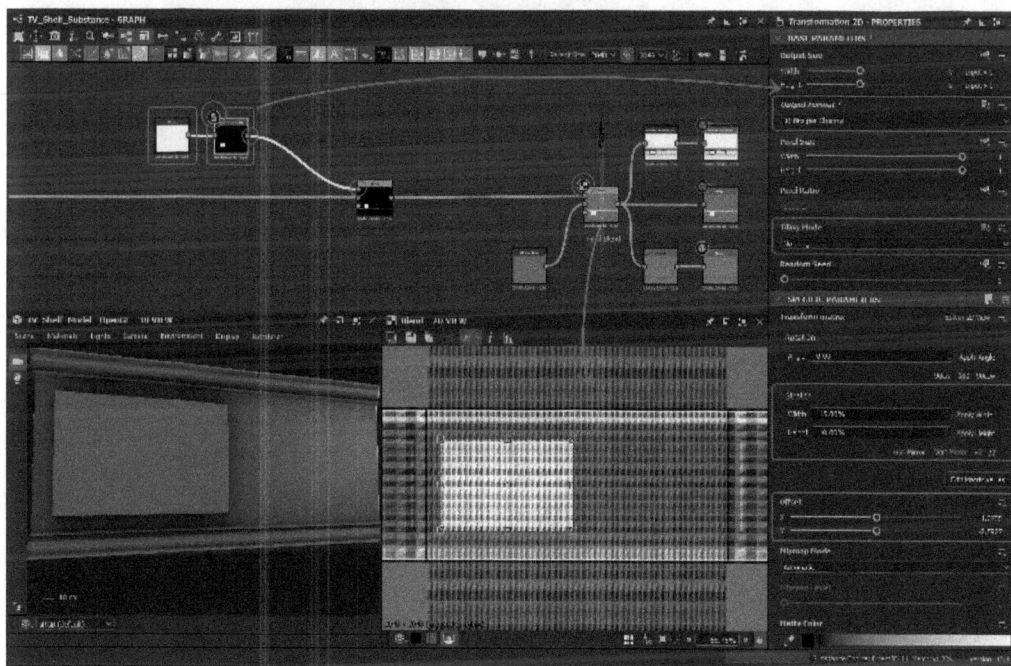

Figure 10.31 – Adding the Shape and Transformation 2D nodes to the new Blend node

5. Now, we need another copy of the door beside the original door. So, add **Mirror Grayscale** between the **Transformation 2D** and **Blend** nodes, then change the following parameters of **Mirror Grayscale** from the **PROPERTIES** window:

    - **Mode**: Mirror Axis X

    - **Axis X Offset**: 0.328

6. Now, select all the nodes related to the front door, right-click with the mouse and choose **Add Frame**, and rename the title of the frame Front Doors, as shown in *Figure 10.32*:

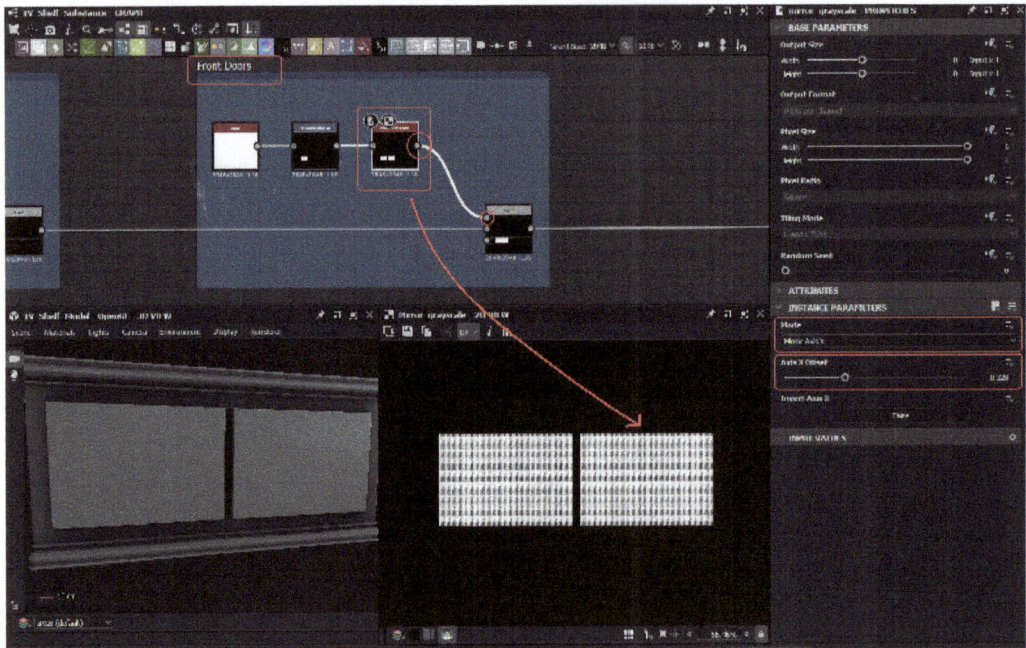

Figure 10.32 – Mirroring the front door with the help of the Mirror Grayscale node

We are done creating the base for the television shelf doors. In the next section, we will learn how to add frame details to the shelf doors.

Creating the door vents and knobs of the television shelf

It is now time to add a second node to the front part of the television shelf. This will be for the door vents. The best node to use to create the vents is **Gradient Linear 1**:

1. Add some space between the **Front Doors** and **Final Blend** nodes, then select the link connection between them and add the **Blend** node to it. Change its **Blending Mode** setting to **Add (Linear Dodge)**, as shown in *Figure 10.33*:

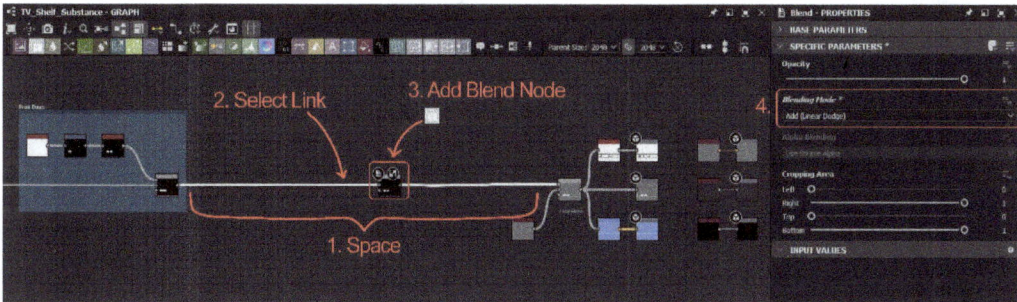

Figure 10.33 – Adding space and a new Blend node with Add (Linear Dodge) Blending Mode

2. Now, add the **Gradient Linear 1** node in the **GRAPH** window, then double-click **Gradient Linear 1** and change its **Tiling** setting to 16, as shown in *Figure 10.34*:

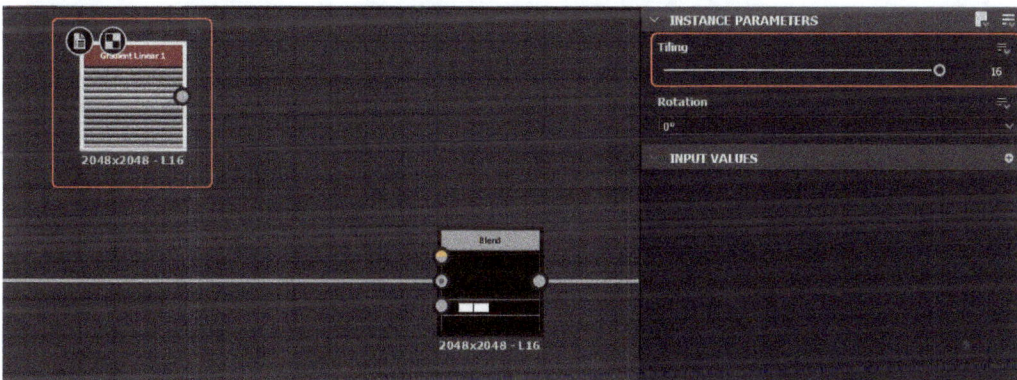

Figure 10.34 – Adding the Gradient Linear 1 node and changing its Tiling parameter

3. Now, add the **Transformation 2D** node to the **GRAPH** window and connect the **Gradient Linear 1** node's **Simple Grad** output to the **Transformation 2D** node's **Input**.

4. Connect the **Transformation 2D** node's **Output** to the previously added **Blend** node's **Background** input and then change the following parameters of the **Transformation 2D** node:

    *   **Tiling Mode: No Tiling**

    *   **Stretch Width**: 12%; **Stretch Height**: 9%

    *   **Offset X**: 2.1595; **Offset Y**: -2.8724

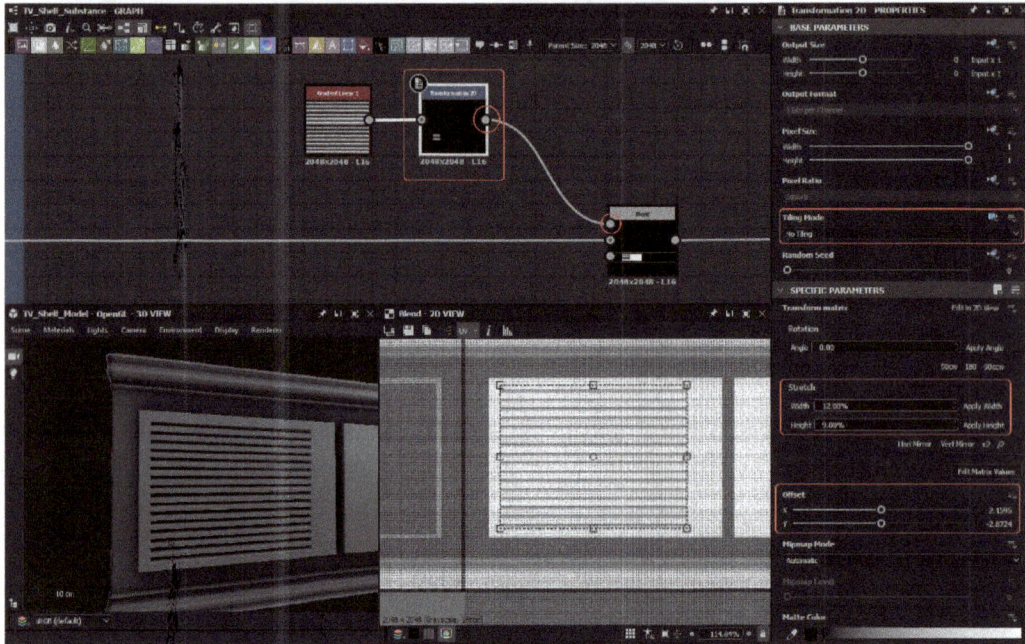

Figure 10.35 – Adding the Transformation 2D node and adjusting its parameters

5. Now, we need another copy of the door vents beside the original door vents. So, add **Mirror Grayscale** between the **Transformation 2D** and **Blend** nodes, then change the following parameters of **Mirror Grayscale** in the **PROPERTIES** window:

   • **Mode**: **Mirror Axis X**

   • **Axis X Offset**: 0.328

Figure 10.36 – Mirroring the door vents

6. Now, double-click on the **Blend** node with the mouse and change its **Blending Mode** setting to **Subtract**, as shown in *Figure 10.37*:

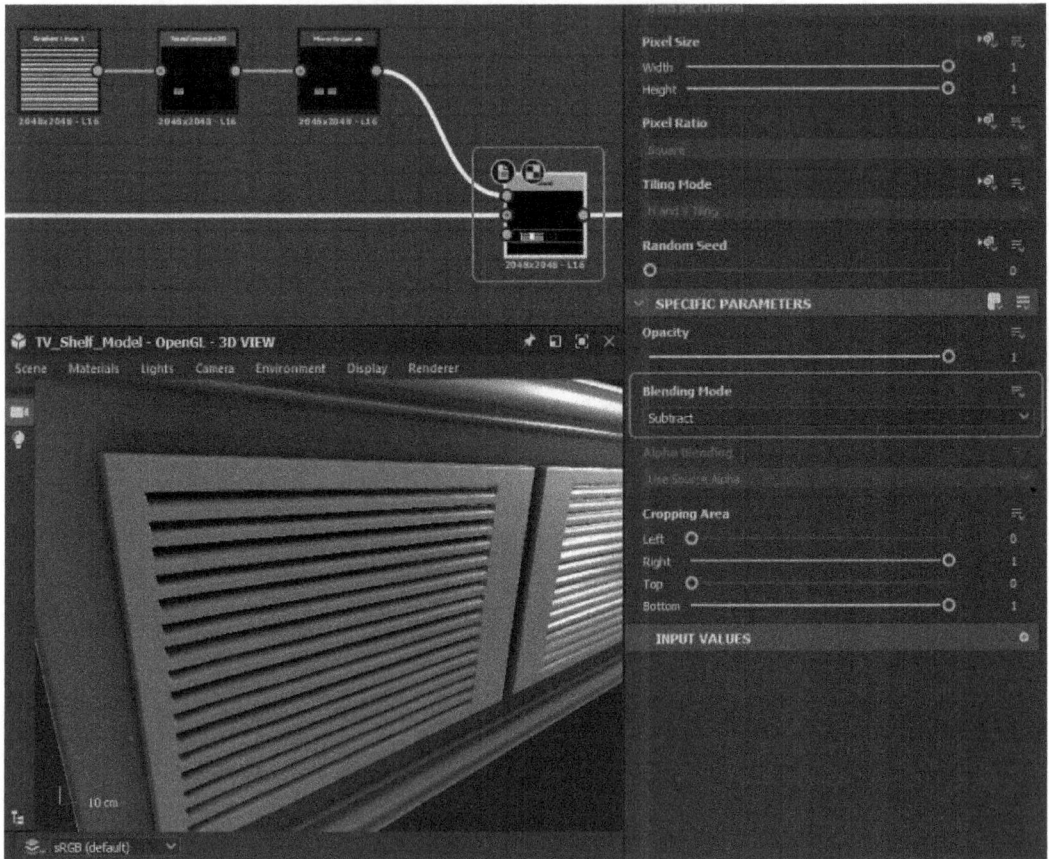

Figure 10.37 – Changing the Blend node's Blending Mode

7. Select all the nodes associated with the door vents, right-click them with the mouse, and choose **Add Frame**. Change the title of the frame to Door Vents, as shown in *Figure 10.38*:

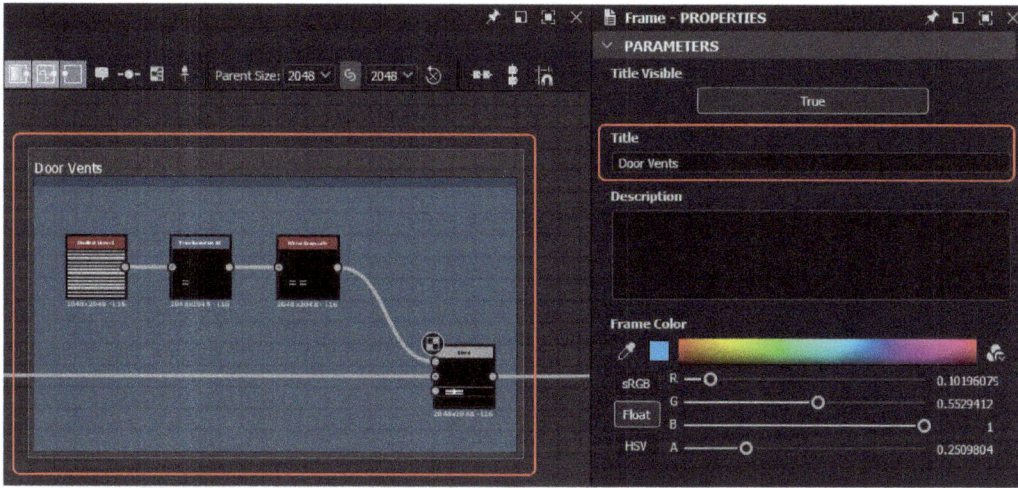

Figure 10.38 – Adding a frame to the Door Vents nodes

8. Add another **Blend** node between the **Door Vents** nodes and the **Final Blend** nodes and change its **Blending Mode** setting to **Add (Linear Dodge)**.

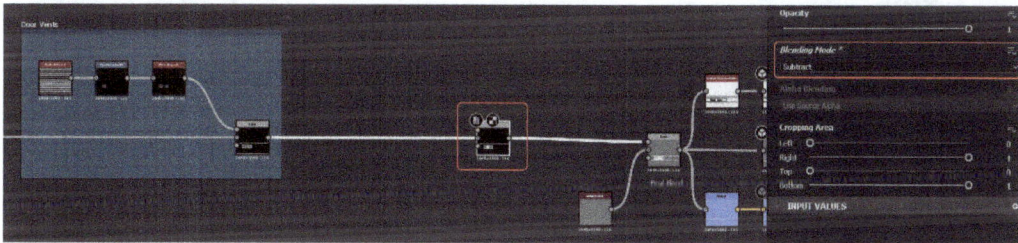

Figure 10.39 – Adding a new Blending node and changing its Blending Mode setting to Add

9. Now, add a **Shape** node and a **Transformation 2D** node and connect both of them.

10. Then, connect the **Transformation 2D** node's **Output** to the previously created **Blend** node's **Foreground** input and change its parameters as shown in *Figure 10.40*:

- **Tiling Mode: No Tiling**

- **Stretch Width:** 1%; **Stretch Height:** 4%

- **Offset X:** 18.7947; **Offset Y:** -6.4644

Figure 10.40 – Adding door knobs to the doors

11. The **Shape** node that we created is a sharp-edged rectangle. To make it a rounded rectangle, we need to add **Blur HQ Grayscale** and change its **Intensity** parameter to 0.5, as shown in *Figure 10.41*:

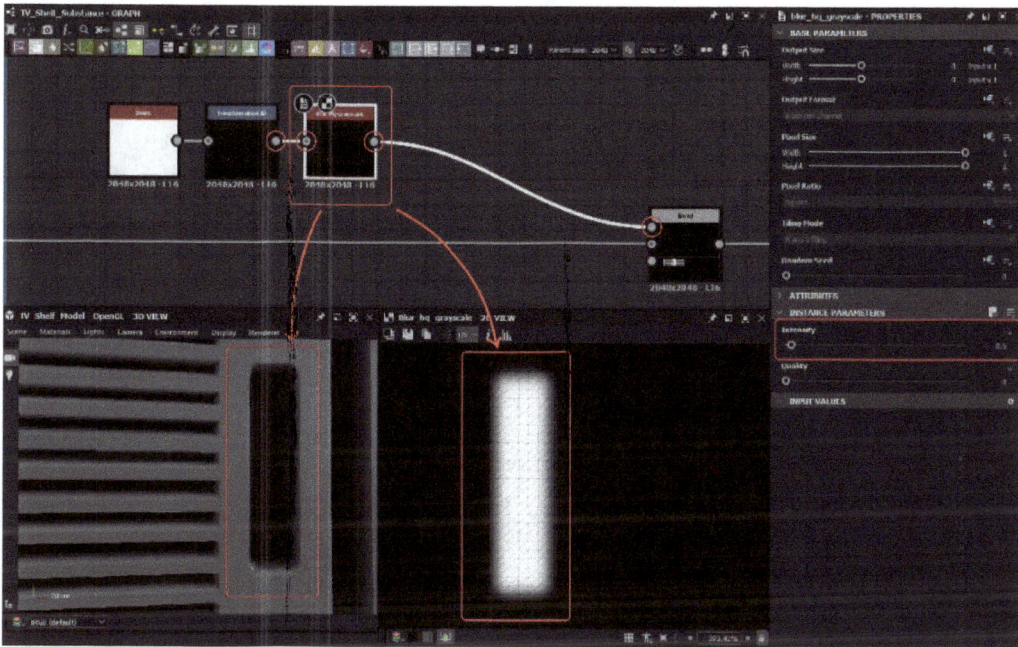

Figure 10.41 – Adding a Blur HQ Grayscale node to Transformation 2D

12. Once you add **Blur HQ Grayscale**, you will notice that the rectangle-shaped knob has now become a rounded rectangle. However, it has also become blurred. To fix this, we can add the **Levels** node to **Blur HQ Grayscale** and change its histogram, as shown in *Figure 10.42*:

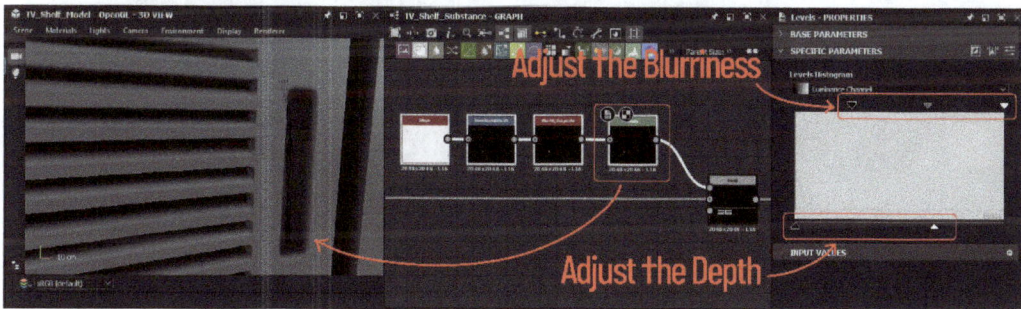

Figure 10.42 – Adjusting the shape and depth of the door knob with the Levels node

13. Now, we need another copy of the door knob beside the original door knob. Add the **Mirror Grayscale** node between the **Levels** node's **Output** and the **Blend** node's **Foreground** input and change its **Mode** setting to **Mirror Axis X** and **Axis X Offset** to 0.3275 under **INSTANCE PARAMETERS**, as shown in *Figure 10.43*:

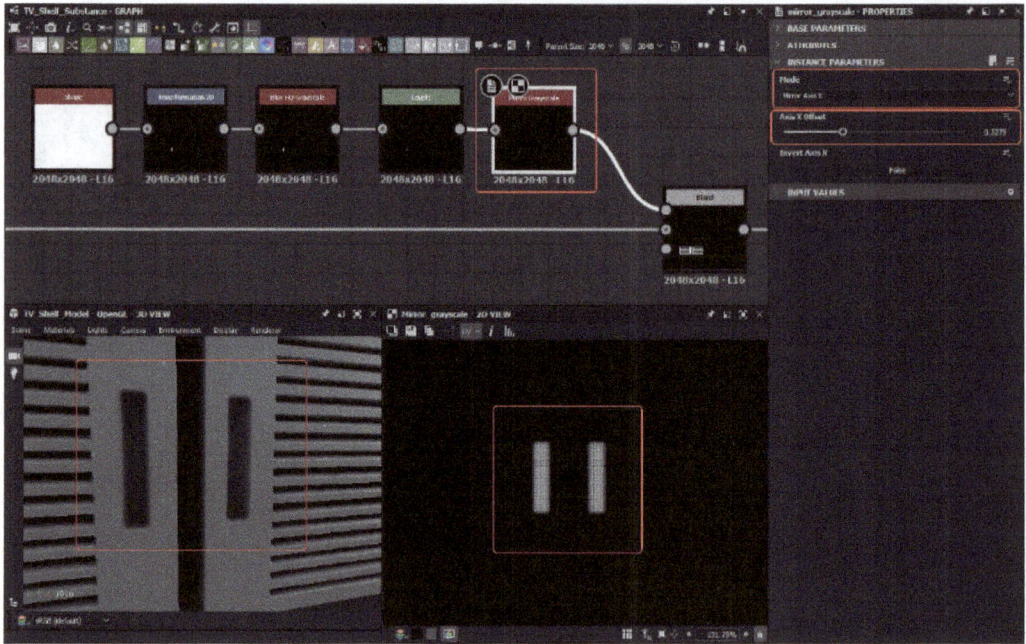

Figure 10.43 – Mirroring the door knob using Mirror Grayscale

14. Now, select all the nodes associated with the door knobs, right-click on them with the mouse, and select **Add Frame**, then change the frame title to Door Knobs, as shown in *Figure 10.44*:

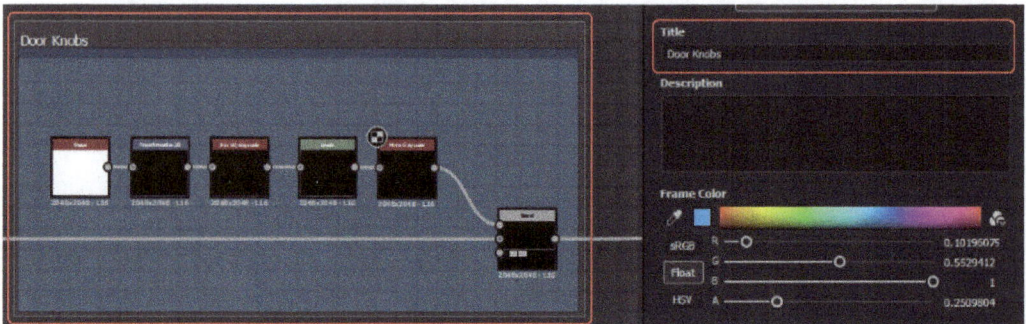

Figure 10.44 – Adding a frame to the Door Knobs nodes

Finally, we are done with the front part of the shelf. In the next section, we will start creating the Albedo Maps and texture for the television shelf.

Creating Albedo Maps, Roughness Maps, and other textures for the television shelf

Now, as we have our television shelf ready, we need to give it some realistic colors with Albedo Maps and Roughness Maps. Also, we will add some dent and scratch effects to the shelf and cover it with some dirt and dust to make it look old.

So, let us start with the steps:

1. First, we will put the **Metallic** map next to the **Final Blend** nodes, the **Base Color** nodes after it, and the **Roughness** nodes after **Base Color**, as shown in *Figure 10.45*:

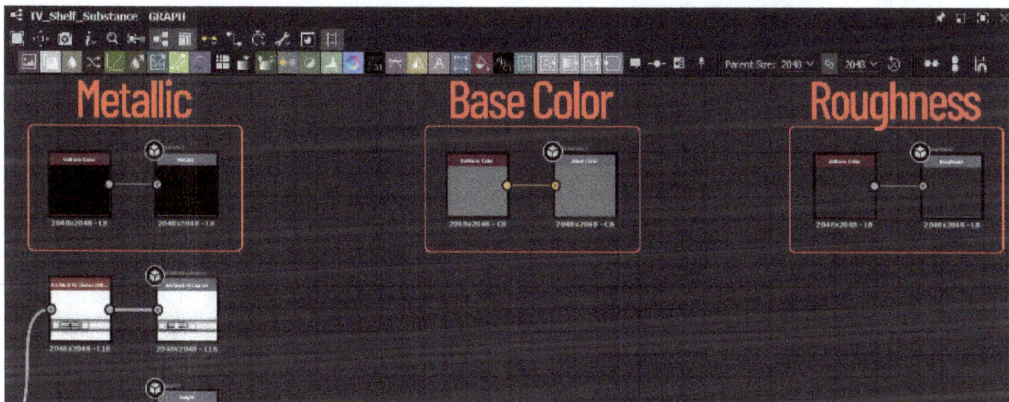

Figure 10.45 – Moving around the Texture nodes in the GRAPH window

2. First, we need to add a wooden texture base material. This is quite a complex material to implement. First, we need to add the **Anisotropic Noise** node and under **INSTANCE PARAMETERS**, change **X Amount** to 2, **Y Amount** to 184, and **Smoothness** to 1 (if it is not already at 1).

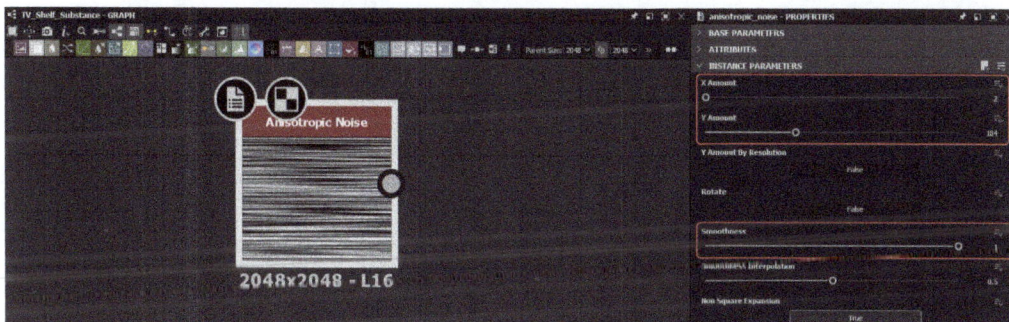

Figure 10.46 – Adding the Anisotropic Noise node and adjusting its parameters

3. Now, add and connect the **Blur HQ Grayscale** node to the **Anisotropic Noise** node and change its **Intensity** setting to 1.73.

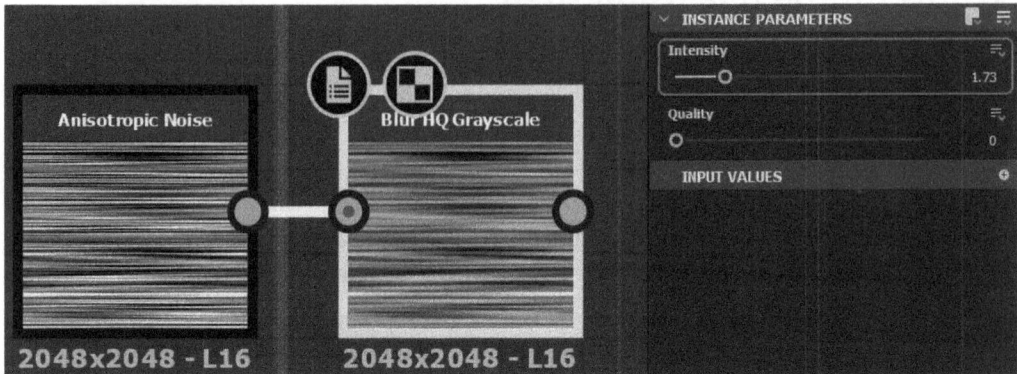

Figure 10.47 – Blurring the Anisotropic Noise node with Blur HQ Grayscale

4. Now, add the **Tile Generator** node, go to its **PROPERTIES** window, and apply the following changes to its parameters:

    -   **X Amount**: 2; **Y Amount**: 4
    -   **Pattern**: **Bell**; **Size | Scale**: 0.6
    -   **Position | Position Random | X Random**: 1; **Y Random**: 1
    -   **Color | Luminance Random**: 1

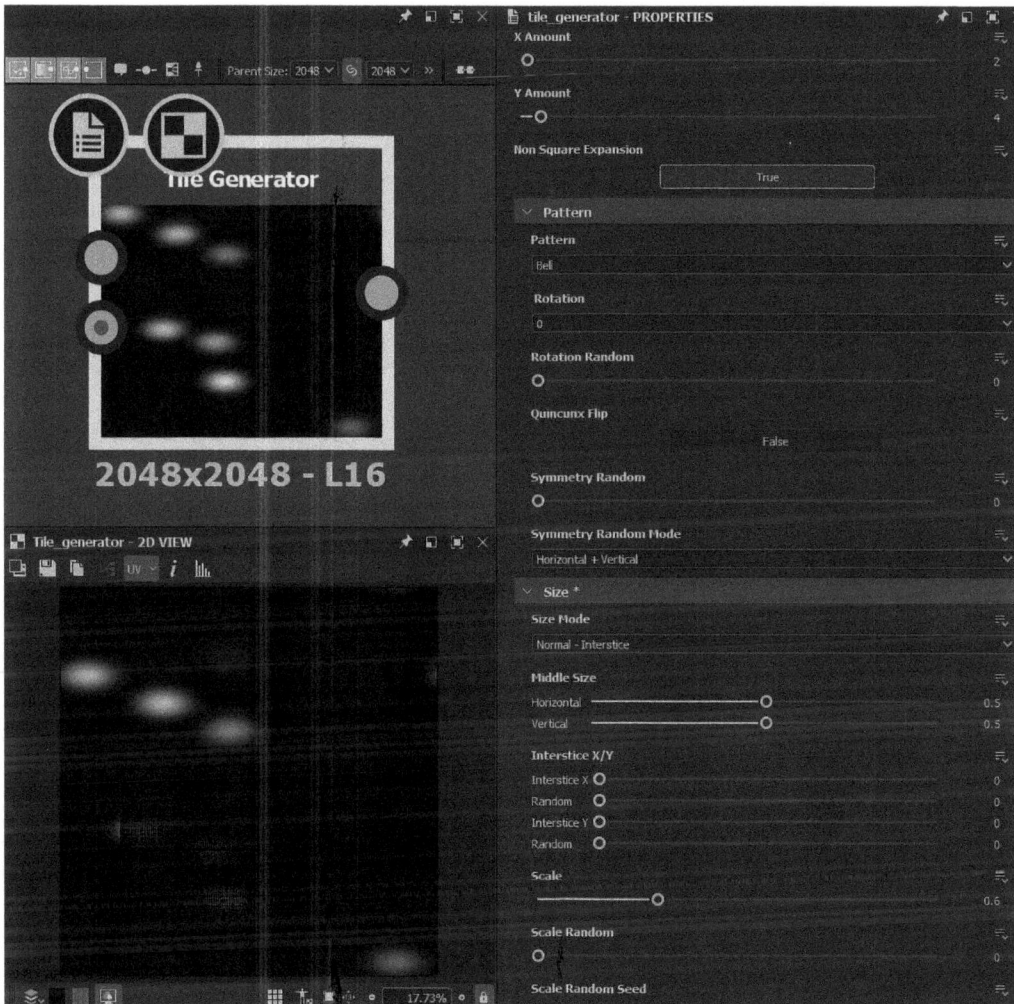

Figure 10.48 – Adding the Tile Generator node

We have added an **Anisotropy Noise** node to create wood grains and a **Tile Generator** node to create wood knots. However, we need to join them together to create a wooden texture.

To blend their shapes, we will add a different kind of node, a **Warp** node.

5. Add a **Warp** node and connect the **Blur HQ Grayscale** node's **Output** to the **Warp** node's **Input**, and the **Tile Generator** node's **Output** to the **Warp** node's **GradientInput**. Then, change the **Warp** node's **Intensity** setting to 0.1.

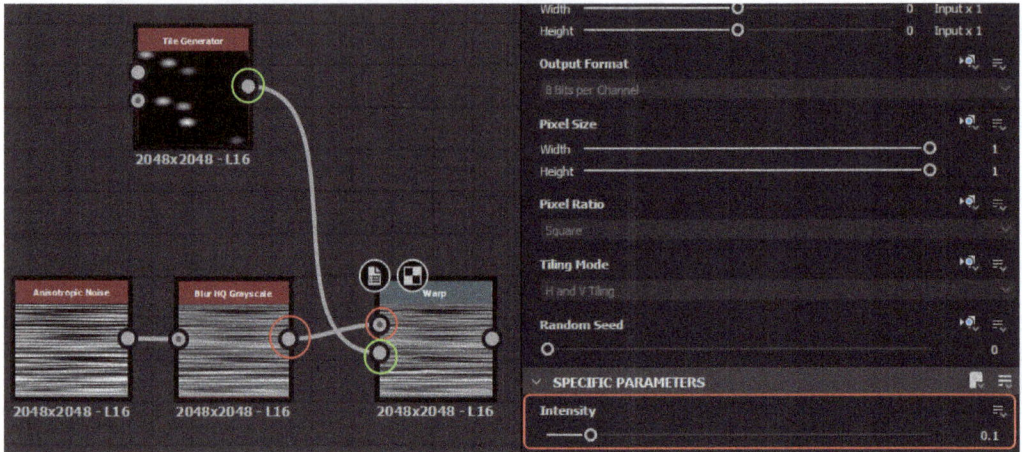

Figure 10.49 – Blending Anisotropic Noise and Tile Generator with the Warp node

The basic shape has been created; however, it needs to be brighter to produce clear results.

6. Add a **Blend** node, connect the **Tile Generator** node's other **Output** to the **Blend** node's **Foreground** input and the **Warp** node's **Output** to the **Blend** node's **Background** input, and change the **Blend** node's **Blending Mode** to **Max (Lighten)**.

7. If you want to customize the connection links, you press *Alt* (Windows) + left mouse button or *Option* (Mac) + left mouse button. This will create a link dot that can be repositioned, and other connections can also be either output from it or input to it, as shown in *Figure 10.50*:

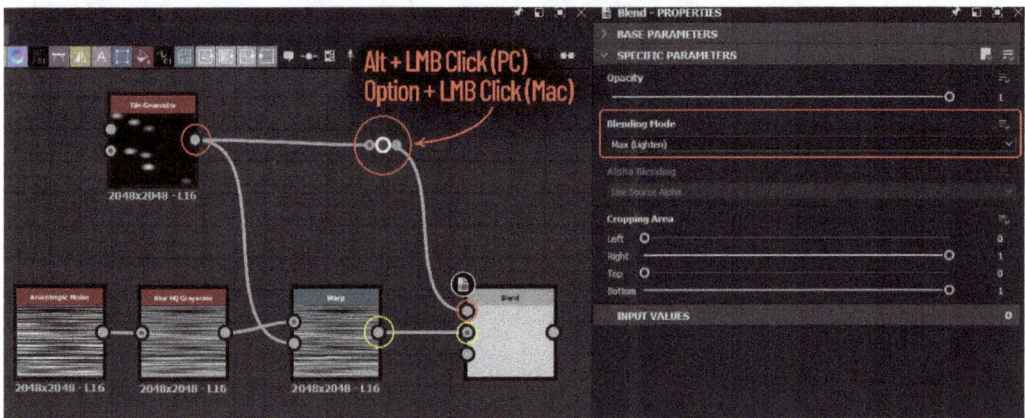

Figure 10.50 – Blending the Tile Generator and Warp nodes

8. Now we have the basic details of the wooden texture, but we need to create finer detail. First, we need a grainy material. Add **Grunge Map 002**. This node will help us to create finer details.

9. However, it's quite large for the television shelf's UV maps. So, to reduce the size of **Grunge Map 002**, connect the **Transformation 2D** node to it and change its **Stretch Width** and **Height** settings to 70%, as shown in *Figure 10.51*:

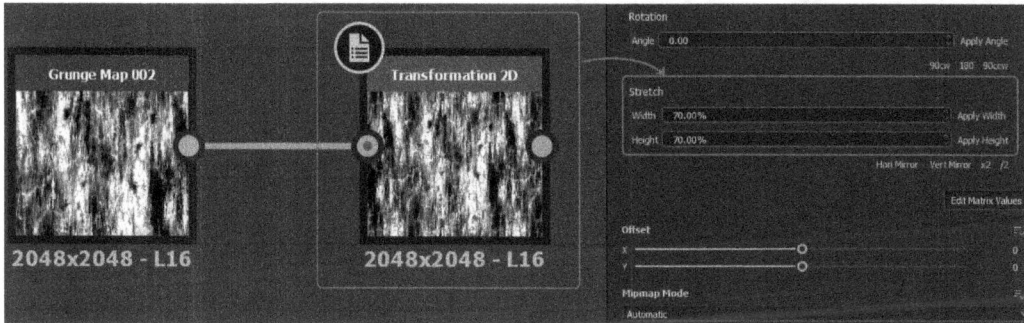

Figure 10.51 – Adding Grunge Map 002 and reducing its size with the help of Transformation 2D

Grunge Map 002 has quite a high dynamic range. Let us reduce its range.

10. To do that, add and connect the **Histogram Range** node to **Transformation 2D** and change the **Histogram Range** node's **Range** setting to 0.2, keeping the default value of 0.5 for **Position**.

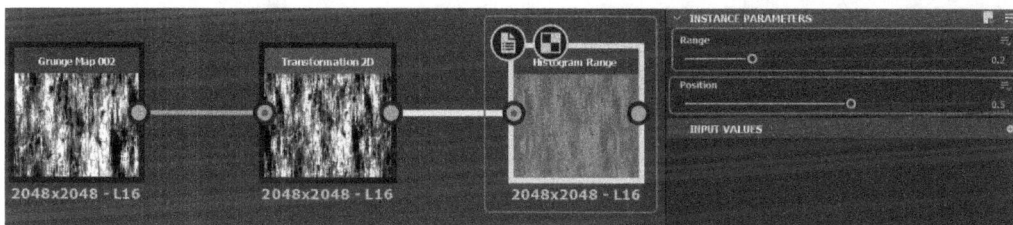

Figure 10.52 – Adding and adjusting the Histogram Range node

Now we have two output nodes to create finer details, as shown in *Figure 10.53*:

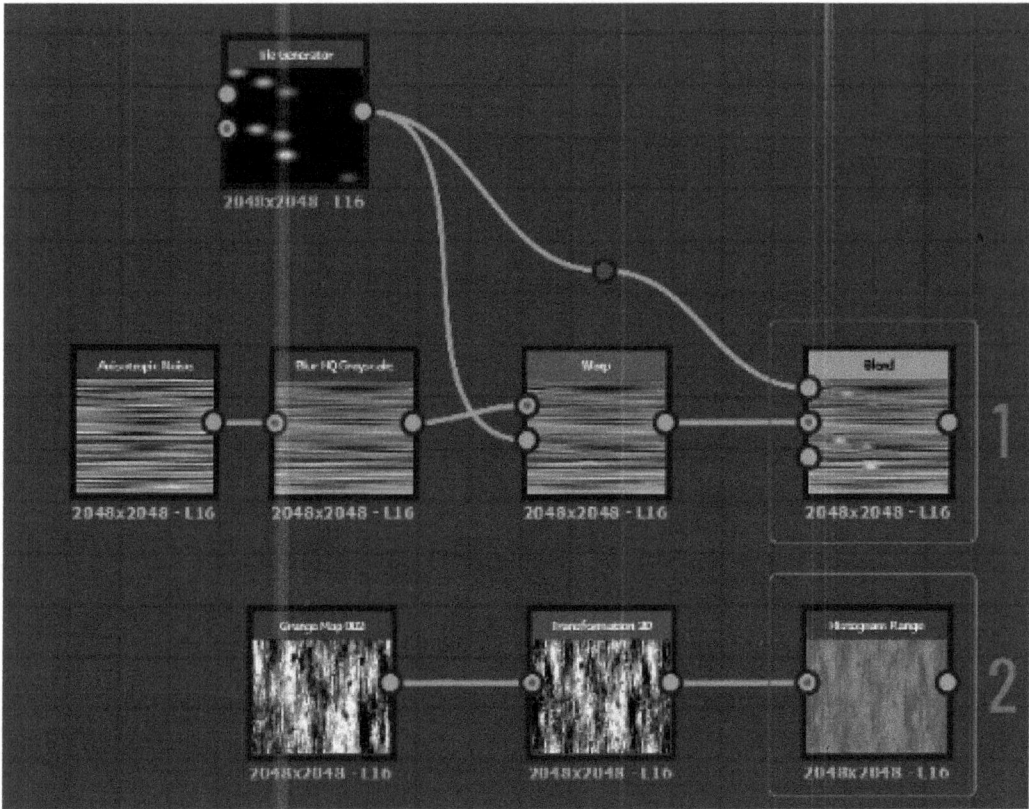

Figure 10.53 – Two output nodes

11. These two nodes must join together. So, add a **Gradient (Dynamic)** node and connect the **Blend** node's **Output** to the **Gradient (Dynamic)** node's **Grayscale Input** and the **Histogram Range** node's **Output** to the **Gradient (Dynamic)** node's **Gradient Input**, as shown in *Figure 10.54*:

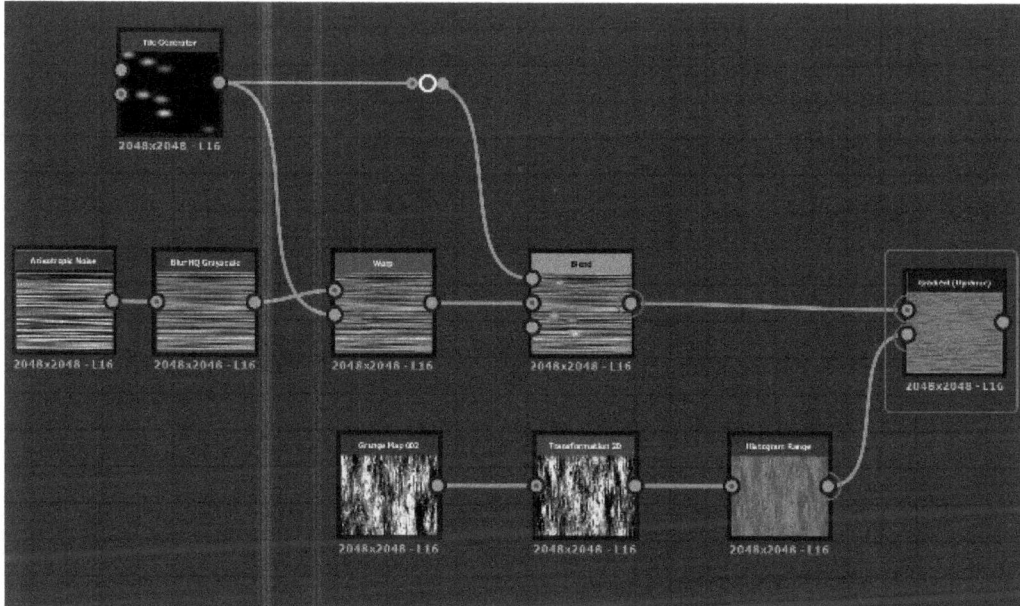

Figure 10.54 – Joining two output nodes with Gradient (Dynamic)

12. Now we finally have our wooden texture, but it's too large to fit on the television shelf's UVs. So, add the **Transformation 2D** node and connect the **Gradient (Dynamic)** node's **Output** to the **Transformation 2D** node's **Input**, then change the **Transformation 2D** node's **Stretch Width** and **Stretch Height** settings to 50%, as shown in *Figure 10.55*:

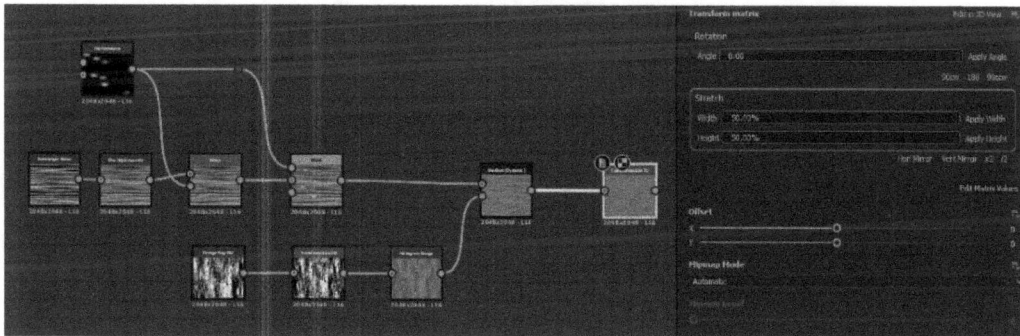

Figure 10.55 – Scaling down the wooden texture material

13. Let's now remove the **Grayscale Uniform Color** node from the **Base Color** node and connect the **Final Wooden Texture** node to it, as shown in *Figure 10.56*. This will create a black-and-white texture on the television shelf.

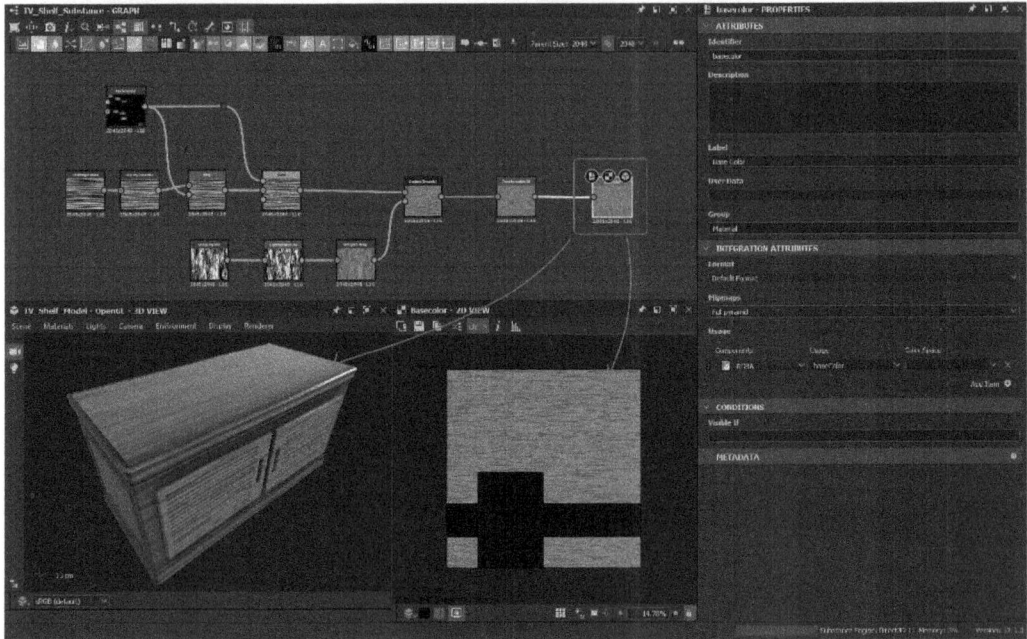

Figure 10.56 – Connecting the final wooden texture to the base color

14. We need a colored wooden texture, so add **Gradient Map** between the wooden texture's **Transformation 2D** node and **Base Color**, as shown in *Figure 10.57*:

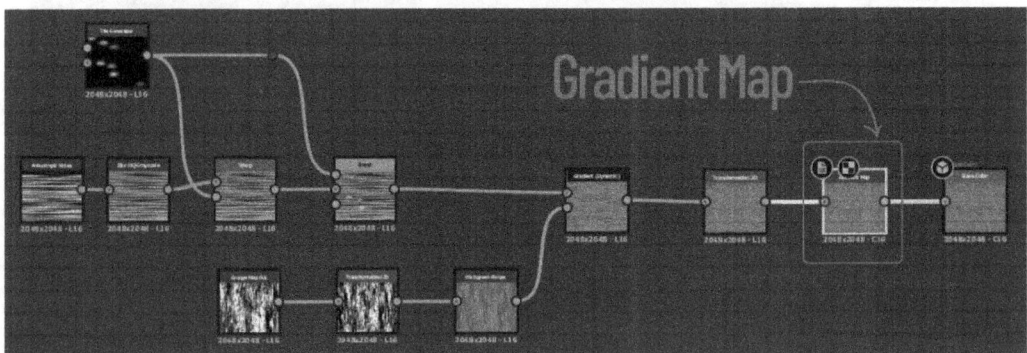

Figure 10.57 – Adding Gradient Map before Base Color

15. To create a colored wooden texture, open wood.jpg given in **Substance Designer Exercise Files** and Substance Designer side by side. Go to the **Gradient Map** node's **PROPERTIES** window and select **Gradient Editor**. Once the **Gradient Editor** window pops up, choose **Pick Gradient** from it, and finally, drag across wood.jpg with the left mouse button, as shown in *Figure 10.58*.

This will create a photorealistic wooden texture Albedo Map:

Figure 10.58 – Picking a gradient from wood.jpg

16. Now, let us move all **Wooden Texture** nodes under the **Door Knobs** nodes so we can easily create its albedo material. Otherwise, we will have to drag long link connections.

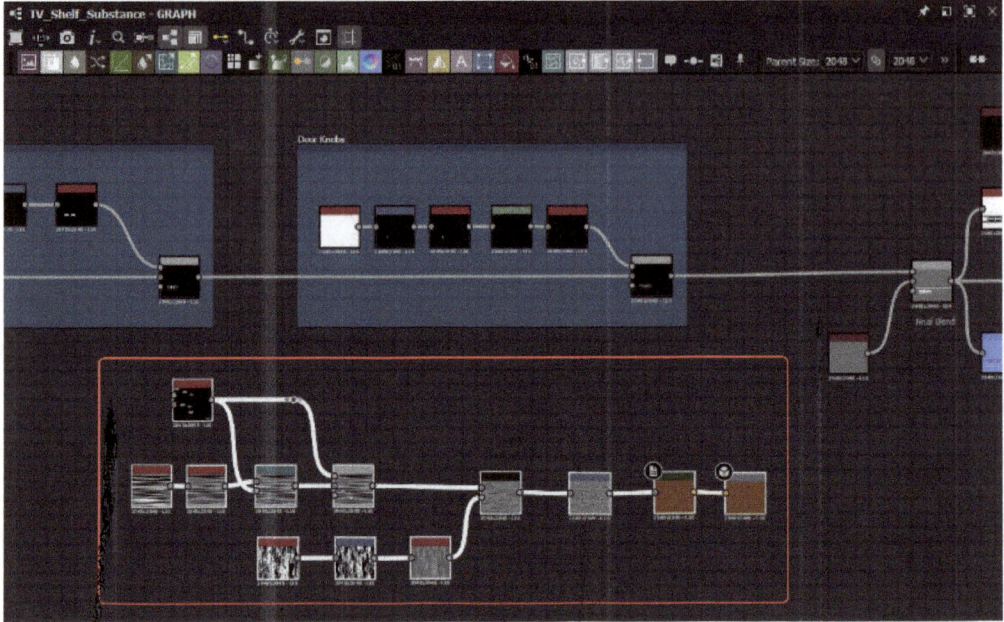

Figure 10.59 – Moving Wooden Texture nodes

17. We will now colorize the door knobs with different colors. Add a new **Blend** node between the **Gradient Map** and **Base Color** nodes, and make sure the **Gradient Map Output** is connected to the **Blend** node's **Background** input.

18. Then, connect the door knobs' **Mirror Grayscale** node's **Output** to the **Blend** node's **Opacity** input, add a new dark-brown **Uniform Color** node, and connect its output to the **Blend** node's **Foreground** input, as shown in *Figure 10.60*. This will colorize the door knobs dark brown:

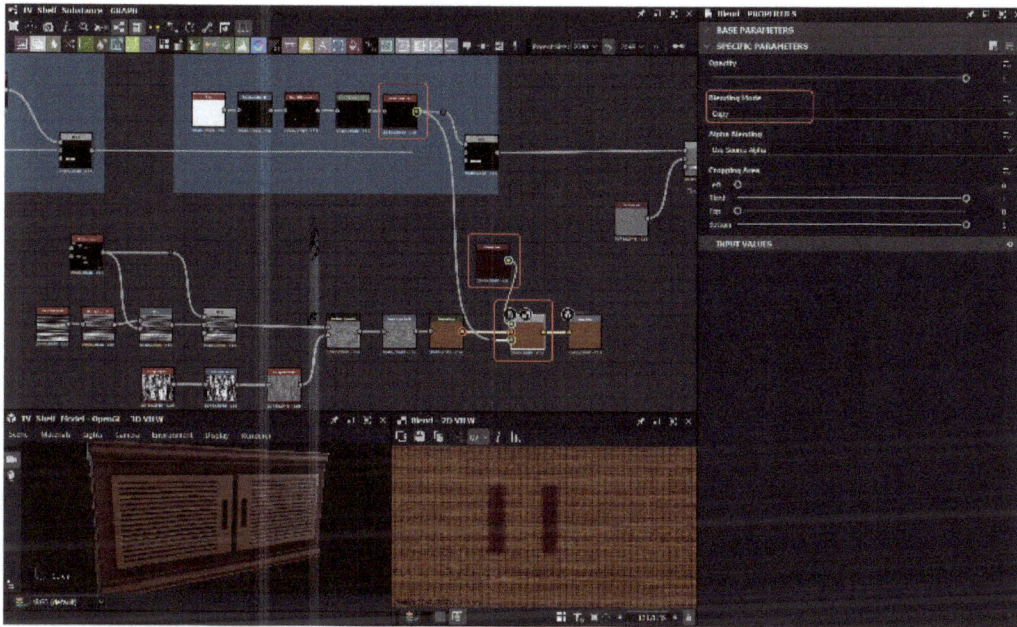

Figure 10.60 – Applying different uniform colors to the door knobs

Now, our **Wooden Texture** Albedo is ready.

19. To organize the albedo map select all the nodes associated with it, right-click with the mouse on it, and choose **Add Frame**. Rename the frame title `Wooden Texture Albedo` and change its color to red so that you can differentiate between the **Height** and **Base Color** maps.

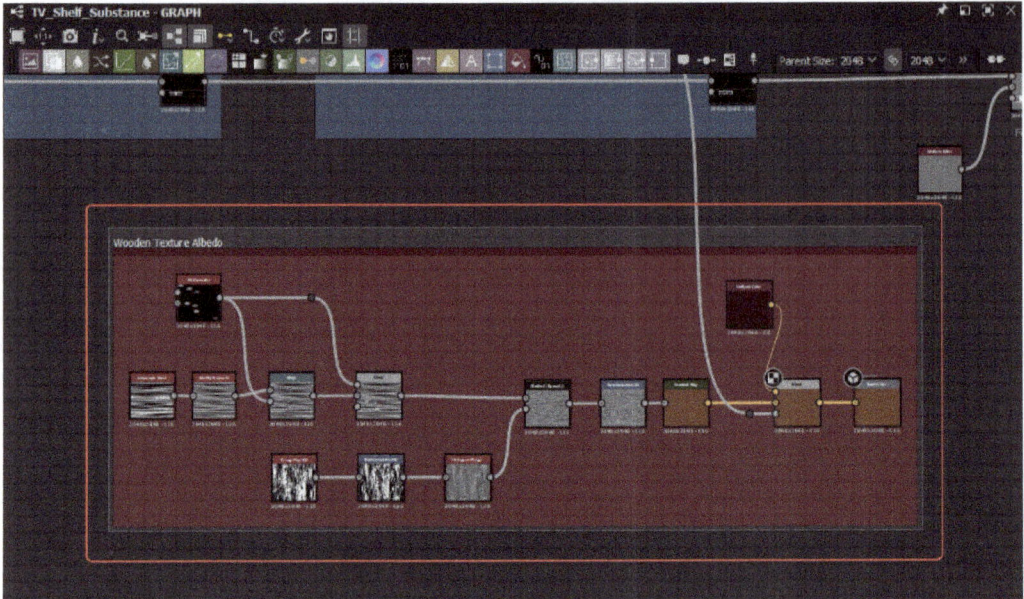

Figure 10.61 – Adding a frame to the Wooden Texture Albedo

20. Let us add a roughness map to the shelf. Remove **Uniform Color** from the **Roughness** map, add the **Grunge Scratch Dirty** node to it, then go to its parameters and change **Balance** to 0.56 and **Contrast** to 0.63.

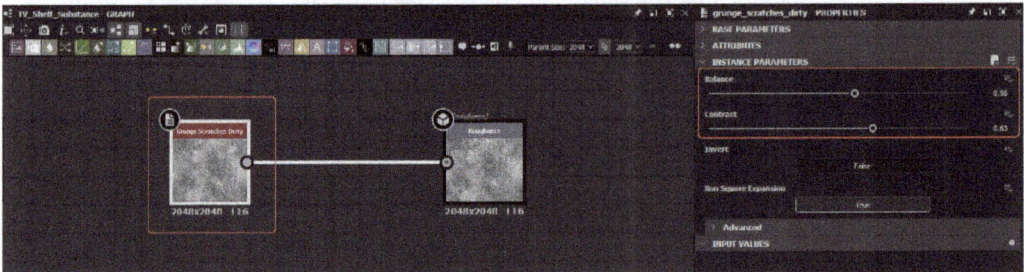

Figure 10.62 – Adding the Grunge Scratch Dirty node to the Roughness map

21. The **Roughness** map is, however, too large to fit in the UVs. So, add the **Transformation 2D** node between the **Grunge Scratch Dirty** and **Roughness** nodes and change the **Transformation 2D** node's **Stretch Width** and **Stretch Height** settings to 25%. Now, the shelf material will look older.

Figure 10.63 – Resizing the Grunge Scratch Dirty node with the help of the Transformation 2D node

22. Now, we will add a **Dirt** map to the shelf. We need more space between the **Wooden Texture Albedo** nodes and **Base Color**, and also need to add a **Blend** node between them so that we can add a **Dirt** node to it.

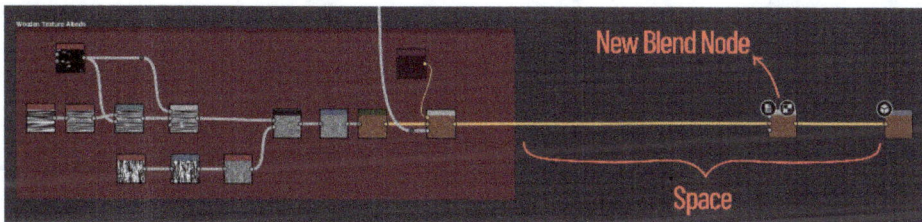

Figure 10.64 – Creating space and adding a Blend node

23. Now, we need three nodes to create the dirt effect, so add a **Dirt** node, a **Uniform Color** node, and a **Curvature Smooth** node.

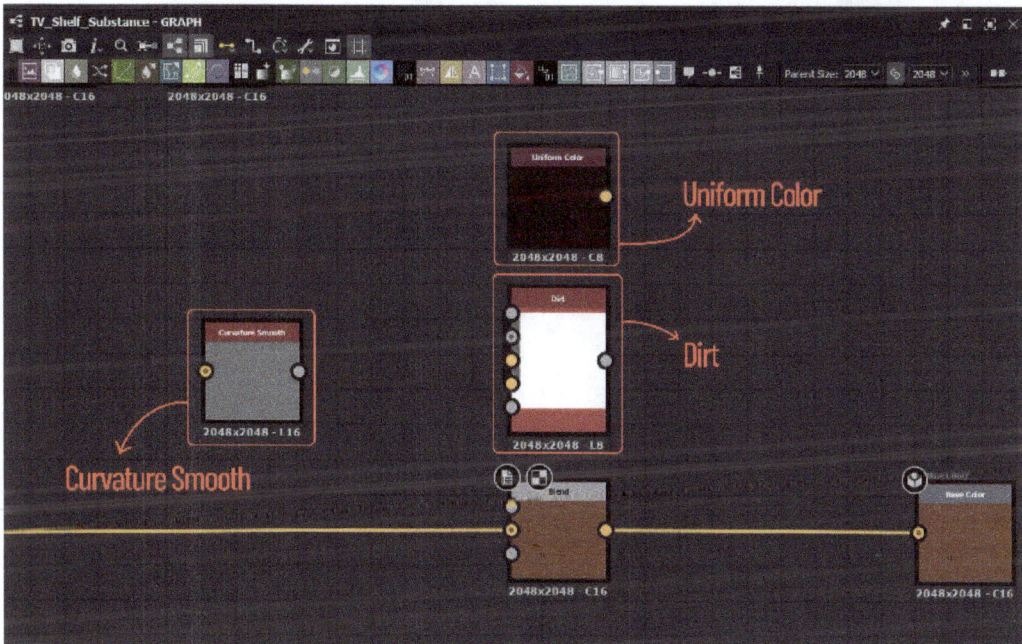

Figure 10.65 – Adding a Dirt node, a Uniform Color node, and a Curvature Smooth node

24. Now, connect the nodes, as follows:

- The **Ambient Occlusion (HBAO)** node's **Output** to the **Dirt** node's **Ambient Occlusion** input
- The **Normal** node's **Output** to the **Curvature Smooth** node's **Normal** input
- The **Curvature Smooth** node's **Output** to the **Dirt** node's **Curvature** input
- The **Uniform Color** (the dirt color that you want) node's **Output** to the **Blend** node's **Foreground**
- The **Dirt** node's **Mask Output** to the **Blend** node's **Opacity** input

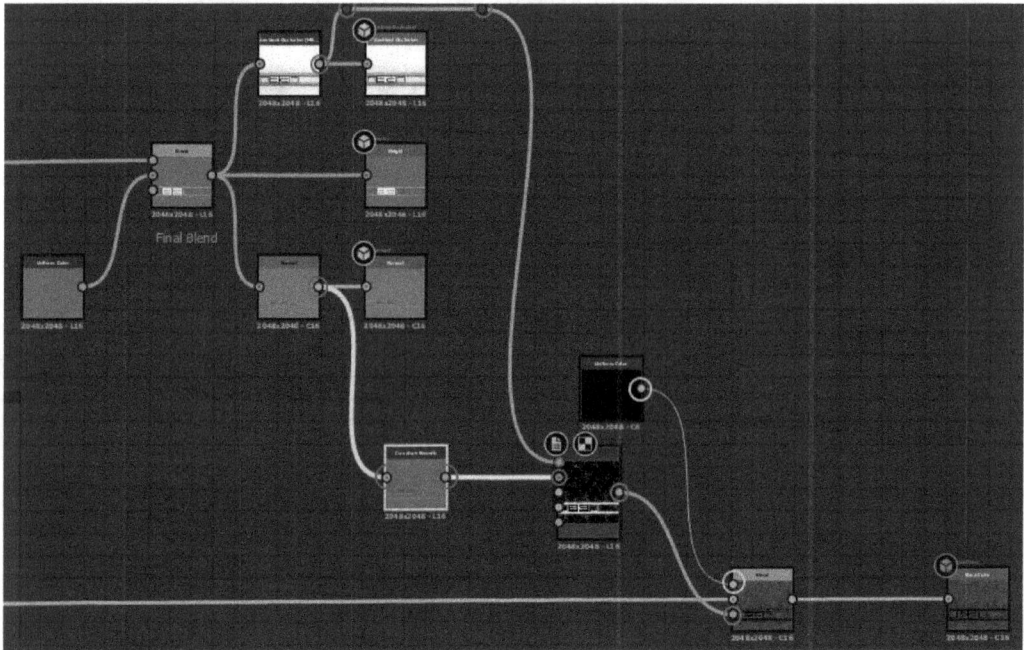

Figure 10.66 – Connecting the nodes

25. As we have connected all the nodes, we need to apply changes to the **Dirt** node's parameter, as follows:

- **Dirt Level**: 0.44; **Dirt Contrast**: 0.13; **Grunge Amount**: 1
- **Edges Masking**: 0.45; **Grunge Scale**: 8

26. Then, change the **Blend** node's **Blending Mode** setting to **Multiply** and **Opacity** to 0.5.

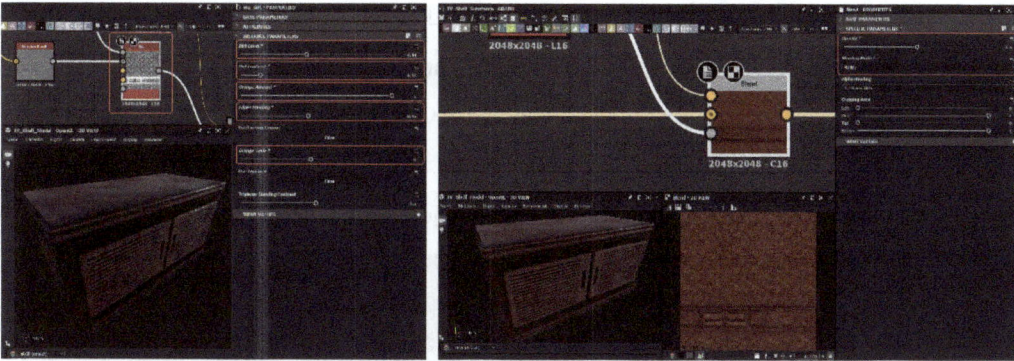

Figure 10.67 – Changing the Dirt and Blend nodes' parameters

Now, we will create a dust effect, so let's create some space for that and add the **Blend** node for it in that space.

27. Once the **Blend** node for the dust effect has been added to the space, add the **Dust** node, the **Fractal Sum 1** node, and the **Gradient Map** node, as shown in *Figure 10.68*:

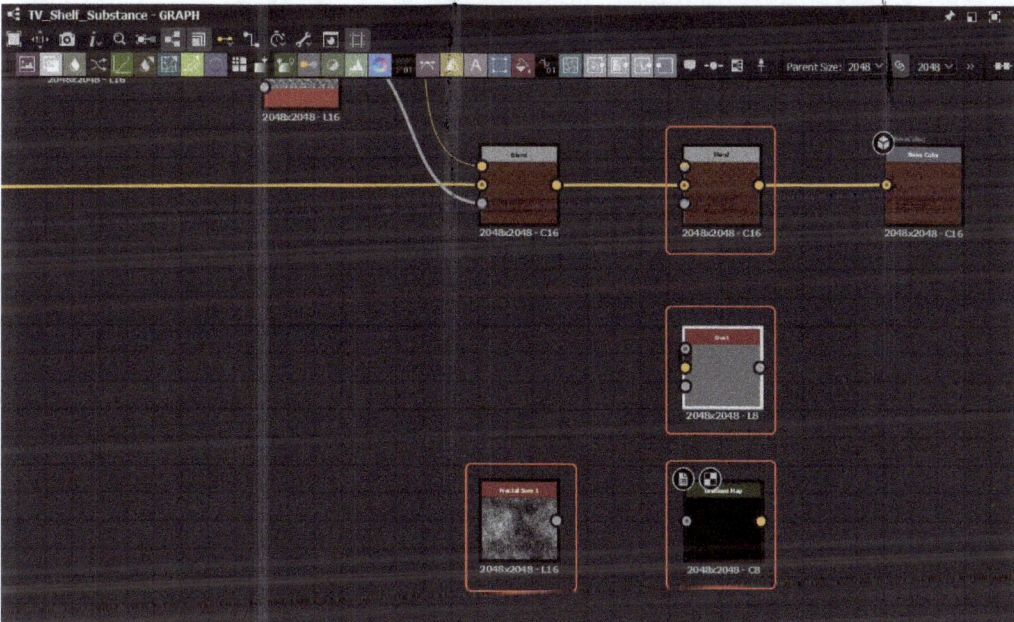

Figure 10.68 – Adding the Blend node, the Dust node, the Fractal Sum 1 node, and the Gradient Map node

28. Now, connect all these nodes, as follows:

- The **Ambient Occlusion (HBAO)** node's **Output** to the **Dust** node's **Ambient Occlusion** input

- The **Normal** node's **Output** to the **Dust** node's **World Space Normal** input

- The **Dust** node's **Mask** output to the **Blend** node's **Opacity** input

- The **Fractal Sum 1** node's **Output** to the **Gradient Map** node's input

- The **Gradient Map** node's **Output** to the **Blend** node's **Foreground** input

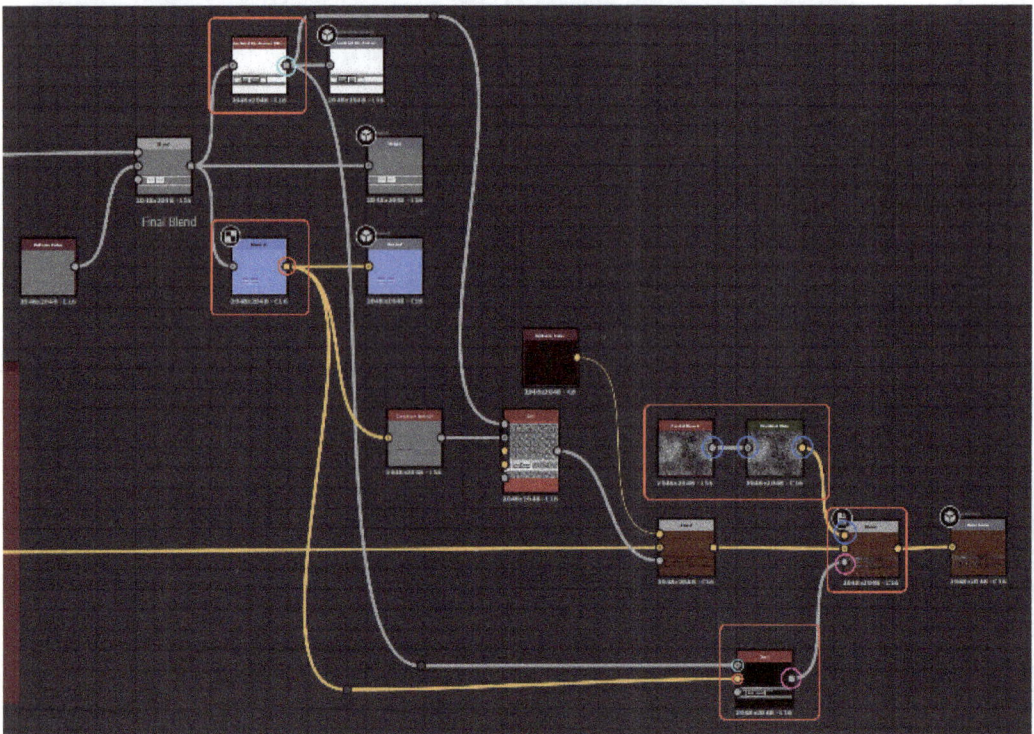

Figure 10.69 – Connecting nodes for the dust effect

Now that you have connected the nodes, you will notice that the dust effect that we have created is in grayscale. You need to select **Pick Gradient** to colorize it.

29. To do that, open the given sand.jpg image and **Substance Designer Gradient Editor** side by side and pick a small gradient by selecting **Pick Gradient**, which you will find in the Gradient Editor, as shown in *Figure 10.70*:

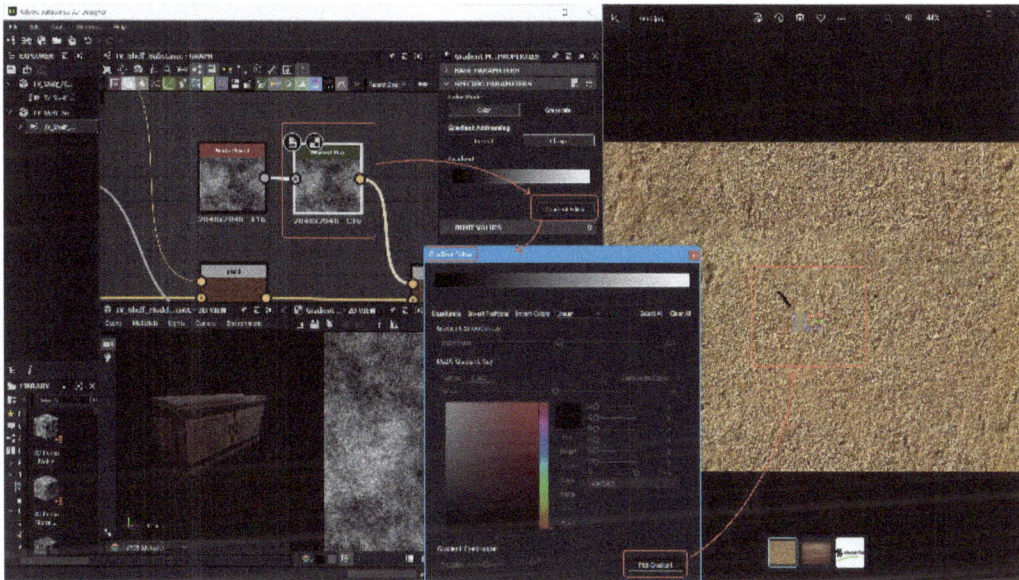

Figure 10.70 – Picking a gradient from sand.jpg

30. To finalize the dust effect, apply the following changes to the **Dust** node:

- **Level**: 0.38; **Contrast**: 0.04; **Occlusion Amount**: 0.6

- **Noise Opacity**: 0.8

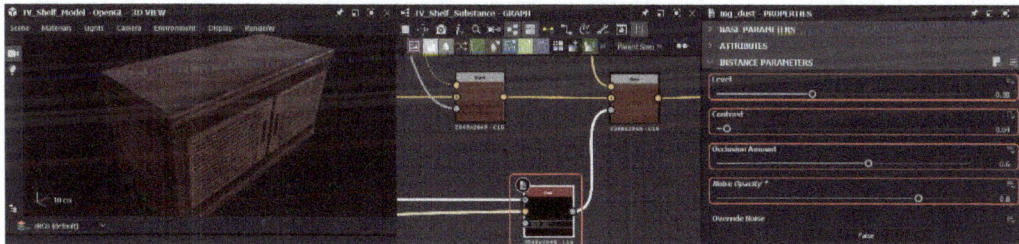

Figure 10.71 – Adjusting the Dust node's parameters

31. Now, you can go to the **Renderer** option inside **3D VIEW** and choose **iRay** to render **3D VIEW**. You can also go to the **Environment** menu in **3D VIEW** and choose **Edit**, then change the **Environment** parameters, such as **Sphere With Ground** under **Dome Type** and the *Y* position to -10.6 under the **Dome Ground** settings.

You can also change the other settings if you want.

Figure 10.72 – Rendering inside Adobe Substance 3D Designer

I hope you have enjoyed learning about Adobe Substance 3D Designer and now you should have total creative control. You can now make whole texture sets as well as patterns and textures that can be tiled.

Summary

You are now fully aware of the non-destructive, node-based environment that Substance Designer provides, enabling you to develop or alter materials from the beginning of your project. You should now be able to change a material's look and make adjustments whenever you want without much effort.

You should now also be able to access and alter any source or filter since you have total knowledge of editing in Adobe Substance 3D Designer. In the next chapter, you will learn how to get started with using Substance 3D Sampler. The chapter begins with an explanation of what this program does, and then moves on to the user interface and panels.

11
Adobe 3D Sampler at a Glance

This chapter delves into the various setup panels and also shows you how to access the ready-to-use components within Adobe Substance 3D Sampler. Then, you'll use what you've learned to create a basic pavement with blending layers and effects. The chapter will conclude with instructions on how to export your material as a smart material and individual maps.

If you're using Adobe Substance 3D Sampler for the first time or if you already know the interface but are unsure of its features, don't worry – we will look at all the fundamentals of the program in this chapter. You may now be wondering what Adobe Substance 3D Sampler is. When you want to create or extract new materials from your scans with single or multiple images, this application is great which can be used for things such as photogrammetry.

You can create the material collection either by tweaking or mixing what you already have, such as the material that is already on your hard drive, polish, and make it ready to be used in other 3D applications. Additionally, Adobe Substance 3D Sampler makes it simple to build your own textures from scratch.

Simply snap some photographs with your camera, convert them to tiles, and construct maps using the images. Photoshop used to be the preferred tool for artists, but with Adobe Substance 3D Sampler, you can now get results of superior quality while spending much less time.

The purpose of this tutorial is to teach you how to use Adobe Substance 3D Sampler if you're interested in texture and want to become a texture artist.

We will cover the following topics in this chapter:

- Getting started with Adobe Substance 3D Sampler
- Settings panels
- The **Assets** panel
- Adobe Substance 3D Sampler viewports
- Right-side panels and the toolbar

- A case study – creating a dirty old pavement
- Sharing

Getting started with Adobe Substance 3D Sampler

Adobe Substance 3D Sampler, also known as **Sampler**, has a quite simple interface. So, let us dissect its user interface. Most of the tools and options in Sampler are toggle-based, so you might have to click on them twice to turn them off. The user interface of Sampler is quite simple, as you can see in *Figure 11.1*.

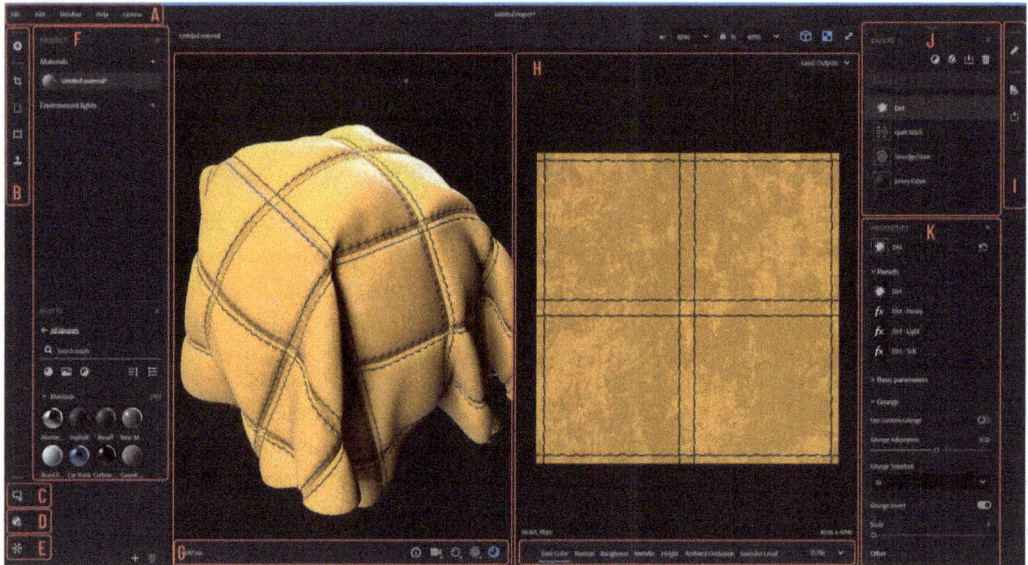

Figure 11.1 – The Adobe Substance 3D Sampler user interface

The following are breakdowns of Sampler's user interface from *Figure 11.1*.

A – the top bar (the application menu bar)

The current project name is displayed in the application's menu bar. The **File**, **Edit**, **Window**, **Help**, and **License** menus are all located there:

- To start a new project, open an existing one, and to save your current project, use the **File** menu
- To undo and redo activities or retrieve your settings, use the **Edit** menu
- Reset your workspace to its default configuration using the **Window** menu
- To find out more about Sampler or how to resolve problems, use the **Help** menu

B – the left sidebar (the 2D modification tool)

The left sidebar holds tools that are used to modify material inside the **2D** view; these are the shortcuts to frequently used 2D modification tools. The following are the tools that are available in this sidebar.

To better understand them, let us work on them practically; therefore, go to **File | Open Project**, and choose `Left_Side_Bar_Practice.ssa` from the given `Substance_Sampler_Exercise_Files` folder. Make sure that both the **3D** view and the **2D** view are active, as shown in *Figure 11.2*.

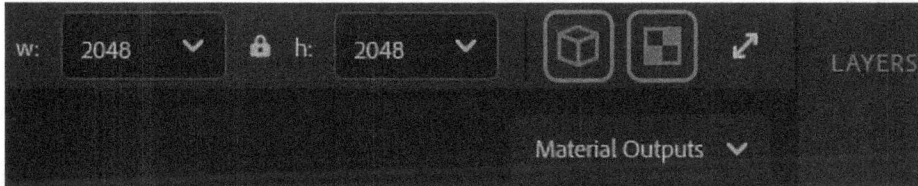

Figure 11.2 – Activating 3D view and 2D view

Get content

This tool allows you to import assets straight into the layer stack or the **Assets** panel. To access additional assets, go to **Substance Source**.

Crop

Images with unusual aspect ratios can be adjusted with the help of the crop tool. For instance, you can utilize the **PROPERTIES** panel's **Input Size** settings and the **Crop** tool to change the scale of an imported picture.

To work with it, you can follow the following steps:

1. Make sure that `Left_Side_Bar_Practice.ssa` from the given `Substance_Sampler_Exercise_Files` folder is open.

2. Select the **Crop** tool from the left sidebar.

3. From the **2D** view, choose the **Layer input** option, which is directly under the 2D/3D view buttons. We are choosing this option so that when you start cropping the material in the **2D** view, the image doesn't change or move; otherwise, it will be quite annoying.

4. There is a toolbar specifically for the **Crop** tool, as you can see in *Figure 11.3*. By changing the scale of the current transformation, the **Make it square** tool makes the resulting square.

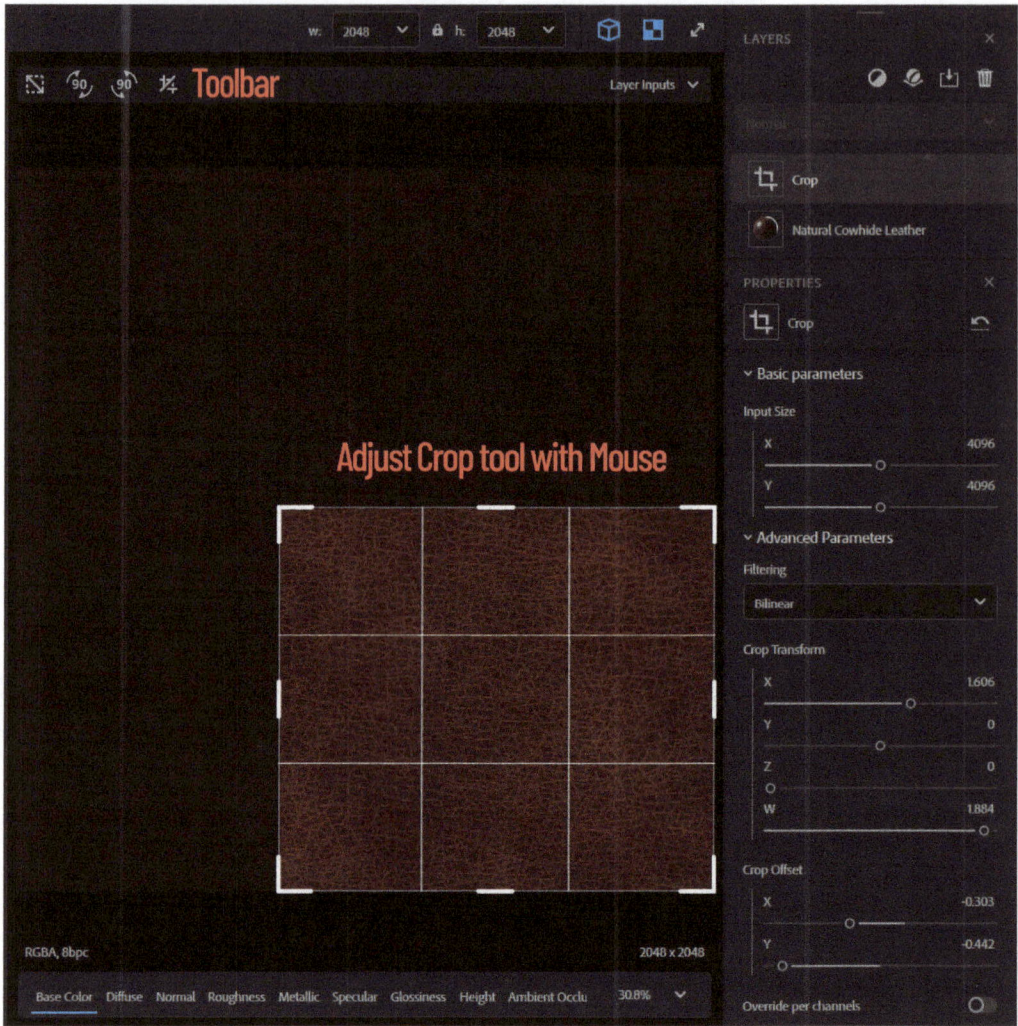

Figure 11.3 – The Crop tool

Rotation by **+90°** (to the right) turns the result clockwise by 90 degrees, rotation by **-90°** (to the left) turns the result counterclockwise by 90 degrees, and **Reset transformation** returns the **Transform** tool to its default settings.

You can use your mouse to adjust the **Crop** tool.

5. Also, you will notice that the **Crop** tool appears inside the **LAYERS** stack; this means that whatever effect, filter, or tool you apply appears in the **LAYERS** stack. This makes the application non-destructive, so you can always go back and adjust any effect, tool, or filter.

6. In the **PROPERTIES** panel, you will see there are a number of parameters that you can adjust to control **Crop** instead of using the mouse.

I hope the **Crop** tool is clear now. Let us move on to **Perspective transform** in the next section.

Perspective transform

To correct perspective errors in an image, use the **Perspective Correction** tool. On materials, perspective correction can also be applied.

To work with it, you can execute the following steps:

1. Make sure that `Left_Side_Bar_Practice.ssa` from the given `Substance_Sampler_Exercise_Files` folder is open, and then select the **Perspective transform** tool.

2. Select **Layer Outputs** in the **2D** view, which will show you the output result while you adjust the image perspective by moving around the circular control points in the **2D** view.

   Also, note that **Perspective transform** is also added to the **LAYERS** stack.

3. In the **PROPERTIES** panel, you can also reset **Perspective transform**; a toolbar is visible at the top of the **2D** view when the **Perspective transform** layer is selected. The handles on the **Perspective transform** layer can be returned to their original positions by clicking the **Reset positions** button. You can see how the perspective transform is applied in *Figure 11.4*.

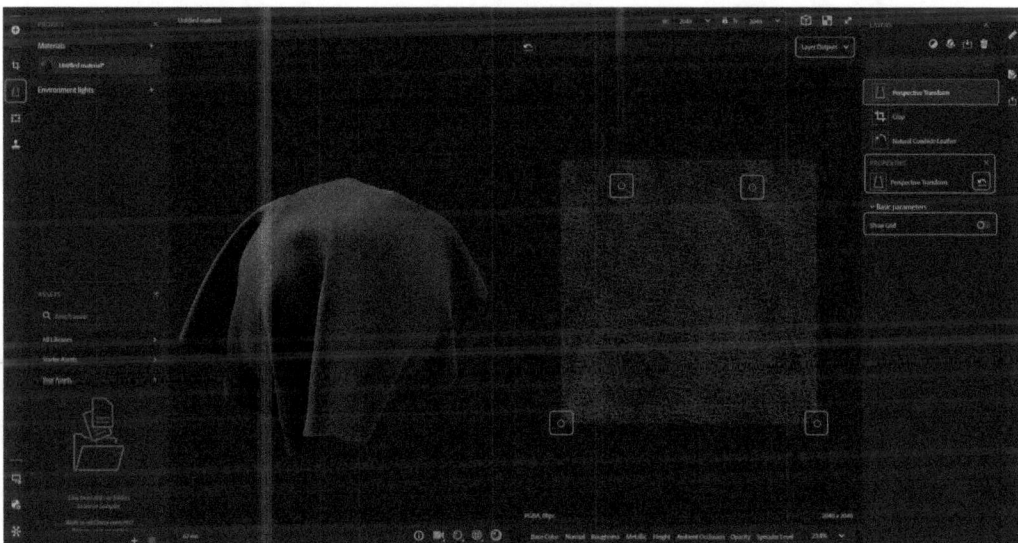

Figure 11.4 – Applying Perspective Transform

4.   You can also toggle the **Show Grid** option; this will display the perspective grid.

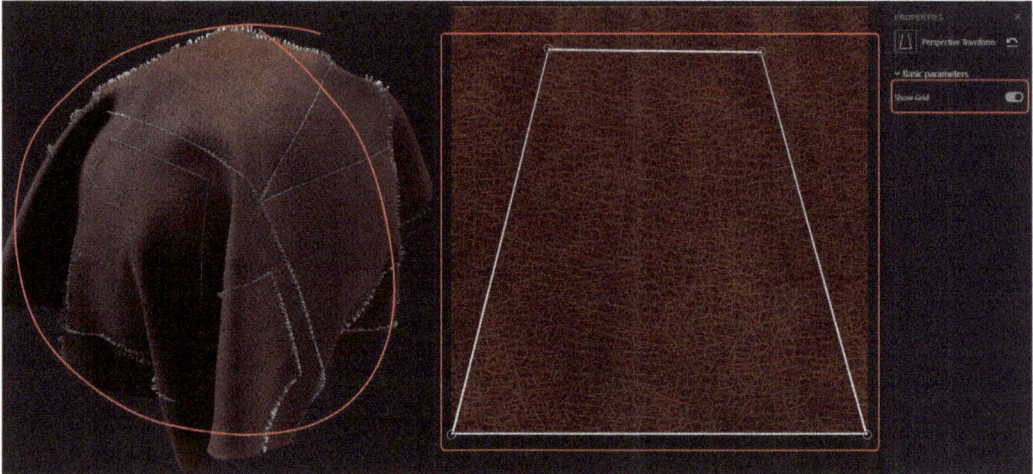

Figure 11.5 – The Show Grid option in Perspective Transform

I hope **Perspective transform** is clear. Now, let us move on to the **Transform** tool in the next section.

Transform

The **Transform** tool in Sampler is used to move, scale, or rotate your image or material.

To work with it, you can follow the following steps:

1.   Make sure `Left_Side_Bar_Practice.ssa` from the given `Substance_Sampler_Exercise_Files` folder is open, and then select the **Transform** tool.

2.   Select **Layer Outputs** in the **2D** view, which will show you the output result while you adjust the image's position, size, or rotation with the control points in the **2D** view by using the mouse.

3.   Also, note that **Transform** is also added to the **LAYERS** stack.

4.   You can transform the material in the **2D** view with the mouse. However, if you do not want to use the mouse, you can adjust the material using the parameters in the **PROPERTIES** panel.

5.   In the **PROPERTIES** panel, you can also reset **Transform**. You can also reset a specific parameter by hovering over its label.

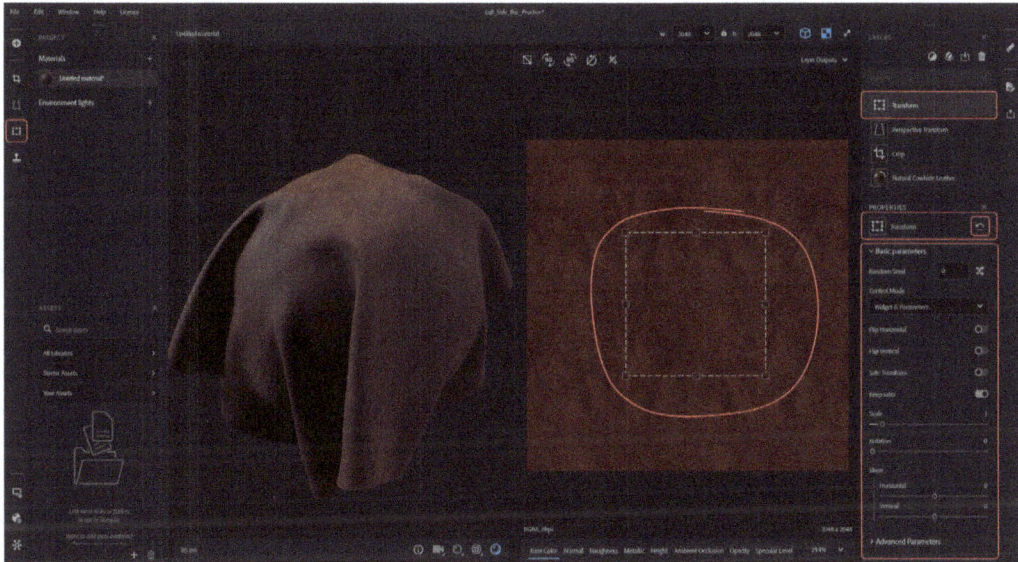

Figure 11.6 – Applying Transform

6. **Safe Transform** can be enabled or disabled. The transform node will retain tiling and prevent the loss of pixel information because of minute offsets and rotations when it is enabled. This limits your ability to manipulate the transformation, and turning on **Safe Transform** will make some parameters invisible.

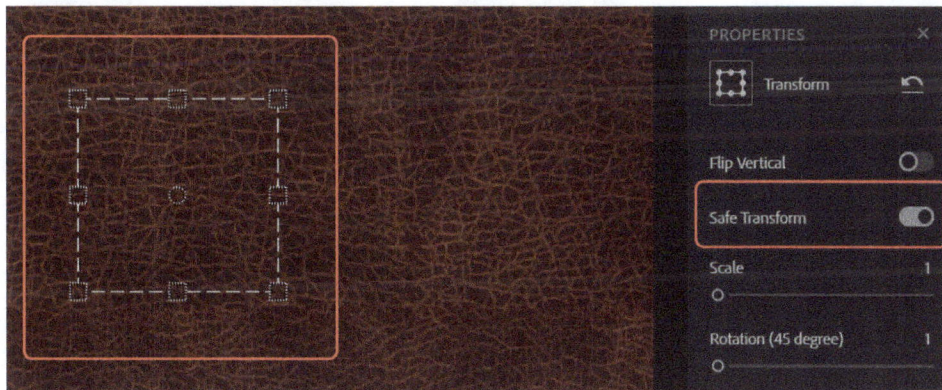

Figure 11.7 – Safe Transform

7. You can enable the **Deactivate Transform per Channel** option in **Advanced Parameters**. This option can deactivate the transformation for each channel.

I hope **Transform** is clear. Now, let us move on to the **Clone Stamp** tool in the next section.

Clone Stamp

One of the tools on the left sidebar is the Clone Stamp filter. You can manually copy or patch up portions of your material with the **Clone Stamp** tool. This is helpful to repair seams or eliminate mistakes in your material:

1. Make sure `Left_Side_Bar_Practice.ssa` from the given `Substance_Sampler_ Exercise_Files` folder is open, and then select the **Clone Stamp** tool.

2. Select **Layer Outputs** in the **2D** view, which will show you the output result.

3. The **2D** view in the viewport is automatically opened when a **Clone Stamp** filter layer is created. As soon as the **Clone stamp** layer is chosen, a toolbar is displayed at the top of the **2D** view.

Figure 11.8 – Clone Stamp in the 2D view

4. In the **2D** view, click and drag over the troublesome region to begin utilizing the **Clone Stamp** tool. Depending on the source, the content will start to update automatically. When you use the **Clone Stamp** tool, the areas where it is used get highlighted, as shown in *Figure 11.9*.

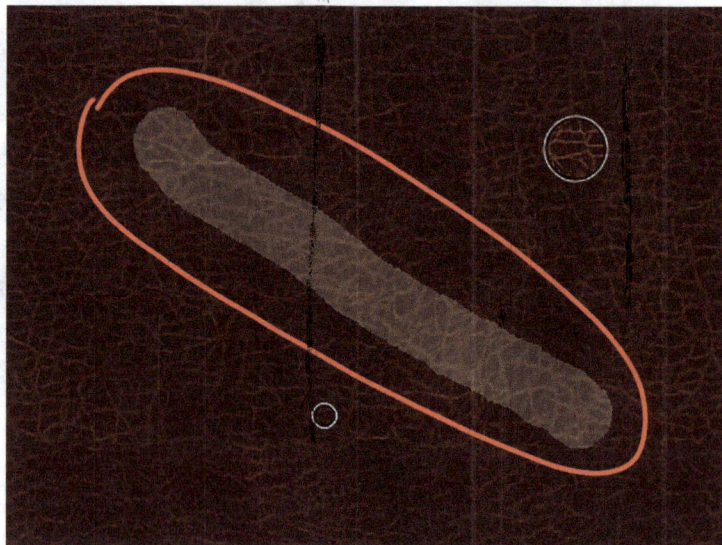

Figure 11.9 – The Clone Stamp tool's highlighted area

5. You can either press *Ctrl* + click the left mouse button (Windows) or *Command* + click the left mouse button (Mac), or drag the source handle from the **2D** view's center to specify the region to replicate. The *Ctrl* + left mouse button (Windows) or *Command* + left mouse button (Mac) adds another stamp layer – for example, **Stamp 2**. The *Ctrl* + right mouse button (Windows) or *Command* + right mouse button (Mac) moves the source dot.

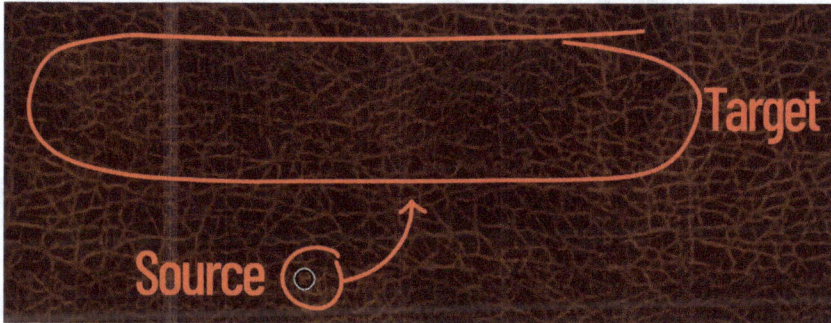

Figure 11.10 – Selecting the source

6. Generally speaking, it's a good idea to attempt to keep the source point away from the region you're copying over. It is possible to clone the troublesome area if the source point is near to it.

I hope the **Clone Stamp** tool is clear. Now, let us move on to **VIEWER SETTINGS** in the next section to discuss the various setting panels available.

Settings panels

There are various settings panels inside Sampler. These panels serve different purposes, so let us go through them in this section.

C – VIEWER SETTINGS

Both the **3D** and **2D** views can be customized from the **VIEWER SETTINGS** panel. The **VIEWER SETTINGS** panel contains different parameters; let us study them one by one.

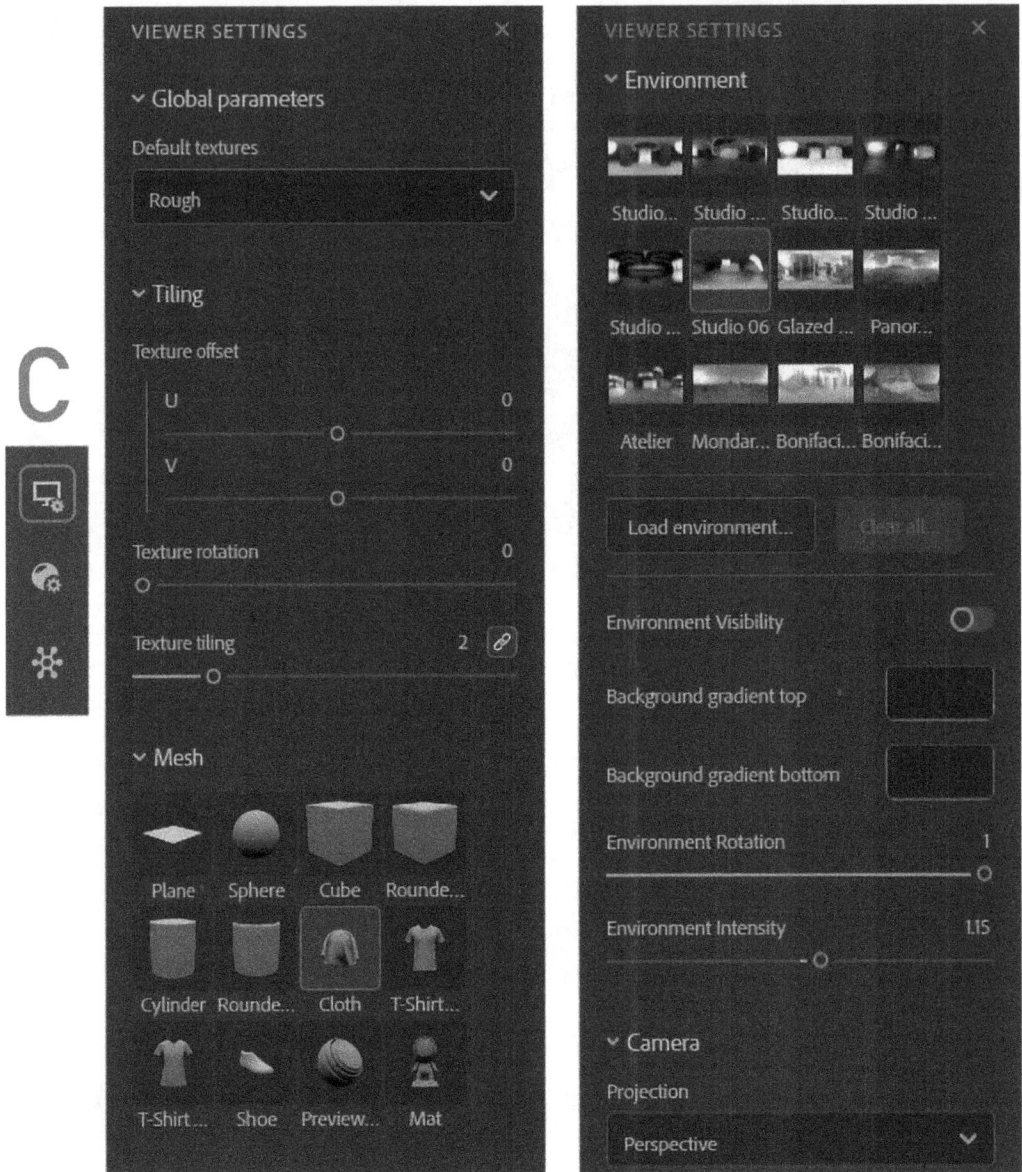

Figure 11.11 – VIEWER SETTINGS

Global parameters

Global parameters is the part of the **Viewer Settings** panel; the global parameters don't impact your project but change aspects of Sampler. Following are the two global parameters settings:

- **Renderer**: This allows you to choose a render engine. The same source materials and surroundings can be used by different render engines to yield somewhat different effects.

- **Default textures**: Here, you can decide whether the building blocks seem rough or reflective.

Tiling

How your textures are tiled on the 3D mesh is determined by the tiling choices. There are three ways to modify tiling:

- **Offset**

- **Rotation**

- **Scale**

You can also create non-square tiles as well. The **Tiling** shortcut is located in the lower-right corner of the **3D** view. It also allows you to change the texture scale.

Mesh

With this parameter, you can add a 3D mesh to the **3D** view, and you can test your materials on the selected mesh. An **AO mixing mode** option will show up if you choose **Preview Sphere** or **Mat mesh**. You can choose to replace or multiply the mesh AO with the material AO in **AO mixing mode**.

Environment

For this parameter, you can choose an HDRI-based environment. EXR and HDR photos for environments are supported by Sampler. For environment maps, it is strongly advised against utilizing low dynamic range photos.

The **Environment Visibility** option allows you to turn the environment on or off in the 3D viewport. The 3D viewport's backdrop can be modified using **Background gradient top** and **Background gradient bottom** when **Environment Visibility** is off. The environment's opacity can be changed when **Environment Visibility** is enabled.

From **Environment Rotation**, you can rotate the environment around the mesh. Using this, you can rapidly change the lighting of your objects. Pressing *Shift* + clicking the right mouse button will also allow you to do that.

Camera

The **Camera** parameters allow you to change the camera's **Projection** settings; there are two different kinds of projections, **Perspective** and **Orthographic**. The **Perspective** settings give distant items a smaller appearance than close-up ones. The **Orthographic** settings, regardless of how far away an object is from the camera, will make the object's size appear to remain the same.

When the **Perspective** projection is chosen, the **Field of view** settings can be changed. You can also find **Field of view** in the **3D** view's lower-right corner.

I hope **VIEWER SETTINGS** are clear. Now, let us move on to **SHADER SETTINGS** in the next section.

D – SHADER SETTINGS

You can customize how the shader displays your elements on the mesh in the 3D viewport through the **SHADER SETTINGS** panel.

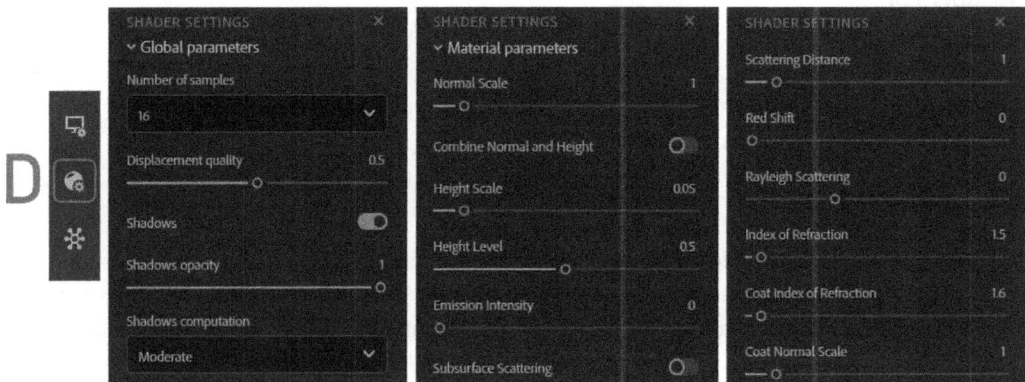

Figure 11.12 – SHADER SETTINGS

Global parameters

Global parameters affect Sampler but have no bearing on our project. **Number of samples** allows you to change how many passes the shader makes. A render with more samples will be of greater quality. **Displacement quality** allows a subdivision depth for tessellation displacement. **Shadows** allows you to turn shadows on or off.

Shortcuts for changing the height scale, displacement quality, and toggling shadows are located in the lower-right corner of the **3D** view.

Material parameters

You can modify the way your current material is rendered by the shader using **Material parameters**. **Normal Scale** allows you to change the usual map's strength or intensity. **Height Scale** allows you to adjust the displacement effect's strength. **Height Level** allows you to modify the height for the displacement's base level. **Emission Intensity** allows you to adjust the emissive maps' intensity.

Index of Refraction is where the angle at which light refracts off surfaces can be adjusted. **Coat Index of Refraction** allows you to angle the surface coat so that light reflects off it at the right angle. **Coat Normal Scale** is used only for the surface coat and to adjust the standard scale.

I hope **SHADER SETTINGS** are clear. Now, let us move on to **CHANNEL SETTINGS** in the next section.

E – CHANNEL SETTINGS

The list of channels computed for your current content is controlled by the **CHANNEL SETTINGS** panel. Here, you can control which channels are visible:

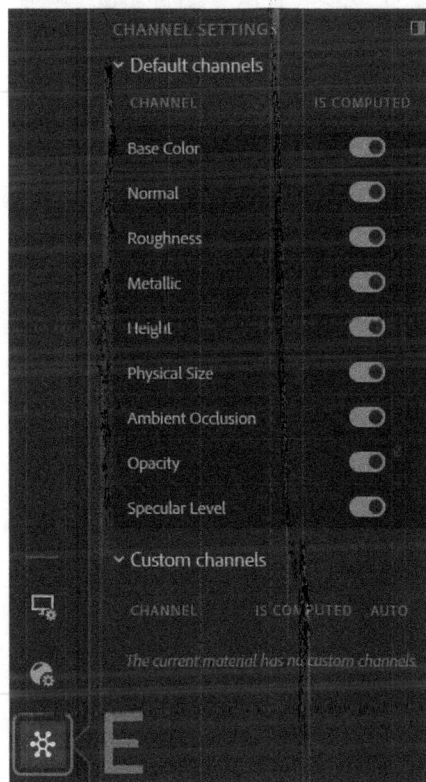

Figure 11.13 – CHANNEL SETTINGS

- **Default channels**: The list of channels that are automatically calculated depending on the process is displayed in this section

- **Custom channels**: Toggle any extra channels that aren't by default a part of the process you've chosen

I hope you are now well aware of settings panels, so let us move on to the **assets panels** and learn about them.

Assets panels

There are mainly two different types of assets panel in Sampler, **PROJECT** and **ASSETS**. Let us explore both of them in this section.

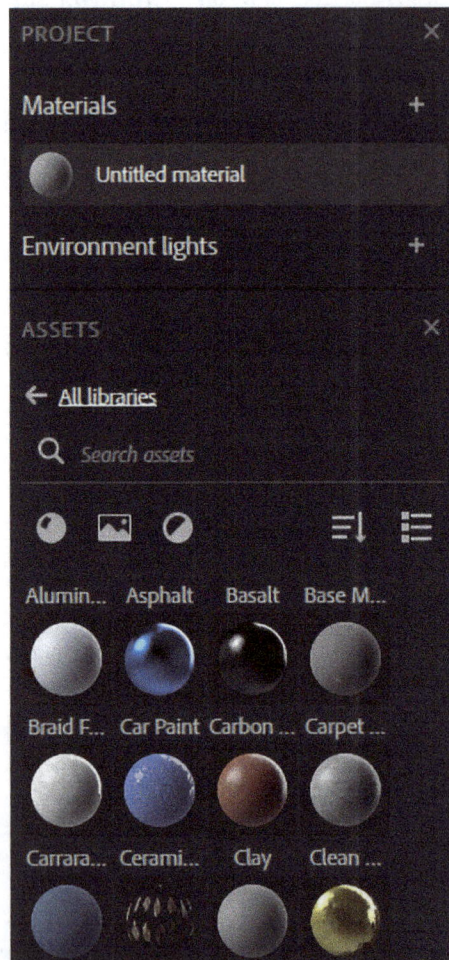

Figure 11.14 – The PROJECT and ASSETS panels

F – The PROJECT and ASSETS panels

In Sampler, the **PROJECT** panel functions like a library that can hold a variety of objects. The assets that make up your current project are shown in the **PROJECT** panel.

Let us open a new Sampler file to see how the **PROJECT** panel works and follow the following steps:

1. Drag an asset into the **Layers** panel to utilize it in other assets, or click one of your assets to load it in the viewport.

2. To add new material to your project, click the plus sign (+) next to **Materials**. To add a new environment light to your project, use the + sign next to **Environment lights**.

3. To rename, remove, or duplicate an asset, use the right-click menu. Some asset metadata is shown in the right-click menu, as you can see in *Figure 11.15*.

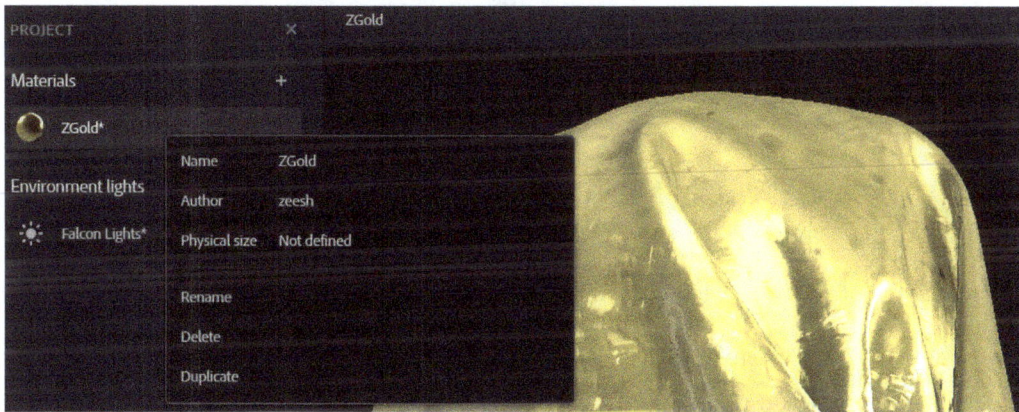

Figure 11.15 – Creating new materials and environment lights in the PROJECT panel

You can utilize the assets in the **ASSETS** tab to develop your masterpieces. You can get started with the resources included in Sampler. Some controls that facilitate organizing and finding assets are located at the top of the **ASSETS** panel:

*   : The search box can be used to locate assets fast

*   : Sort assets based on their kind

*   Organizes assets into distinct kinds and allows users to switch between list and icon views

Adding assets to the ASSETS panel

To create or add your own assets, let us go through the following steps:

1. Click the + in the lower-right corner of the **ASSETS** panel to add your own assets.

2. Select your asset by browsing to its location, and then click **Open**.

3. By dragging them over the trash can, you can delete assets from the **ASSETS** panel. This just works with your own assets, not with starter assets.

Figure 11.16 – Adding new assets

Working with the ASSETS panel

Working with the **ASSETS** panel is a no-brainer; however, we will go through the following steps:

1. Create a new sample file and load the mesh of your choice in the **3D** view; I will load **Preview Sphere**.

2. Now, drag the **Materials** asset of your choice over the mesh.

3. Note that the **Materials** parameters are now appearing in the **PROPERTIES** panel; there are sometimes presets available, as in the case of the **Asphalt** material.

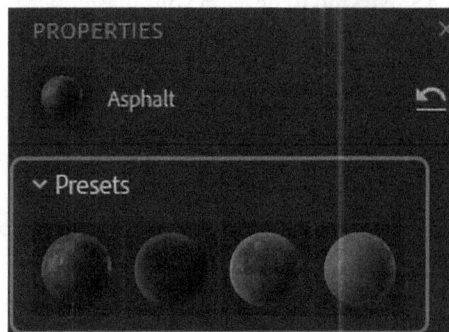

Figure 11.17 – Material presets

4. You can also change various parameters; each material holds different parameters, which are pretty self-explanatory.

I hope you have understood the **PROJECT** and **ASSETS** panels in this section. In the next section, we will move on to viewports.

Adobe Substance 3D Sampler viewports

There are two viewports inside Sampler, the **3D** view, and the **2D** view. In this section, we will dissect these two viewports

The name of your project and choices to customize the viewport's appearance are found at the top of the viewport, as shown in *Figure 11.18*.

Figure 11.18 – The view appearance toolbar

Utilize these choices to do the following:

- Change an asset's width and height in pixels
- Activate or deactivate the **3D** view
- Activate or deactivate the **2D** view
- Change the viewport between the default and full screen

G – the 3D view

Your current asset is shown in the **3D** view. To navigate your mesh inside the **3D** view, you can use the following hotkeys:

For Windows:	For Mac:
Alt + left mouse button = rotate	Option + left mouse button = rotate
Alt + right mouse button = zoom mouse scroll wheel = zoom	Option + right mouse button = zoom
	mouse scroll wheel = zoom
Alt + middle mouse button = pan	Option + middle mouse button = pan
The preceding hotkeys can work without Alt in new versions.	The preceding hotkeys can work without Option in new versions.

To set up your environment and rotate **Environment Lights**, you can use the following hotkeys:

For Windows:	For Mac:
Shift + right mouse button	Shift + Command + right mouse button

The cost of processing your current asset is shown by a counter at the bottom of the viewport.

The interface of the **3D** view bottom bar is shown in *Figure 11.19*.

Figure 11.19 – The 3D view bottom bar

More controls are located at the bottom of the viewport and allow you to do the following:

1. Change the field of view on the camera.
2. Change the materials' quality and displacement amplitude in the viewport.
3. Change how the materials are tiled throughout the viewport model.
4. Turn shadows on and off.

> **Note**
>
> To improve viewport visuals, enable shadows. To increase Samplers' performance, keep the shadows off.

The options in the **VIEWER SETTINGS** panel allow you to alter your viewport.

H – the 2D view

The **2D** view can carry a lot of valuable information and controls for various filters, however, only the **3D** view is accessible by default. Therefore, you must use the **2D view** button in the viewport's top left to launch the 2D view, as shown in *Figure 11.18*.

You can also change the source of the channels by using the menu in the top-right corner of the **2D** view, as shown in *Figure 11.20*.

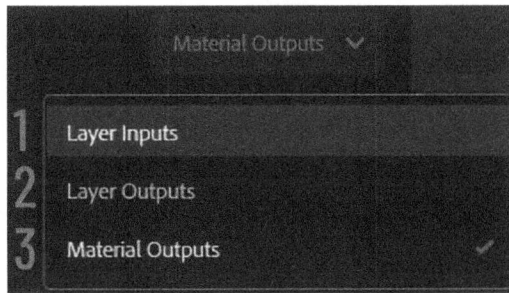

Figure 11.20 – The source of the channels

Let us see how channels are displayed:

1. The channels being input to the chosen layer are displayed in the layer inputs.

2. The output channels from the chosen layer are displayed in the layer outputs.

3. Layer selection has no impact on **Material Outputs**, which displays the channels being output by the top layer.

At the bottom of the **2D** view, you can see a channel bar, as shown in *Figure 11.21*, and with this, you can do the following:

- View the resolution of the channel you have chosen

- See a channel in 2D by choosing it

- View the bit depth and color channels for the chosen material channel

Figure 11.21 – The 2D view of the bottom channel bar

As we are done with Sampler's viewports, it's time to move on to the right side panels in the next section.

Right side panels and the toolbar

There are a couple more panels and one toolbar on the right side of Sampler's user interface. In this section, we will go through them.

I – the right sidebar

The sidebar contains frequently used panel shortcuts. When closed, the following panels fit within the right sidebar:

Figure 11.22 – The right sidebar

1. The **Layers** panel
2. The **Properties** panel
3. The **Physical Size** panel
4. The **Metadata** panel
5. The **Share** panel

By default, the **Exposing parameters**, **Physical size**, **Metadata**, and **Share** options are shown in the right sidebar, and if you close panels like **LAYERS** and **PROPERTIES**, and so on their icons will appear in the right sidebar so that you can reopen them.

J – the LAYERS panel

The layer stack and shortcuts for managing your layers are kept in the **LAYERS** panel. The assortment of components, filters, and other resources that make up your item is known as the **layer stack**. The layer stack functions the same way as Substance 3D Painter and Photoshop, going from the bottom layer first to the top layer last.

The layers above and below each other can therefore have an impact. Select a layer from the **LAYERS** panel to view its properties in the **PROPERTIES** panel, which works closely with the **LAYERS** panel.

There are three primary sections to create and edit layers in the **LAYERS** panel:

*   There are buttons in the tools area that you can use to do the following:

    *   Add a layer

    *   Add a base material

- Import a custom filter

- Remove a layer

- You can modify how a layer blends with layers underneath it using the **Blend mode** picker. Filters don't employ blend modes; the **Blend mode** option is only accessible when a material layer is chosen.

- All of the layers that make up your asset are included in the layer stack.

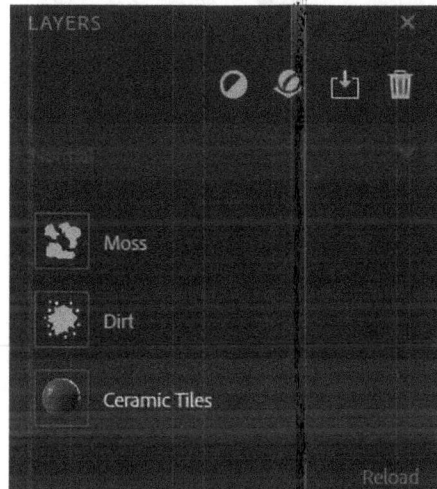

Figure 11.23 – The LAYERS panel

K – the PROPERTIES panel

You can view the settings and characteristics of the layers you choose in the **LAYERS** panel in the **PROPERTIES** panel. Playing around with settings and observing their effects on your asset is the best way to learn what they do.

Depending on what you've chosen in the **LAYERS** panel, different parameters will show up in the **PROPERTIES** panel. One layer may occasionally have several sets of editable attributes. For instance, a **Material** layer that isn't at the bottom of the stack can include blend attributes. When a **Material** layer has both material properties and blend properties, there will be two icons on that layer, since each icon in the layer stack represents a separate set of attributes and parameters.

You can see the interface of the **PROPERTIES** panel interface in *Figure 11.24*.

Figure 11.24 – The PROPERTIES panel

Hopefully, you now understand how the right side panel and toolbar work. In the next section, we will do a practical exercise to create a dirty old pavement.

Case study – creating a dirty old pavement

By now, you might have become familiar with toolbars and panels inside Sampler. It is now time to work on a practical exercise to better understand it. We will first create a dirty old pavement using existing assets and imported assets, so without further ado, let us get started:

1. Go to **VIEWER SETTINGS**, choose **Plane** under **Mesh**, and **Mondarrain 3** under **Environment**.

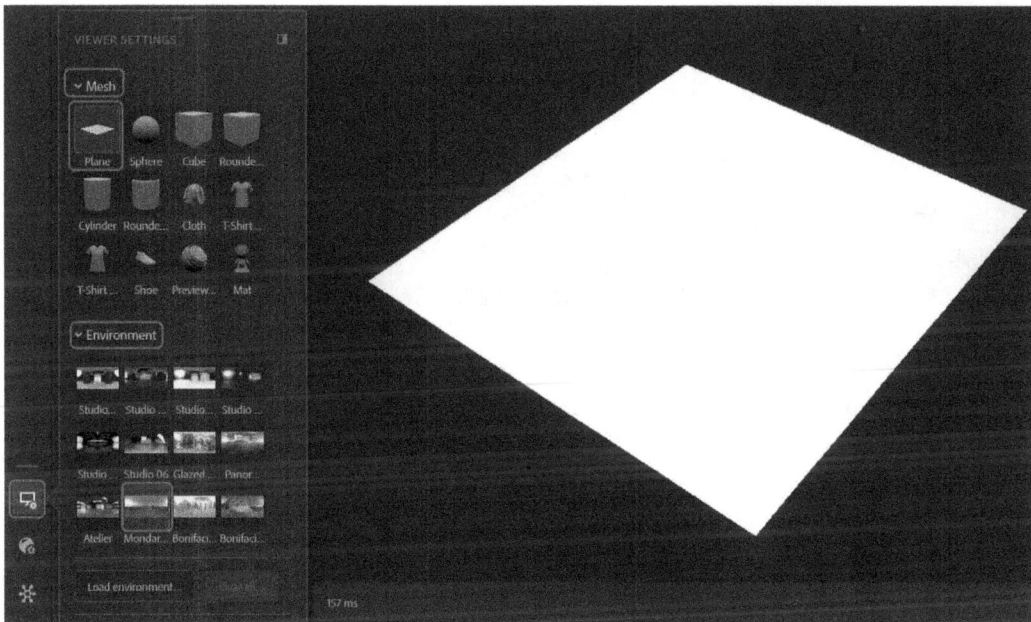

Figure 11.25 – Choosing Mesh and Environment

2. Make sure both **3D** and **2D** views are active.

3. Now, we need a base material on the plane; therefore, choose **Asphalt** and drag and drop it on the **Plane** mesh in the **3D** view.

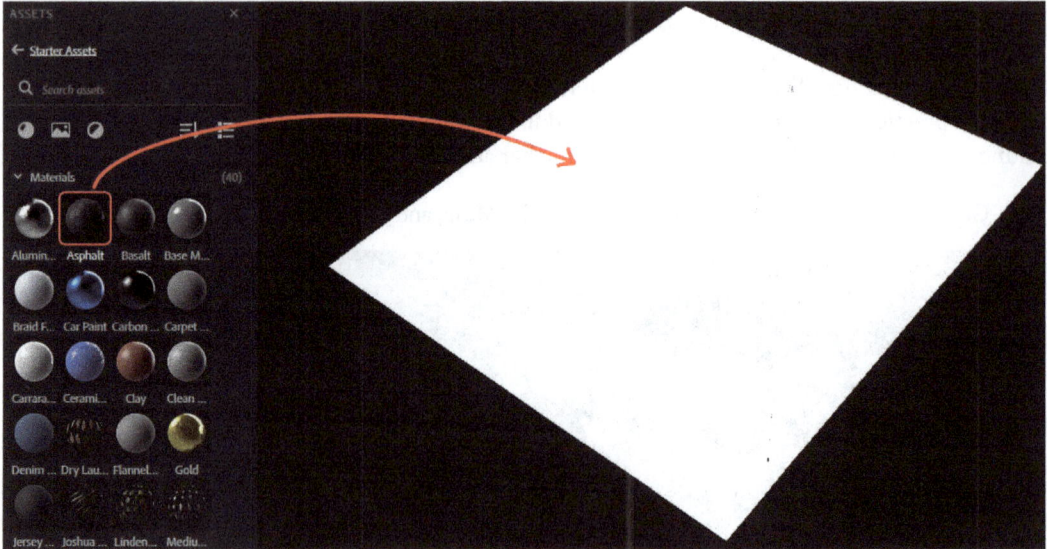

Figure 11.26 – Applying the Asphalt material to the mesh

4. Select **Asphalt** from the **LAYERS** panel, and then change its color to **Maroon** (#440D0D) and **Dirt Amount** to 0.2.

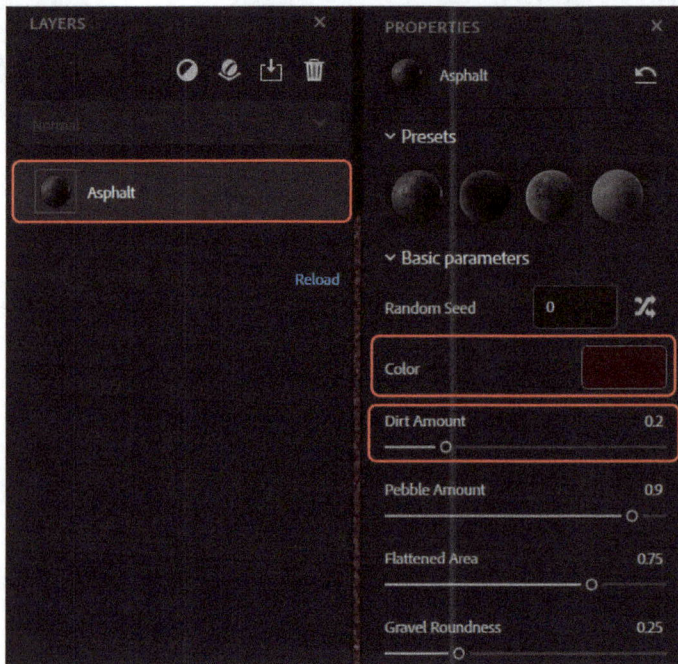

Figure 11.27 – Changing the color of Asphalt and Dirt Amount

5. Now, we need to apply a **Pavement** filter to the **Asphalt** layer. Filters are effects that can manipulate your layers. You can either drag the **Pavement** filter from the **ASSETS** panel or the **LAYERS** panel by clicking **Add a layer** and choosing the **Pavement** filter from the list.

Figure 11.28 – Applying the Pavement filter

6. Now, as we have the **Pavement** filter, it's time to apply the following changes to the parameter:

    - **Basic parameters | Brick Spacing**: 0.15

    - **Random Elevation Intensity**: 0.3

    - **Pattern | Pattern Type: Basket Wave**

    - **Basket Additional Number X**: 0

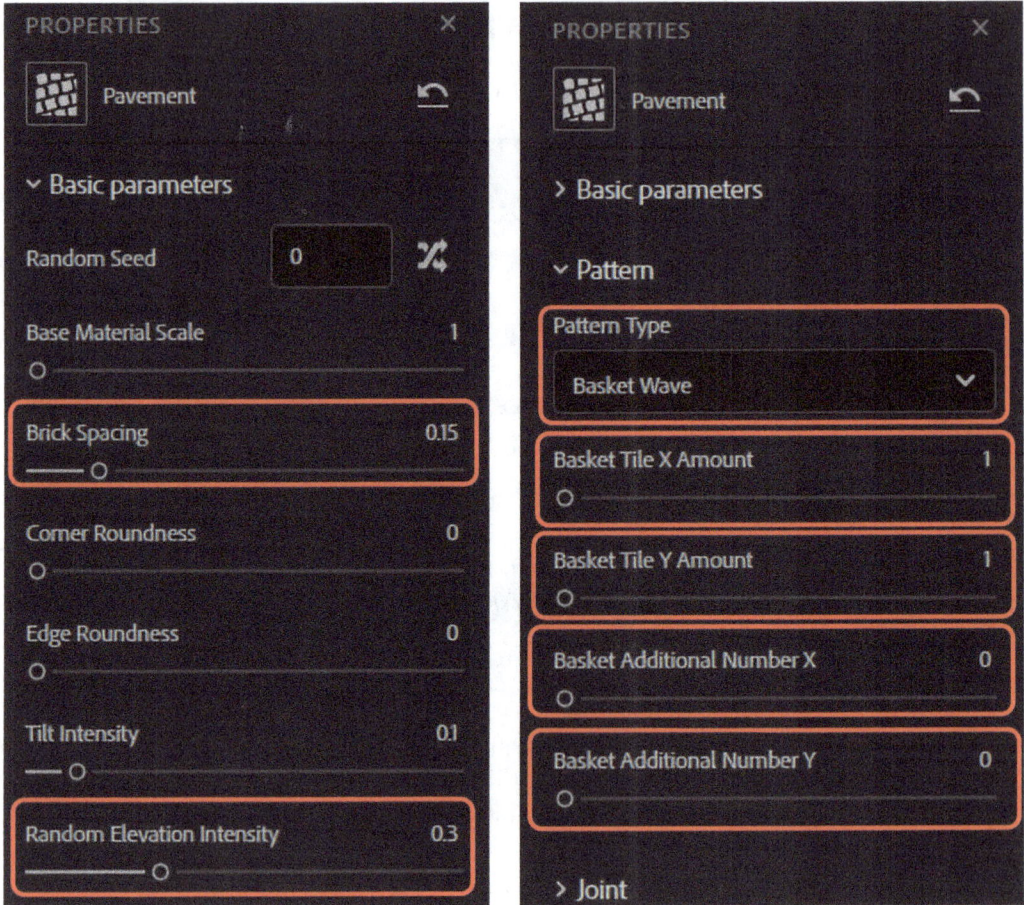

Figure 11.29 – Changing Pavement Parameters

7. You can also increase **Height Scale** to 0.1 and **Displacement quality** to 1 from the shortcut at the bottom of the **3D** view.

Figure 11.30 – Adjusting Height Scale and Displacement quality

8. Now, let us make it look dirty, so add the **Mud** material from the **LAYERS** panel, or drag and drop the **Mud** material from the **ASSETS** panel on the **Plane** mesh.

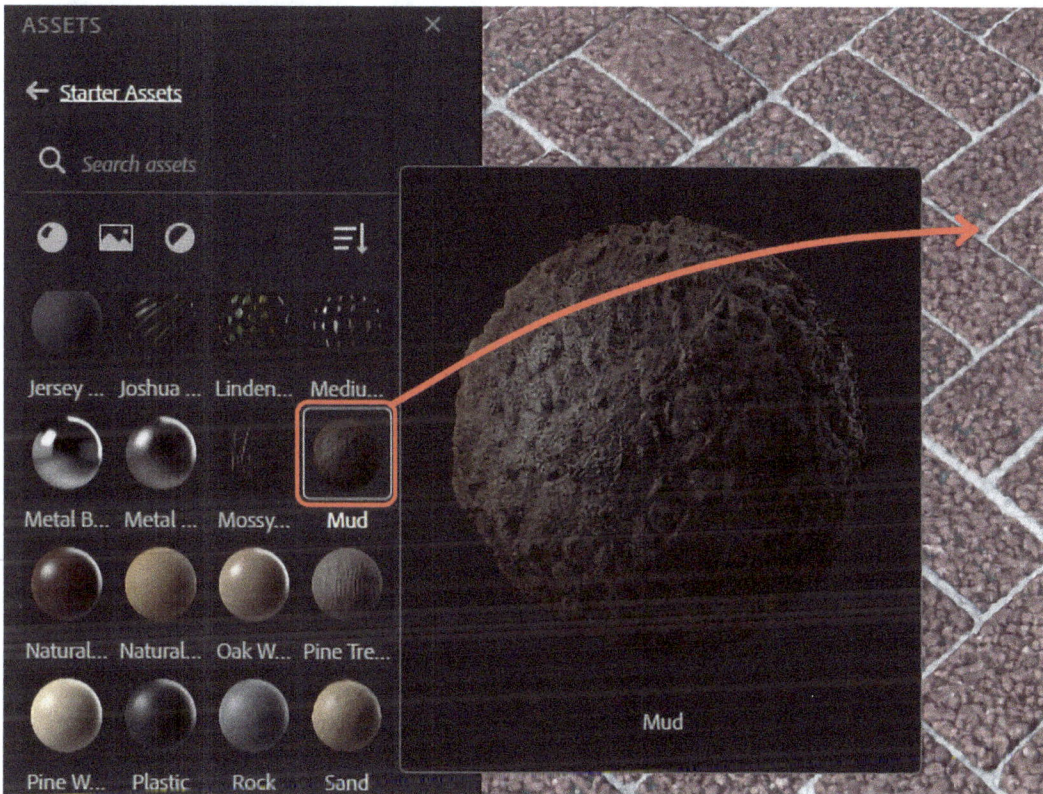

Figure 11.31 – Applying the Mud material

9. Note that the **Mud** material in the **LAYERS** panel is divided into two parts, one being **Mud** and the other being its **Height Blend** parameter. First, select **Mud** and change **Color** to `darker brown` and **Dirt Roughness** to `0.31`.

10. Then, select **Height Blend** and change **Offset** to 0.48 and **Contrast** to 0.83.

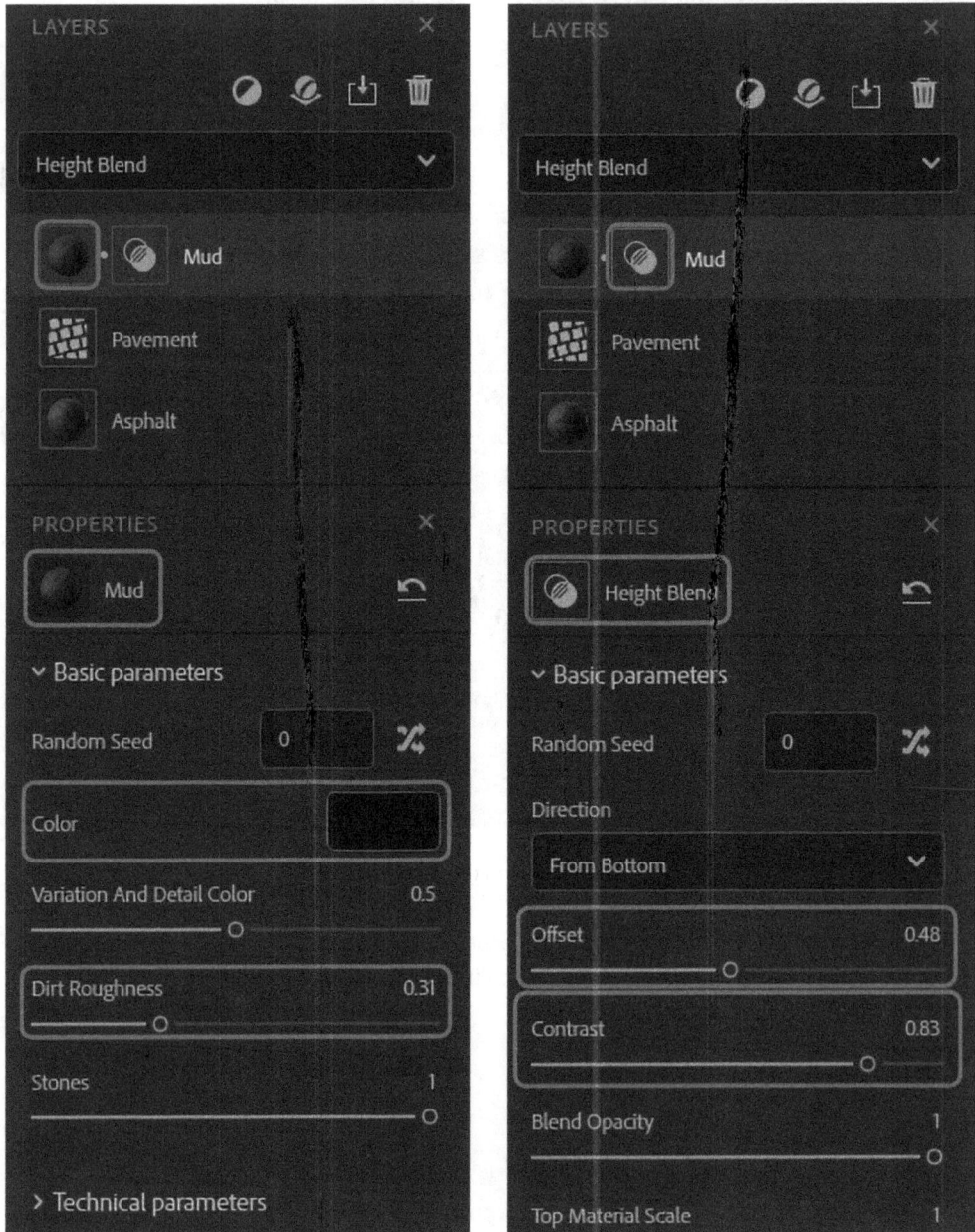

Figure 11.32 – Changing Mud and the Mud Height Blend parameter

11. Now, let us add some dried leaves to the pavement; therefore, add **Atlas Scatter** on the top of the layers.

Figure 11.33 – Adding Atlas Scatter

12. Now, drag **Dry Laurel Leaves** in **INPUT 1** of **Atlas Scatter**, as shown in *Figure 11.34*.

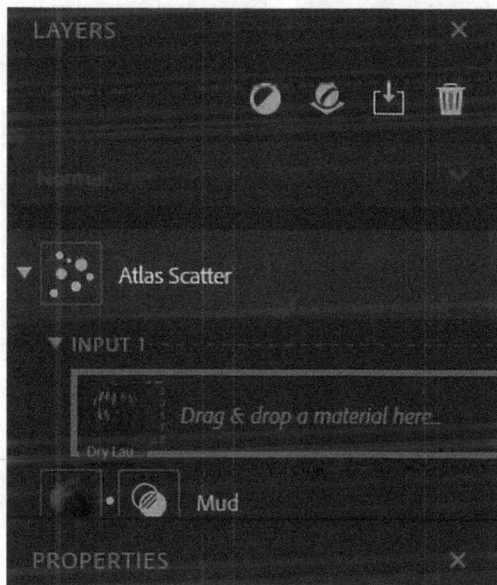

Figure 11.34 – Dragging Dry Laurel Leaves to INPUT 1 of Atlas Scatter

13. If you want, you can change the **Dry Laurel Leaves** settings; however, for this project, we will not change them. We will only change the **Atlas Scatter** parameter, so just click on **Atlas Scatter** and apply the following changes to the **Atlas Scatter** parameters:

  - **Basic parameters** | **X Amount**: 10 and **Y Amount**: 10

  - **Size** | **Scale Random**: 0.5

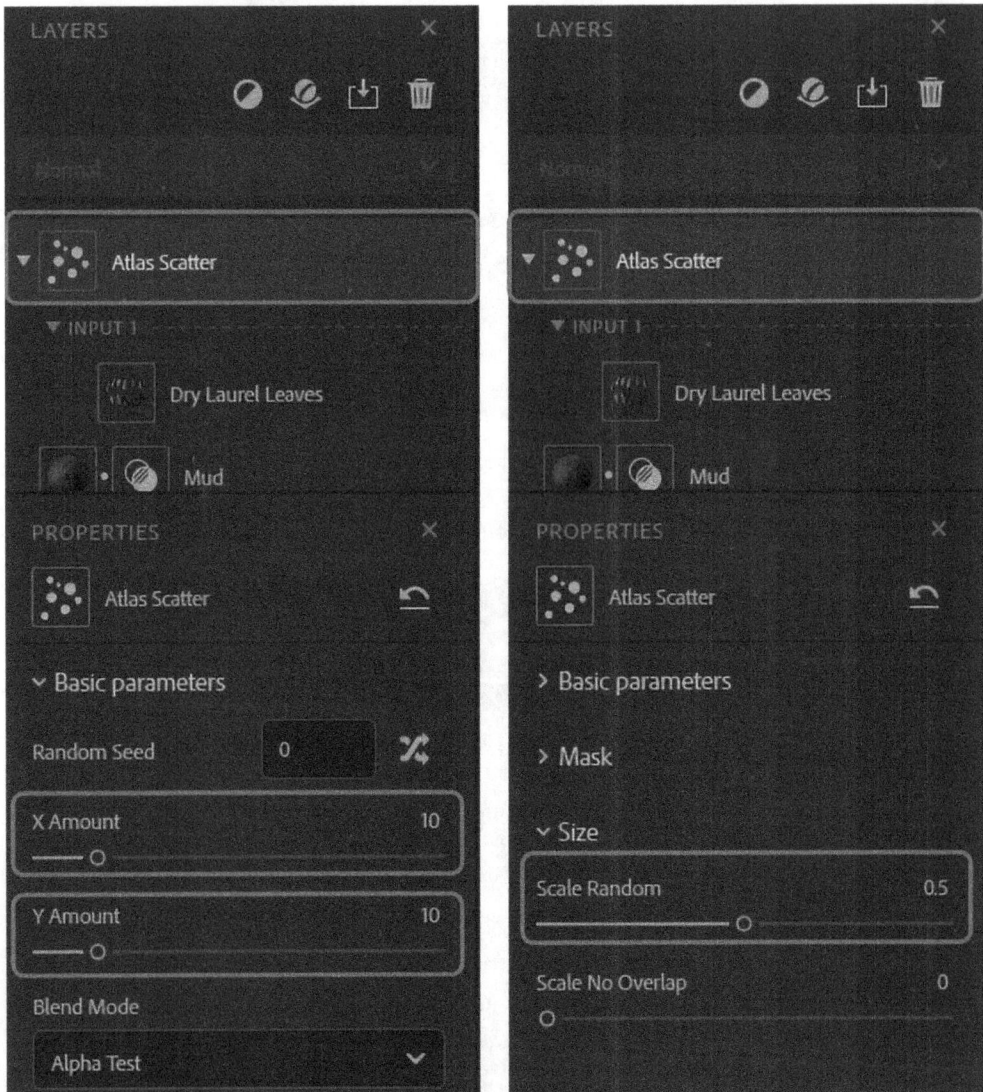

Figure 11.35 – Applying changes to the Atlas Scatter parameters

14. Now, save your file as `Dirty_Old_Pavement.ssa`. If you want, you can apply more materials and filters to your project; there are a total of 40 **Material** options and 80 **Filter** options.

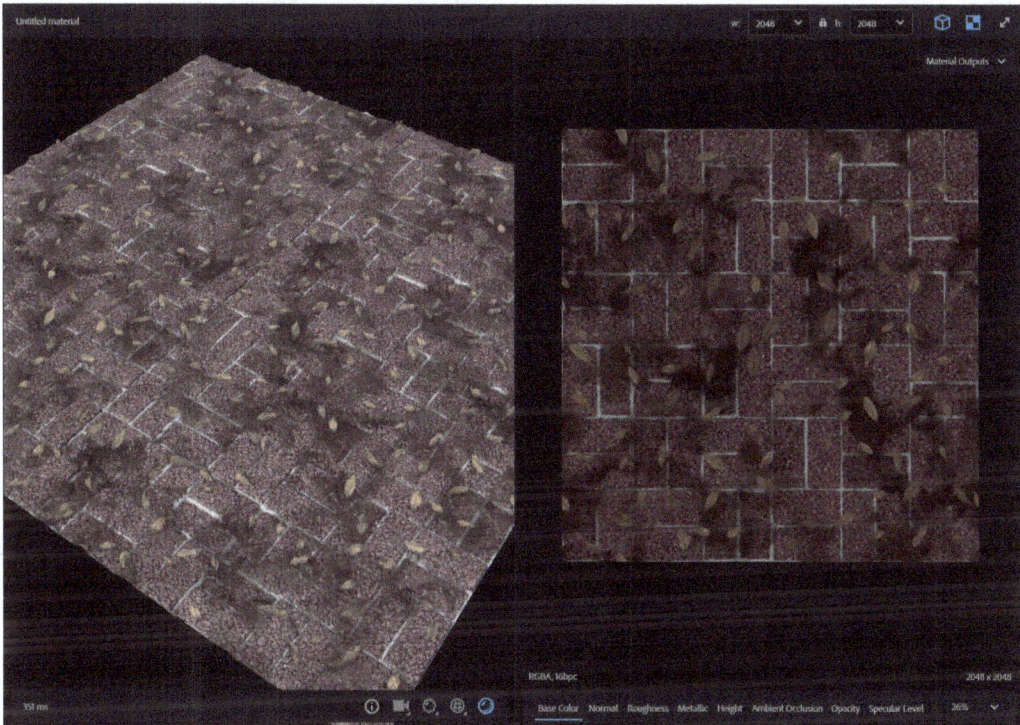

Figure 11.36 – The final output of Dirty Old Pavement

I hope it is now clear how to create materials from existing Sampler materials. Now, let us see how we can create a material from our own images:

1. Open a new Sampler project and drag the given `Rocks.png` file from `Substance_Sampler_Exercise_Files` to Sampler's **LAYERS** panel.

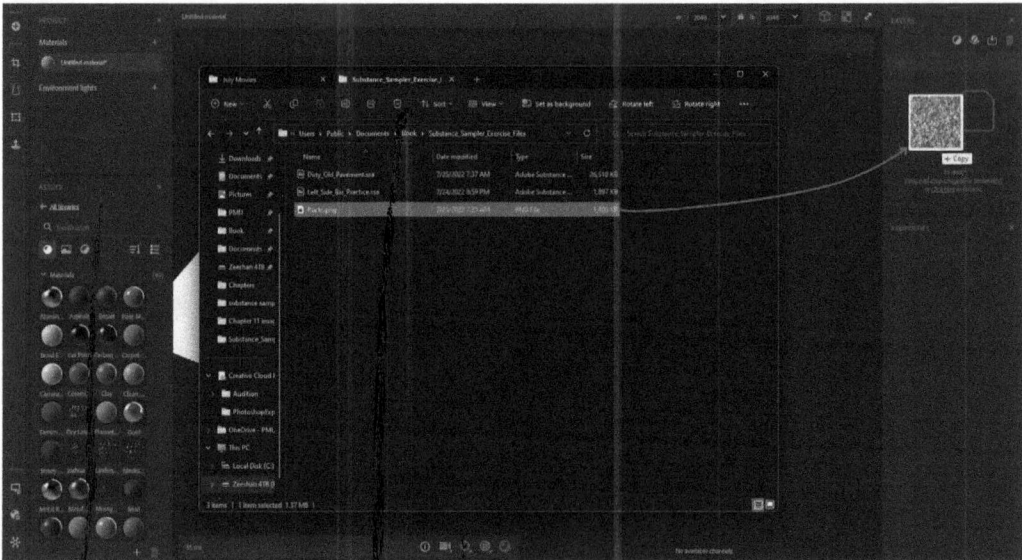

Figure 11.37 – Dragging Rocks.png to the LAYERS panel

2. Now, you will see the **Image to material (AI Powered)** window. Choose **Image to material** and click **Import**.

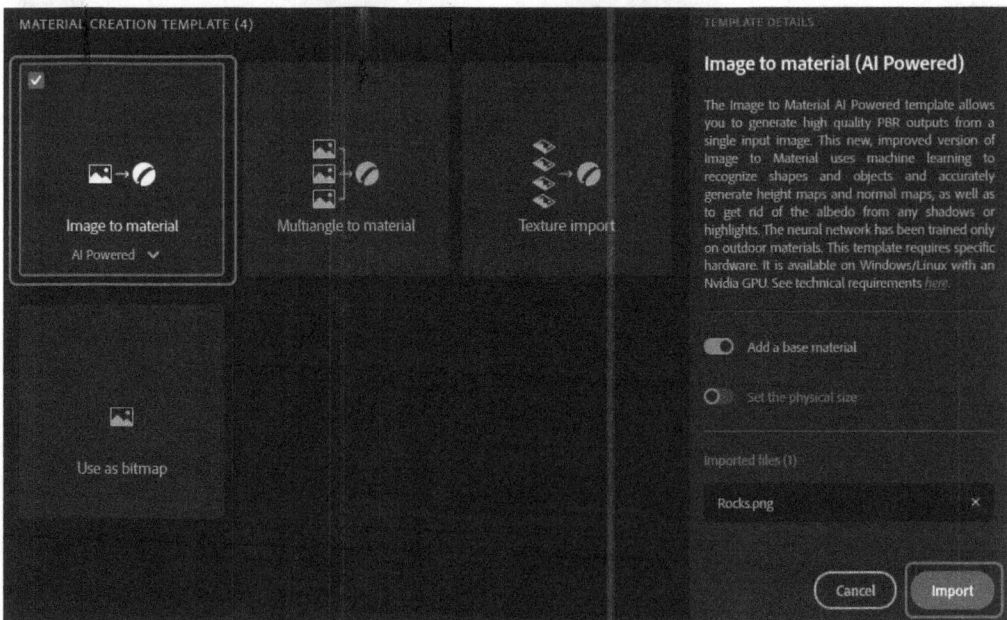

Figure 11.38 – Image to material (AI Powered)

3. You can explore all three layers in the **LAYERS** panel or change their parameters. All the parameters are self-explanatory.

4. Now, you can add the **Equalize** filter. This filter will remove all the irregularities in the colors.

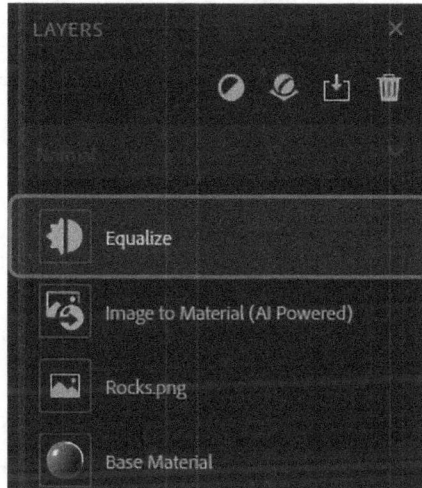

Figure 11.39 – The Equalize filter

5. To remove all the seams and make the material tileable, you can add the **Make It Tile** filter. This filter will make your material clean and tileable, and will remove all the seams.

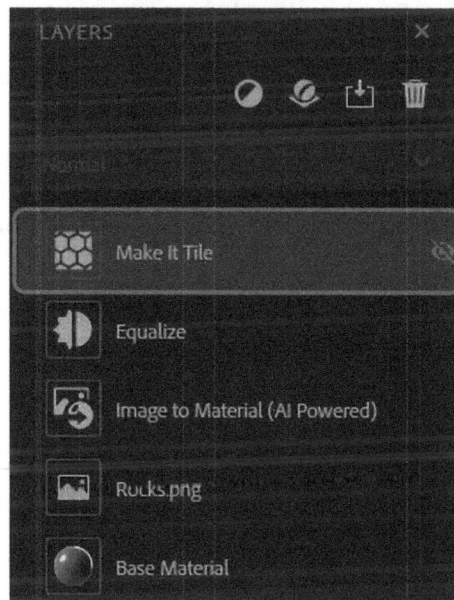

Figure 11.40 – The Make It Tile filter

6. You can add different filters; it is all about trial and error. Just explore what other filters do. For example, use **Water** to add a water effect to your material.

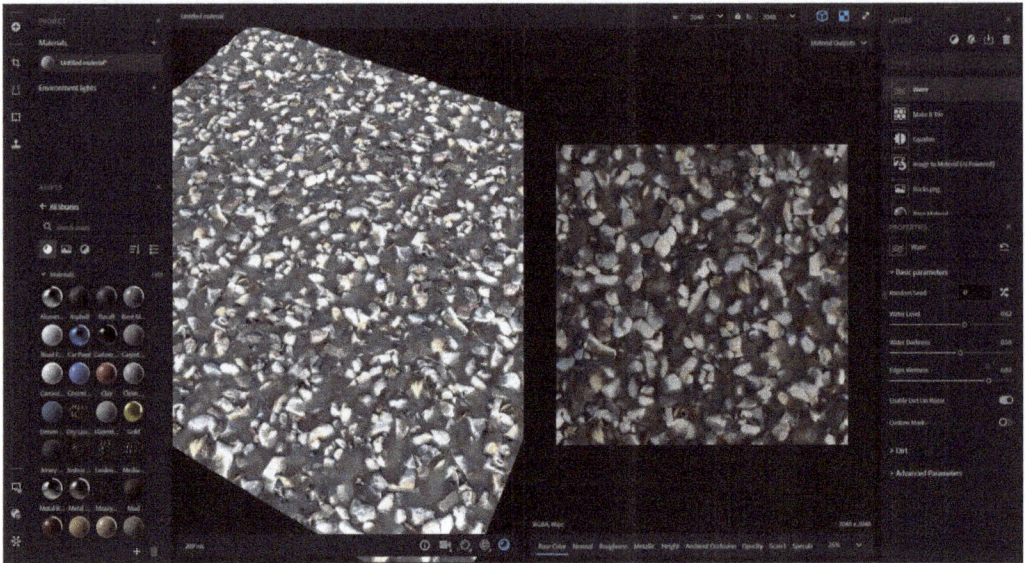

Figure 11.41 – The final output of Image to material (AI Powered)

I hope it is clear now how to create Sampler materials from your own images. Let us now learn how to export them in the next section.

Sharing and exporting materials

You can export Sampler materials using the **Share** option. Once you click the **Share** option, you will get the **Send to** option. You can directly send your material to **Stager**, **Painter**, or **Designer** with a single click.

Using the **Export as** option, you can export Sampler material to third-party 3D tools. You can also choose channels and set other parameters.

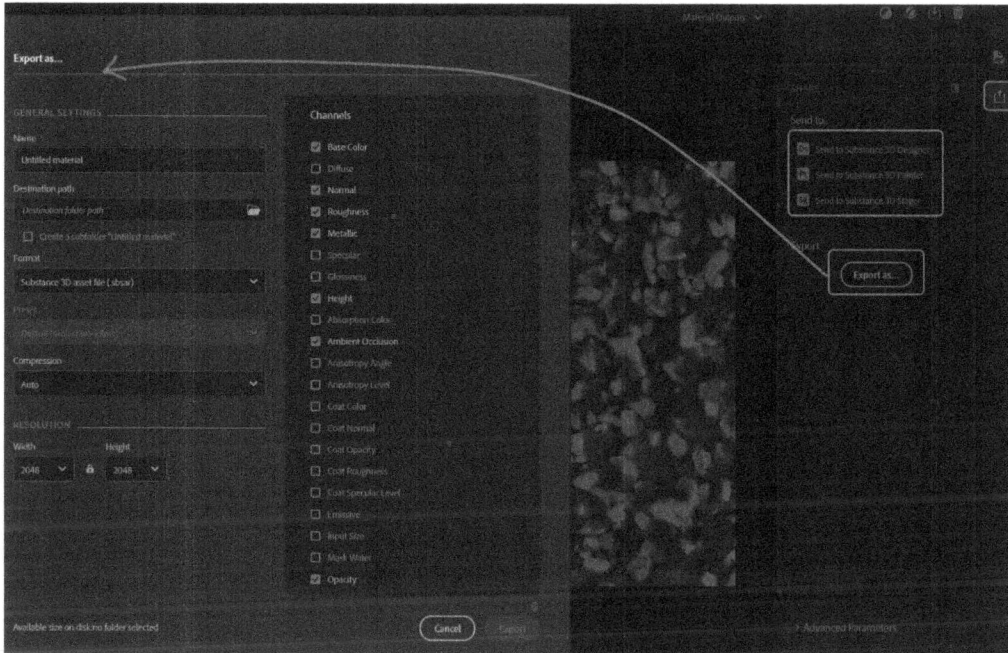

Figure 11.42 – Exporting Sampler materials

Summary

I hope you are now well aware of Adobe 3D Sampler and have comprehended how useful it is to create your own materials very quickly efficiently and quickly. With this chapter, the *Adobe Substance 3D Designer and Sampler* part of this book comes to an end.

In the next part, we will start learning about Adobe Substance 3D Stager, also known as **Stager**. Stager allows you to create scenes and render them using the efficient staging tool, upload materials and set up your scenario, modify the physical lighting, as well as create image-based lighting by adding textures and materials.

12
Getting Started with Adobe Substance 3D Stager

We have reached the last part of this book, which is about Adobe Substance 3D Stager, also known as **Stager**.

Stager allows you to create scenes and render them using the efficient staging tool. You can upload materials and set up your scenario and modify the physical lighting as well as image-based lighting by adding textures and materials. Stager allows you to save cameras with various resolutions and subsequently produce real-time photos.

You are free to make selections based on your imagination in this situation. With real-time alterations and compositional improvements, you can edit and visualize complex materials with intricate lighting and shadows. Using quick, effective, and complex tools, you can create your own realistic 3D scenes.

Stager is an easy-to-use and effective design tool for creating 3D scenes. Staging is a term used to describe the process of creating a 3D scene. Layout, texturing, lighting, and framing are the steps involved.

A wide range of projects can be produced using this staging method. You can organize virtual products and commodity pictures. You can produce many creative 3D designs, such as rooms or other areas, brand images, and drawings.

You can make artistic choices when using Substance 3D Stager. In real time, you can alter and fine-tune your compositions, editing and visualizing sophisticated lighting and shadows in advanced materials. With the aid of effective and speedy smart tools, you can create realistic 3D scenarios.

You can block out forms, snap objects together, turn on physics to prevent model collisions, and create basic lighting. Stager includes models, supplies, and lighting to get things going. Alternatively, browse and utilize the tens of thousands of premium assets that are a part of your Substance 3D Collection plan, which was created by 3D professionals. In this chapter, we will be going through User Interface and Navigation inside Stager so that we can prepare to work on any Stager project.

We will cover the following topics in this chapter:

- Understanding the UI

- Understanding the navigation controls

Understanding the UI

The interface of Stager is quite self-explanatory and a no-brainer. Like other Adobe Substance products, Stager is also simple to use.

When you first launch Stager, you are welcomed with a home screen. You can create, open, and view recent documents on the home screen. To get started producing new papers or discovering recent work, you can go to the home screen.

To practice on Stager's UI, click on **Open** on the home screen, as shown in *Figure 12.1*, and open the `Chapter_12_Practice.ssg` file given in the `Substance_Stager_Exercise_Files` folder.

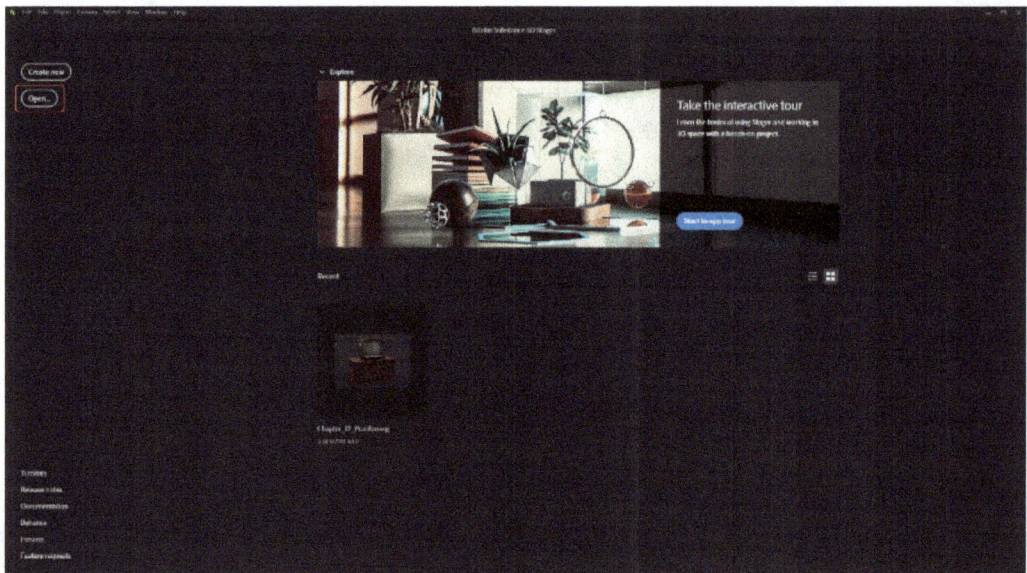

Figure 12.1 – Adobe Substance 3D Stager home screen

There are two different modes in Adobe Substance 3D Stager – Design mode and Render mode.

Design mode

This mode is used to stage 3D scenes and construct compositions. It is used for editing the environment where you add stuff and alter it to create a composition. Let us dissect Stager's **Design** mode interface, as labeled in *Figure 12.2*.

Figure 12.2 – Adobe Substance 3D Stager Design mode interface

A: The Design and Render tabs

You can switch between the **Design** mode and **Render** mode with the help of these tabs. They are located next to the home screen button, as shown in *Figure 12.3*.

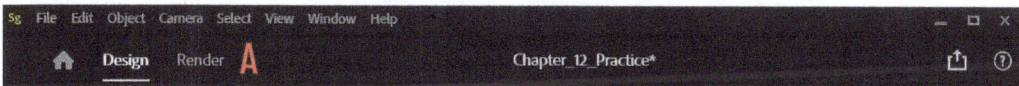

Figure 12.3 – Design and Render tabs

B: The ASSETS panel

Access to the starter assets and Creative Cloud Libraries is provided through this panel. You can access and arrange content for Substance 3D Stager in the **ASSETS** panel.

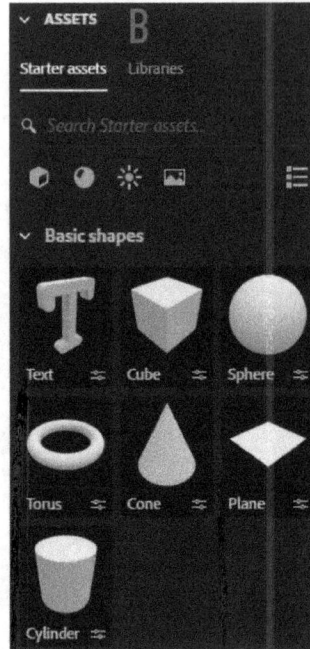

Figure 12.4 – ASSETS panel

The main purpose of using the **ASSETS** panel is to utilize materials from a library or the startup assets. It also shows different types of 3D models, and when you click on them, they are created in the center of the 3D viewport.

Drag the asset into the window to perform the desired action (for 3D models, it will load at the target position).

The **ASSETS** panel is divided into **Starter assets** and **Libraries,** so let us go through them in the next section.

Starter assets

Stager comes with a large selection of models, supplies, lighting, and pictures. These resources are a terrific way to rapidly start learning and working and can all be used royalty-free in any project.

The initial assets allow you to do the following:

- Locate an asset by name, using search
- Discover assets by kind
- Use filters – models, materials, lighting, and photos can all be filtered
- Include an asset in your scene – click or drag it into the viewport

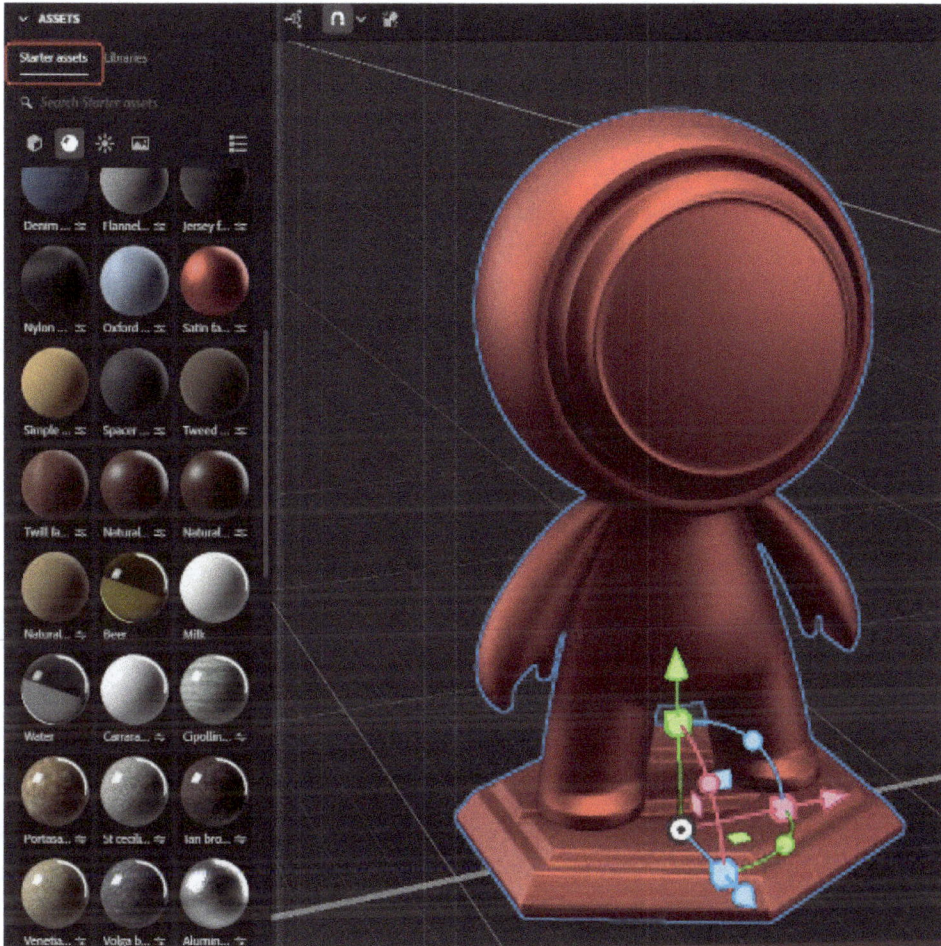

Figure 12.5 – Starter assets

Libraries

We can organize cloud-synchronized information for usage across applications, projects, devices, and teams, and access Creative Cloud Libraries via the **Libraries** panel.

Use Creative Cloud Libraries to do the following:

- Create, remove, and rename libraries
- Add, remove, rearrange, and manage content in a library
- Exchange a library
- Utilize Adobe Stock to find and purchase content

C: Toolbar

Tools for interacting with scene content are available on the toolbar. You can also access the **Render Settings** panel and the **ASSETS** panel from here.

Figure 12.6 – Toolbar

D: Viewport

The viewport in Adobe Substance 3D Stager serves as the main workspace. You can interact with the 3D scene by using the tools and actions available here. It is impossible to describe all of the characteristics of the viewport in one area since it is so essential for editing operations in Stager.

Through the rendering mechanism, the viewport renders the data from the 3D scene. Stager renders material by combining real-time and ray tracing methods. You can alter the look of the viewport (and the finished product) by adjusting the render parameters.

Although you can fine-tune content using the properties panels, engaging the viewport is often faster. A variety of tools and actions are used to interact.

Each tool has a distinct purpose. Objects can be moved and selected using the transformation tools. Cameras are controlled by the camera tools. Throughout the picture, the **Sampler** tool instantly samples and applies elements.

Figure 12.7 – Viewport

E: Viewport control bar

The render engine, current camera, and viewport parameters are controlled by this bar.

Figure 12.8 – Viewport control bar

F: Render settings panel

The settings and active render engine are controlled by this panel. You can manage the rendering quality of both the viewport and the finished product. Modes share the same render settings.

Figure 12.9 – Render settings panel

G: SCENE panel

All of the information in your current scene is summarized in this panel. Models, the setting, cameras, and lighting are all included in this.

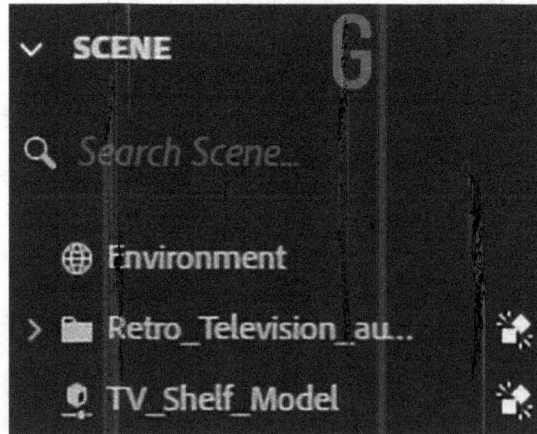

Figure 12.10 – Scene panel

H: ACTIONS panel

The contextual actions for the currently selected option are shown in this panel. Actions are often shorter commands that modify the scene in a specific way. The app has various locations where actions can be found.

The app menus, the viewport control bar, the **Scene** panel, the **Properties** panel, and occasionally the viewport itself all have the Actions panel. Although there are many different kinds of actions, creating objects or changing important characteristics are frequent actions.

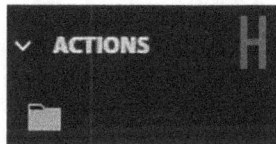

Figure 12.11 – ACTIONS panel

I: PROPERTIES panel

Properties for chosen items are shown in this contextual panel. The thorough control of scene material is managed in this panel. The specific characteristics of an object that determine its position, appearance, and behavior are called its properties.

Hopefully, you are now familiar with the **Design** mode. In the next section, we will study the **Render** mode.

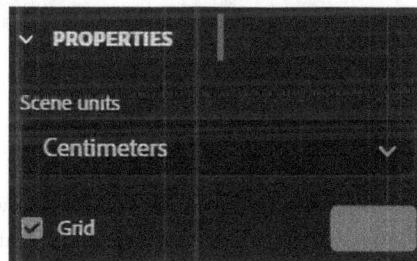

Figure 12.12 – PROPERTIES panel

J. Application menu bar

The application menu bar holds common menus such as the **File** menu, **Edit** menu, **Object** menu, **Camera** menu, **Select** menu, **View** menu, **Window** menu, and **Help** menu.

Figure 12.13 – Application menu bar

Render mode

The render mode is used to produce generated images of the highest quality. You can set up several renderings in the render mode as shown in *Figure 12.4*.

Figure 12.14 – Adobe Substance 3D Stager render mode interface

Each of these renderings is explained here:

1. **Viewport**: It occupies a sizable portion of the application window's center. The viewport in render mode displays the final produced pictures at a pixel-accurate size.

2. **RENDER SETTINGS panel**: The settings and active render engine are controlled by this panel. You can manage the rendering quality of both the viewport and the finished product. Modes share the same render settings.

3. **EXPORT SETTINGS panel**: The format, save location, and cameras to export are all controlled by this panel.

4. **RENDER STATUS panel**: As renders are finished, this panel displays their progress.

Hopefully, you are now familiar with Stager's interface, which makes it easier for you to work with this application, especially when we are moving to navigation, which is important for any 3D application to view your 3D mesh.

Understanding the navigation controls

Navigation controls in Stager are similar to other Adobe Substance 3D tools. The view that you see in the viewport is basically a camera view. Stager comes with a camera by default, so when you are navigating inside Stager, you are actually controlling the camera.

There are three tools that are available for the camera movement:

- The camera spins using the **Orbit** tool

- The camera may be panned up, down, left, and right **Pan**

- Forward and backward camera movement is accomplished with the **Dolly** tool

You may access each of the camera tools from the toolbar or the tool access shortcuts that follow. For a brief period, you may move from your primary tool to the camera controls by using fast access shortcuts. Your initial tool will once again be accessible when you release the shortcut.

Navigating with hotkeys

You can use the following hotkeys to navigate the Adobe Substance 3D Stager camera:

For Windows:

- **Orbit tool:** *Alt* + left click

- **Pan tool:** *Alt* + middle click

- **Dolly tool:** *Alt* + right click

For Mac:

- **Orbit tool:** *Option* + left click

- **Pan tool:** *Option* + middle click

- **Dolly tool:** *Option* + right click

To move the camera to focus on your choice, press *F*. This will focus on every item in your scene if nothing is selected.

Navigating with the toolbar

You can navigate in Stager using hotkeys or buttons in the toolbar as shown in *Figure 12.15*.

Figure 12.15 – Camera navigation tools

A: Orbit tool

View the area with the camera rotating. You can access this tool with hotkey *1*. If you keep the **Orbit** tool button pressed, you will get some extra orbiting tools, as shown in *Figure 12.16*.

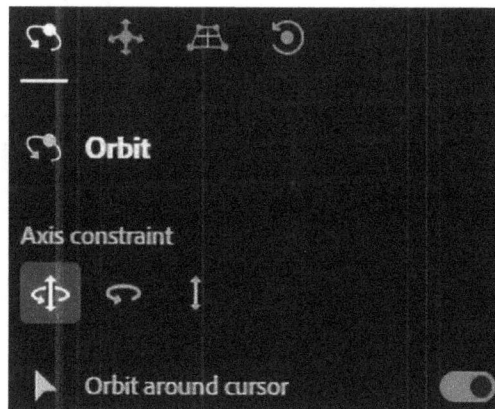

Figure 12.16 – Extra Orbit tools

The extra tools contain the **Tripod** orbit mode, **Horizon** orbit mode, and **Roll** orbit mode.

- **Tripod**: Turn the camera to scan the area.

- **Horizon**: Aim the camera at the horizon of the ground.

- **Roll**: Position the camera at an angle.

Moreover, if you want your camera to be constrained on an axis, you can use **Axis constraint**, and if you want the camera to orbit around where your cursor is, then you can toggle on the **Orbit around cursor** button.

B: Pan tool

This tool allows the camera to move left, right, up, and down. You can access this tool with hotkey *2*. If you keep the **Pan** tool button pressed, you will get some extra **Pan** tools, as shown in *Figure 12.17*. If you want the camera to pan around where your cursor is, then you can toggle on the **Pan under cursor** button.

Figure 12.17 – Extra Pan tools

C: Dolly tool

This tool allows you to move the camera forward and backward. You can access this tool with hotkey 3. If you keep the **Dolly** tool button pressed, you will get some extra **Dolly** tools as shown in *Figure 12.18*.

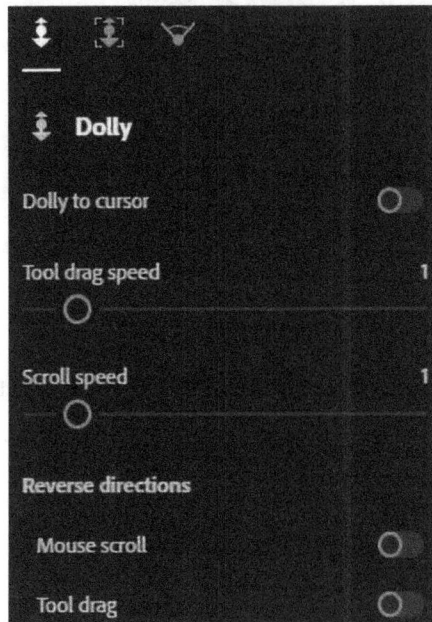

Figure 12.18 – Extra Dolly tools

These tools are explained as follows:

- **Field of View**: Angle the camera's lens differently
- **Dolly Zoom**: To keep the framing, adjust the field of view and move the camera

Summary

Hopefully, you are now familiar with the Adobe Substance 3D Stager's user interface and navigation system. You should know by now that Stager's interface is quite simple but powerful. We have also gone through all its panels and learned how to use them. Moreover, we have learned that the navigation system of Stager is a no-brainer, and studied how to navigate inside Stager in this chapter.

In the next chapter, we will learn how, in real time, a composition is refined and adjusted. The next chapter covers comprehensive knowledge of basic 3D models and materials and how to use them in real-time projects. We will explore **ASSETS Starter** assets and the **Library**, **Geometries**, **Materials**, and **Lights**. We will also learn how to sample materials and work with collision-based transformation.

13

Models, Materials, and Lights in Adobe Substance 3D Stager

You may deal with several types of objects, such as models, lights, and cameras, in Adobe Substance 3D Stager. Each type of object has its own actions and characteristics. In this chapter, we will discuss standard models, which are 3D forms made of points and edges that are imported into Stager from other applications.

3D text is a type of parametric model made with Stager's text and extrusion engines, and parametric models are 3D shapes that are generated and may be changed in Stager with parameters. Typically, these models are created in Adobe Substance 3D Designer.

We can group an organizing aid object without a visual counterpart while using Stager's form tools with the help of shapes, which are a type of parametric model. Moreover, in Stager, materials are used to adjust the appearance of surface-mounted objects such as models and text. Also, cameras are used in Stager to capture images of a scene. This is done by simulating a physical camera lens and its settings.

You can make artistic decisions when using Substance 3D Stager and then refine and alter your composition as you go. You can also visualize and edit sophisticated materials with intricate lighting and shadows. We will cover all these properties in this chapter and work on them in practice.

We will cover the following topics in this chapter:

- Exploring Adobe Substance 3D Stager models
- The Adobe Substance ACTIONS panel
- Exploring Adobe Substance 3D Stager materials and lights
- Collision-based transformation

Exploring Adobe Substance 3D Stager models

There are different types of models inside Stager. Let us explore them all and see how they differ from each other and how to create them:

1. Launch Adobe Substance 3D Stager and create a new empty file.

2. Go to the **ASSETS** panel.

3. Make sure you are in the **Starter assets** tab of the **ASSETS** panel.

4. Select the **Models** asset type, as shown in *Figure 13.1*:

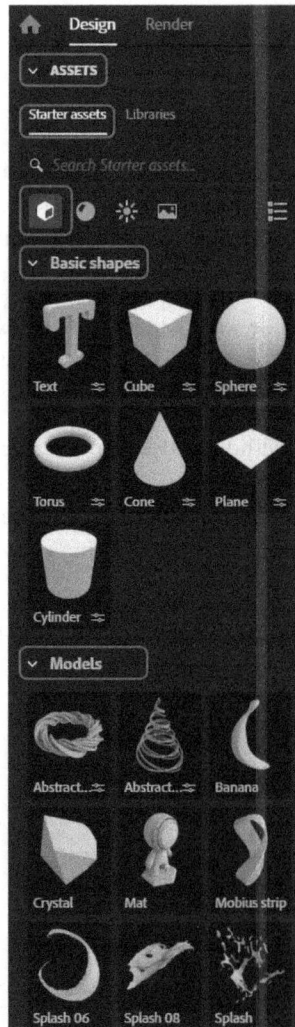

Figure 13.1 – The Models asset type

You will notice that there are two types of **Assets** in Stager – **Shapes** and **Models**.

5. To create a model, you must drag it into the viewport. However, it will not be in the center. Therefore, you have to move it to the center using your mouse with the help of the theGizmo or adjust the necessary parameters in the **PROPERTIES** panel, as shown in *Figure 13.2*:

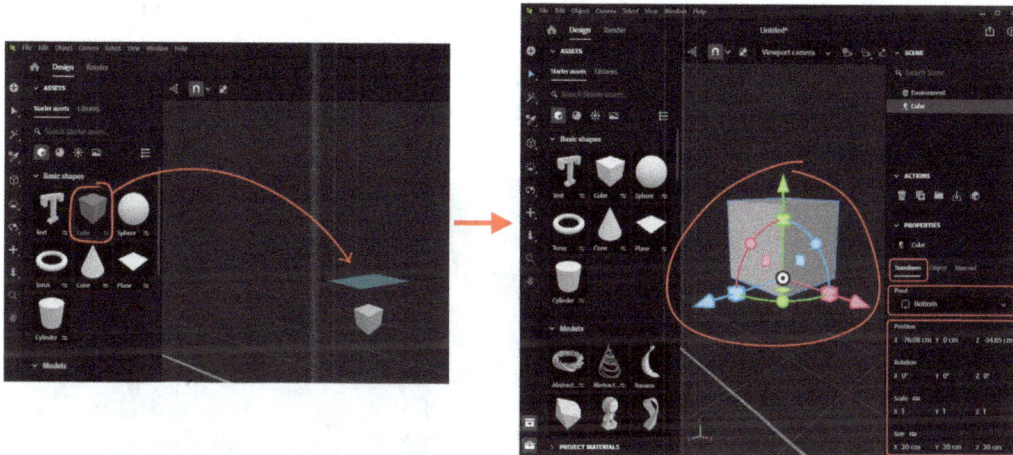

Figure 13.2 – Creating shapes by dragging them

6. If you want the shape or model to be created in the center, then you just need to click on the shape or model in the **ASSETS** panel with the left mouse button. This will put the shape or model in the center of the viewport:

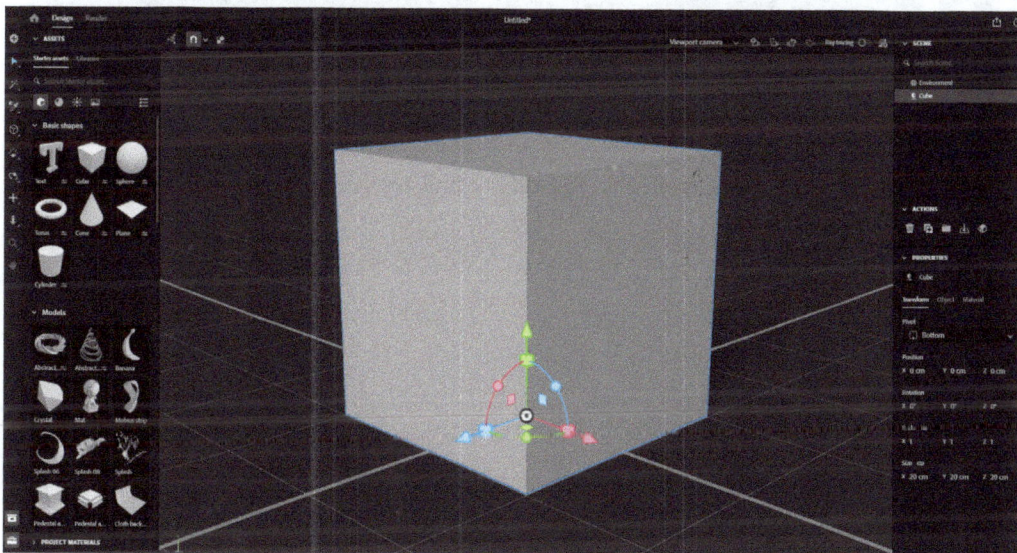

Figure 13.3 – Creating shapes by single-clicking the left mouse button

7. You will notice that some shapes and models are marked with **slider icons**, as shown in *Figure 13.4*. These slider icons indicate that the shapes or models can be modified using the parameters in the **PROPERTIES** panel. If there is no slider icon, this means there are no parameters available to modify in the **PROPERTIES** panel.

8. Create a cube shape and a banana model side by side in the viewport by clicking them in the **Basic shapes** and **Models** lists and moving them using the Gizmo, as shown in *Figure 13.4*. Notice that the cube shape is marked with the slider icon, which means you can adjust the cube through the **Object** tab of the **PROPERTIES** panel. This includes turning on the **Bevel** option for our **Cube**:

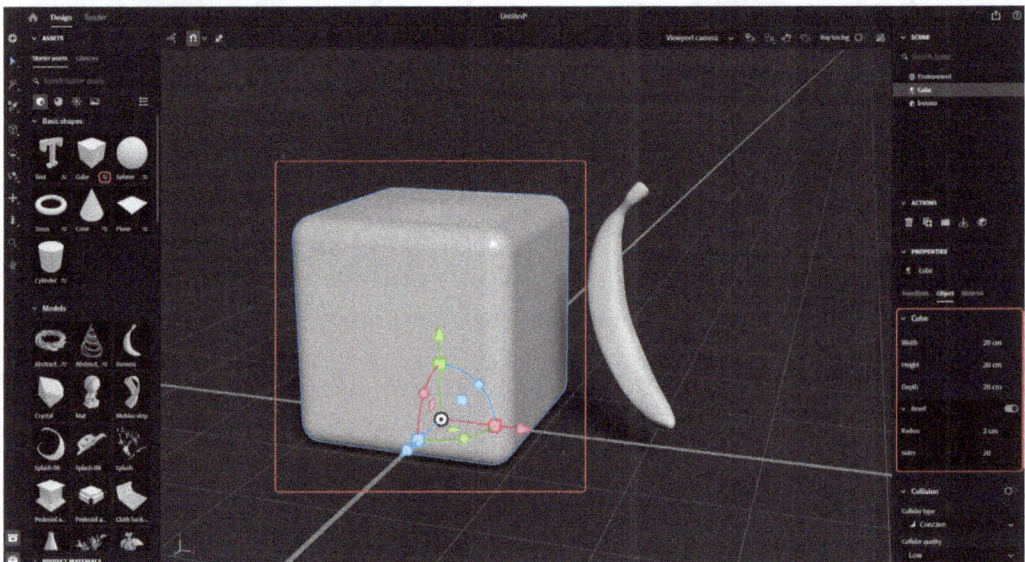

Figure 13.4 – Modifying shapes and models with the slider icon

9. When you select the **Banana** model, you will not find any parameters under the **Object** tab in the **PROPERTIES** panel:

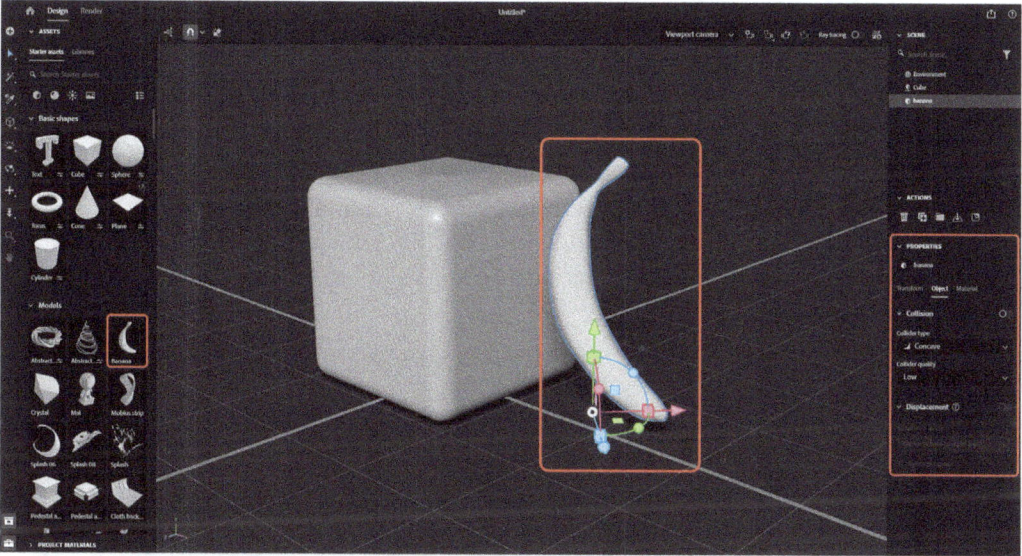

Figure 13.5 – Shapes and models without the slider icon

10. Now, delete the cube and banana so that we can create a **Parametric** model.

The parametric models are located in the **Models** area. Here, select the **Foliage a** parametric model from the **Models** area, as shown in *Figure 13.6*:

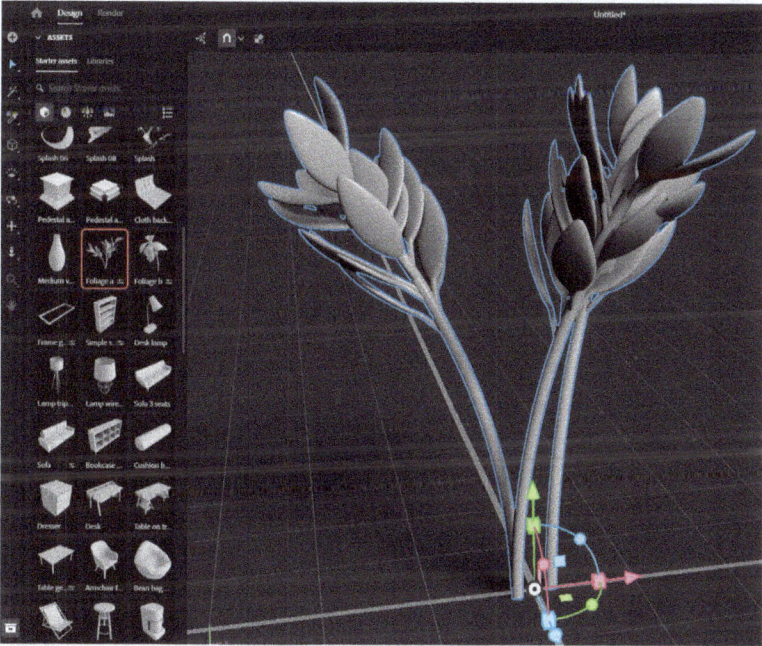

Figure 13.6 – Creating a parametric model

> **Parametric models**
>
> The 3D models known as **parametric models** allow you to change the factors that determine their 3D structure. You can use Adobe Substance 3D Designer to make parametric models that you can import into Stager.

11. Once you create the **Foliage a** parametric model, its highly customizable parameters will appear in the **Object** tab under the **PROPERTIES** panel. You can customize this model in any way you want and change its parameters. This includes adding branches, bending the stem, increasing the leaves, and so on, as shown in *Figure 13.7*:

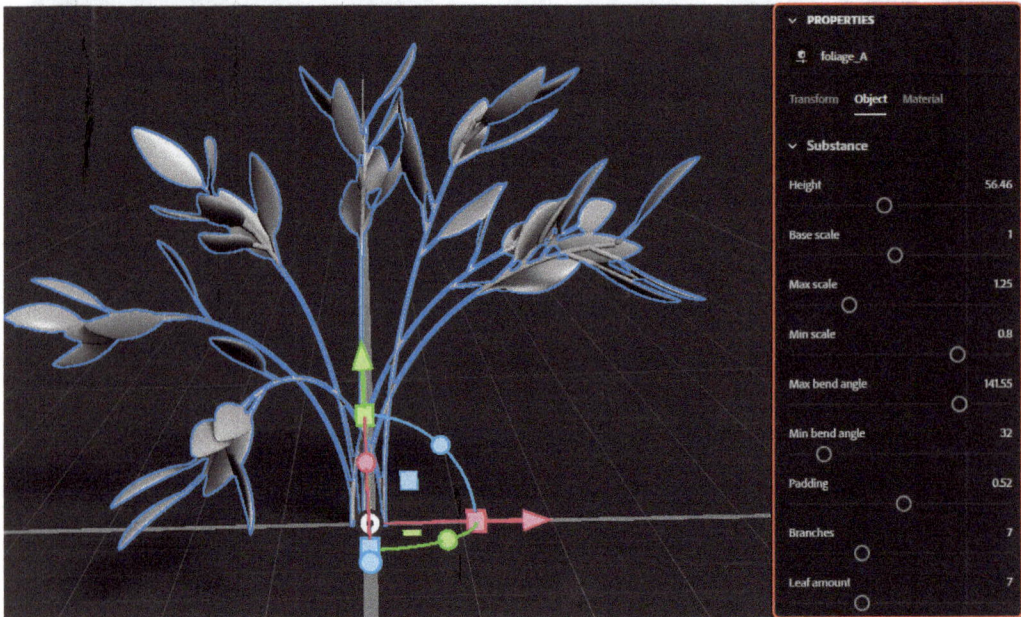

Figure 13.7 – Customizing the parametric model

12. You can manipulate any **Shape** or **Model** using the Gizmo. The default Gizmo is all-in-one: you can move, rotate, or scale with it. The hotkey for the Gizmo is *V*. However, if you only want to move, rotate, or scale, you can use the *W* (**Move**), *E* (**Rotate**), and *R* (**Scale**) hotkeys, respectively, as shown in *Figure 13.8*:

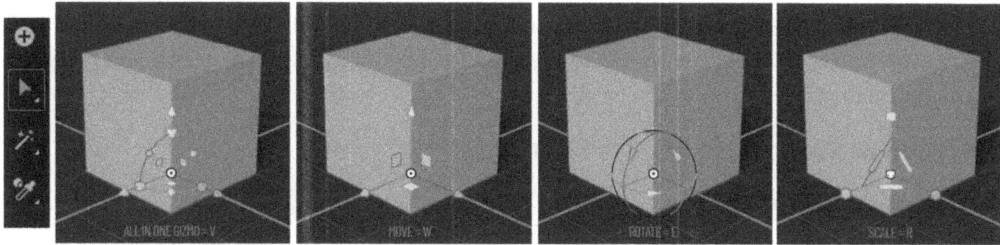

Figure 13.8 – Gizmo controls with hotkeys

If you want to snap while manipulating any **Shape** or **Model**, then use the *Shift* key on your keyboard as you are manipulating with your mouse.

With that, you should be familiar with models and shapes and how to manipulate them inside Stager. Now, let us create a small isometric bedroom inside Stager using its built-in shapes and models.

Creating an isometric bedroom

In this exercise, we will create an isometric bedroom and add a plane, walls, a standing lamp, a coffee table, decoration items, and a sofa. Let us start creating it by following these steps:

1. Create a new empty Stager file, either from the home screen or the **File** menu.

2. Add a **Plane Shape** and **Wall square windows** by clicking on **Plane** in the **Basic shapes** list and **Wall square windows** in the **Models** list. Later, you can transform them using the transform tools, such as move, scale, and rotate. If **Plane** is smaller or bigger than **Wall square windows**, rescale it so that it matches the size of the wall, and move **Plane** next to **Wall square windows** so that their edges meet, as shown in *Figure 13.9*:

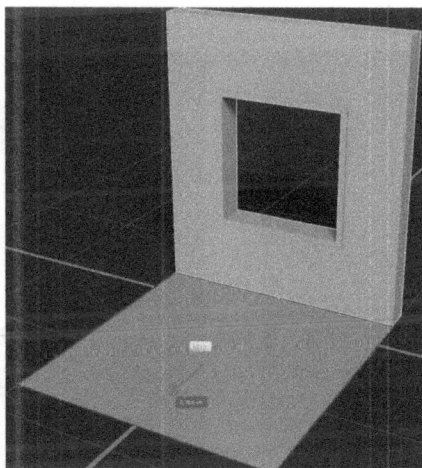

Figure 13.9 – Creating a Plane and Wall square windows

3. Select **Plane**, go to the **PROPERTIES** panel, and change the following settings under the **Transform** tab. Deselect the chain icon next to the labels of the parameters if you want to change them non-uniformly:

    *   **Position**: X: 0 CM, Y: 0 CM, Z: 67 CM

    *   **Rotation**: X: 0, Y: 0, Z: 0

    *   **Scale**: X: 10, Y: 10, Z: 10

    *   **Size**: X: 200 CM, Y: 0 CM, Z: 200 CM

4. Select **Wall square windows**, go to the **PROPERTIES** panel, and change the following settings under the **Transform** tab:

    *   **Position**: X: 0 CM, Y: 0 CM, Z: -44 CM

    *   **Rotation**: X: 0, Y: 0, Z: 0

    *   **Scale**: X: 1, Y: 091, Z: 0.97

    *   **Size**: X: 200 CM, Y: 200 CM, Z: 24 CM

    Then, change the following parameters of **Wall square windows** under the **Object** tab:

    *   **Window height**: 61.65

    *   **Window width**: 77.33

    *   **Window offset**: 0.64

5. You will notice that whatever **Model**, **Shape**, **Material**, or **Environment** you add to the viewport, it will appear in the **SCENE** panel. Therefore, change the name of **Plane** to `Floor` and the name of **Wall square windows** to `Wall with Windows`:

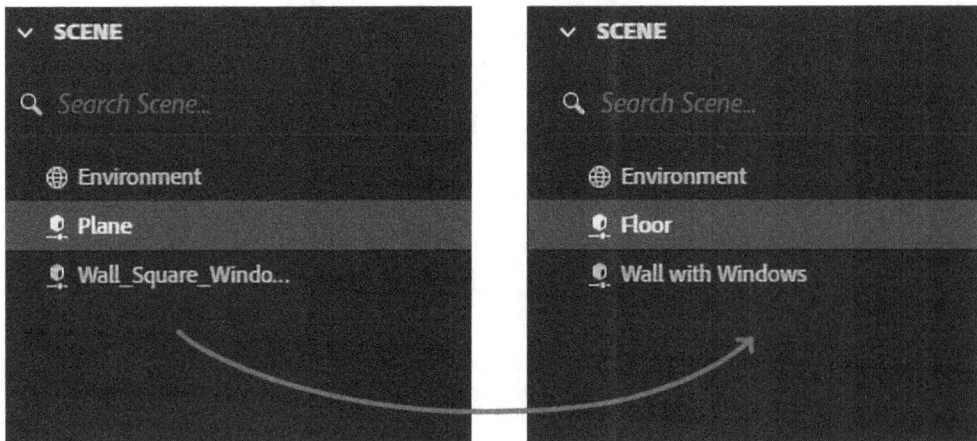

Figure 13.10 – The SCENE panel

6. Now, add a regular cube, then go to the **PROPERTIES** panel of the cube and change the following settings under the **Transform** tab. Deselect the chain icon next to the labels of the parameters if you want to change them non-uniformly:

- **Position**: **X**: `-112` CM, **Y**: `0` CM, **Z**: `67` CM

- **Rotation**: **X**: `0`, **Y**: `90`, **Z**: `0`

- **Scale**: **X**: `10`, **Y**: `10`, **Z**: `1.2`

- **Size**: **X**: `200` CM, **Y**: `200` CM, **Z**: `24` CM

7. Now, add **Sofa 3 seats**, **Frame generator**, **Table generator**, and **Lamp tripod**, and change their parameters the way you want, as shown in *Figure 13.11*:

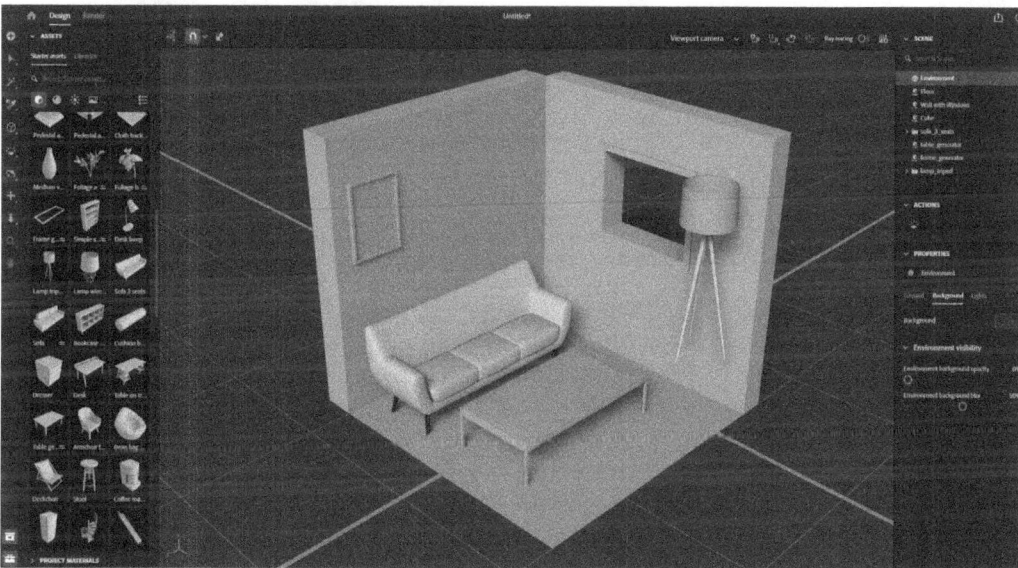

Figure 13.11 – Adding Sofa 3 seats, Frame generator, Table generator, and Lamp tripod

With that, you know how to stage your scenes inside Stager. In the next section, we will further tweak our scene using the **ACTIONS** panel.

The Adobe Substance ACTIONS panel

We have finished staging our scene; however, you will notice that **Frame generator** is out of balance and that **Lamp tripod** is in the air. To fix them, we will use the **ACTIONS** panel:

1. To fix **Frame generator**, we can duplicate it through the **ACTIONS** panel. We will duplicate it twice and arrange the objects, as shown in *Figure 13.12*:

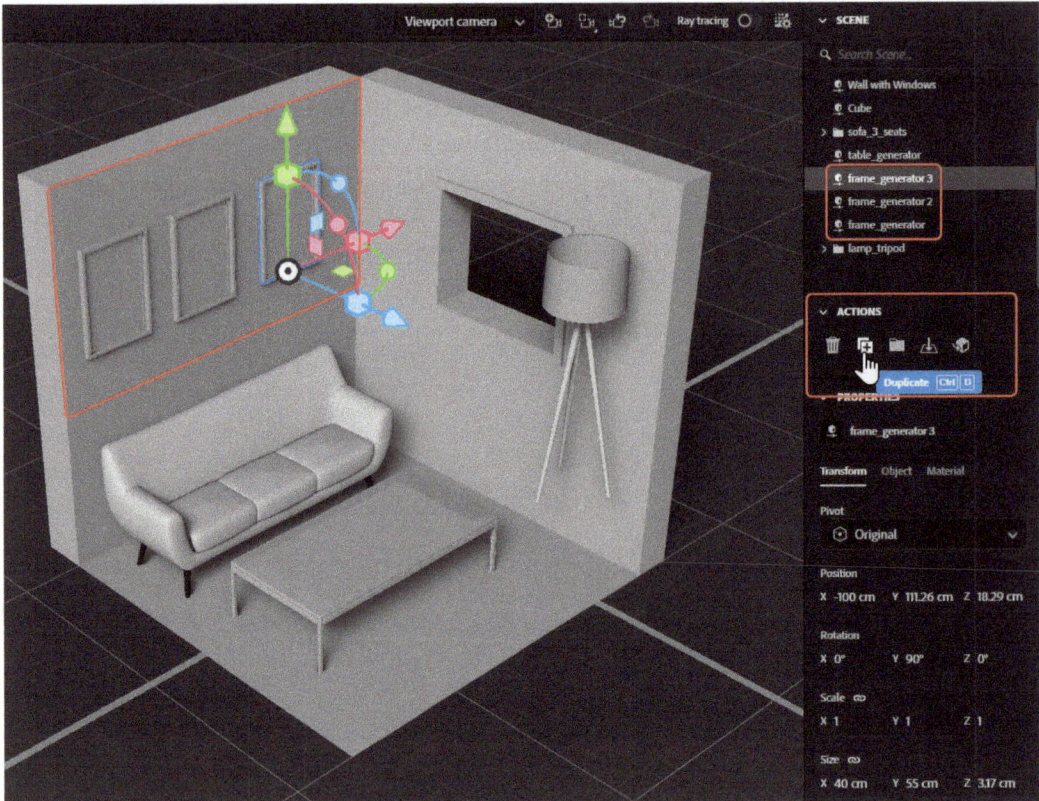

Figure 13.12 – Duplicating Frame generator

2. To fix **Lamp tripod**, select it from the **SCENE** panel, and choose **Move to ground** from the **ACTION** panel. This will bring **Lamp tripod** to the ground:

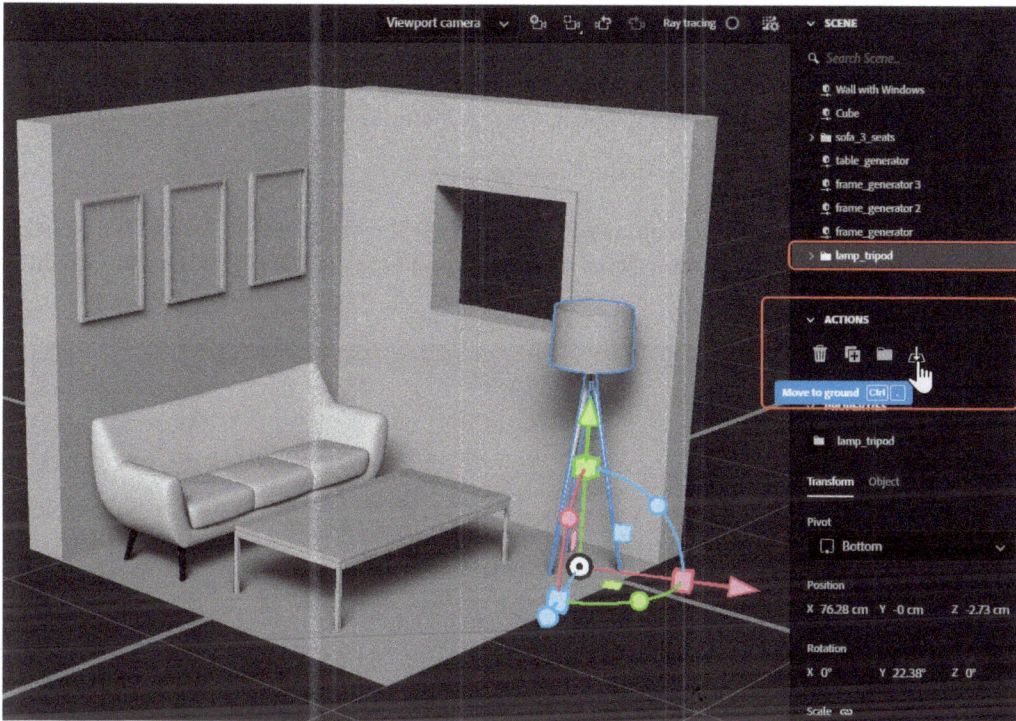

Figure 13.13 – Moving Lamp tripod to the ground

Hopefully, you are now familiar with the **ACTIONS** panel, how to use it, and when to use it. In the next section, you will learn how to apply materials to our scene.

Exploring Adobe Substance 3D Stager materials and lights

Materials are a crucial component of 3D design workflows since they determine how 3D objects look. Strong physically based materials are available in Adobe Substance 3D Stager. This makes setting up materials and obtaining reliable visual results simple.

Moreover, **lights** are the sources of illumination that can be added to the scene physically. Physical lights can mimic several types of lighting, including lamps, traffic signals, spotlights, windows, and sunshine. Four different types of physical lighting exist: **Area**, **Spot**, **Point**, and **Directional**.

So, let us add materials and lights to our scene.

Adding materials

There are two types of materials: **Basic Materials** and **Materials**. **Basic Materials** act as the base of a material, which is usually created from scratch, whereas **Materials** are ready-made SBSAR files.

Let us work on materials with the help of the following steps:

1. First, you must select the 3D objects you want to apply the material to in the **SCENE** panel. Then, drag and drop the **Paint** material onto the wall and use the **PROPERTIES** panel to edit the material's settings, as shown in *Figure 13.14*:

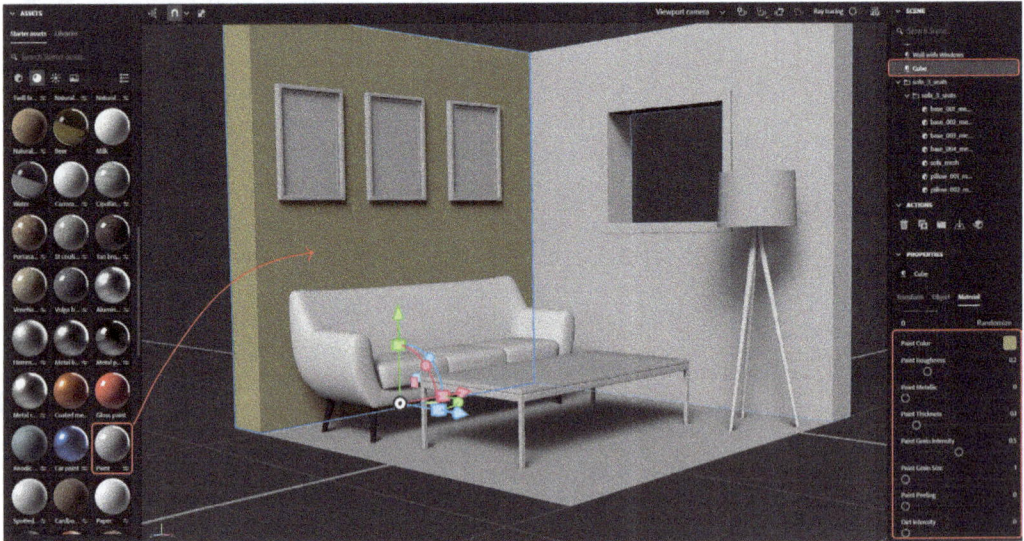

Figure 13.14 – Applying a material

2. You can also sample any **Material** using the **Sampler** tool. The **Sampler** tool allows you to pick any **Material** from the object and apply it to another object:

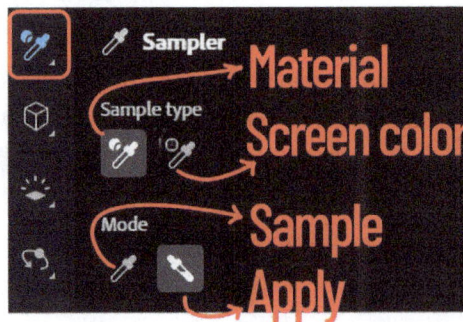

Figure 13.15 – Sampler tool

3. There are two ways you can sample the material; you can try either of the following:

    - Select **Wall with Windows**, press *I* on your keyboard to activate the **Sampler** tool, then click on the wall that we applied the material to previously.

    - Select the wall that we applied the material to previously and press *I* on your keyboard to activate the **Sampler** tool. Then, press *Shift* on your keyboard and click on **Wall with Windows**:

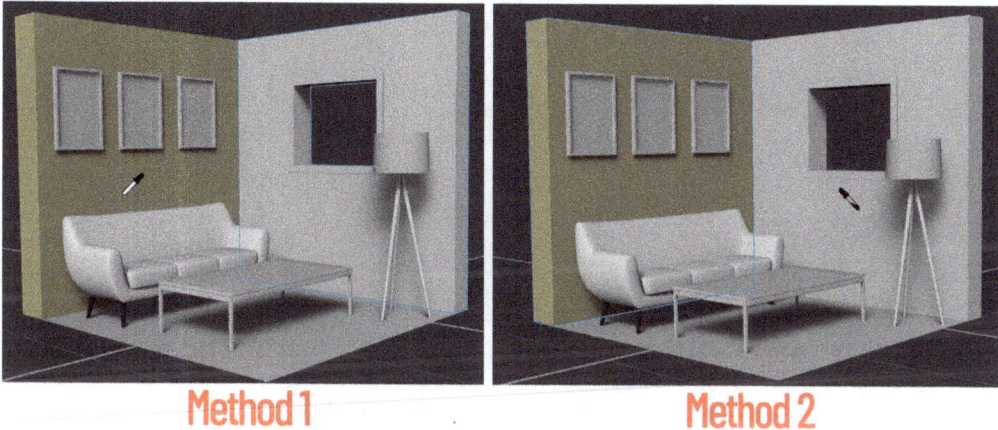

Figure 13.16 – Using different sampler methods

4. The material that you've applied with the help of the **Sampler** tool will become an instance, which means if you change the original **Material**, the sampled **Material** will also change.

5. As you may have noticed, **Sofa 3 seats** was created with a combination of different objects. You can apply different **Materials** to different objects. You just need to select the desired objects from the **SCENE** panel and drag the material you want on them, just like I did to the rest of the objects, as shown in *Figure 13.17*:

Figure 13.17 – Applying materials to multiple objects

6.   If you want, you can select multiple objects in the **SCENE** panel and group them by pressing *Ctrl + G* (Windows) or *Command + G* (Mac) on your keyboard. Alternatively, you can select the **Group** option from the **ACTIONS** panel.

Now that you know how materials work, in the next section, we'll learn how lights work.

Adding lights

As I explained earlier, physical lights help you simulate a photorealistic environment. If you switch to **Lights** in the **ASSETS** panel, you will notice there are a variety of lights. We will cover a few lights here; you can explore the rest by yourself. As you will see, using lights is quite simple:

1.   Go to the **ASSETS** panel and switch to the **Lights** tab. Select **Area light** and place it in the window opening with the help of the Gizmo. An area light's illumination comes from a planar source. Every side of this plane emits light in all directions. By adjusting this plane's settings, you can modify the light that is emitted. You can also turn on **Ray Tracing** to see more photorealistic results, as shown in *Figure 13.18*:

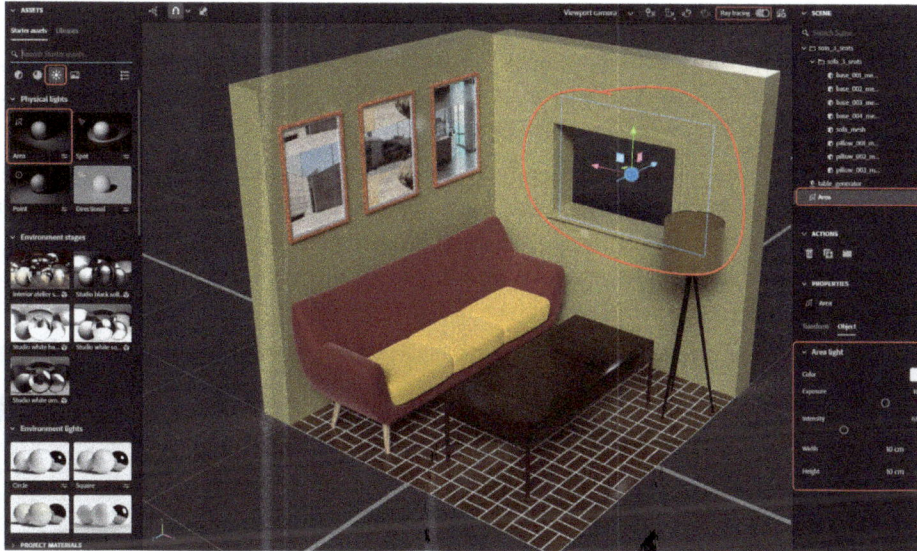

Figure 13.18 – Adding area lights

2. You can also adjust the **Exposure** setting of the lights, which is an exponentially increasing amount of light that the light source emits. You can also adjust the **Intensity** setting, which is a multiplier on a linear scale based on the exposure value.

3. Now, add a **Point light** and place it inside the lampshade, as shown in *Figure 13.19*, with the help of the Gizmo. A **Point light** casts light in all directions from a single point in space:

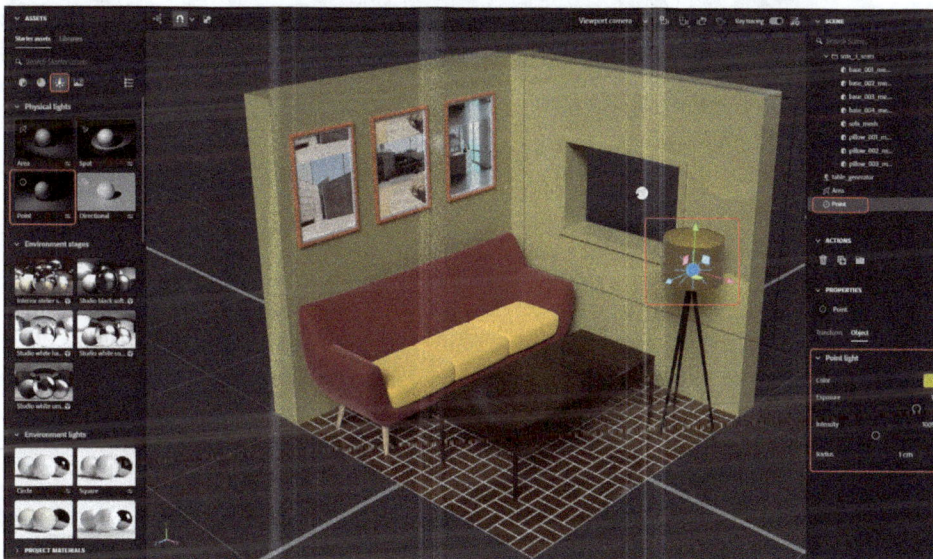

Figure 13.19 – Adding point lights

4. There are two more lights: **Spotlight** and **Directional**. **Spotlight** has an emitter source in a conical area that beams forth light, while a **Directional** light is usually infinitely far away, producing parallel light rays that illuminate the entire scene. This produces an ambiance effect similar to sunlight. However, we are not going to use these lights.

5. Let us make the scene a little dark. Scroll down to **Environmental lights** in the **Lights** tab, then choose **Studio 80s horror flick b** and select **Environment** in the **SCENE** panel. From the **PROPERTIES** panel of the **Lights** tab, reduce its **Intensity** to 34% and change its **Rotation** to 0. You can also rotate the environment by holding *Shift* and pressing the right mouse button. You will get a more realistic result, as shown in *Figure 13.20*:

Figure 13.20 – Changing the environmental lighting

Now that you are aware of the scene setup, material, and lights, we will cover collision-based transformation in the next section.

Collision-based transformation

The mechanism known as collision transformation modifies how the transform tools behave. When two items come into contact, they might interact naturally. Let us see how this works:

1. Open the `Chapter_13_Practice.ssg` file of the `Substance_Stager_Exercise_Files` folder.

    You will notice that **Frame generator**, **Table generator**, and **Medium vase bottle mesh** are in the air instead of in contact with the other objects. If you try to put the objects on top of each other, they will penetrate through each other instead of colliding.

2. To avoid this, we need to activate **global collision**. To do that, go to the **Object** menu and choose **Enable/Disable Collision** or press C. You can do the same from the **contextual** toolbar, as shown in *Figure 13.21*:

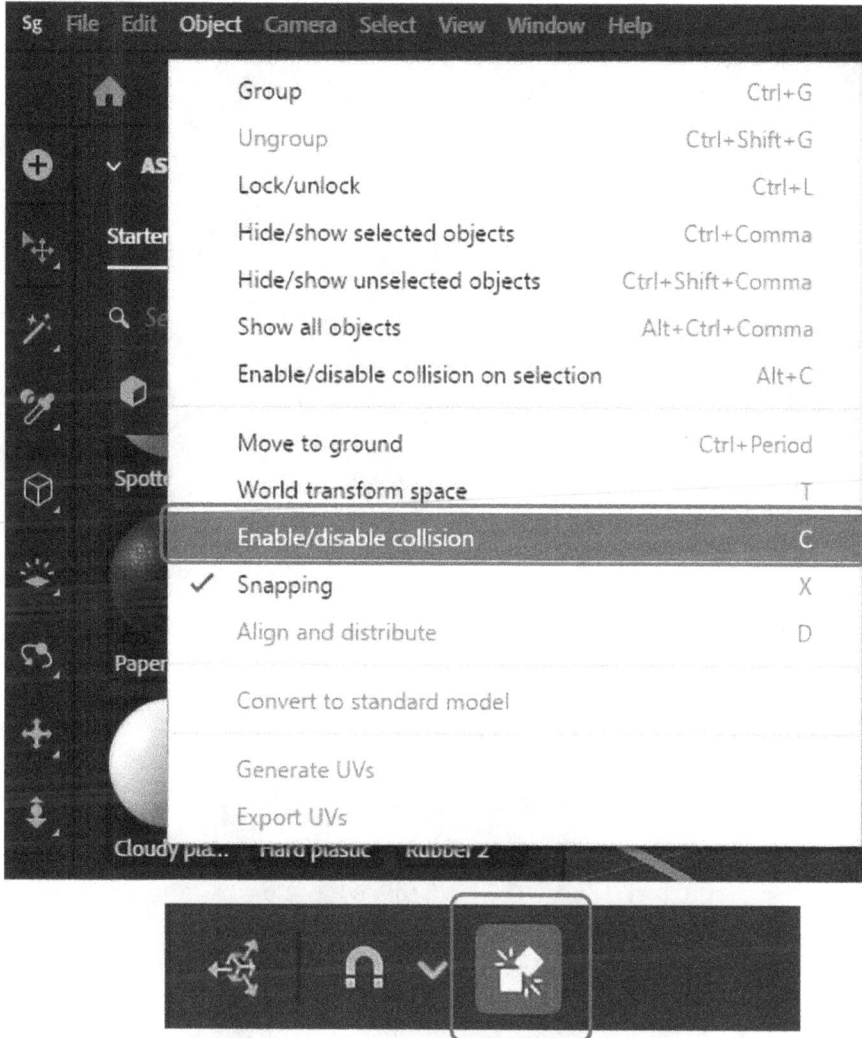

Figure 13.21 – Enabling global collision

3. Now, select the objects you want to collide with each other one by one. While doing so, go to each object's **PROPERTIES** panel, go to the **Object** tab, and toggle the **Collision** option **ON**. It's ideal to turn this **ON** for all objects:

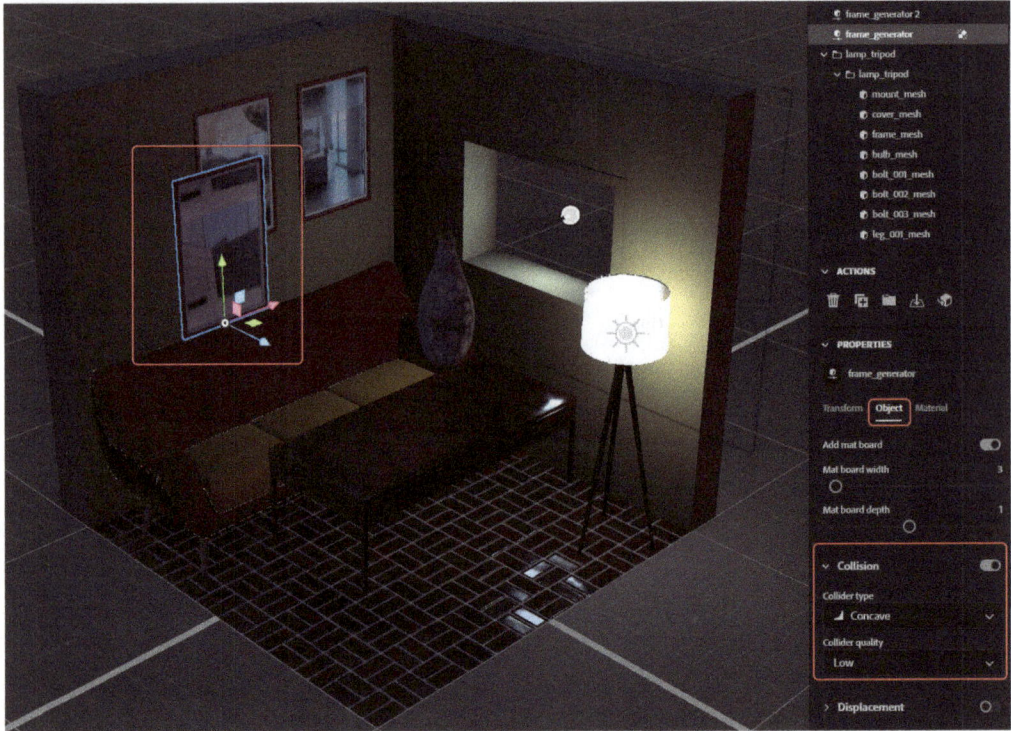

Figure 13.22 – Turning on Collision

4.   Now, if you will try to put the objects on top of each other, they will collide instead of penetrating through each other:

Figure 13.23 – Objects interacting with each other after turning on Collision

There are three types of colliders, as shown in *Figure 13.21*:

- **Convex**: The collider will only follow the object's outside boundaries. The quickest choice is this one.

- **Concave**: The collider will follow the object's crevices and indentations.

- **Surface**: When an item is a thin plane of geometry, it generates a two-sided collider. This works on flat or concave surfaces.

You can also control the collider's quality, although this requires more processing power. Higher quality is more accurate for surfacing features.

Summary

In this chapter, you learned how to stage your scene, arrange your objects, apply materials, and set up lights, which are highly crucial skills you need to know to work with Stager.

In the next chapter, you will learn how to stage a photorealistic scene with the objects we created in Substance Painter and Substance Designer. You will also learn how to set up a 3D camera and render our 3D scene.

Cameras and Rendering inside Adobe Substance 3D Stager

This chapter will cover how the visualization and manipulation of intricate lighting and shadows work when rendering a scene. We will also look at how to set up any camera inside Adobe Substance 3D Stager and how to render the scene with it.

Before setting up the camera and rendering our final scene, we will learn how to import Substance Painter and Designer files into Stager, so let us start with that.

We will cover the following topics in this chapter:

- Importing Substance Painter and Designer files
- Setting up the camera
- Adding spotlights and environment lights
- Rendering the scene

Importing Substance Painter and Designer files

It's a quite cumbersome process when you import or export a file from Substance Painter or Designer and then apply materials to it. However, we will look at an easier way to bring Painter and Designer files into Stager:

1. First, you need to launch Adobe Substance 3D Painter and open `Retro_Television.spp` from the `Substance_Painter_Exercise_Files` folder.

2. Once `Retro_Television.spp` is open in **Painter**, go to the **File** menu, hover over **Send to**, and choose **Send to Substance 3D Stager**:

Figure 14.1 – Sending the Painter file to Stager

3. Once you select the **Send to Substance 3D Stager** option, Stager will automatically open and may load a message as shown in *Figure 14.2*. You just need to select **Skip** because we need to preserve the UVs of our Painter file:

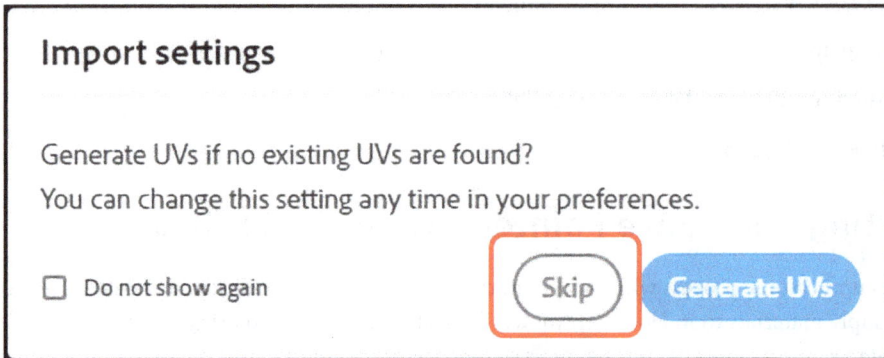

Figure 14.2 – Skipping UV generation

4. The `Retro_Television` model size will be quite small when you import it for the first time, as shown in *Figure 14.3*, and for a realistic lighting effect, the model size should be closer to a realistic size:

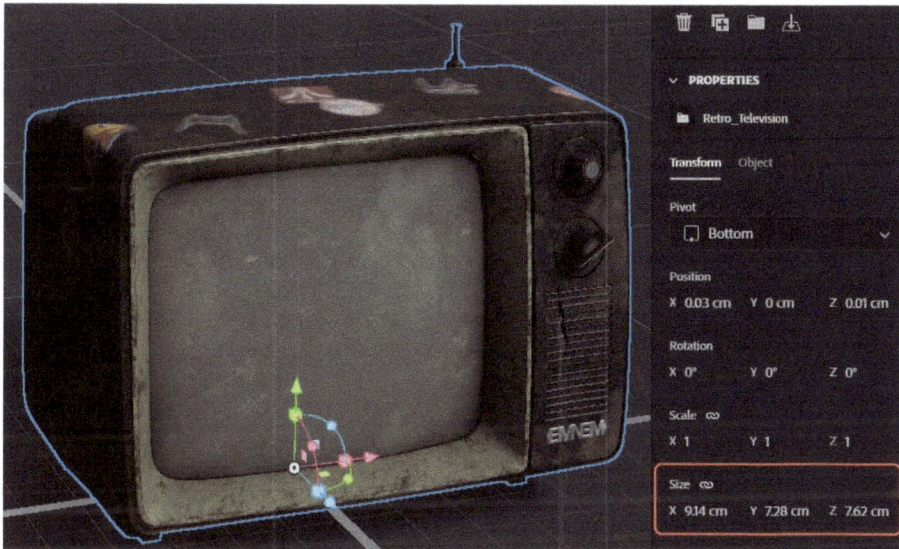

Figure 14.3 – Original dimensions of the Retro_Television model

5. The average size of a CRT television is around 50 to 53 cm wider and higher – therefore, change the size of the `Retro_Television` model to the following:

Size: X = 50 cm, **Y** = 39.81 cm, **Z** = 41.67 cm:

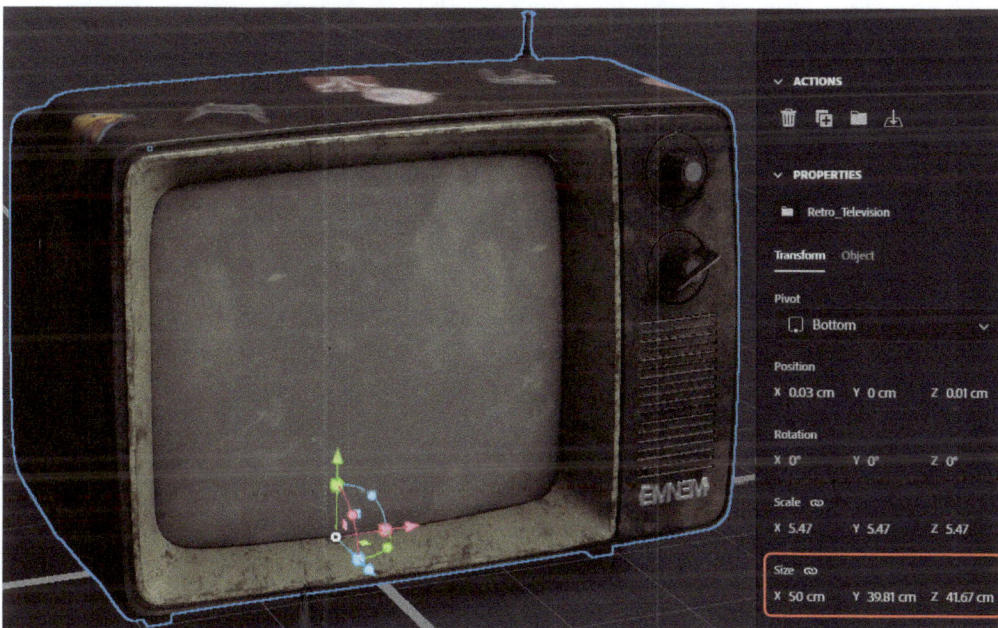

Figure 14.4 – Rescaling the Retro_Television model

6. Now, keep Stager open, launch **Substance Designer**, and open TV_Shelf_Model.sbs from the Substance_Designer_Exercise_Files folder. Once the TV_Shelf_Model.sbs file is open, right-click on it in Designer's **EXPLORER** window and choose **Send to Substance 3D Stager**:

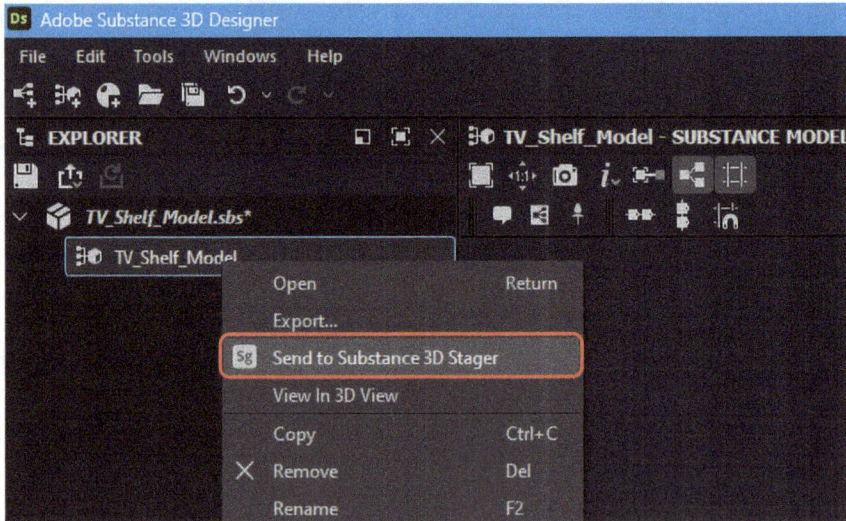

Figure 14.5 – Sending a Designer file to Stager

7. Once TV_Shelf_Model.sbs is imported inside Stager, it will be at exactly the same location as the Retro_Television file. The file is also smaller in size and is blending into the ground as shown in *Figure 14.6*:

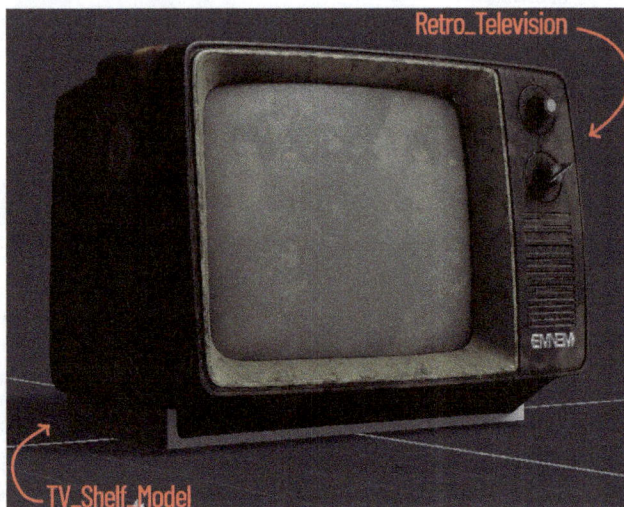

Figure 14.6 – TV_Shelf_Model inside Stager

8. We need to move `Retro_Television` and `TV_Shelf_Model` apart and rescale `TV_Shelf_Model` with the following values:

Size: **X** = `72.98` cm, **Y** = `35.68` cm, **Z** = `34.06` cm

Now, the `Retro_Television` and `TV_Shelf_Model` sizes match as shown in *Figure 14.7*.

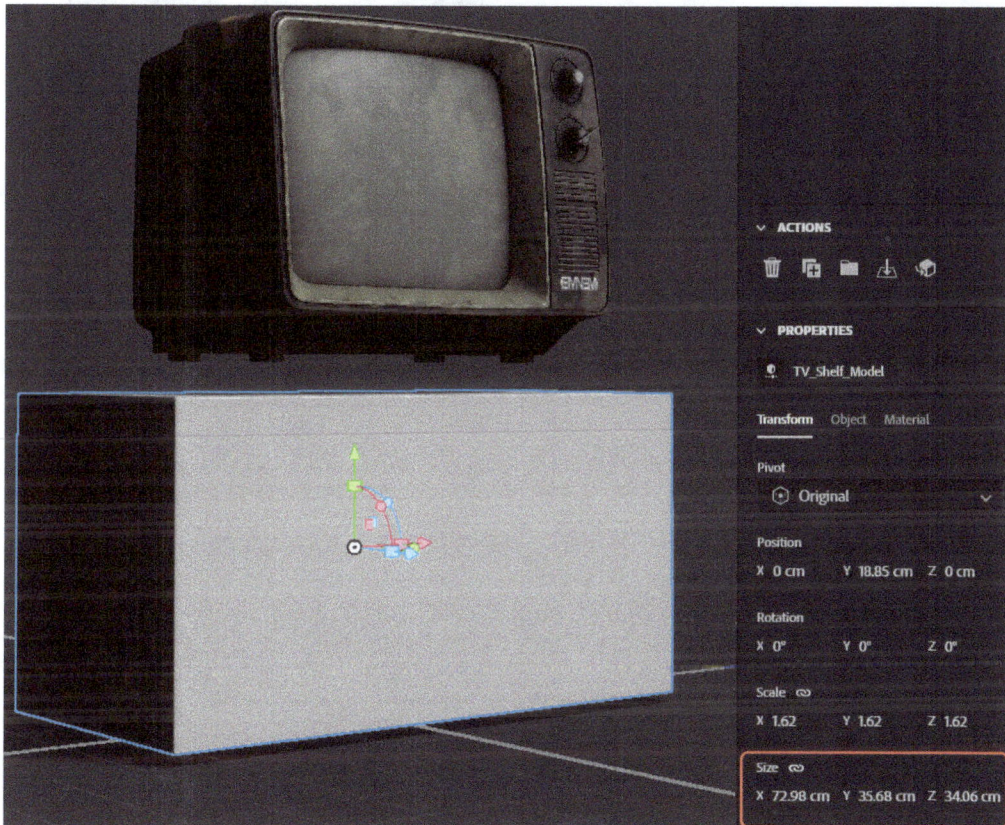

Figure 14.7 – Rescaling TV_Shelf_Model

9. Now, open `TV_Shelf_Substance.sbs` from the `Substance_Designer_Exercise_Files` folder in Designer. Right-click `TV_Shelf_Substance.sbs` and choose **Publish .sbsar file**:

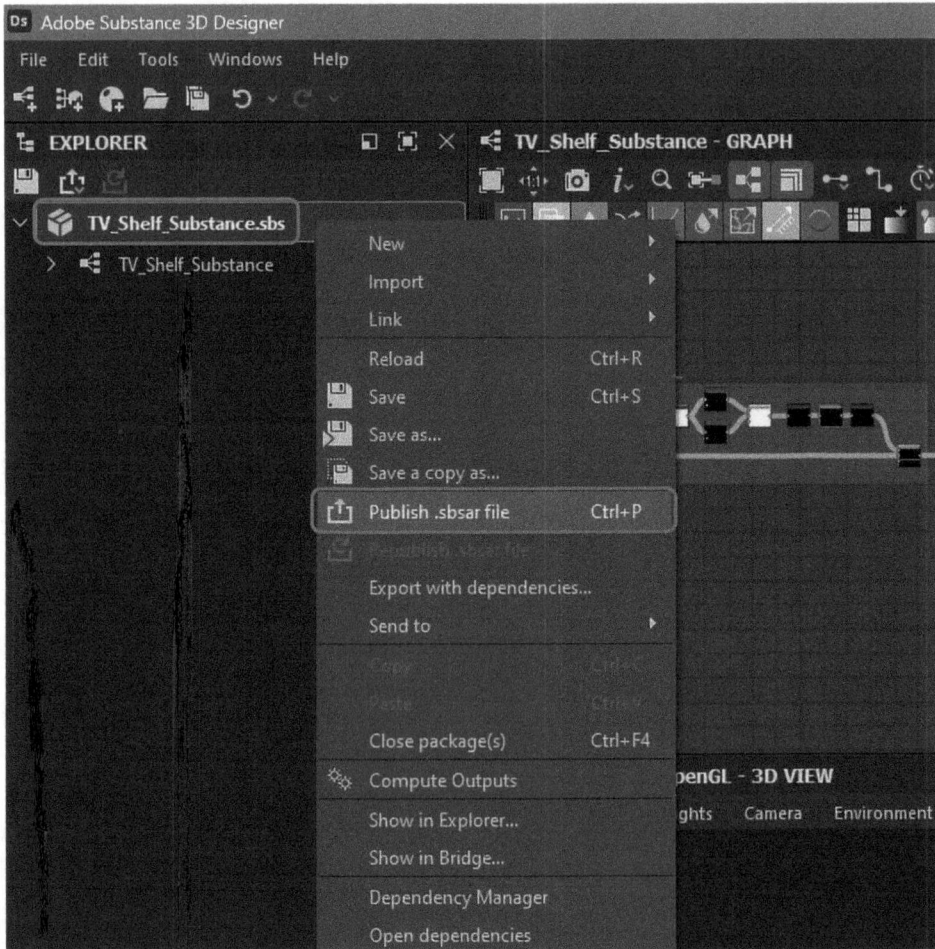

Figure 14.8 – Publishing the .sbsar file for TV_Shelf_Substance.sbs

10. When you select **Publish.sbsar file**, the **Substance 3D asset publish options** window will open. From there, choose the three dots next to **File path**.

11. Choose your `Substance_Stager_Exercise_Files` folder so you can keep all the Stager files in one location for better organization and save it as `TV_Shelf_Substance.sbsar` in that folder:

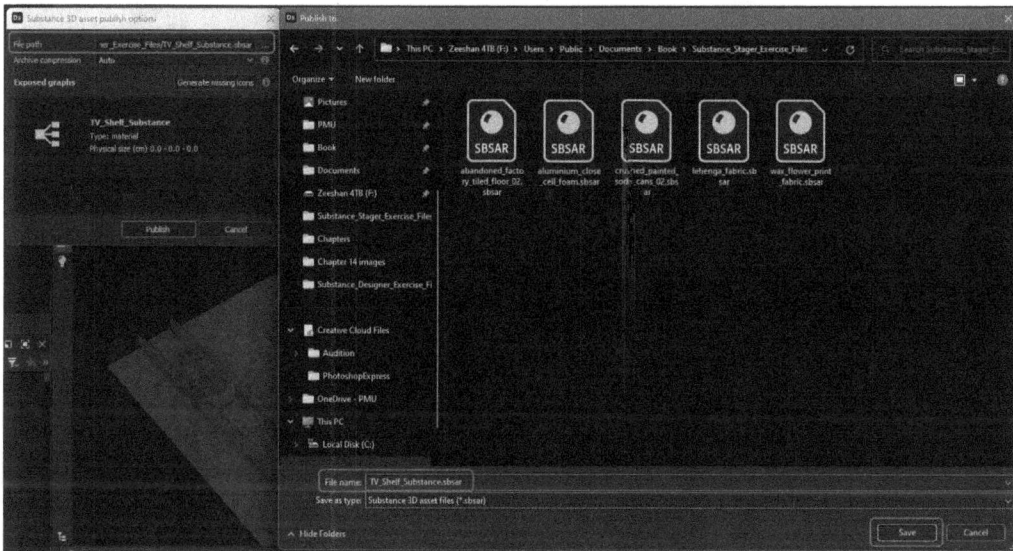

Figure 14.9 – Saving TV_Shelf_Substance.sbsar

12. Now, finally, you can press the **Publish** button and your TV_Shelf_Substance.sbsar file will be saved:

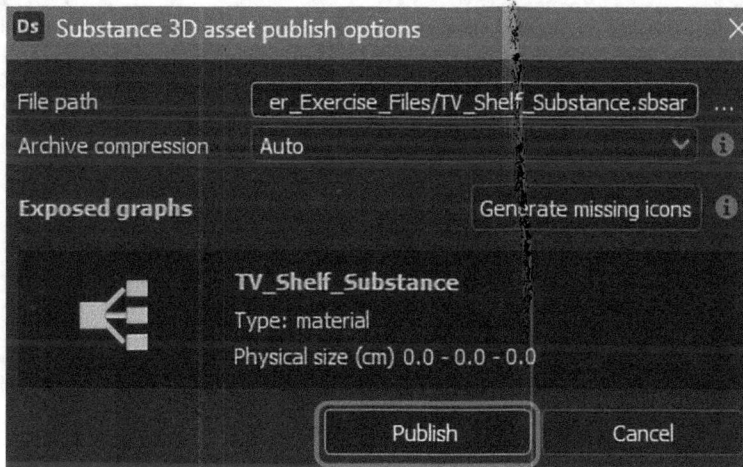

Figure 14.10 – Publishing TV_Shelf_Substance.sbsar

13. Now, you can close Designer and go back to Stager and select **c model** from the **SCENE** panel, go to the **File** menu, hover over **Import**, and choose **Place material on selection**:

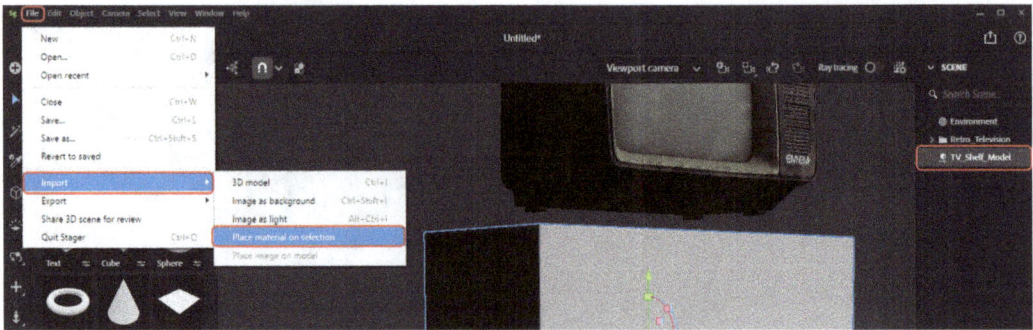

Figure 14.11 – Placing material on selection

14. When the window opens for selecting the material file, choose the TV_Shelf_Substance. sbsar file. You will notice that the material is flat once it is applied to TV_Shelf_Substance. sbsar. Therefore, to give it some depth, you need to select TV_Shelf_Model in the **SCENE** panel, then go to the **PROPERTIES** panel, and turn on **Displacement** under the **Object** tab:

Figure 14.12 – Toggling on Displacement

Now, you can turn on the collision for the TV and the shelf and place them on top of each other. There are a few Painter files inside the Substance_Stager_Exercise_Files folder and we need to import them to Stager to create a scene.

15. Open Room.spp in Painter and repeat *step 2* and *step 3*.

16. Once Room.spp is imported into Stager, it will be significantly bigger. Scale it down to match it with the TV and the shelf and rearrange the scene as shown in *Figure 14.13*:

Figure 14.13 – Arranging Room, Retro_Television, and TV_Shelf_Model

17. The following are the Painter files given in the Substance_Stager_Exercise_Files folder and you need to import them into Stager using the same steps as *step 2* and *step 3*:

- Book_Shelf.spp and Room.spp (if not imported before)
- Carpet.spp and Sheet.spp
- Coffee_Table.spp and Sofa.spp
- Curtain.spp and Wall_Clock.spp
- painting.spp and Wall_Light.spp

18. Now, you can finally arrange the room the way you want – by moving, rotating, and scaling the models. For example, you can see the room arranged as in *Figure 14.14*. You can also toggle the **Displacement** option **ON** for all the models as you did in *14*:

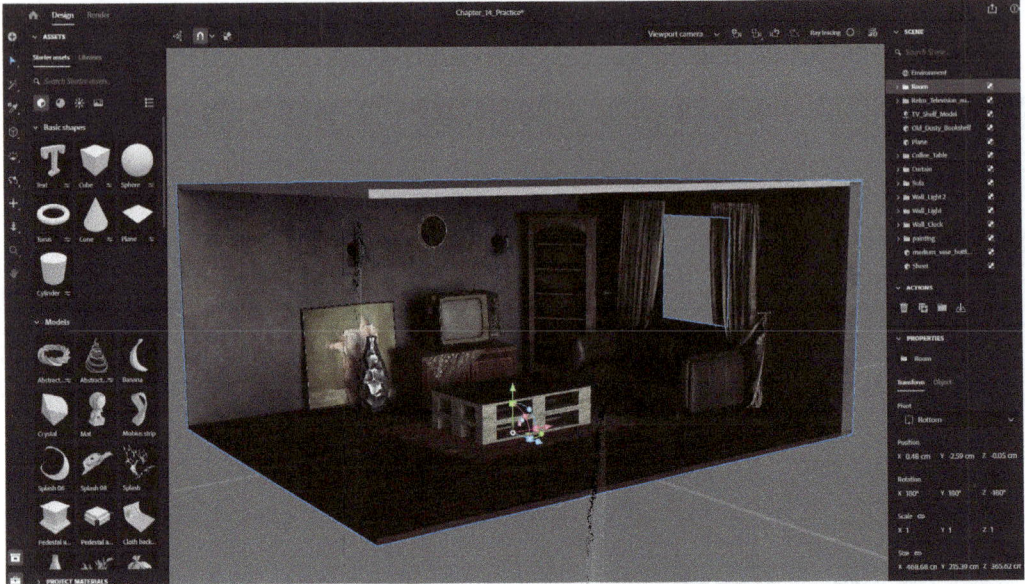

Figure 14.14 – Creating the scene inside Stager

Now, as we have our scene ready, it is time to set up a camera, so let us create some cameras inside Stager in the next section.

Setting up the camera

To create and render a good scene, you need a good camera angle. You can create more than one camera in Stager depending on what you need. There are two different kinds of cameras – the viewport camera, which is the default camera, and the one you create with the **Add camera** option:

1. To create your own camera, go to the top bar on the viewport and choose **Add camera** as shown in *Figure 14.15*. You can also frame a focal point and undo and redo the camera movement or any other changes that you have applied to it:

Figure 14.15 – Adding a camera to the scene

2. Now, you can set up the camera using the same navigational control that we learned about in *Chapter 12, Getting Started with Adobe Substance 3D Stager*. You can set up two different kinds of cameras as shown in *Figure 14.16*. One camera can show the whole room and the other camera can show a closeup of the TV and the shelf:

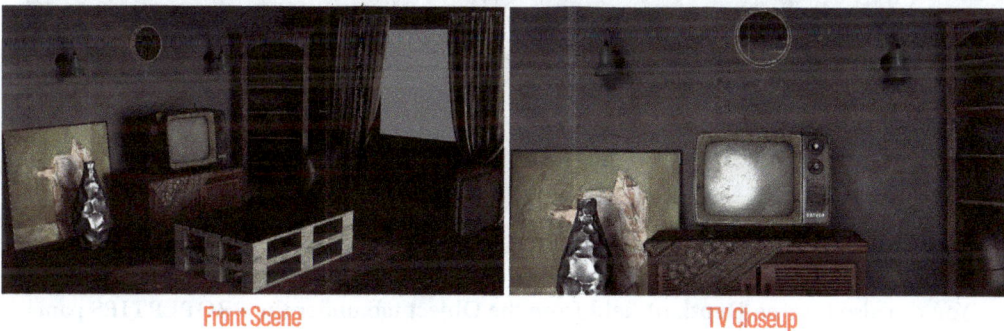

Front Scene   TV Closeup

Figure 14.16 – Creating two cameras

3. You can now rename the cameras in the **SCENE** panel `Front Scene` and `TV Closeup`. You can switch between the cameras from the **View camera** option as shown in *Figure 14.17*:

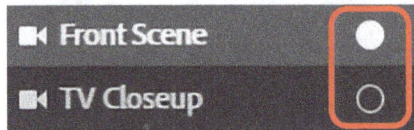

Figure 14.17 – Switching between cameras

4. Now, select the `Front Scene` camera from the **SCENE** panel and go to the **Object** tab under the **PROPERTIES** panel. Set up **Output size**. You can also add your own **Background** if you want by placing a simple **Plane** basic shape that is visible from the window and applying any background texture of your choice to it. Change **Focal length** to `35` mm, which is an ideal focal length for a full-frame scene:

Figure 14.18 – Setting up the camera

5. You can also turn on **Depth of field** from the **Object** tab under the **PROPERTIES** panel. Moreover, you can set the focus point by selecting the **Set focus point** option, clicking on the focus point as shown in *Figure 14.19*, and configuring the **Blur amount** settings. However, you can only see **Depth of field** when you are in the **Ray tracing** option:

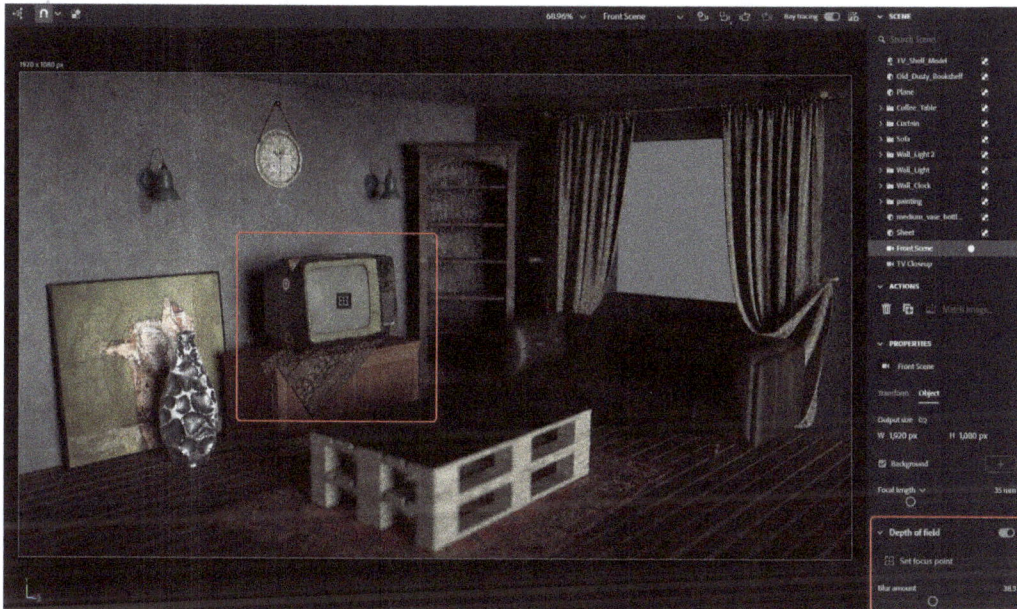

Figure 14.19 – Setting Depth of field

Now, as we have set up our cameras, it's time to learn about adding some spotlights and environment lights in the next section.

Adding spotlights and environment lights

A scene is not complete unless it has good lighting. There are spotlights in Stager that provide realistic photometric light effects and environment lights that provide high-definition, realistic lighting effects, so let us add some lighting in this section:

1. Add two spotlights from **Physical lights** in the **ASSETS** panel and place one in each wall light model in the viewport.

2. Move the lights inside the wall light folders in the **SCENE** panel as shown in *Figure 14.20*. You can change the **Object** parameters of the spotlights as shown in *Figure 14.20*:

Figure 14.20 – Adding and setting spotlights

3. To add realistic ambient lighting to your scene, you can go to **Environment lights** under the **ASSETS** panel and choose **Small apartment**.

4. Select **Environment** in the **SCENE** panel, go to its **PROPERTIES** panel, and switch to the **Lights** tab.

5. Under the **Lights** tab, you can adjust the **Environment light** settings for **Intensity** and **Rotation**.

    For this exercise, I will change the **Intensity** settings to 50% and **Rotation** to 0 degrees:

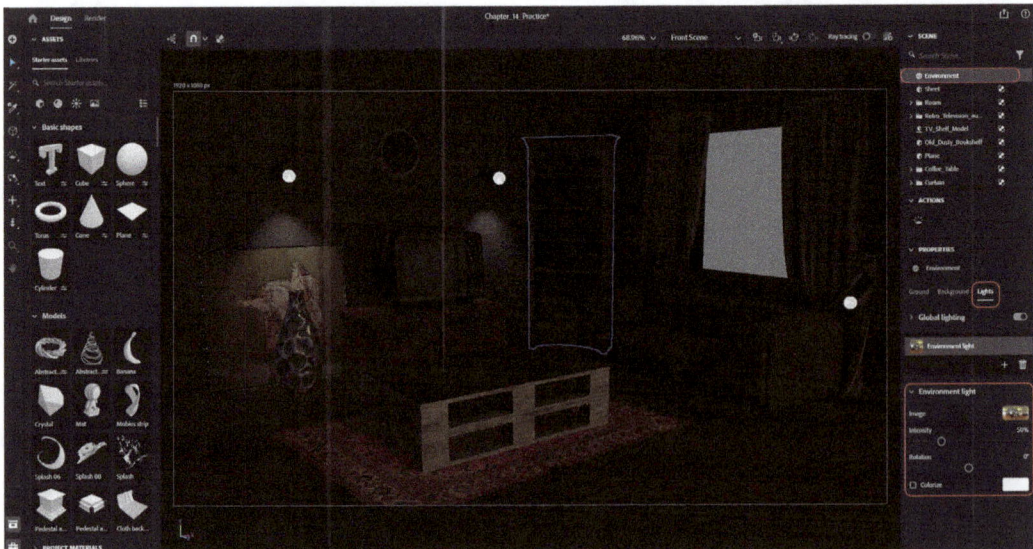

Figure 14.21 – Adjusting the Environment light settings

6. To check the light setup, you can toggle on the **Ray tracing** button.

Once you are satisfied with your light setup, you can start rendering the scene, which we will learn how to do in the next section.

Rendering the scene

To finalize the output of any 3D scene inside Stager, we need to render it. We need to switch to the **Render** mode and go through the following steps:

1. Go to the **Render** mode.

2. Toggle **ON** the **Render with GPU** settings if you have a good GPU. Otherwise, leave it **OFF**.

3. In the left-hand sidebar, under the **Preset** option, you can choose the preset that suits your system. For this exercise, I will choose **High**, which will set **Samples** to 2,048 and **Noise Level** to 1 while keeping **Resolution** set to **Full**.

4. On the right-hand sidebar, choose your camera. I chose **Front Scene,** and rename the export filename Chapter_14_Practice.

5. Go to **Export formats**, select **PSD (32 bits/channel)**, and choose the output folder in the **Save to** option. I will choose the Substance_Stager_Exercise_Files folder for this exercise.

6. Once you have set everything up, click **Render**. The render will save the output file automatically to your **Save to** folder:

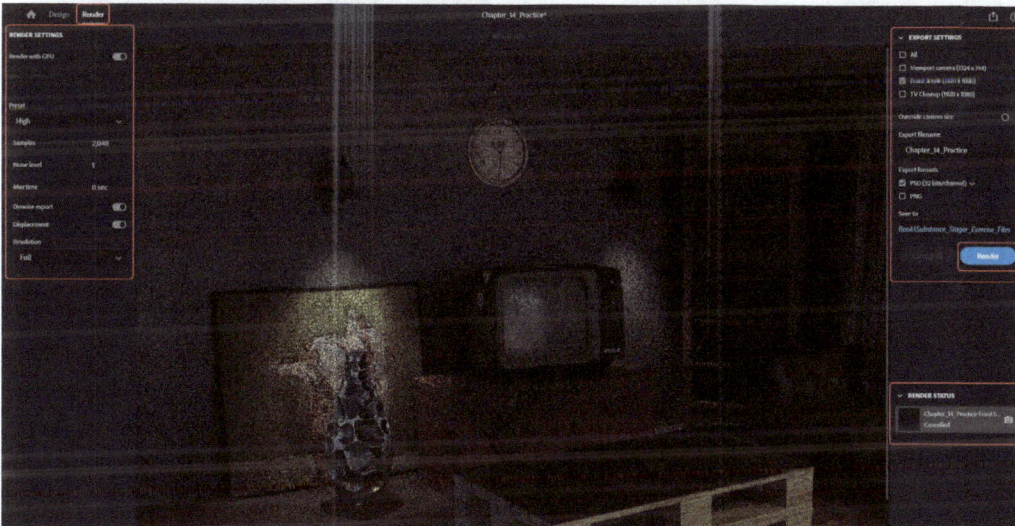

Figure 14.22 – Rendering in Adobe Substance 3D Stager

7. Since you have saved the 3D render as a 32-bit PSD file, you can open it in Adobe Photoshop. Once you open it in Photoshop, you will find that Stager has stored the Render passes as Photoshop layers, which you can tweak for more creative and high-quality effects:

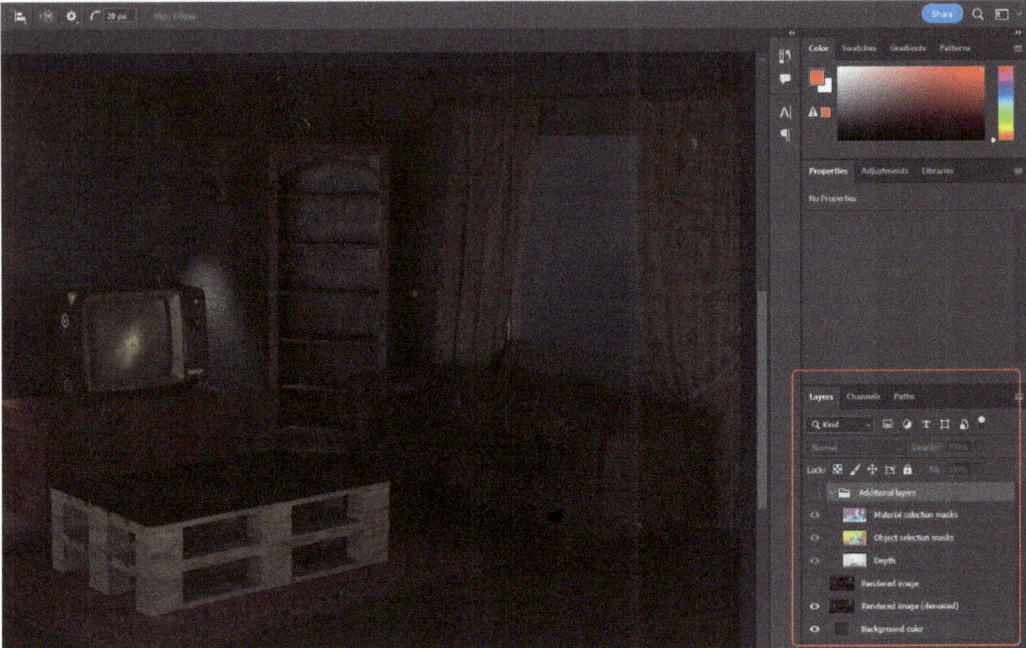

Figure 14.23 – Opening the final render in Adobe Photoshop

I hope that now you have acquired the skill sets to produce hyper-realistic renders inside Adobe Substance 3D Stager.

Summary

We covered all the skills for using Adobe Substance 3D Stager in this last part of the book. We have gone through all the skills for importing 3D assets from different Adobe Substance 3D applications, setting up cameras, and adding light inside in this chapter.

Hopefully, you are now able to produce realistic assets inside Adobe Substance 3D Painter, Adobe Substance 3D Designer, and Adobe Substance 3D Planar, and stage them creatively and professionally inside Adobe Substance 3D Stager.

I hope you have learned a lot about this amazing Adobe Substance 3D suite and acquired the skill sets to produce astonishing creative projects. You can now present your 3D object in context with Adobe Substance 3D Stager and set up the ideal shot by dragging and dropping 3D objects, materials, lighting, and cameras.

You can choose a brush in Substance 3D Painter and texture your asset quickly and easily, utilizing layer manipulation to quickly alter current paint strokes. You can now also quickly turn a representation of the real world into a 3D substance using Adobe Substance 3D Sampler.

Finally, thanks to Substance 3D Designer's node-based approach, you will now have total authoring control over material development. For more tutorials and videos, you can visit my YouTube channel at www.youtube.com/@zinteractive.

Index

‹packt›

Packt.com

Subscribe to our online digital library for full access to over 7,000 books and videos, as well as industry leading tools to help you plan your personal development and advance your career. For more information, please visit our website.

Why subscribe?

- Spend less time learning and more time coding with practical eBooks and Videos from over 4,000 industry professionals

- Improve your learning with Skill Plans built especially for you

- Get a free eBook or video every month

- Fully searchable for easy access to vital information

- Copy and paste, print, and bookmark content

Did you know that Packt offers eBook versions of every book published, with PDF and ePub files available? You can upgrade to the eBook version at packt.com and as a print book customer, you are entitled to a discount on the eBook copy. Get in touch with us at customercare@packtpub.com for more details.

At www.packt.com, you can also read a collection of free technical articles, sign up for a range of free newsletters, and receive exclusive discounts and offers on Packt books and eBooks.

Other Books You May Enjoy

If you enjoyed this book, you may be interested in these other books by Packt:

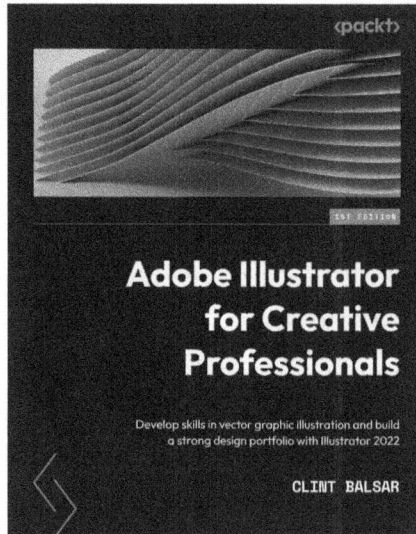

Adobe Illustrator for Creative Professionals

Clint Balsar

ISBN: 978-1-80056-925-6

- Master a wide variety of methods for developing objects
- Control files using layers and groups
- Enhance content using data-supported infographics
- Use multiple artboards for better efficiency and asset management
- Understand the use of layers and objects in Illustrator
- Build professional systems for final presentation to clients

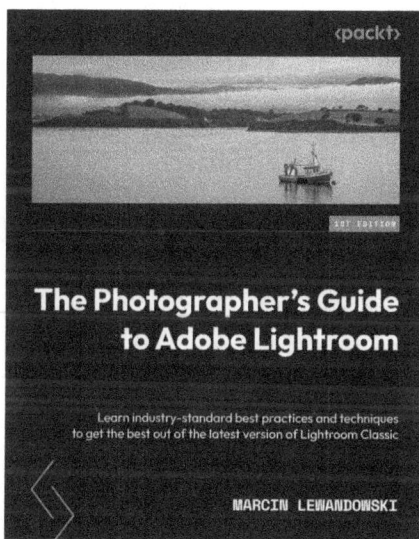

The Photographer's Guide to Adobe Lightroom

Marcin Lewandowski

ISBN: 978-1-80107-010-2

- Import photographs from different sources
- Understand how to create and refine edits
- Use and manage catalogs, folders, and collections
- Develop photographs using all available tools
- Prepare files for print and online viewing
- Create slideshows and book layouts

Packt is searching for authors like you

If you're interested in becoming an author for Packt, please visit `authors.packtpub.com` and apply today. We have worked with thousands of developers and tech professionals, just like you, to help them share their insight with the global tech community. You can make a general application, apply for a specific hot topic that we are recruiting an author for, or submit your own idea.

Share Your Thoughts

Now you've finished *Realistic Asset Creation with Adobe Substance 3D*, we'd love to hear your thoughts! Scan the QR code below to go straight to the Amazon review page for this book and share your feedback or leave a review on the site that you purchased it from.

`https://www.amazon.in/review/create-review/error?asin=1803233400`

Your review is important to us and the tech community and will help us make sure we're delivering excellent quality content.

Download a free PDF copy of this book

Thanks for purchasing this book!

Do you like to read on the go but are unable to carry your print books everywhere?

Is your eBook purchase not compatible with the device of your choice?

Don't worry, now with every Packt book you get a DRM-free PDF version of that book at no cost.

Read anywhere, any place, on any device. Search, copy, and paste code from your favorite technical books directly into your application.

The perks don't stop there, you can get exclusive access to discounts, newsletters, and great free content in your inbox daily

Follow these simple steps to get the benefits:

1. Scan the QR code or visit the link below

https://packt.link/free-ebook/9781803233406

2. Submit your proof of purchase
3. That's it! We'll send your free PDF and other benefits to your email directly